D1219374

Fundamentals of

Aquacultural Engineering

Fundamentals of Aquacultural Engineering

Thomas B. Lawson

Department of Biological Engineering
Louisiana State University

CHAPMAN & HALL

 An International Thomson Publishing Company

New York • Albany • Bonn • Boston • Cincinnati • Detriot • London • Madrid • Melbourne • Mexico City
Pacific Grove • Paris • San Francisco • Singapore • Tokyo • Toronto • Washington

Cover photo courtesy of Dr. James W. Avault, Jr., Louisiana State University Agrilcultural Center

A division of International Thomson Publishing Inc.
I(T)P The ITP logo is a trademark under license

Printed in the United States of America

For more information, contact:

Chapman & Hall
One Penn Plaza
New York, NY 10119

International Thomson Publishing
Berkshire House 168-173
High Holborn
London WC1V 7AA
England

International Thomson Editores
Campos Eliseos 385, Piso 7
Col. Polanco
11560 Mexico D.F. Mexico

Thomas Nelson Australia
102 Dodds Street
South Melbourne, 3205
Victoria, Australia

International Thomson Publishing Gmbh
Konigwinterer Strasse 418
53228 Bonn
Germany

Nelson Canada
1120 Birchmount Road
Scarborough, Ontario
Canada M1K 5G4

International Thomson Publishing Asia
221 Henderson Road #05-10
Henderson Building
Singapore 0315

International Thomson Publishing - Japan
Hirakawacho-cho Kyowa Building, 3F
1-2-1 Hirakawacho-cho
Chiyoda-ku, 102 Tokyo
Japan

1 2 3 4 5 6 7 8 9 10 XXX 01 00 99 98 97 96 95

Library of Congress Catloging-in-Publication Data
Lawson, Thomas B., 1943-
Fundamentals of aquacultural engineering / Thomas B. Lawson
p. cm.
Includes bibliographical references and index.
ISBN 0-412-06511-8
1. Aquacultural engineering. I. Title
SH137 ,L38 1994 93-38073
639 ' . 8--dc20 CIP

British Library Cataloguing in Publication Data available

Please send your order for this or any other Chapman & Hall book to
Chapman & Hall, 29 West 35th Street, New York, NY 10001, Attn: Customer Service Department.
You many also call our Order Department at 1-212-244-3336 or fax you purchase order to 1-800-248-4724.

For a complete listing of Chapman & Hall's titles, send your request to
Chapman & Hall, Dept. BC, One Penn Plaza, New York, NY 10119.

CONTENTS

PREFACE

Aquaculture is the science and technology of producing aquatic plants and animals. It is not new, but has been practiced in certain Eastern cultures for over 2,000 years. However, the role of aquaculture in helping to meet the world's food shortages has become more recently apparent.

The oceans of the world were once considered sources of an unlimited food supply. Biological studies indicate that the maximum sustainable yield of marine species through the harvest of wild stock is 100 million MT (metric tons) per year. Studies also indicate that we are rapidly approaching the maximum sustainable yield of the world's oceans and major freshwater bodies. Per capita consumption of fishery products in the United States increased by about 25% during the 1980s, and it has increased considerably in many other nations, for which current consumption figures are lacking. Experts predict that, at the present rate of population growth, by the year 2000, we will be hard-pressed to fill the food and fiber needs of the world's masses. Two means by which we may meet the world's future food needs are: (1) greater production of traditional agronomic crops; and (2) aquaculture.

Commercial aquaculture production lies in the realm of biotechnology, which requires a balanced support from the biological and engineering sciences. However, commercial aquaculture has become so complex that, in order to be successful, one must also draw upon the expertise of biologists, engineers, chemists, economists, food technologists, marketing specialists, lawyers, and others. The multidisciplinary approach to aquaculture production became apparent during the early 1990s. It is believed that this trend will continue as aquaculture production becomes more and more intensive in order for the producer to squeeze as much product as possible out of a given parcel of land.

Although many aquaculture books exist, few explore the engineering aspects of aquaculture production. This book is a humble attempt to fill that void and bridge the gap between the traditional fishery biologist and the engineer. The primary aim is to concentrate on the technical engineering principles that are applicable to aquaculture systems. Aquacultural engineering is such a broad field that it is impossible to do justice to all facets of the discipline in one text. Therefore, I keep to the basic engineering fundamentals that apply to aquaculture production and do not cover such complex topics as harvesting gear, harvesting vessels, processing, storage, transportation, etc. Biological cultural practices are not discussed in detail since this information is available from so many other sources.

Although this book is written primarily for students and practicing aquacultural engineers, my intention was to write in basic enough terms that it can also be of benefit to fishery biologists, aquaculturists, seafood technologists, entrepreneurs, and others interested in aquaculture. I have touched upon a variety of subjects, covering some in greater depth than others, but still addressing each major topic in basic terms. It was my intention that the reader could use this book to develop an understanding of aquacultural engineering fundamentals that would enable him or her to solve more complex technical problems. The reader should understand that, where a description of a system or process is presented, a generalized method is being described, and there is no *one* way to culture a particular species, nor is there *one* system that works best to the exclusion of all others.

The book begins with an introductory chapter that describes the commercialization of several important species in the United States. This chapter discusses the importance of commercial aquaculture and finishes with a brief summary of the future of aquaculture. Chapters that follow concern topics such as water quality aspects, siting considerations, water supply, pumping, water flow measurement, open culture systems, pond and raceway systems, closed recirculating systems, aeration, oxygenation, and disinfection.

The last chapter includes an extensive bibliography from which citations were obtained and are used liberally throughout the book. This bibliography is a starting place for obtaining additional information, but it is by no means complete. I cited sources that, for the most part, are easy to locate and avoided those that are very difficult or impossible to obtain. I drew heavily upon the expertise of many experts in their respective fields: Dr. Fred Wheaton, University of Maryland; Dr. Claude Boyd, Auburn University; Dr. John Colt, James M. Montgomery Consulting Engineers; Dr. Stephen Spotte; Dr. Ron Malone, Louisiana State University; and many others to whom I am eternally grateful.

Grateful thanks are extended to my loving wife, Charlotte, to my friends and peers, to the Louisiana State University Department of Biological and Agricultural Engineering, and to the Louisiana State University Agricultural Center for their support and encouragement throughout the preparation of the manuscript.

Fundamentals of

Aquacultural Engineering

CHAPTER 1
INTRODUCTION

The term *aquaculture* as used in this book refers to the culture of finfish, shellfish, other aquatic animals, and aquatic plants in either freshwater or salt water. The term has supplanted *mariculture*, which was once used to indicate the culture of animals and plants in marine or brackish water environments. Mariculture is no longer a valid term, emphasized by the changing of the name of the World Mariculture Society to the World Aquaculture Society several years ago.

Aquaculture is often equated to *water farming* or *underwater agriculture*, and is the production of aquatic organisms for human consumption. Most aquaculture takes place in land-based farm ponds. In this respect, aquaculture can easily be thought of as agriculture, and the emphasis of this book will be on aquaculture for the production of human food. However, aquaculture also encompasses the production of fish to be used in restocking programs, bait-fish production, tropical- and ornamental-fish production, and aquatic plant culture.

Many scientists feel that maximum production through traditional agriculture is rapidly approaching around the world. New human food sources are being sought to prevent food shortages for future generations. The oceans, long considered to be unlimited food sources, are logical environments for exploitation since they cover about 71% of the earth's surface. However, studies indicate that the maximum sustainable yield of marine species through capture fisheries is limited to 100 million metric tons (MT)[1] per year. Controlled aquaculture production, on the other hand, can ensure a steady and regular supply of food. Thus, fish production worldwide can be increased considerably with the planned input of modern technology. The growth of aquaculture is likened to the development of land-based food crops, and aquaculture has the potential to be the second largest means of food production next to traditional agronomic food crops.

The worldwide demand for fish and seafood is steadily rising. By the year 2000, aquaculture is projected to account for 25% of the world's fishery production (Beach 1989). But aquaculture must compete with the domestic wild catch and imports for a share of the market. How much of the market aquaculture will be able to capture depends upon price competitiveness and consumer demand. Aquaculture producers in the United States face fierce competition from foreign producers.

Continued aquaculture growth in the United

1. One metric ton = 2,205 lb.

States could help reduce the demand for imported fishery products and the trade deficit in fishery products. It is estimated that, if per capita consumption of fishery products continues on its present rate of growth, and if the U.S. population increases by approximately one million people per year, then U.S. consumers would require an additional 680,000–907,000 MT (1.5 billion–2 billion lb) of fishery products by 1995. If domestic landings follow present trends, they will be able to supply only 25–30% of that demand. Thus, the remaining 70–75% must be supplied either through imports or domestic aquaculture production.

ROLE OF THE ENGINEER IN AQUACULTURE

Aquaculture was once viewed as only a minor part of the fish and seafood industries. However, in recent years its impact has become significant if both the public and private aquaculture sectors are considered. The artificial propagation of fish has become very complex over the years. Today's aquaculture technology demands a good understanding of the physical, chemical, and biological processes necessary for successful production. Because success depends on a blend of expertise from many disciplines, many government research institutions and the larger private companies take the "teamwork" approach to aquaculture production. A team may consist of a fishery biologist, a chemist, an economist, a marketing specialist, a legal advisor, and an engineer. Other specialists may be brought in as the need arises.

Aquacultural engineering can simply be defined as the application of engineering principles to the production of food and fiber from aquatic environments. The aquacultural engineering specialty developed as a spin-off from the agricultural engineering programs in several major universities in the United States in the early 1970s. Aquaculture continues to be closely aligned with the agriculture sector, and agricultural engineers long ago recognized the need for blending the engineering sciences into the aquaculture sector if the industry was to survive and grow. Other engineering disciplines have filled niches as needs arose. Today many chemical, civil, sanitary, environmental, and other specialty engineers are involved with the aquaculture industry. Any of these engineers may call themselves *aquacultural engineers*.

The aquacultural engineer may find himself involved in endeavors other than those for the production of food and fiber. The functions of the industry also include the propagation of fish for restocking programs; ornamental pool and aquarium fish; aquatic fish and/or plants used in the pharmaceutical industry; aquatic organisms for production of industrial products such as oil, jewelry, or animal feed; and sport fish. Technological advances in any of these subareas may require an engineer's input.

NATURAL FISHERIES

Fishery production from natural waters comes mainly from commercial and sport fisheries. This text focuses mainly on commercial food-fish production, therefore sport fisheries are not addressed. Data on world commercial fishery landings are tabulated by the Food and Agricultural Organization (FAO) of the United Nations. Figure 1-1 shows the world commercial landings by continent in 1990. Total world landings of freshwater and marine fish and shellfish were 97.2 million MT (214 billion lb) in that year.

Figure 1-2 ranks the top 10 fishing nations in 1990. Japan led all nations, supplying 12% of the world's total commercial fishery landings. Fishing in the United States has not received the same emphasis as in other countries. Ranked fifth, the United States supplied only about 6% of the world's total landings. Data for the U.S. commercial landings from 1980–1991 indicate a very gradual, insignificant overall increase, whereas the total world catch increased dramatically over the same period. U.S. landings for the 50 states in 1991 totaled 4.3 million MT (9.5 billion lb), excluding the weight of mollusk shells, for a total value of over $3 billion (USDA 1992b). At present, domestic aquacul-

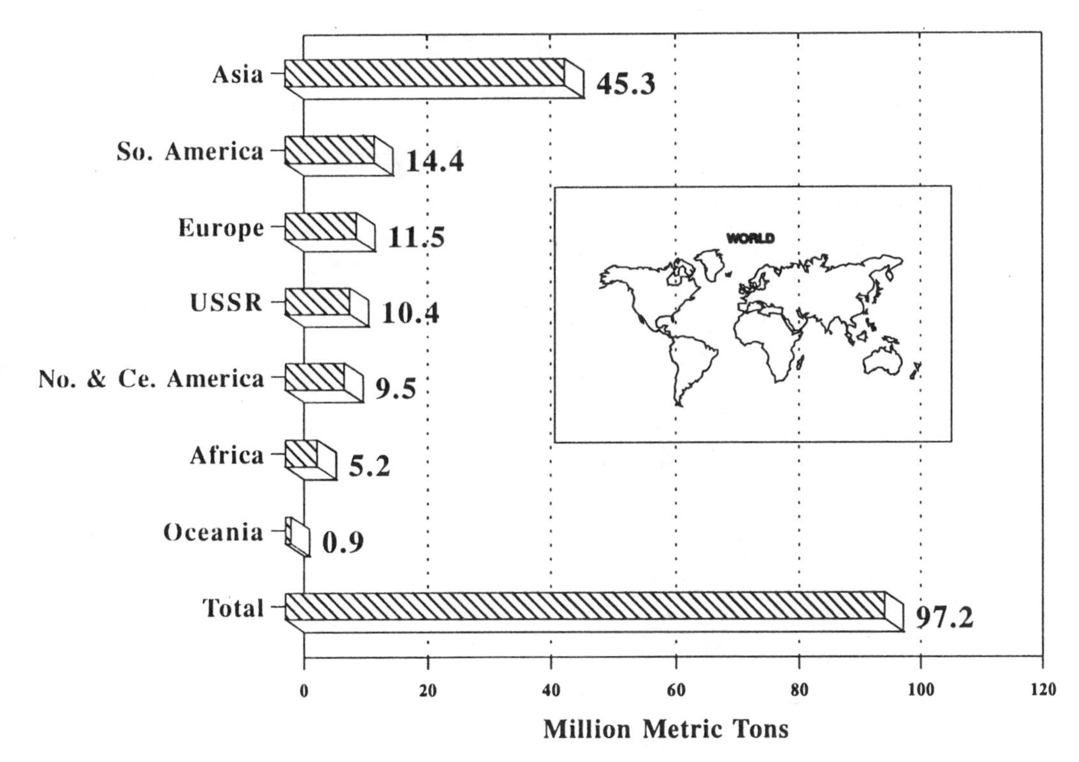

Figure 1-1. World landings by continent in 1991 (USDA 1992b).

ture provides only about 11% of all fishery products consumed in the United States, relatively little compared to domestic capture fisheries or imports.

PER CAPITA CONSUMPTION

U.S. consumption of all edible fishery products for 1975–1991 is shown in Figure 1-3. U.S. per capita consumption[2] of all edible fishery products in 1991 was 6.8 kg (14.9 lb) compared to the worldwide average per capita consumption of about 13 kg (FAO 1989). U.S. citizens consume only about half as much fish and seafoods as their counterparts in other countries. Domestic landings of edible fishery products have remained relatively stable, but consumer demand for fish and shellfish is increasing.

2. Derived by dividing the total amount of edible fishery products by the total civilian population.

U.S. imports for the years 1980–1991 indicate that stable commercial landings and a strong growth in demand have led to a rapid increase in imports of fish and seafood products over the past five years. U.S. imports grew from 0.95 million MT (2.1 billion lb) in 1980 to 1.37 million MT (3 billion lb) in 1991. Total dollar value in 1991 was $5.7 billion, up $438 million from 1990 (USDA 1992b).

U.S. AQUACULTURE

Fish culture began in America in 1853 when Theodatus Garlick and H.A. Ackley were the first to fertilize the eggs of brook trout (*Salvelinus fontinalis*). Until the early 1960s, commercial fish culture in the United States consisted principally of rainbow trout (*Salmo gairdneri*), bait fish, and several warm-water species (Parker 1989). A rapid growth in aquaculture production began in the 1970s and, although at

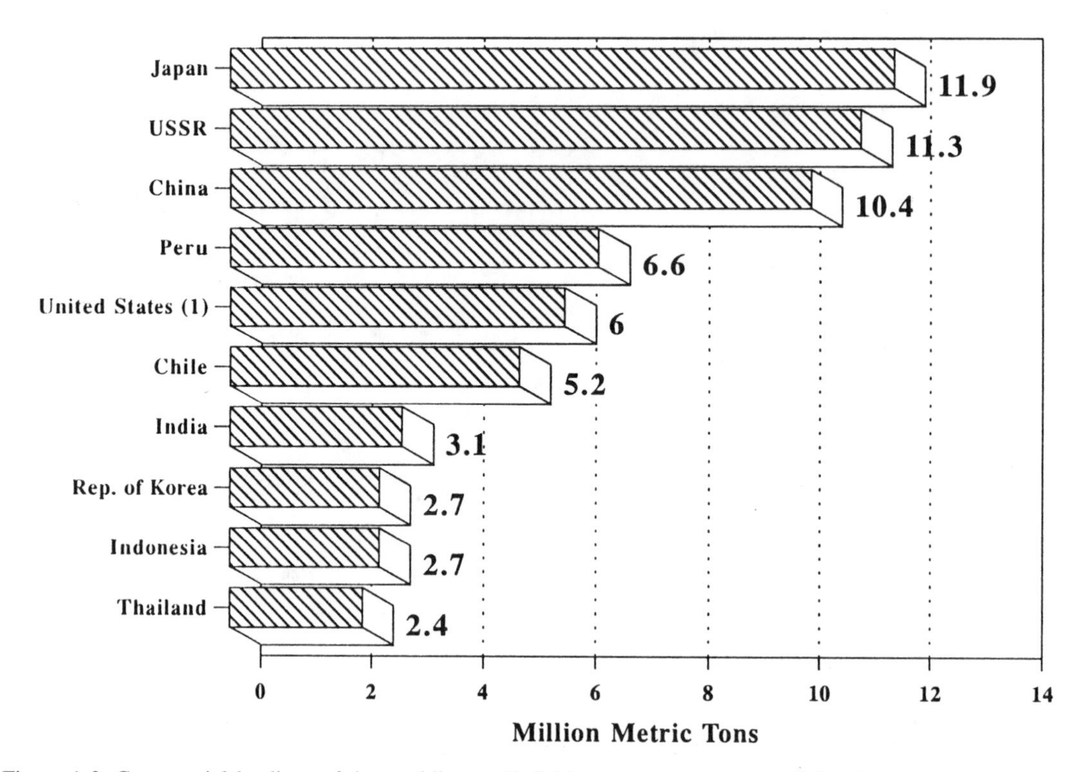

Figure 1-2. Commercial landings of the world's top 10 fishing nations in 1989 (NMFS 1990).

a slower pace, continued into the 1980s. Most of the increase in U.S. aquaculture production during this period came from the catfish industry. A portion of the growth in the industry from the late 1980s through the early 1990s was due to the establishment of new species for culture including tilapia, striped bass, walleye, and red drum and the introduction of specialty items like soft-shell crawfish and alligator.

SPECIES FOR AQUACULTURE

In the recent past catfish and salmonids were the only two species groups that were reared profitably in large numbers in the United States. Many other species now appear to have potential or have recently begun to be reared profitably. Total U.S. private aquaculture production for 1983–1987 is shown in Figure 1-4. The species are categorized into four groups: finfish, crustaceans, mollusks, and other species. Aquaculture production increased from 177,600 MT (391.5 million lb), worth an estimated $270.7 million in 1983, to 364,400 MT (803.4 million lb), worth $589.9 million in 1987. The bulk of U.S. private aquaculture production was in finfish, with catfish making up about half of all aquaculture species produced.

Finfish

Catfish. Catfish aquaculture began in the United States in the late 1950s. The principal species cultured commercially is the channel catfish (*Ictalurus punctatus*). During the 1980s and early 1990s the industry experienced tremendous growth, increasing over 600% (Fig.

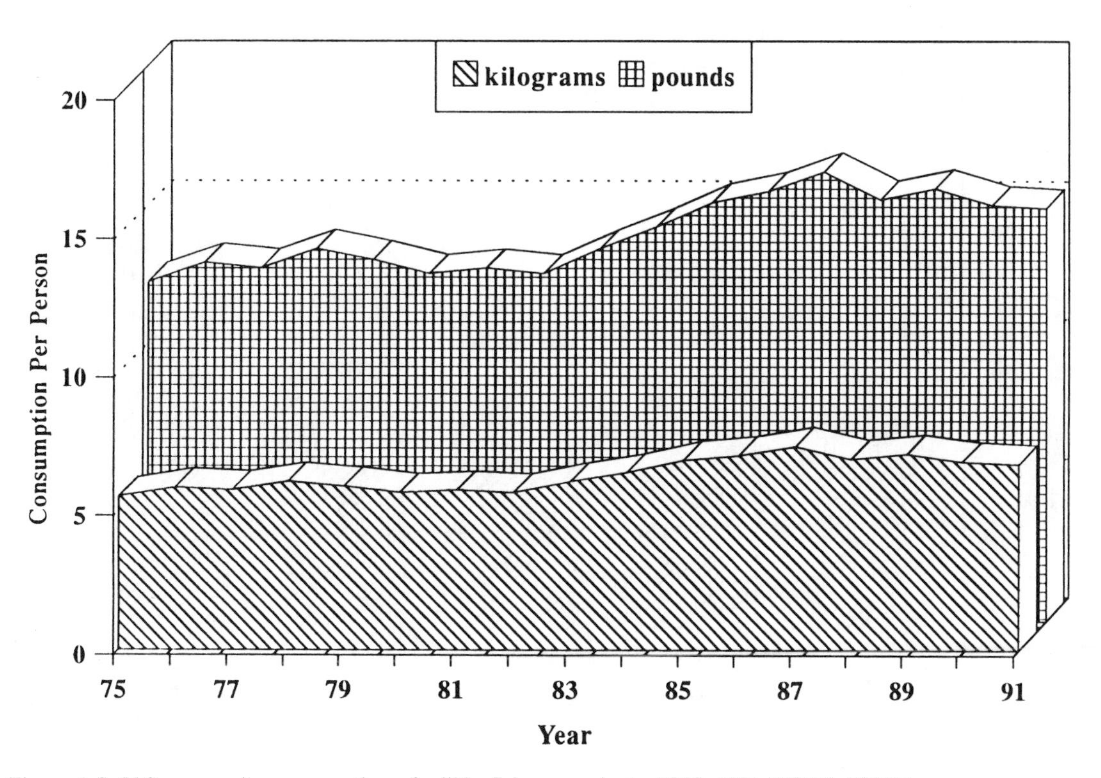

Figure 1-3. U.S. per capita consumption of edible fishery products, 1975–1991 (USDA 1992b).

1-5). In 1991, catfish farmers reported selling a total of 186,000 MT (410 million lb) of food-size fish, up 4% from 1990 (USDA 1992b). The value of sales was $213 million, down 13% from the previous year. One factor contributing to the loss in total value was a drop in catfish prices.

In the states surveyed in 1992, over 64,000 ha (158,400 ac) of catfish ponds were in production. The highest production was in Mississippi with 60% of the total U.S. water surface area. The three next highest states in terms of total water surface in catfish production were Arkansas, Alabama, and Louisiana. These four states together account for over 90% of all U.S. catfish aquaculture production. Most production is centered in the Southeast and Midwest.

Trout. Trout culture is the oldest aquaculture industry in the United States, dating to the 1800s. Trout production was originally begun to replenish wild stock in natural waterways by individuals and sportsmen's groups. These private hatcheries eventually evolved into the current state and federal hatchery system. State and federal hatcheries produce a number of species for restocking programs, but private commercial food-trout producers concentrate almost exclusively on the rainbow trout (*Salmo gairdneri*).

Trout growers are found in 15 states, but the majority of food-trout producers are established in the Snake River Valley in Idaho. This area provides exceptional water volume and quality. Limitations in future water resources could be offset by culture in recirculating systems.

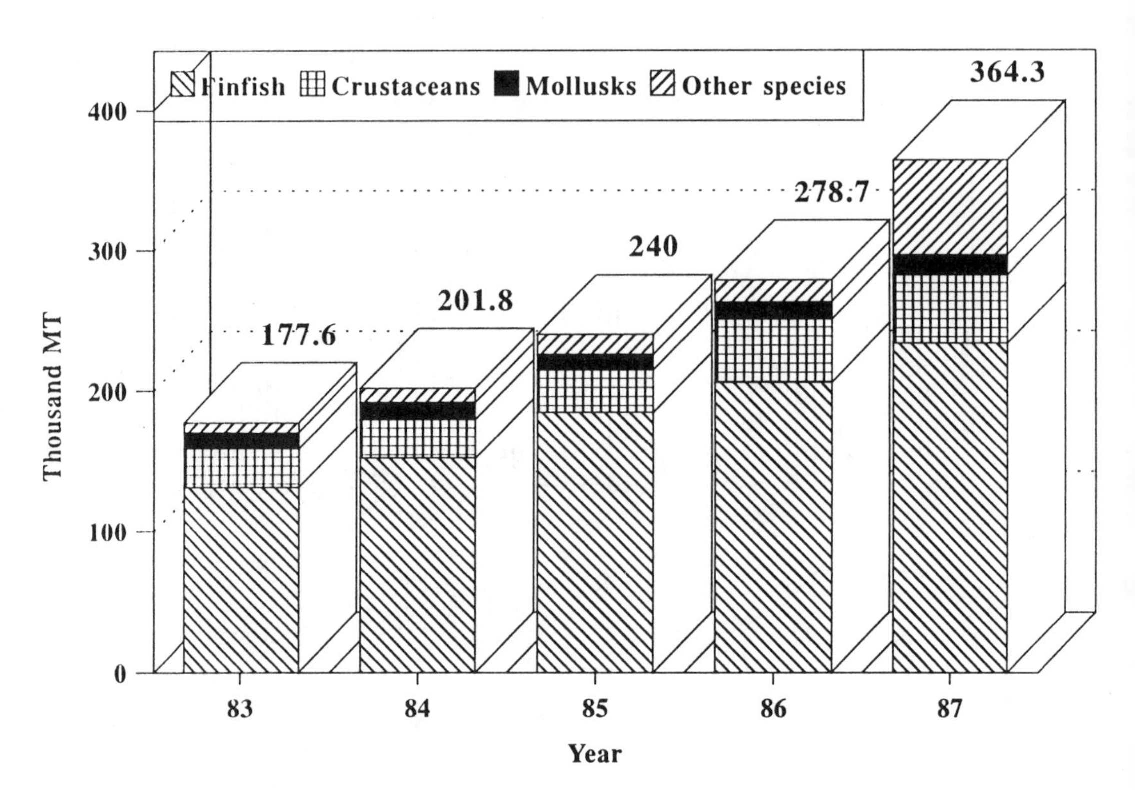

Figure 1-4. U.S. private aquaculture production by species group, 1983–1987. Data shown are live weights except for mollusks, which are meat weights. Eggs, fingerling, and other intermediate products are excluded (USDA 1989b).

Smaller production centers exist in other western states and along the mid-Atlantic coast.

Food-size trout production totaled about 27,000 MT (59.5 million lb) in 1991, valued at over $59 million (USDA 1992a). Although production increased by 1,800 MT, overall value decreased by about $5.5 million from 1990 prices. The decrease resulted from lower prices paid for food-size fish and stockers. The lack of adequate water supplies for the trout industry outside of Idaho continues to be the major obstacle to an expanding industry.

Salmon. Although facing strong competition from the domestic capture fishery and from foreign imports, the salmon aquaculture industry in the United States continues to grow. The wild salmon fishery is dependent almost entirely on five species: chinook (*Oncorhynchus tschawytscha*), chum (*O. keta*), pink (*O. gorbuscha*), sockeye or red salmon (*O. nerka*) and coho or silver salmon (*O. kisutch*). Farmed salmon in the United States are almost exclusively Pacific species, while in the rest of the world the industry consists of primarily Atlantic (coho and chinook) salmon. They are cultured in net pens and raceways.

The top five salmon farming nations in 1991 in descending order of production are Norway, Scotland, Chile, Japan, and Canada. The United States ranked seventh, with a production of over 7,000 MT (15.4 million lb) (Folsom and Sanborn 1992). Maine was the largest producing state with 4,500 MT (9.9 million lb).

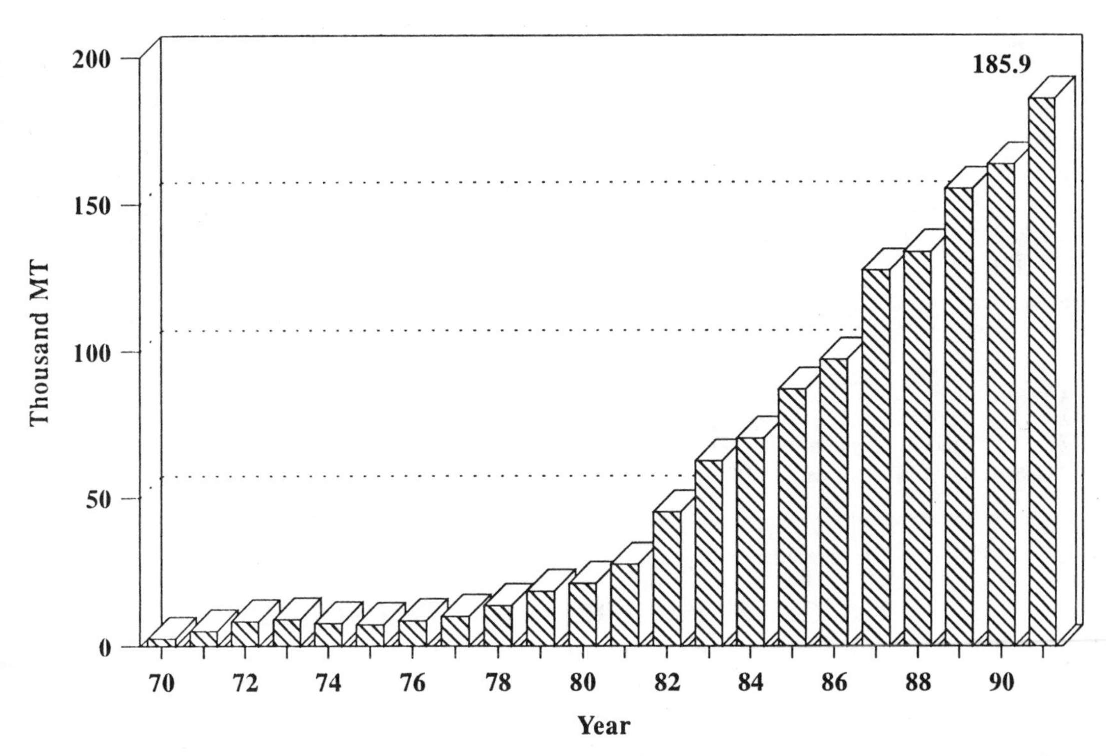

Figure 1-5. U.S. private catfish aquaculture production, 1970–1991. Eggs, fingerling, and other intermediate products are excluded (USDA 1989b, 1990, 1992a).

The outlook for the salmon aquaculture industry in the United States into the near future is for relatively slow growth. Norway, the United States' largest supplier of imported salmon, maintains a large stockpile of frozen salmon, which has depressed market prices. Large fluctuations in wild stocks also keep the market price in a cycle. Since the United States, Japan, and Europe are the three largest consumers of salmon, any disruptions to these markets will affect prices.

Tilapia. Tilapia culture has the potential to become one of the fastest growing segments of the aquaculture industry. Tilapia are important food fish throughout the Middle East, Africa, and some parts of Southeast Asia and South America. Unfamiliar to most Americans, they are sometimes marketed as the *Nile perch*, the *African perch* or *St. Peter's fish*. They are not native to the U.S., but at least one species, *Tilapia aurea* (the blue tilapia), has established itself in the wild in Florida (McLarney 1984). Several other species are of commercial importance: *T. nilotica* (the Nile tilapia), *T. mossambica*, *T. zillii*, *T. rendalli*, *T. hornorum*, and several hybrids. Each has unique skin coloration, but all are basically similar. The ''golden hybrid'' is the preferred species, and consumers prefer large fish (0.3 kg or 3/4 lb) over smaller fish.

One hindrance to expansion of this species is its inability to survive water temperatures much below 10°C (50°F). Tilapia can be grown in cooler regions in ponds supplied by geothermal wells, but in most of the United States production is restricted to indoor tank culture in heated water system or buildings that can be heated during the winter months. Commercial production has been identified in Arizona, California, Florida, Idaho, Louisiana, New Jersey, and Texas. Reported production figures are minimal. In early 1989, less than 4,500 MT (9.9 million lb) of production was confirmed (USDA 1989a).

The future outlook for tilapia culture is bright providing that the market expands. In many Southern states wildlife authorities are concerned about the potential for this species to establish itself in the wild and displace native species, which is a hindrance to industry expansion.

Crustaceans

Total crustacean aquaculture production in the United States in 1980 was 11,000 MT (24.2 million lb), valued at $14.2 million. By 1987 production had increased to 48,400 MT (106.5 million lb), with a dollar value exceeding $57 million (USDA 1989b). Marine shrimp and freshwater prawn (*Macrobrachium rosenbergii*) aquaculture is very small in comparison to crawfish. The promise that these species showed in the late 1970s and early 1980s has diminished somewhat in the United States since it was demonstrated that these species can be produced more cheaply in other countries.

Crawfish. The crawfish aquaculture industry in the United States is dominated by farm-raised crawfish. Production nationwide increased from less than 11,000 MT (24 million lb) in 1980 to over 40,000 MT (88.2 million lb) in 1990, down by about 7,000 MT (15.4 million lb) from the peak of 1987 (Fig. 1-6). Two species of com-

Figure 1-6. U.S. crawfish aquaculture production, 1980–1990 (Lorio 1992).

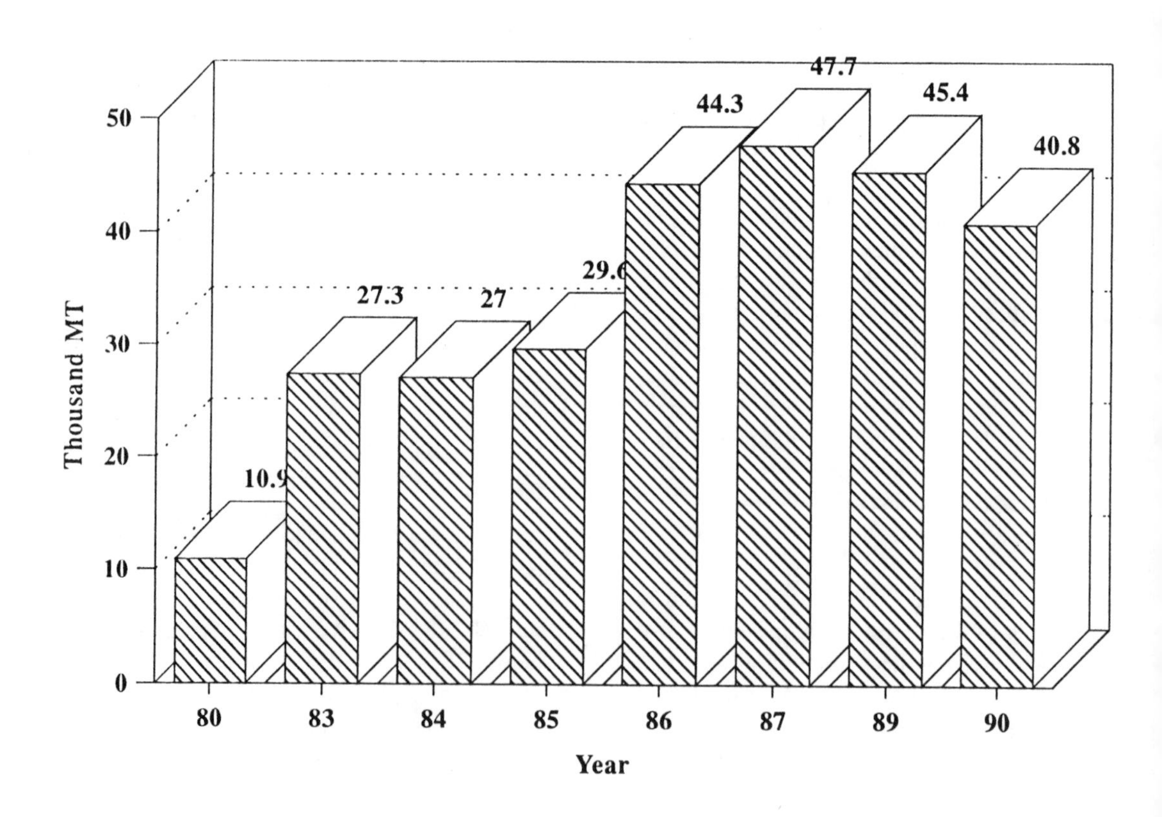

mercial importance in the United States are *Procambarus clarkii* and *P. zonangulus*. About 90% of production is in Louisiana. Smaller amounts are produced in Texas, Arkansas, Mississippi, Florida, South Carolina, and California.

A spin-off from the crawfish industry is softshell crawfish. They are produced by holding premolt crawfish in tanks of filtered water and harvesting soon after they molt. The yield is about 80% compared to 20% for hard crawfish. They are promoted as gourmet items and are sold whole, either frozen or fresh, primarily to upscale restaurants. The industry had its beginnings in Louisiana and evolved around *Procambarus clarkii* since molting procedures are well defined for this species (Culley and Duobinis-Gray 1990). About 36 MT (80,000 lb) of soft crawfish were produced in 1988 with an average farm price of \$3.60–\$4.50 per kg (\$8–\$10 per lb) (USDA 1989a).

Expansion of the crawfish industry is dependent upon acceptance of the product in nontraditional crawfish-consuming areas. Although the industry has spread to many states throughout the United States in recent years, most crawfish produced are still consumed within limited geographical regions such as the southeastern United States. A European market has developed for crawfish, particularly in Scandinavian countries. In 1991, the United States exported 2,950 MT (6.5 million lb) of crawfish, worth over \$5.2 million, to Europe, mostly to Sweden (USDA 1992a).

Shrimp. The U.S. wild shrimp harvest in 1991 was 145,000 MT (320 million lb) (USDA 1992b). By contrast, the U.S. shrimp aquaculture industry is very small. Production in 1991 was estimated at 1,590 MT (3.5 million lb). Viable production systems have developed in Hawaii, Texas, and South Carolina, but producers will remain at the mercy of the fluctuations in production of the much larger shrimp-farming countries and the domestic wild harvest.

Mollusks

It is difficult to define mollusks in terms of whether or not they are aquaculturally produced. Some operations marginally fit the commonly accepted definitions of aquaculture.

The U.S. molluskan aquaculture industry focuses on three major species groups: oysters, clams, and mussels. Abalone and scallops are also farmed, but there are no reliable production figures for these species. Harvest of the three major groups increased from 11,700 MT (25.8 million lb) in 1983 to 14,200 MT (31.3 million lb) in 1987 (USDA 1990). The molluskan aquaculture industry is still very small compared to the commercial wild harvest, which was 157,000 MT (346 million lb) of meats in 1991 valued at \$421 million (USDA 1992b).

The future outlook for the U.S. molluskan aquaculture industry is somewhat cloudy, although modest growth is expected. Diseases in the oyster industry have decimated wild stocks in recent years. Much research is centered on developing disease-resistant strains. Numerous media warnings about the dangers of consuming raw oysters and clams have hurt the industry by making the consumer cautious about the product.

Other Species

Hybrid Striped Bass. The hybrid striped bass receiving most attention in the United States is a cross between a female striped bass (*Morone saxatilis*) and a male white bass (*M. chrysops*). The reverse cross is also being looked at for its culture potential. The culture of striped bass or hybrids may be illegal in those states where the fish is not a native. The Louisiana State Legislature changed the status of the hybrid from a game fish to a domestic fish in 1988, thus opening the door for aquaculture of this species (Fitzgerald 1990). Other states may soon follow suit.

Many ventures to culture hybrid striped bass have developed around intensive culture in tanks since the animals have a faster growth rate in these systems. Early results indicate that 2.2–6.6 kg (1–3 lb) fish can be produced in one to two years. However, during the 1990s, it is anticipated that tank systems will be challenged by open pond growers since developmental and operating costs are less in ponds. The culture of this species is largely still in an experimental phase.

The U.S. government does not provide production figures for hybrid striped bass, but the Hybrid Striped Bass Growers Association estimated that 1990 domestic production was 634 MT (1.4 million lb) live weight (USDA 1992a). The long term potential for this species is good because it doesn't face direct competition from production in other countries, and the wild harvest is relatively small. Markets outside the United States are small, and fingerling production remains a relatively difficult matter. These two factors must be overcome to enhance growth of the industry.

Red Drum. Aquaculture of the red drum (*Sciaenops ocellatus*) generated much excitement in the United States in the mid-1980s when several Gulf Coast states banned the commercial harvest of this species due to severely declining populations. But the jury is still out on this one. Several drawbacks make culture of this species difficult. Highly cannibalistic, the species requires constant culling to separate the larger aniamls from the smaller ones. The species is sensitive to sudden freezes, which prevents its culture in cold climates. Attempts to culture this species in recirculating systems at Louisiana State University in the mid-1980s achieved only moderate success. Pond culture of red drum is the subject of much research in both Texas and Louisiana.

Alligators. The U.S. alligator (*Alligator mississippiensis*) industry experienced rapid growth in the late 1980s and early 1990s. Culture of this species is limited to just a few states, including Louisiana, Texas, Florida, Georgia, and California. Louisiana is by far the leading producer, where 150,000 hides, worth an estimated $33.3 million, were produced in 1990 (USDA 1992a). The industry exists mainly for the hides, but there is a rapidly growing market for the alligator meat, and a small tourist industry has grown near large metropolitan areas, where tourists pay to view the animals.

THE FUTURE OF U.S. AQUACULTURE

The aquaculture industry in the United States has grown rapidly over the past decade, but it remains an immature and unstable industry. The next 10 years could be more important in terms of growth. Several forces should contribute to continued growth. First, world population increases will increase the demand for fish and seafood items. The catch for many species is approaching the maximum sustainable yield, and many countries are placing harvesting quotas on species taken from their territorial waters. Thus, the increased demand can only be met through an increase in aquacultural production.

Second, many aquaculture products are produced under controlled conditions. This should increase their appeal to consumers concerned about pollution and seafood safety. Third, there is increased emphasis on nutrition and health in the United States and certain other countries. Considerable marketing effort went into demonstrating the health benefits derived from fish consumption.

Finally, the continued expansion of the aquaculture industry could be keyed to several other issues: the changing organizational structure of the industry in the production and marketing areas, the emergence of new species for aquaculture, increasing competition from foreign growers, and possible trade conflicts. Continued growth in the established industries will result from better production technologies, including net pen, cage culture, and closed recirculating systems; genetic manipulation; better control of diseases and off-flavor; continued increases in feed conversion and stocking densities; development of new processing technologies; and development of more efficient harvesting systems.

There are three bills currently before the U.S. Congress that could have a strong impact on the aquaculture industry. Two deal with seafood inspection. The first, the Fish Safety Act of 1992 (S.2884), was written to establish a mandatory seafood inspection program with the U.S. Department of Agriculture (USDA) as the lead agency for the program. Other government agencies involved are the National Marine Fisheries Service (NMFS) and the Food and Drug Administration (FDA). Under this bill, the seafood inspection program would be based on the Hazard Analysis Critical Control (HACCP) method, which is a process control system which emphasizes prevention rather than correction for already contaminated seafood products. The second bill, the Consumer Seafood Safety Act of 1992 (S.2538), names the FDA as the lead agency in charge of the seafood inspection program. It also emphasizes the development and implementation of a shellfish safety program.

The third bill before Congress, the National Aquaculture Development Act of 1992 (H.R. 2798), focuses on more clearly defining the role of the federal government in aquaculture and identifying one agency to take the lead role for interacting with the aquaculture industry (USDA 1992a). Major provisions of the bill are: (1) Designation of the USDA as the lead agency for establishing aquaculture policy; (2) establishment of an Office of Aquaculture Policy Coordination within the USDA; (3) requirement of the departments of Commerce and Interior to name aquaculture coordinators within their respective agencies; (4) official definition of aquaculture as ''the controlled cultivation of aquatic plants and animals;'' (5) instruction of the USDA to treat aquaculture products as agricultural commodities; and (6) mandate that the USDA spend a portion of its funds on aquaculture research, development, promotion, and demonstration projects. The highest priority would be given to research programs for therapeutic compounds.

How these and other issues are dealt with in the near future will determine how far and how fast the aquaculture industry will expand.

CHAPTER 2
WATER QUALITY AND ENVIRONMENTAL REQUIREMENTS

Aquatic ecosystems comprise a diverse assemblage of organisms whose interactions between each other and the environment cause a series of complex physicochemical reactions, many of which have strong influences on fish and invertebrate culture. These interactions are demonstrated by the web illustrated in Figure 2-1. Water properties and environmental parameters often affect the choice for an aquaculture site and the cultured species. Thus, it is important that the aquaculturist have an understanding of water chemistry and possess the skills necessary to provide a suitable environment for the venture to be successful.

This chapter explores those environmental parameters, both physical and chemical, having a direct impact on fish and invertebrate culture. The focus is on the major parameters and does not dwell on lesser parameters, the effects of which are not well defined. A discussion of every chemical reaction that takes place in an aquaculture system would be too voluminous for inclusion into a single text. Therefore, the reader is referred elsewhere for more detailed discussions on water chemistry. Several good references on aquaculture water chemistry are Boyd (1982, 1990, 1992), Tucker and Robinson (1990), and Brune and Tomasso (1991).

ENVIRONMENTAL CRITERIA FOR AQUACULTURE

The production of fish and invertebrates is influenced and narrowly confined by the hydrobiological characteristics of the environment and by the various microorganisms and toxic substances found in the environment. The physical, chemical, and biological environmental parameters are interrelated in a complicated series of physicochemical reactions that affect every aspect of fish culture (survival, growth, and reproduction). Fish culture also has a pronounced reverse effect on the environment. For example, fish consume oxygen and produce metabolic by-products, such as ammonia and carbon dioxide, which react with the already present constituents of the environment to drive an ever-changing chain of reactions. The effects of these interactions are to produce a type of "chemical soup" in which the fish must live, as depicted in Figure 2-2. These interactions will be discussed in detail in the remainder of this chapter.

Quality standards for aquaculture are presented in Table 2-1. These standards are derived from years of practical experience and the efforts of numerous research projects, therefore, they are not arbitrarily selected. They are meant to be used as a guide only, since certain param-

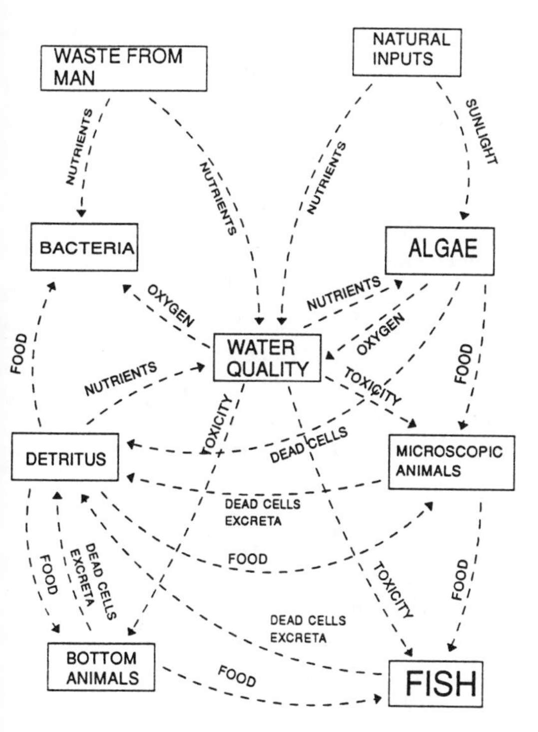

Figure 2-1. Physico-chemical interactions in an aquatic ecosystem.

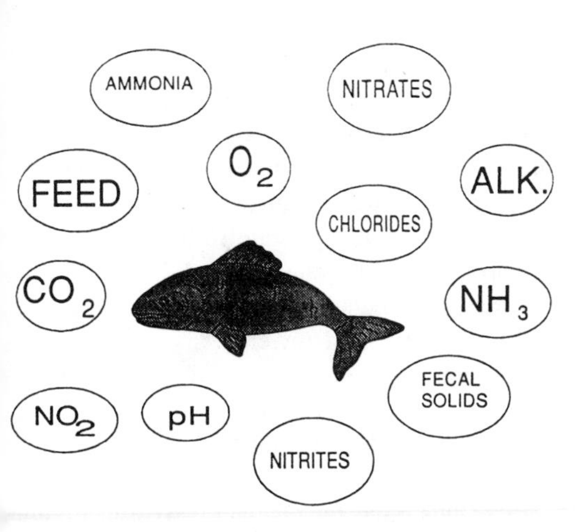

Figure 2-2. Chemical soup in which fish live.

eters can vary significantly from one species to another. A detailed discussion of every parameter shown is not possible. Therefore, the discussion that follows will be limited to several of the more important parameters.

Table 2-1. Water quality standards for aquaculture.

Parameter	Concentration
Alkalinity	10–400
Aluminum	<0.01
Ammonia (NH_3)	<0.02
Ammonia (TAN)	<1.0
Arsenic	<0.05
Barium	5
Cadmium	
Alkalinity <100 mg/L	0.0005
Alkalinity >100 mg/L	0.005
Calcium	4–160
Carbon dioxide	0–10
Chlorine	<0.003
Copper	
Alkalinity <100 mg/L	0.006
Alkalinity >100 mg/L	0.03
Dissolved oxygen	5 to saturation
Hardness, Total	10–400
Hydrogen cyanide	<0.005
Hydrogen sulfide	<0.003
Iron	<0.01
Lead	<0.02
Magnesium	<15
Manganese	<0.01
Mercury	<0.02
Nitrogen (N_2)	<110% total gas pressure <103% as nitrogen gas
Nitrite (NO_2)	0.1 in soft water
Nitrate (NO_3)	0–3.0
Nickel	<0.1
PCB's	0.002
pH	6.5–8
Potassium	<5
Salinity	<5 percent
Selenium	<0.01
Silver	<0.003
Sodium	75
Sulfate	<50
Sulfur	<1
Total dissolved solids (TDS)	<400
Total suspended solids (TSS)	<80
Uranium	<0.1
Vanadium	<0.1
Zinc	<0.005
Zirconium	<0.01

Source: Meade (1989) with permission.

Note: Concentrations are in mg/L except for pH.

Physical Variables

Temperature. Water temperature is the environmental parameter having the greatest effect on fish. It can be thought of as a primary factor affecting the economic feasibility of a commercial aquaculture venture. It is impractical to control temperature in ponds; therefore, large-scale pond culture of any species must be conducted in geographical regions having sufficiently long growing seasons so that a market-size fish can be produced in a reasonable amount of time. Temperature also affects oxygen solubility and causes interactions of several other water quality parameters. How these parameters interact will be discussed throughout this chapter.

Water temperature greatly influences physiological processes such as respiration rates, efficiency of feeding and assimilation, growth, behavior, and reproduction (Meade 1989; Tucker and Robinson 1990). A temperature increase of 10°C (18°F) will generally cause rates of chemical and biological reactions to double or triple. For example, fish will consume two-three times as much oxygen at 30°C (86°F) than they would at 20°C (68°F), and their biochemical reactions will double or triple. Because of this, dissolved oxygen requirements are more critical in warm water than cold water.

Fish can be grouped according to their temperature requirements as *cold-water*, *cool-water* or *warm-water* species. Cold-water species are those preferring water temperatures of 15°C (59°F) or less; cool-water species prefer 15–20°C (59–68°F); and warm-water species prefer waters above 20°C (68°F) (Romaire 1985). A fourth classification, *tropical*, is recognized by some biologists. These classifications are used mainly for convenience. However, there is much overlap, and often species of different classifications are found in the same place at the same time.

Fish are *poikilothermic*, meaning that they have essentially the same body temperature as their surroundings and that temperature is governed by external influences, particularly solar radiation. Each species has a characteristic growth curve and an optimum growth range that changes with temperature and fish size. Every species also has upper and lower temperature limits beyond which it cannot survive. Within a species' tolerable temperature range, growth will continue until a peak is reached and then decline just before reaching the upper lethal limit (Figure 2-3). The optimum temperature for growth is that which the fish would select for itself, given the choice. The fish's ability to ward off diseases is best near the optimum growth temperature. Temperatures on either side of optimum induce stress in the animal, which affects feeding, growth, reproduction, and disease inhibition. The probability for culture success is also greatest near the optimum growth temperature. Optimum temperature ranges for many species of fish and invertebrates are shown in Table 2-2.

Acclimatization is the process by which fish physiology is slowly altered so that it can adapt to environmental changes such as temperature and pH. Fish and invertebrates have a very low tolerance for sudden changes in temperature, that is, often rapid changes of as little as 5°C (9°F) will stress or kill them. Temperature changes must therefore be brought about gradually, a few degrees per day. Very rapid temperature changes of more than 0.9°C (1°F) per minute can cause thermal shock and death. Thus, many aquatic animals can survive broad ranges in water temperature so long as changes are brought about gradually.

Once acclimated to a specific temperature, there are upper and lower limits to which the water temperature can be raised or lowered beyond which the fish cannot survive. The higher the acclimation temperature, the higher will be the upper and lower lethal limits. This relationship is illustrated for channel catfish in Figure 2-4. Catfish acclimated to 16°C (60°F) will survive if the water temperature is raised to about 33°C (91°F). Above 33°C (91°F) the fish will die. The figure was developed using a temperature rise of 1.7°C (2°F) or less per hour. Acclimation temperature becomes less important as the rate of temperature change becomes slower. The area within the lethal limits in Fig-

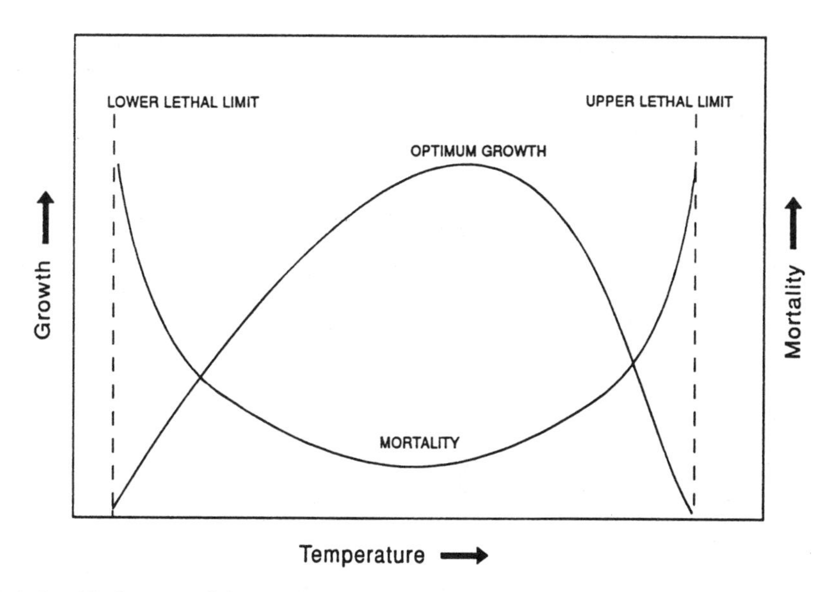

Figure 2-3. Relationship between fish growth and temperature (from Meade (1989) with permission).

ure 2-4 is called the *zone of tolerance*, and each species has its own zone of tolerance.

Water temperature is responsible for the development of thermal stratification in fish ponds. Thermal stratification is the separation of water into distinct layers. The surface layer, the *epilimnion*, is warmer and less dense than the lower layer, the *hypolimnion*. A thin layer of water of rapidly changing temperature, the *thermocline*, exists between the epilimnion and the hypolimnion, as shown in Figure 2-5. In temperate regions ponds may stratify in the spring and remain stratified until fall. "Overturns" experienced in temperate ponds in spring and fall tend to temporarily break up the stratification. In shallow ponds in temperate and tropical regions the stratification may exhibit a diel pattern. When this happens surface and bottom layers stratify during the day but mix at night when surface layers cool (Romaire 1985). Thermoclines produce many effects, one of which is the inhibition of oxygen exchange between surface and lower layers. Consequently, fish are concentrated in the upper layer above the thermocline, and the full capacity of the water column is not utilized. Water mixers and aerators are effective at breaking up thermal stratification in ponds.

Chemical treatments of fish ponds are also affected by water temperature. In warm water, fertilizers dissolve and herbicides react more rapidly, fish toxicants are more effective and react quicker, and the rate of oxygen consumption by decaying manure or vegetation is greater (Boyd 1979).

Density. Pure water is most dense at approximately 4°C (39°F). Below this temperature water becomes less dense and freezes at 0°C (32°F). The variation of density with temperature is shown in Table 2-3. Adding impurities and salts to water increases its density. Thus, seawater reaches its maximum density at −1.4°C (29.5°F) and a salinity of 24.7 g/L (24.7 ppt) (Wheaton 1977). One would think that as seawater freezes ice would form and accumulate on the bottom of the oceans. However, this is not the case. As seawater freezes the ice and salt separate, and the ice formed floats since it is less dense than seawater. The salinity of the water immediately surrounding the ice increases above that of the surrounding water. An increase in pressure increases water density. However, the change is so slight that, for all practical purposes, water is considered to be an incompressible fluid.

Table 2-2. Optimum rearing temperatures for selected species.

Species	Temperature (°C)	Reference
Brook trout	7–13	Piper et al. (1982)
Brown trout	12–14	Petit (1990)
Brown trout	9–16	Piper et al. (1982)
Rainbow trout	14–15	Petit (1990)
Rainbow trout	10–16	Piper et al. (1982)
Atlantic salmon	15	Petit (1990)
Chinook salmon	10–14	Piper et al. (1982)
Coho salmon	9–14	Piper et al. (1982)
Sockeye salmon	15	Petit (1990)
Sole	15	Petit (1990)
Turbot	19	Petit (1990)
Plaice	15	Petit (1990)
European eel	22–26	Petit (1990)
Japanese eel	24–28	Petit (1990)
Common carp	25–30	Petit (1990)
Mullet	28	Petit (1990)
Tilapia	28–30	Petit (1990)
Channel catfish	27–29	Tucker and Robinson (1990)
Channel catfish	21–29	Piper et al. (1982)
Striped bass	13–23	Piper et al. (1982)
Red swamp crawfish	18–22	Romaire (1985)
Freshwater prawn	30	Romaire (1985)
Brine shrimp	20–30	Romaire (1985)
Brown shrimp	22–30	Romaire (1985)
Pink shrimp	>18	Romaire (1985)
American lobster	24	Romaire (1985)
American lobster	20–22	Hedgecock et al. (1976)
American oyster	>8	Romaire 1985

Freezing. The freezing characteristics of water are more unusual than those of other liquids. The hydrogen bonding in water produces a crystalline structure upon freezing that has a larger total volume than the original liquid. The ice that forms has a density less than that of water and floats. In this manner lakes and ponds freeze from the top down. If the reverse were true, water bodies would freeze solid. Salt addition lowers the freezing point of water. Pure water freezes at 0°C (32°F) while seawater freezes at −1.4°C (29.5°F).

Salinity. Salinity is a measure of the concentration of dissolved ions in water expressed in grams per liter (g/L) in the SI system of units or parts per thousand (ppt) in the English system. The symbol o/oo is often used to denote parts per thousand. The major dissolved ions are sodium (NA^+) and chloride (Cl^-). Magnesium (Mg^{+2}), calcium (Ca^{+2}), potassium (K^+), sulfate SO^{-4}) and bicarbonates (HCO_3) are also present in significant amounts (Romaire 1985). The level of salinity reflects geological and hydrological conditions in the geographical region. Surface waters in areas of high rainfall usually have low salinity, whereas waters in arid regions with high evaporation tend to have high salinity. Seawater varies from about 33 to 37 g/L, with 34 g/L being about average. Estuarine waters have varying salinities from near full-strength seawater down to about 3 g/L.

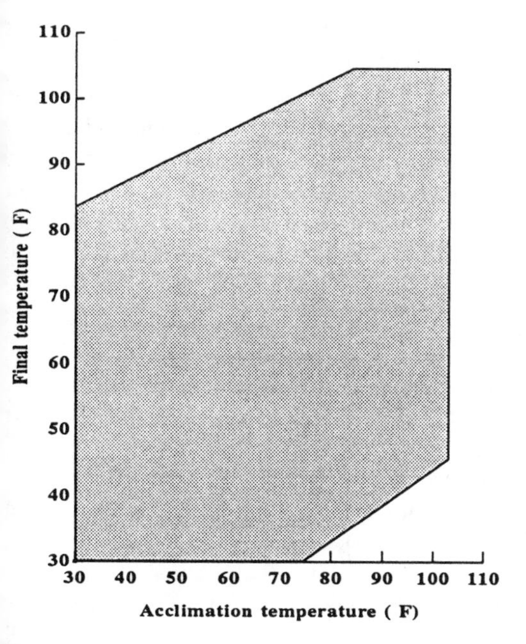

Figure 2-4. Thermal tolerance diagram for channel catfish (from Tucker and Robinson (1990) with permission).

Most inland waters have a salinity less than 2–3 g/L (McLarney 1984; Boyd 1990).

Chlorinity is defined as the total weight of chlorine, bromine, and iodine contained in 1 kg of seawater after all of the bromine and iodine have been replaced by chlorine (Stickney 1991). Since chloride accounts for about 55% of salinity and is easy to measure, chlorinity can be determined and salinity calculated with Equation 2–1 (Ross 1970):

$$C = 1.8066\ [Cl] \qquad (2\text{-}1)$$

where S = salinity, and Cl = chlorinity, both in units of g/L.

The composition and concentration of dissolved salts in the body fluids of fish and invertebrates must be maintained within fairly narrow limits to buffer against changes that can cause physiological disruptions. This process, called *osmoregulation*, requires expenditure of metabolic energy. Some species are *euryhaline*, meaning that they can tolerate a wide range in salinity. Other species have a limited tolerance to salinity changes and are referred to as *stenohaline*. No clearly defined line separates the two groups, and there is no commonly accepted criteria that places a species in one group or the other (Romaire 1985). Every species has an optimum salinity range, and when forced out-

Figure 2-5. Thermal stratification in a fish pond.

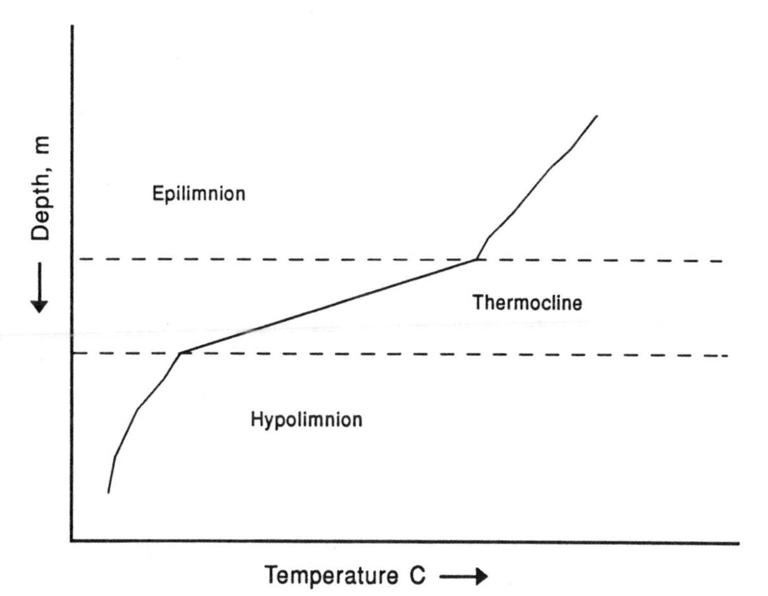

Table 2-3. Density of water at different temperatures.

Temperature (°C)	Density (g/cm^3)	Temperature (°C)	Density (g/cm^3)
0	0.99987	16	0.99897
1	0.99993	17	0.99880
2	0.99997	18	0.99862
3	0.99999	19	0.99843
4	1.00000	20	0.99823
5	0.99999	21	0.99802
6	0.99997	22	0.99780
7	0.99993	23	0.99757
8	0.99988	24	0.99733
9	0.99981	25	0.99707
10	0.99973	26	0.99681
11	0.99963	27	0.99654
12	0.99952	28	0.99626
13	0.99940	29	0.99597
14	0.99927	30	0.99568
15	0.99913		

Source: Boyd (1990) with permission.

side of this range, metabolic energy is spent on osmoregulation at the expense of other functions. If salinity deviates too far from optimum, the animal cannot maintain homeostasis and dies.

The blood of freshwater fish has an osmotic pressure equal to that of a 7-g/L sodium chloride solution (McKee and Wolf 1963). Many freshwater fish can survive in saline waters up to 7 g/L, but growth is poor. Therefore, 2 g/L is recommended as an upper salinity limit for freshwater fish. Euryhaline freshwater species like trout can be cultured in a broad range of salinities from fresh- to seawater. However, even through trout can survive high salinities, survival and growth decreases at about 20 g/L (McKay and Gjerde 1985). Grass carp have an upper salinity tolerance of about 10–14 g/L (Maceina and Shineman 1979). Two tilapia species, *Tilapia aurea* and *T. nilotica*, grew well in salinities ranging from 0–10 g/L (Stickney 1986). The red hybrid (*T. mossambica* × *T. aurea* × *T. nilotica*) does well in salinities up to about 17 g/L.

Adult channel catfish can tolerate salinities up to 11–14 g/L, but growth is poor at salinities greater than about 6–8 g/L (Perry and Avault 1970). The optimal salinity for channel catfish is unknown, but lies in the range of 0.5–3 g/L. Good growth is achieved in salinities lower than 0.5 g/L. For optimal egg and fry development salinities should be lower than 3 g/L.

The egg and larval forms of the freshwater prawn, *Macrobrachium rosenbergii*, require saline water. Tansakul (1983) recommended a salinity of about 12 g/L for prawn hatcheries. Postlarval stages can be grown to market size in freshwater.

The red swamp crawfish, *Procambarus clarkii*, is widely cultured in the United States and Europe. This species can survive salinities up to 20 g/L, but they are seldom found in water having salinity greater than 8 g/L (Huner and Barr 1984).

As with temperature, it is wise to gradually expose fish to changes in salinity. Small fish are more susceptible than adults of the same species. When salinity is changed by more than about 10% within a few minutes or hours, fish and invertebrates may not be able to compensate (Boyd 1990). Sodium chloride (NaCl) in the form of common rock salt can be used to raise salinity in culture systems. Numerous brands of artificial sea salts are commercially available, but these are too expensive to be used in ponds and should be limited to use in small recirculating systems.

Electrical Conductance. Electrical conductance is a measure of the dissolved mineral content (salinity) of water and changes in direct proportion to salinity. The greater the proportion of ions in water, the higher the conductivity. Because water ionizes so slowly, it acts as an insulator and is a poor conductor of electricity. The unit of conductivity is the micromho per centimeter. Distilled water has a conductivity of about 1 μmho/cm while natural freshwaters have conductivities ranging from 20–1,500 μmho/cm (Boyd 1990). Conductance can be used to obtain reliable estimates of salinity or total dissolved solids.

Turbidity. Turbidity is a measure of light penetration in water. It is produced by dissolved and suspended substances, such as clay particles, humic substances, silt, plankton, colored compounds, etc. The more dense these substances, the higher the turbidity. Excessive turbidity can be troublesome in fish ponds and flow-through systems, but seldom is a problem in properly managed recirculating systems.

Turbidity caused by phytoplankton is usually desirable in fish ponds since it enhances fish production. It also limits light penetration, preventing the growth of rooted aquatic plants, which are undesirable in fish ponds. However, turbidity caused by suspended clay and other colloidal particles is undesirable. Clay turbidity that restricts visibility to 30 cm (12 in.) or less can inhibit the development of good phytoplankton blooms (Romaire 1985). Excessive runoff from the surrounding watershed can often cause clay and silt loads to exceed 20,000 mg/L. This is cause for alarm since these particles can clog the gills of small fish and invertebrates, settle onto and smother fish eggs, shield food organisms, settle in tanks, and clog filtration equipment. Fish seem less affected at concentrations below 20,000 mg/L for short periods.

Turbidity caused by suspended solids appears to affect fish more than clay turbidity. Cold-water fish have been killed as a result of exposure to 500–1,000 mg/L suspended solids for three to four (Alabaster and Lloyd 1982). Good to moderate fish production can result at suspended solids concentrations between 25 and 80 mg/L, but 80 mg/L is recommended as a maximum. Channel catfish seem to be more tolerant. Fingerling and adults can survive long-term exposure to 100,000 mg/L suspended solids, but behavioral changes were noticed at 20,000 mg/L (Tucker and Robinson 1990).

There are corrective measures that can be taken to reduce clay turbidity and turbidity caused by other substances in fish ponds. It is not the intent of this text to provide extensive procedures to remedy turbidity problems, therefore, the reader is referred to Boyd (1982, 1990) and McLarney (1984).

Water turbidity in fish ponds is commonly measured with the Secchi disk. A *Secchi disk* is a round disk having a diameter of 30 cm (12 in.). The disk is divided into quadrants, two opposite quadrants painted white and the other two black. The disk is attached to a rope or cable marked in increments. A measurement is taken by lowering the disk into a water body until it just disappears from sight. The depth at which the disk disappears is the *Secchi disk visibility*, and for aquaculture purposes, is usually expressed in centimeters. Other measurement techniques are discussed at the end of this chapter.

Color. Color is a result of the interaction of incident light and impurities in the water. Pure water appears blue in white light since the blue colors of the spectrum travel farther in water than others and is scattered more (Wheaton 1977). The addition of humic substances in water imparts a tea-colored or reddish hue. Heavily manured ponds or ponds in wooded areas or swamplands often are high in dissolved humic substances. Iron associated with humic substances can impart a yellow color. Certain algae impart a color dependent upon the species (i.e., the presence of green algae makes water appear green in color). Water color in highly productive waters like fish ponds is largely dependent on the color of the predominant species of phytoplankton. Unproductive waters generally have a bluish color and are very transparent since color is caused by light scattering as it hits dissolved particles in the water. Impending oxygen shortages in the water can often be detected by changes in color.

Total Gas Pressure. *Total gas pressure* (*TGP*) is the sum of the partial pressures of all gases dissolved in water. The difference between *TGP* and barometric pressure (*BP*) is a term called ΔP. At equilibrium, *TGP* is equal to *BP*, and $\Delta P = 0$. If *TGP* is greater than *BP* ($\Delta P > 0$), the water is supersaturated and gases come out of solution. When *TGP* is less than *BP* ($\Delta P < 0$), the water is undersaturated and gases will diffuse into water from the atmosphere. Gas saturation is often reported as a percentage of local *BP*, as shown in the following relationship:

$$\% \, TGP = \frac{(BP + \Delta P)}{BP} \times 100 \qquad (2\text{-}2)$$

Example problem 2–1. As an example, assume that BP = 760 mm Hg[1] and ΔP = 38 mm H in Equation 2-2. Calculate the total gas pressure.
Solution:

$$\% \, TGP = \frac{760 + 38}{760} \times 100 = 105 \, \% \text{ of saturation}$$

Total gas pressure is important to aquaculturists because supersaturated waters can cause a condition called *gas bubble trauma* or *gas bubble disease*. Supersaturation is an unstable condition, and the gases form bubbles when they come out of solution. Bubbles formed in the blood and tissues of fish restrict circulation and oxygen supply to tissues. A TGP over 105% saturation is undesirable in fish culture systems. Ideally, TGP should be less than 100% saturation in aquaculture systems. Chronic gas bubble trauma can occur in channel catfish after long-term exposure to positive ΔP values less than 76 mm Hg (about 110% saturation) (Tucker and Robinson 1990).

Gas bubble trauma is affected by submersion depth. Fish located deeper in the water column are less affected than those at shallower depths. It is the hydrostatic pressure of the overlying water that keeps the gases in solution. Thus, the deeper a fish swims beneath the surface, the greater the hydrostatic pressure. As an example, in freshwater at 21°C (70°F), ΔP is reduced by 67 mm Hg for every meter (3 ft) of submersion. Fish confined to shallow water, such as in tanks, cages and raceways, are particularly susceptible because they cannot dive to deeper water to reduce the effects of high ΔP.

Gas bubble trauma can be a problem in flow-through and closed, recirculating systems. In those systems, supersaturated water rapidly flows through, and little time is afforded for degassing. Source water for these systems must often be degassed by gravity flow through a packed column, or water can otherwise be vigorously agitated to drive out the gases. Gas bubble trauma almost never occurs in fish ponds because added supersaturated water rapidly reaches equilibrium with the atmosphere.

Chemical Variables

Dissolved Oxygen. Dissolved oxygen (DO), along with temperature, controls the metabolism of fish and invertebrates. Together, these two environmental variables are the dominant determinants of every fish's environmental niche. Fish seek an environmental niche and maintain that niche through a complicated series of behavioral enviroregulation and physiological acclimation processes (Neill and Bryan 1991). Fish can acclimate to low DO and other physico-chemical stimuli, but these adjustments are slow, taking anywhere from a few hours to several weeks.

Although the atmosphere contains nearly 21% oxygen gas, oxygen is only slightly soluble in water. Thus, water contains only small amounts of oxygen available for fish respiration. Fish must spend much more energy breathing water than air-breathing animals of similar size since water is much denser and more viscous than air. Small differences in the metabolic rate of an aquatic community can cause rapid changes in DO concentrations. The effect is greater as water temperature increases. Oxygen solubility decreases with increasing temperature and increasing salinity. Figure 2-6 illustrates the relationship between oxygen, temperature, and salinity. Decreasing barometric pressure, increasing altitude, and increasing impurities decrease oxygen solubility. Oxygen solubility is given in Table 2-4.

Fish oxygen-consumption rates vary with water temperature, environmental DO concentration, fish size, level of activity, time after feeding, and other factors. Metabolic rates vary

---

1. Instruments used to measure barometric pressure are usually calibrated in millimeters of mercury (mm Hg). To convert to the English system, 760 mm Hg = 29.92 in. of water. Standard atmospheric pressure at sea level = 760 mm Hg.

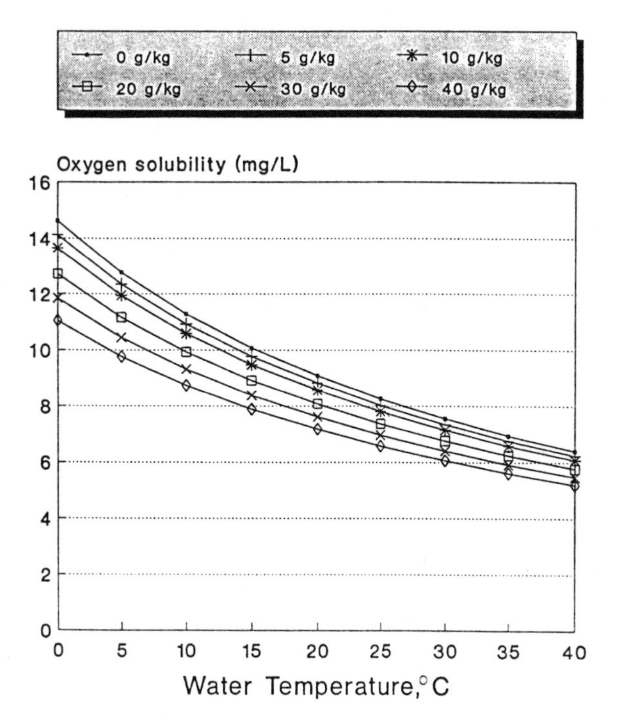

Figure 2-6. Oxygen solubility variation with temperature and salinity.

by species and are limited by low DO conditions. Small fish consume more oxygen per unit weight than large fish of the same species. The effects of temperature and size on the oxygen consumption of channel catfish is shown in Table 2-5. Oxygen consumption increases when fish are forced to exercise, and metabolic energy demands can cause oxygen consumption to double from one to six hours after feeding (Tucker and Robinson 1990).

Critical oxygen demands for a given species are difficult to assess since there is such a broad spectrum of effects as DO changes. However, the published literature contains considerable information on the oxygen requirements of freshwater species. In general, warm-water species tolerate lower DO conditions than cold-water species. Swingle (1969) developed a DO scale (Figure 2-7) for warm-water fish. Warm-water fish in ponds die after short-term exposure to less than 0.3 mg/L DO. To support life for several hours, a minimum of 1.0 mg/L is required, and 1.5 mg/L is necessary to support fish for several days.

Adult channel catfish can survive short-term exposure to 0.5 mg/L DO, and fingerlings can survive for brief periods at even lower concentrations (Tucker and Robinson 1990). Healthy catfish can survive for extended periods at 2–3 mg/L DO, but catfish eggs begin to die at these concentrations. Catfish feed poorly, grow slower, and are more susceptible to disease infections below 5 mg/L DO.

Crustaceans are also sensitive to low DO conditions. Young red swamp crawfish will die at prolonged exposure to DO concentrations below 1.0 mg/L (Avault et al. 1974). Adult crawfish will crawl out of ponds onto the levees when DO concentrations fall below 2 mg/L. Shrimp are probably as tolerant of low DO conditions as freshwater fish (Boyd 1990). The lethal concentration for many Peneaid species ranges from 0.7 to 1.4 mg/L.

Generally, a minimum DO concentration of 5 mg/L is recommended for warm-water fish and 6 mg/L for cold-water species. Huguenin and Colt (1989) recommend a minimum concentration of 6 mg/L for all marine species. Fish

Table 2-4. Solubility of oxygen (mg/L) in water at different temperatures and salinities from moist air at 760 mm Hg.

		Salinity (g/L)							
(°C)	(°F)	0	5	10	15	20	25	30	35
0	32.0	14.60	14.11	13.64	13.18	12.74	12.31	11.90	11.50
1	33.8	14.2	13.72	13.27	12.82	12.40	11.98	11.58	11.20
2	35.6	13.81	13.36	12.91	12.49	12.07	11.67	11.29	10.91
3	37.4	13.44	13.00	12.58	12.16	11.76	11.38	11.00	10.64
4	39.2	13.09	12.67	12.25	11.85	11.47	11.09	10.73	10.38
5	41.0	12.76	12.34	11.94	11.56	11.18	10.82	10.47	10.13
6	42.8	12.44	12.04	11.65	11.27	10.91	10.56	10.22	9.89
7	44.6	12.13	11.74	11.36	11.00	10.65	10.31	9.98	9.66
8	46.4	11.83	11.46	11.09	10.74	10.40	10.07	9.75	9.44
9	48.2	11.55	11.18	10.83	10.49	10.16	9.84	9.53	9.23
10	50.0	11.28	10.92	10.58	10.25	9.93	9.62	9.32	9.03
11	51.8	11.02	10.67	10.34	10.02	9.71	9.41	9.12	8.83
12	53.6	10.77	10.43	10.11	9.80	9.50	9.21	8.92	8.65
13	55.4	10.52	10.20	9.89	9.59	9.29	9.01	8.73	8.47
14	57.2	10.29	9.98	9.68	9.38	9.10	8.82	8.56	8.30
15	59.0	10.07	9.77	9.47	9.19	8.91	8.64	8.38	8.13
16	60.8	9.86	9.56	9.28	9.00	8.73	8.47	8.21	7.97
17	62.6	9.65	9.36	9.09	8.82	8.55	8.30	8.05	7.81
18	64.4	9.45	9.17	8.90	8.64	8.39	8.14	7.90	7.66
19	66.2	9.26	8.99	8.73	8.47	8.22	7.98	7.75	7.52
20	68.0	9.08	8.81	8.56	8.31	8.07	7.83	7.60	7.38
21	69.8	8.90	8.64	8.39	8.15	7.91	7.68	7.46	7.25
22	71.6	8.73	8.48	8.23	8.00	7.77	7.54	7.33	7.12
23	73.4	8.56	8.32	8.08	7.85	7.63	7.41	7.20	6.99
24	75.2	8.40	8.16	7.93	7.71	7.49	7.28	7.07	6.87
25	77.0	8.24	8.01	7.79	7.57	7.36	7.15	6.95	6.75
26	78.8	8.09	7.87	7.65	7.44	7.23	7.03	6.83	6.64
27	80.6	7.95	7.73	7.51	7.31	7.10	6.91	6.72	6.53
28	82.4	7.81	7.59	7.38	7.18	6.98	6.79	6.61	6.42
29	84.2	7.67	7.46	7.26	7.06	6.87	6.68	6.50	6.32
30	86.0	7.54	7.34	7.14	6.94	6.76	6.57	6.39	6.22
31	87.8	7.41	7.21	7.02	6.83	6.64	6.47	6.29	6.12
32	89.6	7.29	7.09	6.90	6.72	6.54	6.36	6.19	6.03
33	91.4	7.17	6.98	6.79	6.61	6.43	6.26	6.10	5.94
34	93.2	7.05	6.86	6.68	6.51	6.34	6.17	6.01	5.85
35	95.0	6.93	6.75	6.58	6.40	6.24	6.07	5.91	5.76
36	96.8	6.82	6.65	6.47	6.31	6.14	5.98	5.83	5.68
37	98.6	6.72	6.54	6.37	6.21	6.05	5.89	5.74	5.59
38	100.4	6.61	6.44	6.28	6.12	5.96	5.81	5.66	5.51
39	102.2	6.51	6.34	6.18	6.02	5.87	5.72	5.58	5.44
40	104.0	6.41	6.25	6.09	5.94	5.79	5.64	5.50	5.36

Source: Colt (1984).

Table 2-5. Oxygen consumption rates (mg oxygen/lb fish/hour) for channel catfish of different weights.

	Average Fish Weight (lb)					
°F	0.05	0.1	0.25	0.5	1.0	1.5
35	50	48	42	35	25	22
50	92	88	77	63	47	41
65	167	160	140	114	86	75
80	326	311	254	224	168	147
95	589	365	493	405	306	267

Source: Tucker and Robinson (1990) with permission.
Note: Table was left in its original English unit form.

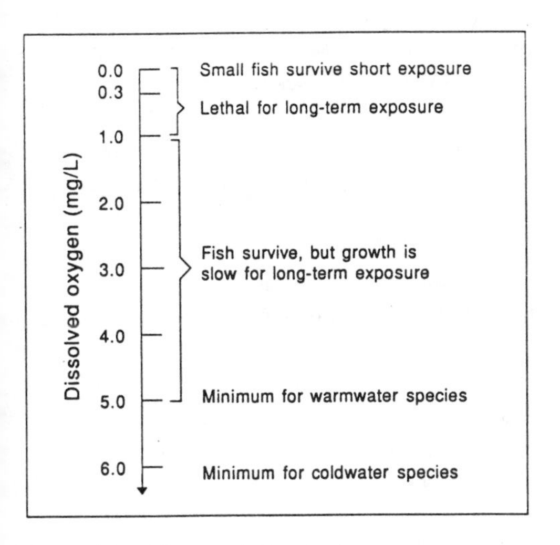

Figure 2-7. Effects of dissolved oxygen on warmwater pond fish (modified from Swingle (1969)).

do not grow well when DO concentrations remain below 25% of saturation for long periods (Romaire 1985). It is commonly accepted by aquaculture researchers and producers that fish perform better and are healthiest when DO concentrations are near saturation. Some authors (Colt and Orwicz 1991) recommend that the DO concentration in aquaculture systems be kept at about 90% of saturation, as a minimum, at all times for optimum performance.

If DO concentrations are adequate, fish meet the increased demand for oxygen during active periods or just after feeding by increasing the volume of water passing over their gills. They do this by using two physiological responses to oxygen need: (1) By increasing the ventilation (breathing) rate; and (2) by taking in larger gulps with each breath. The effect is overall increased DO intake. When DO declines, the behavioral and physiological responses differ. At first, the fish will seek zones of higher DO concentrations. Failing this, they will become less active in order to conserve energy and remaining metabolic oxygen. They often will stop feeding in response to their need to reduce their metabolism. Thus, fish should not be fed when the DO concentration decreases to 3–4 mg/L or less (Tucker and Robinson 1990).

As DO concentrations continue to decline, a *compensatory point* is reached where the oxygen demand of the tissues is greater than that which can be supplied by behavioral and physiological responses. At this point fish gape at the surface in an attempt to remove oxygen from the thin surface film. Small fish are more efficient at doing this than large fish. Hence, during oxygen depletions, large fish usually succumb first, soon followed by the small fish. When fish gulp at the surface, the DO concentration is near the lethal level. They can survive short periods under these conditions since their metabolic energy demands can be partially met by glycolysis or anaerobic metabolism. However, glycolysis produces acidic end products that lower blood pH, decreasing the ability of the blood hemoglobin to extract oxygen. Tissues are thus further starved of oxygen, and death is imminent.

Dissolved oxygen is usually the first limiting factor in pond aquaculture. The rate of diffusion of atmospheric oxygen into water is slow. Hence, photosynthesis is the major source of oxygen in static fish ponds. Phytoplankton (algae) must produce at least as much oxygen as consumed by the fish, zooplankton, chemical reactions, and bottom muds or oxygen will become depleted. A diel oxygen fluctuation is observed in fish ponds. The magnitude of daily changes in DO concentration is influenced by phytoplankton density. Oxygen is lowest at sunrise, before photosynthesis becomes active, increases during the daylight hours to a peak in late afternoon or early evening, and declines at night. This cycle is illustrated in Figure 2-8. The most successful fish farmers monitor the DO in their ponds throughout the night, and, if oxygen depletion appears imminent, emergency aeration equipment can be brought into action.

When suitable plankton blooms are present the water should have a Secchi disk visibility of 30–60 cm (12–24 in.) (Romaire 1985). As Secchi disk visibility decreases to less than about 30 cm (12 in.) the probability of a low DO problem developing increases. In water bodies where the Secchi disk visibility due to algae blooms is only about 10–20 cm (4–8in.) in depth, DO concentrations may be reduced so low at night as to cause fish mortalities (Romaire and Boyd 1978).

The topics of measuring DO concentrations and mitigative measures to correct for low DO conditions are discussed at the end of this chapter.

Total Alkalinity. Total alkalinity is defined as the total amount of titratable bases in water expressed as mg/L of equivalent calcium carbonate ($CaCO_3$) (Romaire 1985). By expressing alkalinity as $CaCO_3$ the contribution of a variety of ions that compose alkalinity can be standardized. The principle ions which contribute to alkalinity are carbonate (CO_3^-) and bicarbonate (HCO_3^-) and, to a lesser degree, hydroxides,

Figure 2-8. Typical diel fluctuation of dissolved oxygen in ponds.

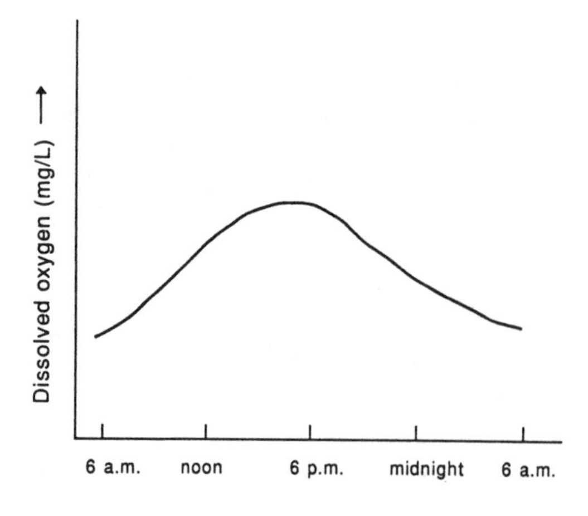

ammonium, borates, silicates, and phosphates. Alkalinity is determined in fish ponds by the quality of the water source and the nature of the bottom muds.

Alkalinity is a measure of the pH-buffering capacity or the acid-neutralizing capacity of water. When a source of calcium is added to water to adjust pH, in effect the alkalinity is being altered. The higher the alkalinity measure, the more stable the water is against pH changes. In natural freshwater systems alkalinity ranges from less than 5 mg/L in soft water to over 500 mg/L in hard water (Boyd 1990). Natural seawater has a mean total alkalinity of about 116 mg/L. Thus, alkalinity is rarely of concern in brackish and seawater aquaculture systems.

Total alkalinity does not have a direct effect on fish but, generally speaking, waters having a total alkalinity below 30 mg/L are considered poorly buffered against rapid pH changes. Fish ponds having low total alkalinity often have pH values of 5–7.5 at daybreak. Also, dissolved metals, such as copper, are more toxic in waters having low alkalinity and hardness. One should be cautious about using copper sulfate in waters having a total alkalinity less than 50 mg/L as $CaCO_3$. A total alkalinity range of 20–400 mg/L is considered satisfactory for most aquaculture purposes (Meade 1989; Tucker and Robinson 1990).

Total alkalinity has a marked effect on the biochemical processes in recirculating systems. Nitrification, the conversion of ammonia to nitrate, is an alkalinity-consuming process. Paz (1984) observed that natural waters with an alkalinity less than 40 mg/L as $CaCO_3$ affected the nitrification process independent of pH. Thus, enough alkalinity must be available to the bacteria in biological filters or the filter may fail. Nitrification is discussed in a later section.

Pond waters having a total alkalinity of less than 15–20 mg/L usually contain little available CO_2, and phytoplankton production is low (Boyd 1974). Waters with total alkalinities of 20–150 mg/L contain sufficient CO_2 for photosynthesis to occur. The removal of CO_2 results in rapidly rising pH (Sawyer and McCarty 1978). At a given alkalinity and temperature, the ratio of total alkalinity to carbon dioxide decreases very rapidly with increasing pH.

Total Hardness. Total hardness is the total concentration of metal ions expressed in terms of mg/L of equivalent calcium carbonate ($CaCO_3$). The ions are primarily calcium (Ca^{2+}) and magnesium (Mg^{2+}), but hardness also includes iron and manganese. As a rule, the most productive waters for fish culture have a total hardness and total alkalinity concentration of about the same magnitude (Romaire 1985). In some waters alkalinity may exceed hardness and vice versa. If total alkalinity is high and hardness is low, pH may rise to above 11 during periods of rapid photosynthesis (Boyd 1982).

Waters are classified with respect to the degree of hardness as follows (Sawyer and McCarty 1978):

Soft	0–75 mg/L
Moderately hard	75–150
Hard	150–300
Very hard	> 300

This classification is used by some aquaculturists, but in reality it has no biological significance.

Calcium is required for osmoregulation, but it is also important for bone formation in fish and exoskeleton formation in crustaceans. Some fish species often exhibit bone deformities and cease to grow in water that is too soft. Pursley and Wolters (1989) recommend a minimum hardness of 100 mg/L for grow-out of juvenile red drum, *Sciaenops ocellatus*. Crawfish cease to molt and their exoskeleton begins to soften in water that is too soft. De la Bretonne et al. (1969) recommend a minimum hardness of 100 mg/L as $CaCO_3$ for optimum crawfish production. Desirable concentrations of total hardness for fish culture generally fall within the range of 20–300 mg/L (Boyd and Walley 1975).

Generally, calcium concentrations increase with increasing salinity. The total hardness of seawater averages about 6,600 mg/L (Boyd 1990). Obviously, hardness will not be a problem in seawater and brackish water systems.

pH. The *pH* of water expresses the intensity of its acidic or basic character. In equation form, pH is the negative logarithm of the hydrogen ion activity:

$$pH = \log \frac{1}{[H^+]} = -\log[H^+] \qquad (2\text{-}3)$$

Water ionizes very slightly, producing only 10^{-7} moles of hydrogen ions and 10^{-7} moles of hydroxyl ions per liter. At neutrality there are just as many hydrogen ions as hydroxyl ions, therefore the neutrality pH is 7 at 25°C (77°F). Below 7 the $[H^+]$ ion predominates and the water is said to be *acidic* and above 7 the $[OH^-]$ ion predominates and the water is basic or *alkaline*. The pH scale ranges from 0 to 14.

The pH of most natural waters ranges between 5 and 10 (Boyd 1990). Most often, the pH of natural waters falls between 6.5 and 9, but there are many exceptions. In recent years, because of acid rain and other pollution problems, the pH of many natural water bodies has fallen several points below these values. Especially at peril are lakes and streams near large metropolitan areas or near plants which burn fossil fuels. The atmospheric fallout in these areas is typically very acidic. Acid rain occurs almost over the whole world but is especially troublesome in the Northeastern United States, Eastern Canada, and Northern Europe. The mean pH of ocean surface water is about 8.3, and remains fairly constant because of the great buffering capacity of the oceans.

Hydrogen ion concentration (pH) changes according to the influence of many factors: acid rain, man-made modifications, pollution, and CO_2 from the atmosphere and fish respiration are but a few. The decay of organic matter and oxidation of compounds in bottom sediments also alter pH in water bodies. In ponds phytoplankton and other aquatic plants use up CO_2 during photosynthesis, so the pH of a water body rises during the day and drops at night. In poorly buffered pond waters the pH can be as low as 5–6 in the morning to 9 or more in the afternoon. In waters with a high alkalinity, pH typically ranges from 7.5–8.0 at daylight to 9–10 in the afternoon. In some instances the pH may rise to over 11 during periods of rapid photosynthesis (Boyd 1990). High afternoon pH values are often responsible for fish kills. Waters in which the pH remains below 5.0 usually contain sulfuric acid resulting from the oxidation of sulfide-containing minerals in bottom soils, most commonly from the oxidation of iron pyrite by sulfur-oxidizing bacteria under anaerobic conditions.

Fish and invertebrates should be gradually acclimated to broad changes in pH. Very rapid changes can result in pH shock and mortality. The relationship of pH to pond fish culture is illustrated in Figure 2-9. The ranges shown are intended as a guide, since the pH effects vary among species. However, in general, fish are intolerant to pH extremes outside of the range 5–9 (Randall 1991). The toxic values for maximum and minimum pH vary with species and life stage at which the fish are exposed (Carrick 1979). For example, healthy populations of crawfish have been observed in ponds with a pH range between 5.8 and 10.0 (Huner and Barr 1984). Hymel (1985) observed no mortalities among crawfish exposed to waters with a pH range of 4.0–9.0 over a 96-hour period.

Carbon Dioxide. Carbon dioxide (CO_2) is a normal component of all natural waters. It enters surface waters by diffusion from the atmosphere and is also produced through fish respiration and the biological oxidation of organic compounds. It is carried in fish blood mainly as bicarbonate. Bicarbonate is converted to CO_2 at the gills by enzymic action and then diffuses into the water. As the CO_2 concentration in the water increases, the water pH shifts toward acid. The concentration gradient necessary for diffusion through the gills is reduced, and blood CO_2 levels increase, lowering blood pH. The amount of oxygen that the blood hemoglobin can carry is reduced, and respiration distress occurs, even though sufficient oxygen exists in the water. This phenomenon is known as the *Bohr-Root effect* (Tucker and Robinson 1990). Carbon dioxide also interferes with oxygen uptake and use by fry and eggs. Ideally, hatchery waters should be free of CO_2.

Carbon dioxide is highly soluble in water, but it is only a minor constituent in the atmosphere. Therefore, concentrations in pure water are small (Table 2-6). Optimal pH for an aquaculture system is defined as the pH of the water when it is in equilibrium with atmospheric CO_2. The pH of water saturated with CO_2 depends on temperature, salinity and alkalinity. Given values for these three parameters one can calculate the CO_2 concentration in water by using Table 2-7. The factor in the table is multiplied by the

Figure 2-9. Effects of pH on warm-water pond fish (modified from Swingle (1969)).

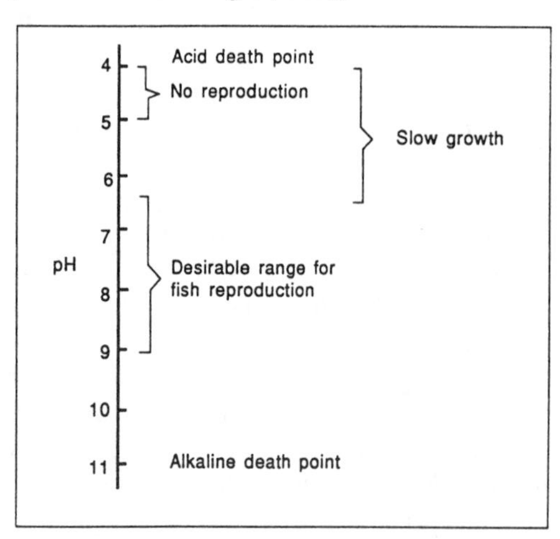

Table 2-6. Solubility of CO_2 (mg/L) in water at different temperatures and salinities from moist air at 760 mm Hg.

	Salinity (g/L)						
°C	0	5	10	15	20	30	40
0	1.09	1.06	1.03	1.00	0.98	0.93	0.88
5	0.89	0.87	0.85	0.83	0.81	0.77	0.73
10	0.75	0.73	0.71	0.69	0.68	0.64	0.61
15	0.63	0.62	0.60	0.59	0.57	0.54	0.52
20	0.54	0.53	0.51	0.50	0.49	0.47	0.45
25	0.46	0.45	0.44	0.43	0.42	0.41	0.39
30	0.40	0.39	0.39	0.38	0.37	0.35	0.34
35	0.35	0.35	0.34	0.33	0.33	0.31	0.31
40	0.31	0.30	0.30	0.29	0.29	0.28	0.27

Source: Boyd (1990) with permission.

Table 2-7. Factors used to calculate approximate CO_2 concentrations in dilute freshwaters from pH, temperature, and alkalinity.

	Temperature °C (°F)						
pH	5(41)	10(50)	15(59)	20(68)	25(77)	30(86)	35(95)
6.0	2.915	2.539	2.315	2.112	1.970	1.882	1.839
6.2	1.839	1.602	1.460	1.333	1.244	1.187	1.160
6.4	1.160	1.010	0.921	0.841	0.784	0.749	0.732
6.6	0.732	0.637	0.582	0.531	0.495	0.473	0.462
6.8	0.462	0.402	0.367	0.335	0.313	0.298	0.291
7.0	0.291	0.254	0.232	0.211	0.197	0.188	0.184
7.2	0.184	0.160	0.146	0.133	0.124	0.119	0.116
7.4	0.116	0.101	0.092	0.084	0.078	0.075	0.073
7.6	0.073	0.064	0.058	0.053	0.050	0.047	0.046
7.8	0.046	0.040	0.037	0.034	0.031	0.030	0.030
8.0	0.029	0.025	0.023	0.021	0.020	0.019	0.018
8.2	0.018	0.016	0.015	0.013	0.012	0.012	0.011
8.4	0.012	0.010	0.009	0.008	0.008	0.008	0.007

Source: Tucker and Robinson (1990) with permission.

Note: Carbon dioxide concentrations (mg/L as $CaCO_3$) are estimated by multiplying the total alkalinity of the water (as mg/L $CaCO_3$) by the factor in the table at a specific pH and temperature. Carbon dioxide is negligible above pH 8.4.

total alkalinity to give the CO_2 concentration. For example, water having a temperature of 25 °C (77°F), pH of 7.6, and total alkalinity of 100 mg/L will contain 100 × 0.05 (from Table 2-7) = 5 mg/L of CO_2.

CO_2 is not particularly toxic to fish provided that sufficient dissolved oxygen is available. Channel catfish have a relatively small Bohr-Root effect compared to most species and can survive in waters with a CO_2 concentration as high as 50 mg/L provided that sufficient oxygen is present (Tucker and Robinson 1990). Many species can survive short-term exposure to slightly higher concentrations, but a CO_2 concentration of 10–15 mg/L is recommended as a maximum for finfish (Boyd 1990). Toxic effects of CO_2 for trout are about 9–10 mg/L (Petit 1990). The upper tolerance limits for invertebrates are unknown. Jaspers (1969) reported CO_2 concentrations of 24–372 mg/L in crawfish burrows, but concentrations in open pond waters averaged less than 6 mg/L.

Groundwater and water in the hypolimnion of fish ponds at times contains high concentrations of CO_2. It can be removed by vigorous mixing and aeration. The concentration of CO_2 in fish ponds is highly variable since it is involved in the metabolism of plants and animals. CO_2 is produced by organisms during respiration and consumed by photosynthesis during the daylight hours. Conditions that favor rapid photosynthetic rates also favor rapid CO_2 removal. Concentration fluctuations on a 24-hour basis are generally opposite that of dissolved oxygen, as illustrated by the curves in Figure 2-10. At night, phytoplankton respiration may produce CO_2 in excess of the buffering capacity of the pond. Thus, pond pH at dawn can often be very acidic (Parker 1984). The pH and DO fluctuations generally follow one another in fish ponds.

High CO_2 concentrations normally occur in the summer months when DO is low, and low concentrations occur in the winter months when DO is high. CO_2 concentrations in channel catfish ponds often range from 0 in the afternoon to 5–10 mg/L at dawn the following day (Tucker and Robinson 1990). During extended cloudy weather, CO_2 may not be depleted in the afternoon and can exceed 10 mg/L by dawn. High CO_2 concentrations are also usually present af-

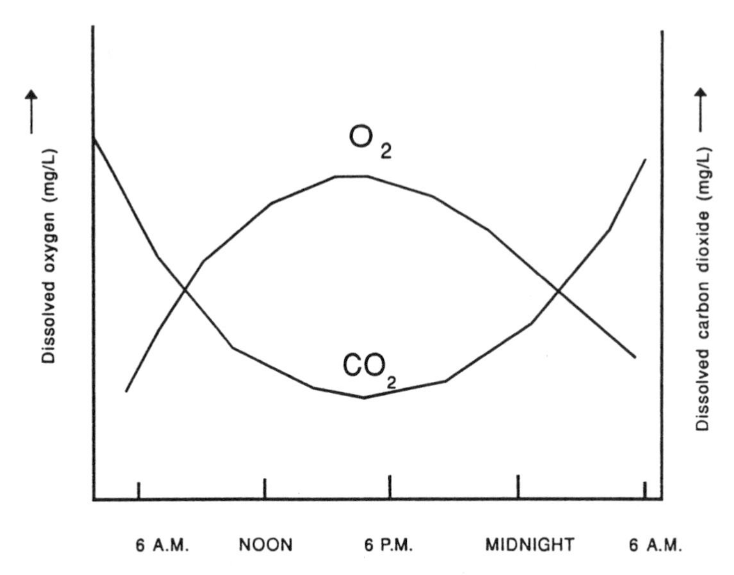

Figure 2-10. Diel cycle of dissolved oxygen and free carbon dioxide in ponds.

ter phytoplankton die-offs and after loss of thermal stratification.

If CO_2 concentrations are high in incoming water, particularly when pumping from wells, it should first be aerated to drive out the free CO_2. In closed tank systems CO_2 produced by fish respiration must be removed by aeration since there is no photosynthesis for removal. CO_2 removal practices in ponds are described later in this chapter.

Nitrogenous Compounds. In aquaculture systems, if all of the oxygen demands are met, the second factor that becomes limiting is the accumulation of nitrogenous compounds. The major source of nitrogen (up to 90%) in aquaculture systems is from fish feeds and is produced through the normal metabolic processes of the fish. Ammonia also results from the decay of organic matter. Most of the nitrogen in organic matter exists as amino acids in proteins.

The chemistry of nitrogen is very complex because of the many states in which nitrogen can exist, i.e., NH_3, NH_4^+, N_2, N_2O, NO, N_2O_3, NO_2^-, NO_3^- and N_2O_5 (Sawyer and McCarty 1978). The oxidative states of many of these compounds have little significance in aquaculture systems. The major forms of nitrogen found in aquaculture systems are shown in Table 2-8. The main nitrogenous compounds of concern are gaseous nitrogen (N_2), un-ionized ammonia (NH_3), ionized ammonia (NH_4^+), nitrite (NO_2^-), and nitrate (NO_3^-).

Molecular nitrogen gas (N_2) readily diffuses in and out of an aquaculture system. It is the major gas in the atmosphere, comprising about 78% of the total. Although N_2 is relatively insoluble in water, equilibrium concentrations are higher than for O_2 because N_2 is the principal gas in air. Similar to O_2 and CO_2, the equilibrium concentration for N_2 declines with increasing temperature and salinity (Table 2-9). At normal fish culture temperatures water contains about 10–20 mg/L N_2 at equilibrium.

Nitrogen gas is considered inert for aquaculture purposes, and normally poses no toxicity threat to fish and invertebrates. However, water supersaturated with nitrogen gas can cause gas bubble trauma. This condition is described elsewhere in this chapter and in Chapter 11. It is occasionally necessary to strip nitrogen gas from water entering flow-through and recirculating aquaculture systems, but N_2 is rarely a problem in fish ponds.

THE NITROGEN CYCLE. Nearly every aquaculturist is familiar with the *nitrogen cycle*, which relates the various nitrogenous com-

Table 2-8. The major forms of nitrogen in aquaculture systems.

Form	Notation	Comments
Nitrogen gas	N_2	Inert gas; transfers in and out from atmosphere; no significance.
Organic nitrogen	Org-N	Decays to release ammonia.
Un-ionized ammonia	NH_3	Highly toxic to aquatic animals; predominates at high pH levels.
Ionized ammonia or	NH_4^+	Nontoxic to aquatic animals except at very high concentrations; predominates at low pH levels.
Total ammonia	NH_3+NH_4^+	Sum of unionized and ionized ammonia; typically measured in the test for ammonia; converted to nitrite by nitrifying bacteria.
Nitrite	NO_2^-	Highly toxic to aquatic animals; converted to nitrate by nitrifying bacteria.
Nitrate	NO_3^-	Nontoxic to aquatic animals except at very high concentrations; readily available to aquatic plants.

Table 2-9. Solubility of nitrogen (mg/L) in water at different temperatures and salinities from moist air at a pressure of 760 mm Hg.

Temperature (°C)	Salinity (g/L)								
	0	5	10	15	20	25	30	35	40
0	23.04	22.19	21.38	20.60	19.85	19.12	18.42	17.75	17.10
5	20.33	19.61	18.92	18.26	17.61	16.99	16.40	15.82	15.26
10	18.14	17.53	16.93	16.36	15.81	15.27	14.75	14.25	13.77
15	16.36	15.82	15.31	14.81	14.32	13.86	13.40	12.97	12.54
20	14.88	14.41	13.96	13.52	13.09	12.68	12.28	11.89	11.52
25	13.64	13.22	12.82	12.43	12.05	11.69	11.33	10.99	10.65
30	12.58	12.21	11.85	11.50	11.17	10.84	10.52	10.21	9.91
35	11.68	11.34	11.02	10.71	10.40	10.10	9.82	9.54	9.26
40	10.89	10.59	10.29	10.01	9.73	9.46	9.20	8.94	8.70

Source: Boyd (1990) with permission.

pounds in aquaculture systems (Figure 2-11). Ammonia is produced through the biological conversion of organic nitrogen through a process called *ammonification*. Ammonia is also produced as the major end product of protein catabolism excreted by fish and invertebrates (Campbell 1973). It is excreted primarily as un-ionized ammonia (NH_3) through the gills. Ammonia is also produced through the decomposition of urea, fish feces, and uneaten food. In aqueous solutions, un-ionized and ionized ammonia exist in an equilibrium as shown by Equation 2-4:

$$NH_3 + H_2O = NH_4OH = NH_4^+ + OH^- \quad (2\text{-}4)$$

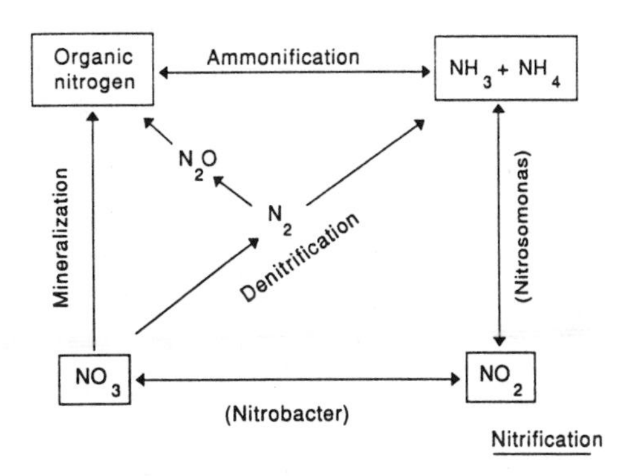

Figure 2-11. The nitrogen cycle.

Ammonia in fish ponds is rapidly assimilated by phytoplankton and rooted aquatic plants. The amount released in excess of plant requirements is oxidized by chemoautotrophic bacteria in a two-step process called *nitrification*. The first step is the oxidation of ammonia to nitrite by *Nitrosomonas* bacteria (Sawyer and McCarty 1978):

$$2NH_3 + 3O_2 = 2NO_2^- + 2H^+ + 2H_2O \qquad (2\text{-}5)$$

Nitrite is further oxidized to nitrate by *Nitrobacter* bacteria:

$$2NO_2^- + O_2 = 2NO_3^- \qquad (2\text{-}6)$$

In systems where oxygen is not limiting, the conversion of ammonia to nitrite is the rate-limiting step in the total process. Nitrite is converted to nitrate fairly rapidly, therefore, in most natural systems nitrite concentrations are usually of no concern, except in special situations discussed later in this chapter.

When plants die and decay, they are broken down, and ammonia is released back into solution through a process called *mineralization*. Ammonia is rarely a problem in fish ponds having good phytoplankton blooms unless the bloom suddenly dies. However, ammonia can be problematic in ponds when feeding rates exceed about 56 kg/ha/day (50 lb/ac/day) (Tucker and Robinson 1990). It is also of concern in serial reuse systems and closed recirculating systems because there are no green plants for assimilation. Ammonia is generally flushed from flow-through systems, but recirculating systems use biological filters for ammonia control.

Nitrification requires oxygen and proceeds most efficiently when DO is near saturation. The process is also most efficient at pH values between 7 and 8 and at temperatures between 27°C (80°F) and 35°C (95°F) (Tucker and Robinson 1990). *Nitrobacter* are less tolerant of low temperatures than *Nitrosomonas*. Thus, nitrite can accumulate at low temperatures.

Nitrites and nitrates are reduced through a process called *denitrification* (Figure 2-11). Nitrates are reduced to nitrites, and nitrites are further reduced. Some bacteria may carry the reduction of nitrites all the way to ammonia, but most of the end products are gaseous forms of nitrogen (i.e., N_2, NH_3, and N_2O), most of which are lost to the atmosphere. Considerable nitrite levels can result from the denitrification of nitrate in the anaerobic bottom mud of fish ponds.

All of the steps in the nitrogen cycle occur even in small ponds and closed systems. The nitrogen cycle does not go to completion in flow-through systems because of the short residence time of the water.

It is common in water chemistry to express inorganic nitrogen compounds in terms of the nitrogen they contain. Thus, the following convention will be used throughout this text: $NH_4^+{-}N$ (ionized ammonia nitrogen); NH_3-N (un-ionized ammonia nitrogen); TAN or $(NH_4^+ + NH_3)$-N (total ammonia nitrogen); NO_2^--N (nitrite nitrogen) and $NO_3^-{-}N$ (nitrate nitrogen). To convert concentrations on a nitrogen basis to concentrations of the specific compound, the following factors can be used: $NH_3{-}N = 0.822$ NH3; $NH_4^+{-}N = 0.776\ NH_4^+$; $NO_2^-{-}N = 0.304\ NO_2^-$, and $NO_3^-{-}{-}N = 0.226\ NO_3^-$. The concentrations of all of the mentioned compounds are normally expressed in milligrams per liter (mg/L) or parts per million (ppm). In aqueous solutions mg/L and ppm mean the same.

AMMONIA. Ammonia exists in two states, ionized ammonia, also called the ammonium ion (NH_4^+) and un-ionized ammonia (NH_3). The sum of the two ($NH_4^+ + NH_3$) is called total ammonia or simply *ammonia*. Total ammonia nitrogen, $(NH_4^+ + NH_3)$-N, is often written as TAN. The toxicity of TAN is dependent on what fraction of the total is in the un-ionized ammonia form since this form is by far the more toxic of the two. In most environments $NH_4^+{-}N$ predominates; however, which species predominates is dependent on pH, temperature, and salinity. Water pH has the strongest influence on which way the equilibrium equation (Eq. 2-4) will shift. As pH lowers, the reaction will shift to the right, and as pH is raised, the reaction will shift to the left. An increase of one pH unit will generally cause a tenfold increase in the proportion of NH_3-N in solution. Thus, use of the term *total ammonia* implies that pH, temperature, and salinity are known so that the concentrations of the two ammonia species can be calculated.

Table 2-10 shows values of the mole fraction of NH_3-N in solution at various temperatures and pH levels. At a pH of 9.4 and a temperature of 20°C (68°F), 50% of the total ammonia is in the NH_4^+-N form and 50% is in the NH_3-N form. At pH 7.0 nearly all (99.6%) of the ammonia is in the ionized (NH_4^+-N) form, and at pH 12 nearly all (99%) is in the NH_3-N form. Systems with high pH are more prone to ammonia toxicity. In seawater, the mole fraction of NH_3-N is about 40% less than that of freshwater at the same pH and temperature. The relationship among NH_3-N, pH, and temperature in pure water is shown graphically in Figure 2-12.

From Table 2-10, the concentration of NH_3-N can be calculated as follows (Huguenin and Colt 1989):

Table 2-10. Mole fraction of un-ionized ammonia in aqueous solution for different temperature and pH values.

Temp. (°C)	6.0	6.5	7.0	7.5	8.0	8.5	9.0	9.5	10.0
0	-	-	-	0.003	0.008	0.025	0.076	0.207	0.453
1	-	-	-	0.003	0.009	0.028	0.082	0.221	0.453
2	-	-	-	0.003	0.010	0.030	0.089	0.236	0.494
3	-	-	0.001	0.003	0.010	0.032	0.096	0.251	0.515
4	-	-	0.001	0.004	0.011	0.035	0.103	0.267	0.535
5	-	-	0.001	0.004	0.012	0.038	0.111	0.283	0.556
6	-	-	0.001	0.004	0.013	0.041	0.119	0.300	0.576
7	-	-	0.001	0.005	0.014	0.044	0.128	0.317	0.569
8	-	0.001	0.002	0.005	0.016	0.048	0.137	0.335	0.614
9	-	0.001	0.002	0.005	0.017	0.052	0.147	0.353	0.633
10	-	0.001	0.002	0.006	0.018	0.056	0.157	0.371	0.651
11	-	0.001	0.002	0.006	0.020	0.060	0.168	0.389	0.668
12	-	0.001	0.002	0.007	0.021	0.064	0.179	0.408	0.685
13	-	0.001	0.002	0.007	0.023	0.069	0.190	0.426	0.702
14	-	0.001	0.003	0.008	0.025	0.074	0.202	0.445	0.717
15	-	0.001	0.003	0.009	0.027	0.080	0.215	0.464	0.733
16	-	0.001	0.003	0.009	0.029	0.085	0.228	0.483	0.747
17	-	0.001	0.003	0.010	0.031	0.091	0.241	0.502	0.761
18	-	0.001	0.003	0.011	0.033	0.098	0.255	0.520	0.774
19	-	0.001	0.004	0.011	0.036	0.105	0.270	0.539	0.787
20	-	0.001	0.004	0.012	0.038	0.112	0.284	0.557	0.799
21	-	0.001	0.004	0.013	0.041	0.119	0.299	0.575	0.810
22	-	0.001	0.005	0.014	0.044	0.127	0.315	0.592	0.821
23	-	0.002	0.005	0.015	0.047	0.135	0.330	0.609	0.832
24	0.001	0.002	0.006	0.016	0.050	0.144	0.346	0.626	0.841
25	0.001	0.002	0.006	0.018	0.054	0.153	0.363	0.643	0.851
26	0.001	0.002	0.006	0.019	0.057	0.162	0.379	0.659	0.859
27	0.001	0.002	0.007	0.020	0.061	0.172	0.396	0.674	0.868
28	0.001	0.002	0.007	0.022	0.066	0.182	0.412	0.689	0.875
29	0.001	0.002	0.007	0.023	0.070	0.192	0.429	0.704	0.883
30	0.001	0.003	0.008	0.025	0.075	0.203	0.446	0.718	0.890

Data excerpted from Emerson et al. 1975.

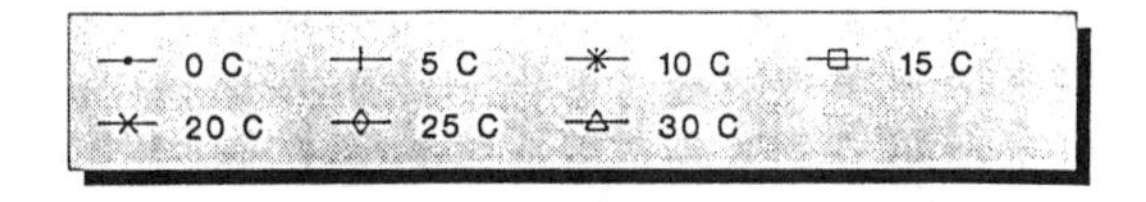

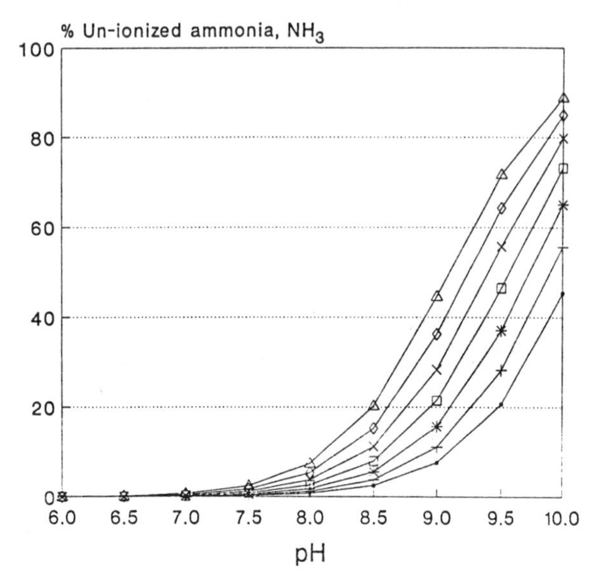

Figure 2-12. Variability of un-ionized ammonia with pH and temperature.

$$\text{Un-ionized ammonia } (mg/L \text{ as } NH_3-N) = (a)\,(TAN) \qquad (2\text{-}7)$$

where (a) = mole fraction of un-ionized ammonia; and TAN = total ammonia nitrogen (mg/L as nitrogen). Table 2-10 is useful because NH_3-N cannot be measured directly. Analytical procedures measure total ammonia nitrogen, and the table can be used with pH, temperature and salinity concentrations to estimate NH_3-N concentrations.

Knowing the system pH and temperature, the mole fraction of ammonia in the un-ionized form can be calculated using the following relationship (Petit 1990):

$$NH_3\text{-}N = \frac{1}{1 + 10^{10.068-0.033T-ph}} \qquad (2\text{-}8)$$

Example problem 2-2. Water in a fish tank has a temperature of 25°C (77°F), a pH of 7.5, a salinity of zero, and a *TAN* of 0.65 mg/L. What is the concentration of un-ionized ammonia in mg/L?

Solution: From Table 2-10, the mole fraction (a) of un-ionized ammonia = 0.0177. The concentration of un-ionized ammonia = NH_3-N = $(a)(TAN)$ = (0.018)(0.65) = 0.011 mg/L.

Un-ionized ammonia, NH_3-N, is very toxic to fish and invertebrates, even in very small quantities. The primary mechanism of ammonia toxicity is unknown, but a number of physiological effects are apparent (Schwedler et al. 1985). As ambient ammonia concentrations increase, the ability of fish to excrete ammonia decreases, and the level of ammonia in the blood and tissues increases. This results in elevated blood pH and adverse effects on enzyme reactions and membrane stability. The effects include gill damage, reduction in the oxygen-carrying capacity of the blood, increased oxygen demand by tissues, and histological damage to the red blood cells and tissues that produce red blood cells. High ambient ammonia concentrations can affect the osmoregulation of fish by increasing the permeability of the animal to water with a corresponding reduction in internal ion concentrations (Lloyd and Orr 1969).

Ammonia concentrations are seldom high enough in fish ponds to be of concern. Concentrations of NH_3-N in channel catfish ponds range from essentially zero to 1 mg/L or more (Tucker and Boyd 1985). The 96-hour LC_{50} of NH_3 to fish was reported to be between 0.5 and 3.8 mg/L (Ball 1967; Colt and Tchobanoglous 1976). In high-density culture situations, such as silos, raceways, and closed recirculating systems, ammonia can increase to toxic levels. Toxic concentrations of NH_3-N for short-term exposure vary between 0.6 and 2 mg/L for many pond fish, and some effects can be seen at 0.1–0.3 mg/L (Boyd 1979). Smith and Piper (1975) reported gill damage in rainbow trout when exposed to 0.021 mg/L of NH_3-N for six months. Table 2-11 lists representative un-ionized ammonia toxicity concentrations for selected species of finfish.

It is easy to see that the maximum recommended concentrations of ammonia are extremely variable. Basically, the literature indicates that warm-water fish are more tolerant to ammonia than cold-water fish, and freshwater fish are more tolerant than marine species. To be safe, ammonia concentrations below 0.05 mg/L as NH_3-N and 1.0 mg/L as TAN are recommended for long-term exposure. Huguenin and Colt (1989) recommend a maximum NH_3-N concentration of 0.01 mg/L for marine fish. Data on ammonia effects on invertebrates is very limited. Maximum concentrations should be the same as for finfish, as a precaution.

Table 2-11. Representative toxicity concentrations of un-ionized ammonia to selected species of finfish.

Species	96-hour LC50 (mg/L NH_3)
Pink salmon	0.08–0.1
Mountain whitefish	0.14–0.47
Brown trout	0.50–0.70
Rainbow trout	0.16–1.1
Largemouth bass	0.9–1.4
Smallmouth bass	0.69–1.8
Common carp	2.2
Red shiner	2.8–3.2
Fathead minnow	0.75–3.4
Channel catfish	0.50–3.8
Bluegill	0.55–3.0

Source: Russo, R.C. and R.V. Thurston. 1991. *Aquaculture And Water Quality*, Advances In World Aquaculture, Vol. 3, The World Aquaculture Society, Baton Rouge, Louisiana (with permission).

The amount of ammonia excreted by fish can be estimated from the net protein utilization (NPU) (the weight protein gain by fish per weight protein added in feed) and the protein level in the feed according to the following formula (Tucker and Boyd 1985):

$$TAN\ (g/kg\ feed) = (1.0 - NPU)\left(\frac{protein}{6.25}\right)(1{,}000) \tag{2-9}$$

where protein is in g protein/g feed and 6.25 is the average value of g protein/g nitrogen. Garling and Wilson (1976) reported an average NPU value of 4.0 for fish fed high protein diets.

Example problem 2-3. Calculate the total ammonia nitrogen excreted from fish that are fed a 25% protein diet.

Solution: TAN = (1.0—0.4)(0.25/6.25) (1,000) = 24 g N/kg feed

The example problem shows that the amount of ammonia reaching the culture water as metabolic waste is proportional to the feeding rate. Doubling the feeding rate will double the ammonia production, tripling the feeding rate will triple ammonia production, etc. Thus, the higher the fish stocking and feeding rates, the more ammonia produced per unit volume of culture water.

Generally, ammonia nitrogen production is assumed to be equal to 3% of the daily feeding rate (Huguenin and Colt 1989; Tucker and Robinson 1990) as shown in Equation 2–10:

$$TAN = 0.03\,R \qquad (2\text{-}10)$$

where R = total ration, kg/day (lb/day). This relationship provides a safety factor over the results obtained with Equation 2-9 when designing for ammonia loading.

Liao (1974) reported that ammonia production could be estimated by the oxygen consumed by the fish:

$$TAN = 0.053 \times kg\ O_2\ consumed\ per\ day \qquad (2\text{-}11)$$

NITRITE. Nitrite (NO_2-N) is the ionized form of nitrous acid (HNO_2), and it can be as lethal as NH_3-N. Nitrite levels in fish ponds typically ranges from 0.5 to 5 mg/L, probably due to the reduction of nitrate in anaerobic mud or water (Boyd 1982). Concentrations of both nitrite and nitrate show distinct seasonal patterns in fish ponds. They both are usually minimal in the summer months and increase in autumn, winter, and spring. The toxicity of NO_2-N is due principally to its effects on oxygen transport and tissue damage. When nitrite is absorbed by fish the heme iron in blood hemoglobin is oxidized from the ferrous to the ferric state. The resulting product is called *methemoglobin* or *ferrihemoglobin* and is not capable of combining with oxygen (Tomasso et al. 1979; Colt and Armstrong 1981; Tucker and Boyd 1985). The result is that the fish suffer from hypoxia (tissue suffocation). Blood containing methemoglobin is brown in color. Thus, fish suffering from nitrite poisoning are said to have *brown blood disease*. A similar condition is thought to occur with the hemocyanin in crustaceans (Colt and Armstrong 1981; Romaire 1985). Brown blood disease occurs most often during the cool months of the year and is linked with a high daily feeding rate of high-protein feeds. Fish deaths increase when low dissolved oxygen is coupled with nitrite. DO levels should be at 6 mg/L or higher where fish are affected by brown blood disease.

The amount of nitrite that moves into the bloodstream of fish is influenced greatly by the concentration of chloride in the water. High nitrite levels in aquaculture systems with low chloride levels can cause reduced feeding activity, poor feed conversions, lowered resistance to diseases, and mortality. Increasing the chloride-to-nitrite ratio will reduce toxicity problems. Chloride competes with NO_2 for active transport across the gills. At least a 20:1 ratio of chloride to nitrite—nitrogen (Cl:NO_2-N) is recommended to lessen the effects of nitrite toxicity (Tucker and Robinson 1990). If nitrite is reported simply as NO_2, then a Cl:NO_2 ratio of 6:1 or greater should be used to reduce nitrite toxicity. Chloride levels can be increased by the addition of common rock salt (NaCl). Calcium chloride can also be used, but it is much more expensive. Nitrite toxicity is not a problem in brackish water systems because of the naturally high chloride content of the water. The lethal concentration of NO_2-N is variable, depending upon fish species and size, nutritional status, and water chemistry. Representative acute toxicity values for nitrite for some species are listed in Table 2-12.

In flow-through systems, ammonia is the

Table 2-12. Representative acute toxicity concentrations for nitrite.

Species	48- or 96-hour LC50 (mg/L NO_2-N)
Rainbow trout	0.19–0.39
Cutthroat trout	0.48–0.56
Chinook salmon	0.88
Common carp	2.6
Channel catfish	7.1–13
Labyrinth fish	28–32
Fathead minnow	2.3–3.0
Goldfish	52
Bluegill	80
Largemouth bass	140
Smallmouth bass	160
Green sunfish	160
Guadalupe bass	190

Source: Russo, R.C. and R.V. Thurston. 1991. *Aquaculture And Water Quality*, Advances In World Aquaculture, Vol. 3, The World Aquaculture Society, Baton Rouge, Louisiana (with permission).

principal toxic metabolite. Water generally does not have a long enough residence time in flow-through systems for nitrite to become a problem. However, nitrite often is a serious problem in recirculating systems where the water is continually reused. In recirculating systems, nitrite is controlled with biological filters, but it can accumulate to toxic levels if the biological filters are not functioning properly or if the system temperature is below the functional range for *Nitrobacter* bacteria.

NITRATE. Like nitrite, nitrate buildup occurs most in the fall in pond systems when water temperatures are cooler. The *Nitrosomonas* bacteria, which convert ammonia to nitrite, function at cool temperatures (16°–20°C (61°–68°F)), but *Nitrobacter*, which convert nitrite to nitrate, do not function well at temperatures this low, hence nitrite will accumulate. Neither species functions well at temperatures below 16°C (61°F). Thus, ammonia accumulates in ponds during cold weather. Nitrate toxicity may be a problem only in recirculating systems due to constant water reuse.

Nitrates are the least toxic of the inorganic nitrogen compounds. The effects on aquatic animals are similar to nitrite having to do with osmoregulation and oxygen transport, but the concentrations at which fish are affected are much higher. The 96-hour LC_{50} value of NO_3-N to many fish and invertebrates lies between 1,000 and 3,000 mg/L (Wickins 1976; Colt and Tchobanoglous 1976). Nitrate toxicity levels for a few species are reported in Table 2-13. Nitrate is controlled in recirculating systems with daily water changes.

Table 2-13. Representative toxicity concentrations for nitrate.

Species	96-hour LC_{50} (mg/L NO_3-N)
Guppy	180–200
Guadalupe bass	1,260
Chinook salmon	1,310
Rainbow trout	1,360
Channel catfish	1,400
Bluegill	420–2,000

Source: Russo, R.C. and R.V. Thurston. 1991. *Aquaculture And Water Quality*, Advances In World Aquaculture, Vol. 3, The World Aquaculture Society, Baton Rouge, Louisiana (with permission).

Hydrogen Sulfide. Hydrogen sulfide (H_2S) is generated by certain heterotrophic bacteria under anaerobic conditions. Production occurs in the hypolimnion of fish ponds and can be produced in oxygen-starved zones in other culture systems, such as in tanks and biological filters. It reaches the zones inhabited by fish through diffusion and mixing. Bacteria use sulfate and other oxidized sulfur compounds as terminal electron acceptors in metabolism and excrete sulfide in turn. This process is demonstrated in the following chemical reaction (Boyd 1990):

$$SO_4^{2-} + 8H^+ \rightarrow S^{2-} + 4H_2O \qquad (2\text{-}12)$$

The sulfide excreted is an ionization product of H_2S and exists in equilibrium with HS^- and S^{2-}:

$$H_2S = HS^- + H^+ \qquad (2\text{-}13)$$

$$HS^- = S^{2-} + H^+ \qquad (2\text{-}14)$$

The distribution of H_2S and other sulfur species is regulated by water pH. The un-ionized form accounts for about 99% of the total sulfide. As the pH rises, the percent of H_2S decreases as shown in Table 2-14.

The net effect of H_2S poisoning is acute oxygen deficiency (hypoxia). Un-ionized hydrogen sulfide (H_2S) is toxic to fish and invertebrates at low concentrations. As little as 0.005 mg/L is toxic to channel catfish sac fry (Tucker and Robinson 1990). The 96-hour LC_{50} for 35-day-old bluegill (*Lepomis macrochirus*) is 0.013 mg/L, and many fish are killed after brief exposure to 0.05 mg/L (Smith et al. 1976). Concentrations less than 0.01 mg/L can inhibit reproduction. Therefore, any detectable concentration should be considered a hazard.

Iron. Iron is found in many forms in aquatic systems. Two of the most common forms are the ferrous (Fe^{+2}) and ferric (Fe^{+3}) ionic states.

Table 2-14. Percent un-ionized hydrogen sulfide in aqueous solution for different pH and temperature values.

	Temperature (°C)				
pH	16	20	24	28	32
5.0	99.3	99.2	99.1	98.9	98.9
5.5	97.7	97.4	97.1	96.7	96.3
6.0	93.2	92.3	91.4	90.3	89.1
6.5	81.2	80.2	77.0	74.6	72.1
7.0	57.7	54.6	51.4	48.2	45.0
7.5	30.1	27.5	25.0	22.7	20.6
8.0	12.0	10.7	8.8	8.0	7.6
8.5	4.1	3.7	3.2	2.9	2.5
9.0	1.3	1.2	1.0	0.9	0.8

Source: Boyd (1990) with permission.

Soil in many areas contains iron mineral deposits. Thus, groundwater can often contain over 10 mg/L of iron. Anaerobic conditions in the soil favor the formation of the ferrous form of iron which is soluble in water. When water is removed from the ground and is exposed to air the iron oxidizes to the insoluble ferric form. This causes the iron to precipitate and causes the orange discoloration that is sometimes seen on pipes, in tanks, or on surfaces where iron-bearing water has come into contact. Also associated with iron in water is iron bacteria, which cause an orange slime to develop, often clogging piping, filters, etc. Tucker and Robinson (1990) recommend that iron concentrations should be less than 0.5 mg/L in channel catfish hatchery water. Huguenin and Colt (1989) recommend concentrations less than 0.1 mg/L in marine aquaculture systems.

Heavy Metals. Natural waters in the modern world contain fairly high concentrations of heavy metals. Environmentalists are concerned about the effects of heavy metals on aquatic life and on organisms higher on the food chain. Table 2-15 shows the toxicity of certain metals to a variety of aquatic species. The safe levels shown are conservative estimates that are 10–100 times lower than the lowest concentrations known to kill aquatic organisms under laboratory conditions.

Table 2-15. Toxicity of metals to aquatic life.

Metal	96-hour LC_{50} (μg/L)	Safe level (μg/L)
Cadmium	80–420	10
Chromium	2,000–20,000	100
Copper	300–1,000	25
Lead	1,000–40,000	100
Mercury	10–40	0.10
Zinc	1,000–10,000	100

Source: Boyd (1990) with permission.

Pesticides. Many of the insecticides and herbicides routinely used on agronomic crops and around fish ponds are toxic to fish and invertebrates. Many insecticides are toxic over the range 0.005–0.1 mg/L. Some are toxic for long-term exposure at much lower concentrations (Boyd 1990). Long-term damage can result even if fish are not killed outright. In certain cases pesticides can affect habitat and food organisms, thus affecting growth and reproduction. The toxicity of selected hydrocarbon insecticides is shown in Table 2-16.

WATER QUALITY MEASUREMENT AND MANAGEMENT

Alkalinity and Hardness

The usual method for determining total alkalinity is by titration to the methyl orange endpoint (pH = 4.5) with sulfuric or hydrochloric acid (APHA 1980). Colorimetric kits that measure methyl orange make field measurements simple and can be purchased for less than $100. Alkalinity and hardness can be increased by adding agricultural limestone. Most ponds with soft, low-alkalinity water require between 2,240 and 4,480 kg/ha (1 and 2 tons/ac) of limestone to raise the alkalinity to 20–50 mg/L.

If total hardness in aquaculture systems is too low it can be raised by liming. Adding lime to culture water increases the total alkalinity and total hardness by the same amount. If it is desired to raise total hardness without affecting total alkalinity, agricultural gypsum (calcium

Table 2-16. Toxicity of selected chlorinated hydrocarbon insecticides to aquatic life.

Pesticide	96-hour LC_{50} (μg/L)	Safe level (μg/L)
Aldrin/Dieldrin	0.20–16	0.003
BHC	0.17–240	4.0
Chlordane	5–3,000	0.01
DDT	0.24–2	0.001
Endrin	0.13–12	0.004
Heptachlor	0.10–230	0.001
Toxaphene	1–6	0.005

Source: Boyd (1990) with permission.

sulfate, $CaSO_4$) can be added. Treatment rates can be found in Boyd (1982). Calcium chloride is also used to increase hardness. It is more expensive than agricultural limestone but reacts quicker.

There is no practical way of lowering total alkalinity or total hardness when they are too high. They can be reduced by water exchanges, but this is impractical in large systems. Hardness can be decreased with the use of water softeners when a small amount of water is required, however, this also is impractical on a large scale.

Carbon Dioxide

Free CO_2 can be removed from culture water by the addition of calcium hydroxide ($Ca(OH)_2$), commonly referred to as slaked or hydrated lime. The addition of 1.68 mg/L of hydrated lime will remove 1 mg/L of carbon dioxide (Boyd 1982). In recirculating aquaculture systems free CO_2 accumulation tends to lower system pH, particularly if alkalinity is low (Malone and Burden 1988a, 1988b). CO_2 accumulations will most likely occur when the system is heavily loaded since both the animals and nitrifying bacteria release CO_2 when they respire. A pH drop inhibits nitrification and toxic ammonia and nitrite may accumulate. CO_2 may be removed by aeration, permitting bicarbonate ions to react with hydrogen ions, thus lowering system pH. Another technique for CO_2 removal in recirculating systems is by air stripping. Air stripping of gases from culture water is discussed in another chapter.

Carbon dioxide can be measured by titration of a water sample to the phenolphthalene endpoint (at pH 8.3) with a standard base (APHA 1980). Small field-test kits are available for less than $100. These kits are not accurate enough for research, but they can be used to spot-check for CO_2 problems.

Conductivity

Conductivity meters are manufactured by several companies in the United States. One of the more popular meters is the YSI model 30 conductivity meter (Yellow Springs Instrument Co., Yellow Springs, Ohio) which retails for about $1,000.

Dissolved Oxygen

A number of procedures have been developed to prevent fish kills when dissolved oxygen concentrations are low. Oxygen demand can be reduced by removal of some of the animals from the system, removal of organic materials in the water, or by partial water replacement. Another treatment is with the addition of 6–8 mg/L of potassium permanganate ($KMnO_4$). Potassium permanganate is reported to oxidize organic matter in the water and lower the DO demand (Romaire 1985). However, this treatment is controversial since results are inconsistent. The only permanently effective treatment of culture water for low DO involves the use of mechanical aerating devices or pure oxygen injection. These processes are discussed in a later chapter.

To properly manage DO in aquatic systems the culturist must learn to measure DO levels. Colorimetric test kits can be purchased for $50–100, but they are not very precise and are time-consuming. Titration gives the most accurate results. Chemicals and apparatus may be purchased for under $100, however, the process is much more time-consuming and requires careful measuring of individual water samples followed by time spent in a laboratory.

The quickest results, also very accurate, are

obtained with battery-operated DO meters. Portable, hand-held DO meters are available that are lightweight and rugged for field use. Their cost ranges from under $250 to over $2,600. The less expensive meters are not as reliable and may not have as long a life span as the more expensive ones. The aquaculturist must decide for himself if the added initial capital investment offsets the potential longer period of use.

Hydrogen Sulfide

Prevention is through vigorous aeration and circulation to eliminate anaerobic zones. Once formed, H_2S can be removed by oxidation with potassium permanganate or by dilution with water exchange.

Iron

For pond fish culture, the most simple method to remove iron from water before it causes problems is to pump the water into a small pond while aerating. The iron will then oxidize to the ferric form and precipitate. A one- to two-day retention time is generally required to allow ferrous iron to oxidize and settle. Water can be vigorously aerated with an aeration tower or with mechanical agitation for more rapid removal. The water can then be passed through a sand filter or settling basin to remove the settled iron before it enters the fish pond. Iron can be removed from small systems with filters and water softeners, however, this is generally not economical on a large scale.

Nitrogenous Compounds

The most accurate way to determine ammonia nitrogen is by wet chemistry methods, the most popular being the nesslerization technique (APHA 1980). Wet chemistry methods also can be used for determining nitrite and nitrate. These methods are time-consuming, however, since distillation of a water sample is necessary. Small field-test kits costing under $100 can be used for determining total ammonia and/or nitrite. They are not extremely accurate, but it may be the only way to determine ammonia and nitrite in the field. Ammonia and specific ion electrodes can be purchased for determining total ammonia, nitrite, and nitrate, but they are expensive, costing upwards of $300, and the membranes require frequent replacement. In addition, nitrite and nitrate electrodes are usually not sensitive enough for accurate analysis of pond water because of the normal low concentrations of nitrite and nitrate present (Boyd 1982).

There is no practical method to reduce concentrations of ionized or un-ionized ammonia in large, commercial fish ponds. In smaller systems dilution by water exchanges and lowering the pH can reduce the TAN concentration temporarily, but the source of the problem will still remain. Green water (water containing a dense phytoplankton bloom) from an adjacent pond can be pumped into a pond with high ammonia as a temporary relief. The only method to reduce TAN with some degree of permanence is to reduce stocking and feeding rates.

pH

Measurements for pH should be made in the early morning and afternoon to determine the diurnal pattern. A pH value of 6.5–9 at daylight is best for fish production. Poor growth will result in waters with a pH of 4–6 or 9–10, although fish may survive. Reproduction diminishes at pH below 6. The acid and alkaline death points are approximately at pH 4 and 11, respectively.

The pH of waters too acid for fish production can be raised by liming. Liming will also increase total alkalinity and total hardness (Romaire 1985). Applications of ammonium fertilizers are recommended to lower pH. The ammonia in the fertilizer is nitrified to nitrate, releasing hydrogen ions and lowering pH. However, at a high pH, a large percentage of the ammonium ions will revert to unionized ammonia.

The pH range for most recirculating aquaculture systems is 7.0–8.0 (Malone and Burden 1988a, 1988b). The optimum pH range is 7.5–8.0, but it is often difficult to hold pH within this range because animal respiration in the system tends to drive pH down. pH in recirculating systems can be regulated by the addition of sodium bicarbonate ($NaHCO_3$), which is common baking soda. This method of pH control is generally the least expensive. Addition of carbonate is necessary when sand filters or plastic filter media are used since these materials have no calcareous compounds for pH buffering.

For aquaculture purposes, the best method for determining pH is with a pH meter. Meters are available with a glass electrode for laboratory use. These units typically cost between $800 and $2000. Small, portable pH meters designed for field use are also available, costing between $200 and $400.

Salinity

Chlorinity can be determined by titration of a water sample with silver nitrate, using a potassium chromate indicator and then converting to salinity (APHA 1980). However, instruments are available which measure salinity directly. Two such instruments are the salinity meter and the salinity refractometer.

Turbidity

Turbidity is measured in samples collected for laboratory analysis in Jackson Turbidity Units (JTUs) or Nephelometric Turbidity Units (NTUs). The Jackson Candle Turbidimeter is used to measure the straight-line path of water while the nephelometer is used to measure the scattering of light at 90% angles to the path of the light beam. Both of these instruments are described in Sawyer and McCarty (1978).

CHAPTER 3

SITE SELECTION FOR AQUACULTURE

Selection of a suitable site for an aquaculture venture determines facility construction costs and strongly influences the ultimate success of the resulting aquaculture enterprise. A number of factors must be considered when selecting a site. These factors should be included in a comprehensive site analysis. Kövári (1984) groups site selection factors into three broad categories: (1) ecological; (2) biological; and (3) economic and social. These and a fourth category, legal matters, are discussed in this chapter.

ECOLOGICAL FACTORS

Location

The aquaculturist is faced with a number of choices for a location for the venture. He may choose an inland, coastal, or offshore site. However, the choice may be somewhat limited if a species for culture has already been decided upon. For example, if it has been confirmed that the venture will produce a brackish water or marine species, it would be a mistake to locate the facility a great distance from a brackish or marine water source unless there are other considerations that take precedence or affect the economic viability of the project. The cost of pumping or hauling seawater to the facility would be prohibitive. Likewise, an aquaculturist would not attempt to produce channel catfish in a marine environment.

A self-imposed limitation occurs where a prospective fish culturist already owns a parcel of land and he is determined to use that land for his aquaculture venture and will consider no alternative. In another situation a landowner may want to convert from a conventional agronomic crop to aquaculture. Conversions of this type are seldom successful, and this somewhat limits species selection since the species must be biologically compatible with local environmental conditions. Additionally, the parcel of land may be too small to produce a commercial-size aquaculture crop, or the shape of the parcel may not be conducive to the proper layout of an aquaculture facility. Finally, the parcel of land may not be near an adequate source of water.

Purchasing land for an aquaculture venture requires a substantial initial capital investment. Obviously, one would not want to invest in prime real estate property for this type of business. It would pay to solicit the aid of local realty agencies in locating property suitable for fish farming. Typically, marginal land that is not suitable for other agricultural purposes is the least expensive to purchase and often makes the best land for aquaculture use.

If a coastal or offshore venture is proposed, the facility should be located in an area away from shipping lanes and local small-craft traffic. The site planner should check with local commercial fishing license authorities to ensure that the site is not planned on leased or privately owned fishing grounds. Local sport fisherman should be queried so that favorite fishing sites can be avoided. Additionally, known migration routes of fish species and marine mammals should be avoided. Offshore aquaculture facilities such as raft or string culture or net pen operations should be located in coves or behind offshore land masses for protection from wind and wave action as much as possible (Figure 3-1). However, the site should be thoroughly evaluated to ensure that water quality problems will not be created. These factors are discussed in greater detail in Chapter 5.

The prospective aquaculturist should be familiar with developmental plans for the area. It would be unwise to select a site where industrial development may cause air and/or water pollution. Also, if the site selected is near a heavily populated area or an area of intense agricultural development, the risk of pollution from human and agricultural wastes or agricultural chemicals is increased. Some agricultural wastes are beneficial and may be incorporated into the aquaculture venture. This possibility should be investigated.

The site should be located near transportation routes or where an access road may be constructed economically and without harming the environment. Routes to local markets must also be planned.

Underground utilities on the site, such as oil or gas pipelines, may render an otherwise good site unsuitable for an aquaculture venture. Ponds should not be constructed so as to contain pipelines, electric power lines, utility poles, radio towers, and similar structures.

Topography

When selecting a site for aquaculture, preference should be given to locations where gravity flow may be used to fill ponds, tanks, raceways, etc. Water can be supplied by pumping, but gravity flow will be more economical. It is also advantageous to drain ponds by gravity flow. Ponds that are drained by pumping are rare in commercial fish farming operations due to the costs involved. To take advantage of gravity flow, the pond bottom should be at a higher elevation than the water table when the pond is drained. Topographic surveys to the scale of 1:500 to 1:5000 with 1-ft contour lines are required to permit designing complete pond filling and drainage schemes and for accurately estimating earthwork volumes (Elekes 1984).

The land slope should be suitable, since too steep a slope will require more excavation and higher facility construction costs. Kövári (1984)

Figure 3-1. Open culture fish farm sheltered from high winds and waves.

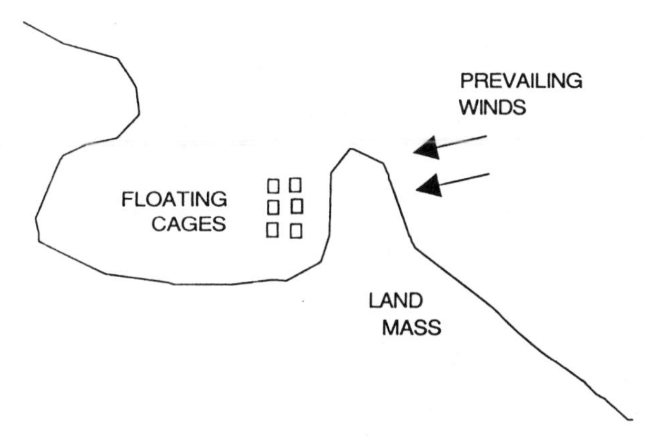

recommends a land slope no greater than 2%. Flat land or land sloping less than 1% is ideal. On land containing depressions or ravines, it may be necessary to haul in soil for filling and smoothing. Both the removal and addition of soil increases facility construction costs. However, in some cases where soil is removed, it may be used at some other location on the site for dam or levee construction.

The type and density of vegetation are partially dependent upon land elevation (Kövári 1984). The size of vegetation and root systems have an influence on the method of land clearing, construction time, and cost. Grassland, open woodlands, and land covered with low-growing shrubs and bushes will be less expensive to clear than heavily forested areas or land with very dense, tall shrubbery.

The shape and size of the land parcel must be considered. Land with a regular shape that is sufficiently large enough to accommodate future expansion is more suited for a fish farm. A parcel of land too small or irregular in shape may inhibit the proper location of ponds, tanks, pumps, drainage canals, roadways, storage structures, and support facilities with respect to one another, decreasing operational efficiency. Facility components requiring the most frequent attention (e.g. hatcheries, rearing and nursery ponds, holding ponds, pumping stations, etc.) should be located as near as possible to the center of the facility (Elekes 1984).

Soil Suitability

Field investigations are used to determine distribution of soil types. Both surface and subsurface soil properties should be determined by a competent soil scientist as early in the site study plan as possible. A soil analysis may reveal features undesirable for aquaculture purposes, thus saving the investors time and money. A soil survey may vary from a simple visual inspection to a detailed subsurface exploration and laboratory analysis. Visual examinations are often employed to determine basic soil properties. Soils have characteristics that can easily be determined by sight and feel. Properties of interest include soil type, pH, clay content, organic matter content, plasticity index, percolation rate (coefficient of hydraulic permeability), grain size, microbial content, and the presence of pollutants, including heavy metals.

Soil cores should be taken in the area where ponds are being considered to ensure that the soil contains adequate clay to hold water. Core samples of surface and subsurface soils should be taken from several locations on the proposed site. Soil properties often change abruptly a short distance away, and by taking samples at several locations a better overall picture of soil properties will be obtained. Subsurface cores are obtained by digging test pits approximately 2 m (6 ft) deep (Figure 3-2). The pits should be deep enough to extend below proposed pond bottom elevations. Core samples should be taken at mid-depth and at the bottom of the pits. Test pits make it possible to visually examine subsurface strata and obtain both disturbed and undisturbed samples of soils in different layers. This detailed analysis may reveal sand veins or other pervious strata through which ponds are to be dug and that may later cause leakage (Figure 3-2).

A sandy clay to clayey loam is the best soil type for fish pond construction. For ponds containing burrowing crustaceans, such as crawfish, soils having a clay content up to 30% are desirable to prevent collapse of the burrows. Too much organic matter in soil is harmful. Land with a layer of organic soil greater than 0.6 m (2 ft) deep is unsuitable for fish ponds because the organic stratum will cause excessive seepage losses when it decays (Kövári 1984). Highly organic soil is not usable for dike or levee construction. Additionally, soils high in organic matter cause rapid oxygen depletion from the water when ponds are filled. Areas with large surface stones or subsurface rock strata may only be suitable for lined ponds or concrete raceways.

A chemical analysis should be performed on the soil by using a composite made from several surface and subsurface core samples. A chemical analysis will determine soil pH, nutrient concentration (nitrogen, phosphorus, potassi-

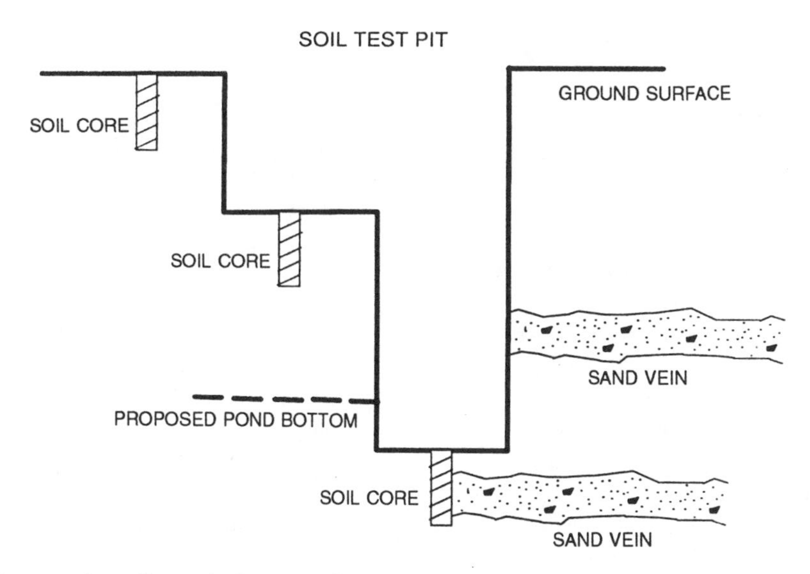

Figure 3-2. Diagram of a soil test pit demonstrating proper method for taking soil core samples.

um), organic matter, and the presence of metals (calcium, magnesium, iron, etc.). Soils with a high or low pH can cause pond waters to become too alkaline or too acidic for fish culture. Ideally, fish pond pH should be between 6 and 9 pH units (see Chapter 2). A chemical analysis should also reveal pesticides and/or other toxic residues in the soil.

Cat's clay soils are those formed from marine sediments containing sulfide compounds and are often found along coastal plains. Ponds receiving drainage from acid soils on the watershed may become too acidic to support fish culture. Sulfuric acid resulting from the oxidation of sulfide-containing minerals in bottom soils may cause the water pH to permanently remain below 5 units (Romaire 1985). Iron pyrites in soils may also cause mineral acidity in pond waters. Cat's clay soils should be avoided for aquaculture if possible; however, there are methods of dealing with this problem. It is not the intent of this text to discuss lengthy mitigative measures for problem soils. Therefore, the reader is referred to Boyd (1990, 1992).

Water Quantity and Quality

A reliable water supply is perhaps the single most important factor to consider when selecting a site for an aquaculture venture. Water of the proper quantity and quality should be available in accordance with the cycles of aquaculture operation. Therefore, when selecting an aquaculture site, the water supply must be thoroughly investigated.

Water can be obtained from an ocean, estuary, river, stream, irrigation canal, reservoir, spring, lake, well, or spring. Well water is usually the preferred source for aquaculture purposes if groundwater is abundant since it is free of undesirable fish, fish eggs, and predators and may be relatively unpolluted. However, it may not always be economically advantageous to use wells as the water source, particularly if the water must be pumped from deep in the ground. Wells must have an adequate yield throughout the year so that the water supply will not fall woefully short at a critical point in the fish production cycle. For channel catfish farming, Boyd (1982) recommends a minimum well yield of 7,600 Lpm (2,000 gpm) for a 32-ha (80-ac) production unit. Consultation with a local well driller should verify if a well with such a yield can be developed on the site.

An analysis of the physical, chemical, and biological properties of the proposed source water must be conducted. The quality of the water must be such that it will support culture of the desired fish species, whether it be a fresh-, brackish, or saltwater variety. The analysis should include potential health hazards for both humans and the aquatic species to be cultured. Although well water is normally the most desirable source, groundwater may contain pollutants such as pesticides, heavy metals, chlorinated hydrocarbons, or other toxic substances. In addition, the salinity of groundwater varies dramatically with location. Some freshwater species may tolerate 1–2 g/L salinity, but groundwater in some areas may contain too much salinity to support culture of some species. In these instances a brackish water or marine species should be considered.

Water test procedures should be in accordance with the relevant standards for the country in which the venture will be located. In the United States, water analyses are conducted according to methods specified by the American Public Health Association (APHA) (APHA 1988). Physical, chemical, and biological properties of water for aquaculture purposes are discussed in Chapter 2.

Hydrological and Meteorological Information

A thorough hydrological analysis should be conducted for the water source and the area immediately surrounding. If the water source is to be a river, stream, or spring, flow records must be examined, and data must be obtained on water stage elevations and flow rates at the anticipated point of removal for filling ponds and for water replacement due to seepage and evaporation.

If the source is to be a lake or reservoir, data must be obtained on water inflows and outflows. Documentation of existing water uses and how they will affect the aquaculture operation should be produced. If a reservoir is to be used as the water source, a minimum catchment area to pond area ratio of 11:1 or 15:1 should be used (Hora and Pillay 1962). A higher ratio is required if the catchment area is woodlands, and a lower ratio is used for land under cultivation. Hydrologic and water use information may be obtained from the local water authority and from state and federal geological survey offices. Other important information that may be obtained from these official sources is data on flooding and water elevations for existing water sources.

Important climatological data may be obtained from the meteorological station nearest the proposed aquaculture site. Data should include mean monthly temperature, rainfall, evaporation, humidity, solar intensity, wind speed, and wind direction for the area. Data should be obtained for the longest period of record available. Data on peak monthly rainfall and evaporation for the area are required for estimating water budgets. Mean extreme temperature information can aid in selecting the species for culture. Temperature data are also valuable for planning feeding rates and schedules and for designing holding and storage facilities for live fish.

Other climatological factors include the incidence of high winds, storms, tornadoes and earthquakes in the area and the amount of damage caused by each. Data on the direction and maximum speed of prevailing winds are used for building design and to design wave protection for ponds. In areas that are characterized by heavy rainfall and frequent floods, the aquaculture facility planner should study flood level and discharge data. Information on the probability of occurrence of design floods may be obtained from the nearest geological survey office. If such information is lacking, large aquaculture facilities should be designed for the 100-year flood (Elekes 1984). For smaller facilities, where dam or dike failure will not cause severe downstream damage, a 25- or 50-year design flood may be used.

Land elevation and flooding are important site considerations. The land should be free from deep flooding. The maximum recommended flood level for the previous 10 years should not be at a higher elevation than the tops of the dikes around the ponds (Kövári 1984). High-water marks left by floodwaters can be observed on tree trunks, fence posts, utility poles, bridge piers, or other nearby structures.

Flood history information can also be obtained by questioning local residents.

BIOLOGICAL FACTORS

Species Selection

Another common self-imposed limitation is that of a prospective aquaculturist restricting himself to a particular species and refusing to consider others. It would be advantageous to select a species that other fish farmers in the same geographic area have found suitable for culture. Once the species is selected, the availability of eggs or stockers (fry, fingerling, or brood fish) must be determined. This information will enable the culturist to determine the desired production level, what size fish will be produced and rearing space requirements. He may also at this point decide on the biomass loading and determine if his operation will be extensive, semi-intensive, or intensive in nature. He may also consider employing polyculture or integrated culture methods rather than the traditional monoculture technique. Inexperienced fish culturists should begin with a single species (monoculture) until operational expertise is gained.

Predator Control

Whatever the source, water used for aquaculture should be free of undesirable fish species, fish eggs, and the egg and larval forms of insect predators. If it is suspected that source water may contain these organisms, it should be screened before use. Small mesh screens for this purpose are available from aquaculture equipment suppliers.

The area should be made as free of predatory birds and mammals as possible. Predatory birds are a particular problem in some areas. For example, the double-breasted cormorant has a reputation as the most serious avian predator on pond-raised catfish in the Mississippi Delta. In a recent study of the predacious effect of these birds on commercially produced catfish, it was estimated that 4.7 fingerling were consumed per cormorant-hour. At a cost of $0.08 per fingerling, 100 birds on a pond for a nine hour day would consume $342 worth of fish (Cormorant 1991). These animals can often be controlled by shooting; however, this practice must be used with caution since a hunting permit may be required, and some species, including cormorants, are protected by law. Other methods used to control predatory birds include using scarecrows, bird netting or noise devices such as small cannons. Some farmers are more creative and use radio-controlled model aircraft to frighten away predatory birds.

Mammal predators can be controlled by shooting or trapping. Again, the farmer should be familiar with local laws and ordinances regarding hunting and the use of firearms. Some mammals and birds are classified as endangered or threatened species and are federally protected in the United States. In some cases animals may be trapped and relocated off the site. In situations where legal hunting or trapping may be used to control predators like mink, otter, racoons, nutria, and muskrats, a second income may be generated from the sale of hides and/or meat. Other predatory animals, such as snakes, frogs, and turtles, can be trapped and sold to biological laboratories or supply houses.

ECONOMIC AND SOCIAL FACTORS

Type of Facility

An important decision must be made regarding how the aquaculture crop will be produced, whether in ponds, raceways, or in tanks, or using some other culture method. Will the facility be used to spawn fish and sell the eggs produced? Will fish be raised to fingerling size or sold as food fish? The choices in these matters will be limited by the amount of working capital available, land availability, operator skill levels, labor requirements, operating costs, and other considerations.

Fish farming operations typically require electric power to operate hatcheries, business offices, pumps, water heaters, chillers, aerators, and other equipment. Therefore, the costs to tie into local public utilities must be considered. It must be known beforehand whether single- or three-phase power will be required. Large electric motors over 10 hp (7.5 kW) in size require three-phase power for operation. The cost of bringing three phase power to a site may be prohibitive. If power lines and utility poles already exist on the property, it must be determined if they will interfere with the daily operation of the farm. In some situations these can be moved at minimal expense. Also, the safety of all personnel on the farm must be taken into consideration. Existing utilities that present a hazard to farm personnel must be corrected.

Marketing

A marketing study should be made as an essential component of a site analysis long before land is purchased and facility construction begins. The aquaculturist must understand marketing strategies and be able to interpret the marketing climate. Several approaches to marketing are identified by Chaston (1983).

Marketing-Oriented Approach. In this approach the marketing analyst begins by identifying potential customers and their needs. Information on product type, price, promotion, distribution, and other variables are then used to develop the marketing strategy. A major fault with this approach is that one occasionally identifies a species for which there is high market demand but for which production techniques are lacking. In this example market demand may be strong, but supply is weak.

Production-Oriented Approach. The production-oriented approach is used by the aquaculturist who produces a specific product before he establishes a market for the product. This approach is typified by the producer who is accustomed to selling fish at the pond bank. When customers stop coming to him directly he must then develop a marketing strategy to move his product.

Environmental-Oriented Approach. Aquaculturists who select a species for culture that will adapt to environmental conditions in a given geographical location are using the environmental approach. This is a self-imposed limitation, as discussed earlier in this chapter. Often culturists will select a species that is popular in a given area but for which production technology is lacking.

Species-Oriented Approach. The species approach is where the aquaculturist selects a given species and will not consider others. This is another self- imposed limitation, and the culturist may find that his product has appeal only in a small geographical region.

Safety and Security

When aquaculture ventures are located in close proximity to densely populated areas, public health and the necessity of guarding against poaching and vandalism must be considered. Facilities must be kept appropriately clean so that noxious odors or vapors, which may affect local populated areas, are not produced. Effluent from ponds, settling tanks, filters, etc. must be properly treated before disposal. In particular, culture water containing chemicals used for treating diseases or for cleaning and disinfecting culture system components must be properly disposed of so as not to jeopardize the safety and health of others.

Poaching and vandalism can be controlled by hiring full- or part-time security personnel or by fencing off the property. Warning signs should be posted conspicuously on the property. Locals can often be hired for a small fee to watch the facility. Many facilities, such as closed recirculating production systems, are protected by being constructed indoors.

Social Considerations

It is wise to precede construction of any aquaculture venture with a publicity campaign to educate the locals on the benefits of having the venture in their community. Aquaculture undertakings can provide jobs and economic opportunities for local residents.

Local engineering firms can be used to design the facility, and local contractors can be used for construction. Professionals like competent biologists and skilled operators are essential for any major aquaculture enterprise. These personnel should be hired locally whenever possible, but they can be brought in from some other geographic locations.

LEGAL MATTERS

When considering a site for aquaculture development, the site plan should contain an analysis of local, state, and national legal requirements. The potential aquaculturist must be familiar with these laws before beginning any aquaculture venture. Special permits are usually required to operate in coastal waters of the United States, for example. Construction of ponds and other facilities in wetland areas are also restricted. Some states have laws prohibiting the use of net pens, cages, rafts, racks, or similar structures in navigable wasters. Local zoning ordinances may prohibit aquaculture operations in certain areas. Restrictions may also prohibit the culture of certain species. Permits may be required before certain species may be transported into and/or across certain states. Some states may have particular labeling requirements and quality standards for products to be marketed within their boundaries. The potential aquaculturist must be prepared to abide by all local laws if he wishes to be welcomed into a given area.

If products are to be marketed internationally, the problem is often compounded. Many countries prohibit the import of certain species for various reasons. Others may impose trade barriers to protect domestic production. Many countries have labeling requirements and quality standards just as in the United States.

All of these examples illustrate the importance of making a thorough analysis of all factors affecting the marketing of aquaculture products. Costly mistakes can be avoided by checking out these factors beforehand. Setting up a test-market program is a safer and cheaper alternative to plunging headlong without knowing how marketing factors are affected.

CHAPTER 4

WATER SUPPLY

Aquaculture is water-intensive, generally requiring more water per unit area or per unit of product than most other agronomic crops or animal production. Reliable sources of good-quality water often determine the success or failure of any aquaculture enterprise. In many geographic regions the availability of water may limit the feasibility of aquaculture production.

Some system types are much more water-consumptive than others. For example, water requirements for recirculating system aquaculture are far less than for pond or raceway culture. Pond and raceway culture methods consume great quantities of water, and the aquaculturist must have a thorough knowledge of his water source (i.e., quantity, quality, and seasonal variability). The study of the water supply should be an integral first part of any site analysis, as discussed in Chapter 3.

WATER QUANTITY

The quantity of water required is dependent on the type of aquaculture system, species cultured, management practices, culture densities, skill of the culturist, and so on. Therefore, it is difficult and foolhardy to generalize about water needs. Wheaton (1977) identified four functions that a water supply must satisfy: evaporative losses, seepage losses, oxygen depletion, and waste disposal. In addition, water may be required for other uses on the property such as fire protection, livestock watering, facility cleaning, irrigating crops, and residential use. The water budget should also be sufficient to take into account future plans for expansion.

Tank and Raceway Culture

Water requirements for fish culture in either tanks or raceways can be estimated from what is known about the incoming water, effects of the fish on the water, and water requirements of the fish (Westers and Pratt 1977). The water budget calculations are based on a mass balance of some limiting water quality factor in the culture unit, usually oxygen or ammonia. In some cases carbon dioxide can be the limiting factor (Colt and Orwicz 1991). The expression for flow requirement in a fully mixed flow-through tank is given by Piedrahita (1991) as:

$$Q = \frac{(PROD_x - CONS_x)}{(C_{x,in} - C_{x,out})} \qquad (4\text{-}1)$$

where Q = steady-state flow rate based on the limiting concentration of water quality factor X.

m^3/hr (ft^3/hr); $PROD_X$ = rate of production or introduction of factor X within the culture unit, g/hr (lb/hr); $CONS_X$ = rate of consumption or removal of factor X from the culture unit, g/hr (lb/hr); $C_{X,in}$ = concentration of factor X in the incoming water, g/m^3 (lb/ft^3); and $C_{X,out}$ = concentration of factor X in the outgoing water, g/m^3 (lb/ft^3).

For recirculating systems the inputs into Equation 4-1 are somewhat more complex. The production and consumption terms are influenced by the physico-chemical processes taking place within the culture system (e.g., aeration and biofiltration) or by the physiological processes of the fish (e.g., respiratory consumption of oxygen and metabolic production of ammonia and carbon dioxide).

Alternatively, when detailed information on the physiology of the fish or the processes affecting the concentration of factor X are not known the flow requirements can be estimated based on the number of water exchanges per hour.

$$Q = RV \qquad (4\text{-}2)$$

where Q = flow rate, m3/hr (ft3/hr); R = number of volume replacements or exchanges; /hr and V = volume of culture unit, m^3 (ft^3). When estimating the total water use for the facility, water used for filter backwashing, washdown, and other water uses should be included in the total.

Pond Culture

Many stock watering ponds and recreational fishing ponds rely on rainfall for filling. However, a commercial aquaculture production pond cannot be filled and its water depth maintained by rainfall alone. Rainfall is not always dependable, and there may be long periods between rainfall events. The water supply should be sufficient to fill the pond in a reasonable amount of time. Once the pond is filled the water source should be sufficient to maintain the desired water level in the pond while compensating for water losses. What is considered to be a reasonable amount of time to fill a pond is based on economical as well as technical considerations. For catfish production it is generally accepted that the water supply should be sufficient to fill ponds in about two weeks (Tucker and Robinson 1990). Seven days is more desirable but may not be economically practical for some operations. If ponds fill too slowly production time is lost, and weed growth in the ponds may become troublesome. The minimum water requirement for levee ponds in the Southern United States is 0.14 m^3/min/ha (15 gal/min/ac). This is about twice the maximum daily evaporation rate in the region and compensates for periods of drought plus moderate seepage losses. Ideally, 0.28 m^3/min/ha (30 gal/min/ac) is considered more desirable. As an example, if the higher flow rate were used to fill a 4-ha (10-ac) pond to a depth of 1.2 m (4 ft), a water source capacity of about 4.6 m^3/min (1,200 gal/min) would be required. It would take 7.5 days to fill the pond.

Once the pond is filled, a regulated inflow of water must be supplied to maintain the desired water depth and to compensate for seepage and evaporation losses. The water budget for a levee pond should include the following terms (Tucker and Robinson 1990):

Inflow:	Outflow:
Regulated inflow	Regulated discharge (draining)
Precipitation	Overflow
Runoff	Evaporation
	Seepage

The runoff portion of the inflow from the surrounding levees is usually considered negligible. Therefore, a typical water budget for a levee pond is (Boyd 1990):

$$P + INFLOW = (S + E) \pm \Delta V \qquad (4\text{-}3)$$

where P = precipitation; Inflow = water pumped from a well or other controlled source; S = seepage; E = evaporation; and ΔV = change in storage volume. All terms in the equation use millimeters or inches as units. For a static pond situation, where the only inflow is precipitation and the only outflow is evaporation and seepage, the equation simplifies to

$$P = (S + E) \pm \Delta STORAGE \qquad (4\text{-}4)$$

or, rearranging terms,

$$\Delta STORAGE = (S + E) - P \qquad (4\text{-}5)$$

In Equation 4-5 the change in storage (Δ STORAGE) is the change in water level in the pond over a period of time and is equal to the amount of water required to maintain the desired water level.

For watershed ponds, input water comes from rainfall and runoff from the surrounding watershed. The water budget for a typical watershed pond is:

$$P + R = (S + E) \pm \Delta V \qquad (4\text{-}6)$$

where R = surface runoff. If water is intentionally drained from or added to the pond, or if the water supply for a watershed pond is supplemented by well water, or if a levee pond has a larger than normal watershed, then the water budget equations must be adjusted to reflect the changes.

For a dynamic pond situation, where water is discharging from the pond, the water budget is given by Equation 4-7 (adapted from Piedrahita 1991):

$$Q_r = Q_s - Q_p + Q_e + Q_o \qquad (4\text{-}7)$$

where Q_r = regulated flow rate needed to balance seepage, precipitation, evaporation, and outflow; Q_s = seepage rate; Q_p = precipitation rate; Q_e = evaporation rate; and Q_o = overflow rate from pond. All terms are in m^3/h or ft^3/h.

Seepage, precipitation and evaporation can be calculated by use of the following relationships (Piedrahita 1991):

Seepage
$$Q_s = \frac{(A \times V \times R_i)}{24{,}000} \qquad (4\text{-}8)$$

where A = pond area (m^2); R_i = infiltration rate (mm/day); and 24,000 = a unit conversion constant.

Precipitation
$$Q_p = \frac{(A_r \times R_r)}{24{,}000} \qquad (4\text{-}9)$$

where A_r = area which drains into pond (m^2); and R_r = rainfall rate (mm/day).

Evaporation
$$Q_e = 2.241 \times A \times WV_2 \times (e_s - e_a) \times 10^{-6} \qquad (4\text{-}10)$$

where WV_2 = wind velocity 2 m above pond surface (km/hr); e_s = saturated vapor pressure (mm Hg) at the temperature of the water, T_w (K), so that

$$e_s = 25.374 \times exp\left(17.62 - \frac{5{,}271}{T_w}\right) \qquad (4\text{-}11)$$

e_a = vapor pressure (mm Hg) above the pond surface, at air temperature T_a (K), so that

$$e_a = R_h \times 25.374 \times \frac{exp\left(17.62 - \frac{5{,}271}{T_a}\right)}{100} \qquad (4\text{-}12)$$

and R_h = relative humidity (%).

Accurate estimates of seepage, precipitation, and evaporation are essential for calculating flow requirements. Precipitation and evaporation can be estimated from local climatological data. In the absence of good evaporation data, pan evaporation tests can be conducted at the site, or evaporation can be calculated from methods discussed by Boyd (1985a, 1985b) and Piedrahita (1991). Pond evaporation can be estimated as 0.81 times the evaporation obtained with a U.S. Weather Service Class A evaporation pan. Information on soil seepage rates can be obtained from local well drillers or by conducting soil infiltration tests at the site.

Water requirements to compensate for seepage losses can be estimated for different sites in the United States with Table 4-1. The table assumes four different seepage rates. It is obvious that seepage rates are a significant factor for

Table 4-1. Water required (in./year) to maintain water levels in ponds with different seepage rates at various U.S. locations.

Location	Seepage rate (in./day)			
	0.05	0.1	0.3	0.5
Tallahassee, Fla.	1	19	92	165
Augusta, Ga.	20	38	111	184
Raleigh, N.C.	20	38	111	184
Columbia, S.C.	17	35	108	181
Knoxville, Tenn.	12	30	103	176
Auburn, Ala.	7	25	98	171
Stoneville, Miss.	15	33	106	179
Little Rock, Ark.	17	35	108	181
Baton Rouge, La.	13	31	104	177
Houston, Tex.	25	43	116	189
Wichita Falls, Tex.	55	73	146	219
Wichita, Kans.	40	58	131	204
Bakersfield, Calif.	81	99	172	245
Sacramento, Calif.	58	76	149	222

Source: Tucker and Robinson (1990) with permission.

regulating water requirements. The table also points out that water requirements are much higher in arid regions.

If ponds are drained annually, the water required for refilling must be added to evaporation and seepage losses. Most levee ponds require about 122 cm (48 in.) of water when filling (Tucker and Robinson 1990). Discharge can be reduced by harvesting the fish without draining or by partially draining the pond. Many ponds go for years without being completely drained. However, it is sometimes necessary to drain ponds to make repairs to levees or drainage structures or to completely remove the total fish population.

The requirement for water can be reduced by increasing inflows or by decreasing outflows. In actuality there are only so many methods by which one may conserve water. Obviously, the culturist has no control over precipitation, therefore, this water input cannot be increased at will. The water source must therefore be carefully planned so that adequate water can be provided during dry phases of the production cycle. During rainy periods precipitation is an added bonus that must be taken advantage of. However, precipitation should be thought of as a supplement to the normal water supply supplement and not as the main water source.

The pond water inflow rate can be increased by increasing runoff from surrounding levees. Runoff for levee ponds can be increased by initially constructing the pond with larger levees, but this practice increases construction costs and reduces the land available for fish production. Properly constructed watershed ponds make better use of the surrounding land area and should function with little pumped water requirements. However, the capability to add water to watershed ponds is essential during dry spells.

The aquaculturist has no control over evaporative losses from ponds. To practice water conservation in levee ponds the only other course of action is to reduce water outflow by either reducing seepage losses, overflow losses, and/or by reducing the need to exchange pond water. Several steps can be taken to reduce seepage losses. First, seepage losses can be mitigated by careful site selection, including a thorough soil analysis, and by utilizing proper pond construction methods. Ponds should not be constructed where highly permeable or sandy soils are present. Seepage rates in properly engineered ponds should not be greater than 5 mm/day (0.2 in./day) (Tucker and Robinson 1990). Often, however, the size and shape of the land parcel and land slope limit choices for a site. Second, where ponds are to be constructed in highly permeable soils, liners can be used to limit seepage losses. Various types of pond liners are available, and costs vary with type and thickness. Third, seepage losses can be reduced by tilling pure clay or chemical soil amendments into the soil at the pond bottom and sides. These materials limit seepage by aiding soil compaction. They are discussed in greater detail in Chapter 9.

Pond overflow can be reduced by designing the pond with sufficient storage capacity. Pond storage is the vertical distance from the water surface to the top of the overflow pipe. If ponds are constructed with little or no storage capacity the water could overflow levees or dams during rainstorms, potentially causing extensive damage. Tucker and Robinson (1990) recommend a minimum water storage of 76 mm (3 in.) below overflow structures.

Excessive overflow is also caused by water exchange or *flushing*. Flushing is the practice of adding oxygen and/or removing metabolites by pumping freshwater into the pond and allowing stale water to flow out at the same rate. Fish pond management recommendations in past years included flushing as a management tool. However, more recently, it has been demonstrated that water quality is not significantly improved, and fish production is not increased by flushing. The same holds true for shrimp and crawfish ponds. The biological processes controlled by temperature, feed inputs, and microbes in the water column have a more rapid influence on water quality than flushing (Tucker and Robinson 1990). In emergency situations aerators can be used to improve dissolved oxygen conditions as effectively as flushing.

WATER QUALITY

Good-quality water is essential for aquaculture production. In industrial societies it is becoming increasingly difficult to locate adequate quantities of good quality surface water. It therefore becomes necessary to use groundwater sources or resort to water reuse.

Good-quality water is difficult to define since, what is good quality for one species may be of marginal quality for another. Basically, good-quality water can be defined as water in which its chemical and physical characteristics are within defined limits for the cultured species. The degree to which water quality conditions can be allowed to deteriorate is related to the amount of fish waste put into the system. Fish waste, in turn, is related to the intensity of culture and the amount and characteristics of the feed added to the system. The key to any successful aquaculture venture is to maintain the highest stocking rate possible without degrading the water quality to the point where excessive management becomes uneconomical.

WATER SOURCES

Aquaculture water can come from one source or a combination of several sources, each having its own distinct advantages and disadvantages. The two most likely water sources are groundwater (well water) and surface water. Other sources are rainwater and city water. Once the quality and reliability of the water source is established, the choice of a source then becomes strictly a matter of economics.

Groundwater Sources

Groundwater is the preferred water source for aquaculture purposes. It is usually more dependable and more uniform over time than surface water. Groundwater is free from wild fish, fish eggs and the larval and adults forms of predatory insects. In addition, groundwater is usually less polluted than surface waters, although this is becoming less true over time. Groundwater sources have a very constant water temperature throughout the year, which can be used to advantage in certain situations. The temperature of shallow groundwater is dependent on the surrounding atmosphere (i.e., deep sources remaining relatively constant). Beginning at about 15.6 m (50 ft) below the ground surface, the temperature of groundwater increases by about 1°C (34°F) for each 32 m (102 ft) of depth in the United States.

There are several disadvantages to groundwater, which may limit its use in certain geographical areas. Shallow aquifers are increasingly becoming threatened by contamination from such sources as septic fields, chemical and radioactive waste dumps, landfills, and agricultural chemicals. All of these pollutants are capable of traveling great distances through the soil from aquifer recharge areas at the surface. Groundwater may also contain toxic gases like hydrogen sulfide (H_2S), methane (CH_4), and carbon dioxide (CO_2).

A major disadvantage to groundwater is that it is devoid of oxygen and must be aerated before use. Additionally, groundwater may contain high concentrations of dissolved iron, which can often be removed with aeration.

The hardness of groundwater is determined by the geology of the area. In limestone areas groundwater will be hard and high in calcium. Hard waters are relatively noncorrosive but sometimes cause problems with scale buildup in plumbing fixtures. Groundwater from granite rock areas will be soft and corrosive. In addition, these waters may contain high concentrations of CO_2. Dissolved minerals are sometimes added to soft waters to increase the hardness and alkalinity before use (refer to Chapter 2).

A thorough chemical analysis should be performed on any potential groundwater source before it is used for aquaculture purposes in order to determine the type and degree of treatment required, if any. The potential culturist should seek out and speak to other fish farmers in the area and use their experiences to access potential problems with the water supply. Local well drillers can be contacted for well log and water quality and quantity information. State and local water resources agencies can be contacted as well. If well drillers or agencies do not have sufficient information on a potential water supply, a test well should be drilled to obtain quality and yield data. As an extreme measure, a pilot-scale study could be conducted if insufficient data exists for the water supply.

The two principal sources of groundwater normally available in sufficient quantity for aquaculture use are springs and wells.

Springs. *Springs* occur when an aquifer is exposed at the ground surface (Figure 4-1) or when a crack or fault occurs in an upper confining layer. Springs can occur as a point source or spread over a broad region. Generally, only point sources provide sufficient flow to use for aquaculture. Springs have all of the advantages of groundwater and usually do not require pumping, which saves on energy costs.

A determination should be made on the yield and reliability of a spring before its use. It is a good idea to consider a spring only as a supplemental source of water since it may not be reliable during dry spells. Rightful ownership of the spring must be established, and permits may be required before use.

Figure 4-1. Sources of ground- and surface water.

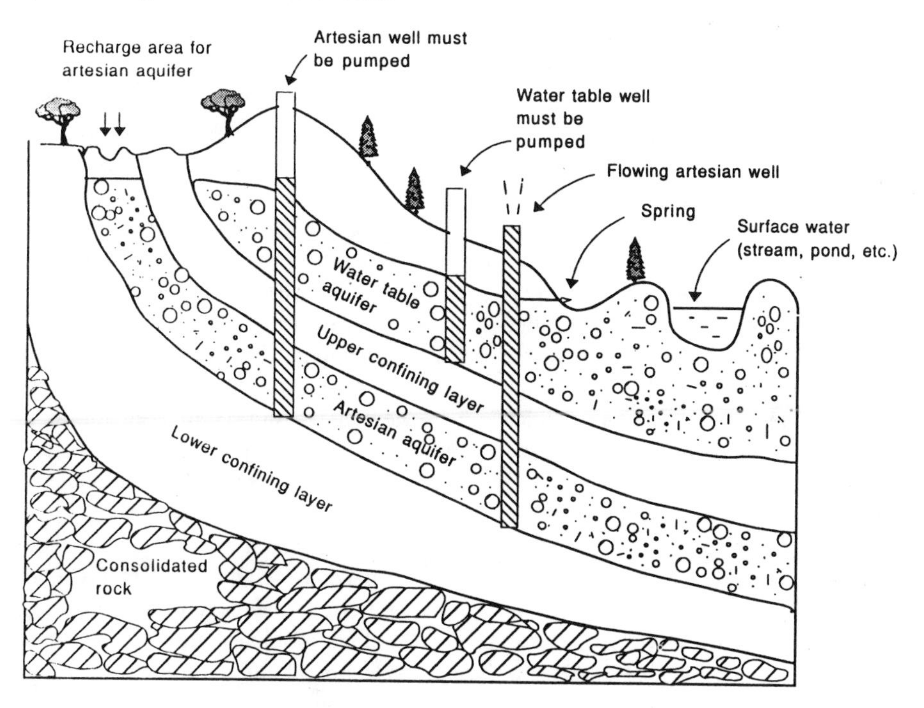

Wells. Well water is usually the best source of water for aquaculture. The water is typically of better quality than surface water, but wells must be pumped in most cases. The costs associated with pumping, particularly from deep wells, are often a large component of the facility operating budget. Well construction costs can also be considerable and depend on the depth to the water and the characteristics of the water-bearing geologic formation. For wells that require pumping, shallow aquifers (those less than 78 m (250 ft) deep) are preferred (Tucker and Robinson 1990). In many parts of the United States well permits are required. The local Soil Conservation Service office can supply information on wells and permits.

Wells are basically of three types: water table, flowing artesian, and nonflowing artesian. *Water table wells* are also referred to as *shallow wells* since they are essentially holes dug into the water table aquifer (Figure 4-1). The water table is the uppermost surface of the water layer in the saturated zone of the soil. The pressure exerted on the water layer surface is in equilibrium with the atmosphere. Shallow wells are affected by fluctuating water tables and may totally dry up during severe periods of drought. Yields vary during the course of a year and are dependent upon the well diameter, soil permeability, and vertical thickness of the aquifer.

Flowing artesian wells are those drilled into aquifers confined between two impermeable layers (Figure 4-1). The aquifer recharge area is at a higher elevation than the well outlet, and the hydrostatic head causes water to flow by gravity. The well starts flowing as soon as drilling is complete, so a shutoff valve should be installed to control the flow. If the well outlet is at the same elevation or above the recharge area the well will not flow and must be pumped. Such wells are known as *nonflowing artesian wells.*

Surface Water Sources

Freshwater Sources. Fresh surface water sources include rivers, streams, bayous, sloughs, lakes, ponds, and reservoirs. They almost always require pumping. The cost of pumping may be high, but it is usually less than pumping water from wells. Surface waters should be carefully evaluated and thoroughly analyzed before use since they are subject to contamination and often carry high silt loads. In addition, they may contain wild fish, parasites, waterborne predators, or disease organisms. The water should be filtered with fine mesh screens before use to remove these constituents. Surface waters from areas where agriculture is intensively practiced may contain pesticides or other potentially harmful chemicals.

Rivers and streams are subject to variations in flow, and the availability of water during dry periods may be limited. Historical flow records must be examined to determine when water shortages may occur. It may be practical to shift the production cycle so that periods of reduced water requirements coincide with periods of low flow.

Surface waters are often subject to environmental regulation that may change without much advance warning, presenting an element of risk to the user. As in the case of springs, proper ownership of the water must be established before use. Permits may be required before surface waters are diverted for aquaculture use, and discharge permits may be required.

Brackish Water and Seawater Sources. Like other surface waters, brackish and seawater sources are subject to contamination, usually increasing in severity the closer one is to the shoreline. Coastal pollution is an ever-increasing problem, and the situation will only worsen without strict governmental regulation. Water intakes should be located as far offshore as economically practical and should not be located near industrial or municipal discharges or near areas subject to agricultural runoff. Salt water is very corrosive, therefore, pipes, fixtures, pumps, and other components that come into contact with the water should be fabricated from corrosion-resistant materials. Fouling is also a considerable problem wherever salt water is used. Measures that can be taken to mitigate fouling problems are discussed at length by Wheaton (1977).

The saltwater source should be of constant salinity. This is generally not a problem where the water intake is located far offshore. Intakes located near freshwater inflows from rivers and streams or in shallow water areas that may be affected by evaporation and rainfall are subject to rapid fluctuations in salinity and should be used with caution.

Alternative Water Sources

Rainwater. Rainwater is a source of free water that the aquaculturist should take advantage of at every opportunity. Rainfall can be collected in tanks or impoundments and stored for later use. Rainfall should, however, be considered a supplement to the normal water source since monthly rainfalls are highly variable and cannot always be relied upon to supply water at critical times.

Rainwater is generally of good quality, but it is somewhat acidic and poorly buffered. Depending upon the use, it may be necessary to add lime to increase total hardness and alkalinity. Pond waters prepared with a mixture of rain and well water are usually of good enough quality for aquaculture purposes without much treatment other than aeration.

City Water. City tap water may also be considered as a water source for aquaculture, but on a limited scale. Obviously, it would not be feasible to fill ponds or a series of large tanks with tap water since the cost would be considerable. City water is usually of excellent quality but normally contains a disinfecting agent to make it safe for human consumption. The most common disinfecting agents used in city water are chlorine and fluorine, two substances that are extremely toxic to fish in small quantities. The agents should be removed before use, but this practice on a large scale is not economical. Sodium thiosulfate is used to remove chlorine and chloramines from tap water. Chlorine can also be removed by vigorous aeration for 24–48 hours. Chemical dechlorination is usually too expensive to be used on a large scale.

Artificial seawater is made by mixing tap water (or other good-quality freshwater) with artificial sea salts. Artificial sea salts are available from a number of manufacturers in the United States, but a given mixture may not be suitable for all species or life stages. In addition, not all marine mixtures are suitable for plant culture. The risk can be minimized by using a mixture that is manufactured for broad application. The cost for preparation of artificial seawater is extremely expensive and is not recommended for large-scale culture with the exception of public display aquaria where cost is generally not a factor.

Saltwater Wells. Saltwater intrusion in some coastal regions of the United States makes it possible to extract brackish water from the ground. The quality of such wells depends on site conditions and the geology of the water-bearing strata. Saltwater wells often must be drilled very deep to reach below freshwater aquifers. Typically, the deeper the well, the more expensive the construction and pumping costs. Also, the deeper one drills, the warmer the water. Cooling may be required before use if the well water has too high a temperature. If the system appears to be economical, saltwater wells have all of the same advantages as fresh groundwater. Where salt- or brackish water is required, saltwater wells reduce system complexity and costs by reducing the need for treatment. A disadvantage of salt water wells is that site conditions and geology often produce unfavorable groundwater quality (Huguenin and Colt 1989).

Recycled Water. Waters that have already been used for some other purpose, like irrigation tailwater or pond water, can sometimes be reused for aquaculture purposes. The availability of reuse water is dependent on the individual system and the degree of treatment required before reuse. In most instances the water will not be of the same high quality as it was when first used, but some potential for reuse may exist nonetheless. Obviously, if water can be reused several times before discharge, a savings is realized on water pumping costs, and valuable water supplies are conserved. Disadvantages are that the availability of water for reuse and water needs do not always coincide. In addition, the proximity of reuse water to the location where it is needed may have limitations.

WATER INTAKES

For those aquaculture systems utilizing a surface water source, the location of the water intake may be the most important aspect of the project (Huguenin and Colt 1989). All applicable permits must be secured before construction begins. A Section 404 permit is required from the U.S. Army Corps of Engineers for locating a structure in, excavating in, or discharging dredged or fill materials into navigable waters of the United States.

Intake openings should be located near the water's surface to utilize the better-aerated surface waters, but not so near as to draw in floating pollutants. Floating intakes are commonly used in freshwater lakes and ponds. In this technique the intake opening remains at the same water depth regardless of the elevation of the surface (Figure 4-2).

The most common source of seawater is from nearshore or offshore intakes. Marine intakes present the most challenging problem. They should be located as far offshore as possible to avoid salinity fluctuations and pollutional problems. A big disadvantage to using natural marine water is that quality fluctuates diurnally and seasonally. Submerged water intakes present a hazard to watercraft and should be located away from shipping lanes and out of the way of normal boat traffic. Signs or warning buoys should mark the site, and no anchoring, dredging, or seining permitted in the area. The intake opening should be located some distance beneath the surface but not near the bottom where sediments and low DO water will be drawn in. The design engineer should have knowledge of tidal fluctuations in the area so that the intake opening is not exposed at low tide.

Wind and wave action can exert considerable forces on intake structures. It may be necessary to consult with an ocean or marine engineer so that the intake will not be subject to failure during storms. The structure should be able to withstand 25-, 50- or 100-year storms if it is to be a large, permanent installation. For short-duration installations the 5–20-year storm may be used for design depending on the life expectancy of the system and its intended use (Huguenin and Colt 1989).

Intake openings should be screened to filter out debris, unwanted fish, fouling, and disease organisms, etc. The mesh size of the screen is important. The openings should be as large as possible to protect the pump and to reduce suction-side head losses. Pump manufacturers usually state the largest size object that may safely travel past the impeller without damage. In lieu of screening it may be desirable to bury the intake in the bottom and use the coarse overlying sand as a filter. One such configuration is illustrated in Figure 4-3. Several design options for buried marine intakes are presented in Huguenin and Colt (1989).

Figure 4-2. Floating water intake scheme.

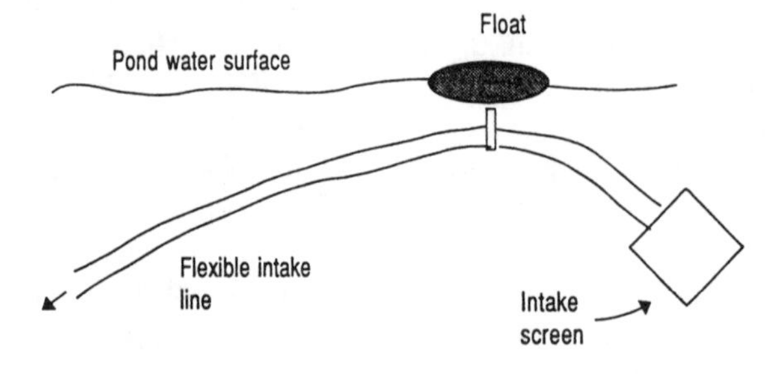

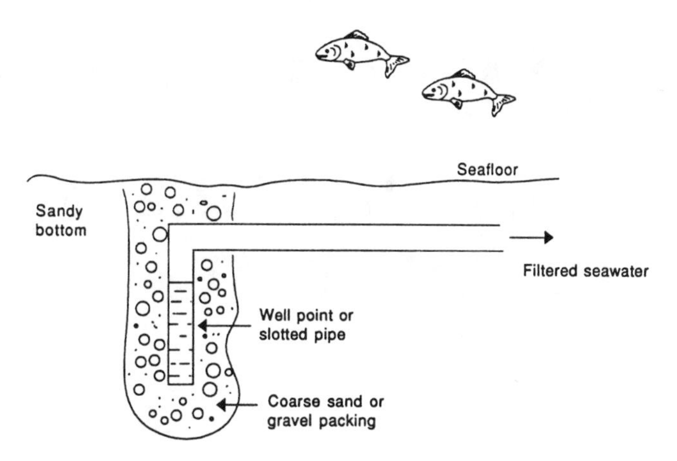

Figure 4-3. Offshore filtered seawater intake.

WATER DISCHARGE

Water must be treated before discharge so that it meets the requirements of local, state, and federal regulatory agencies. The quality of the discharge water is determined by the intensity of culture and the treatment the water receives during and after use. The degree of treatment required before discharge is dependent on the quality promulgated by regulatory agencies. The U.S. Environmental Protection Agency (EPA) and the U.S. Army Corps of Engineers have programs that coordinate the granting of permits for point sources of pollution. Some aquaculture effluents can severely impact the quality of receiving waters. Aquaculture effluents are considered point sources of pollution, and the aquaculture industry is subject to environmental regulation like any other industry. Waste products include fecal matter, nutrients, metabolites, biochemical oxygen demand (BOD), chemical oxygen demand (COD), chemicals, unconsumed feed, low or high pH, and low dissolved oxygen. A permit to discharge is required from the EPA if fish production is over 46,000 kg/year (Tucker 1985). Other state and local permits may be required. Amounts and types of discharges allowed are subject to regulation.

CHAPTER 5
AQUACULTURE IN OPEN SYSTEMS

Aquaculture systems are very diverse in their design and function. The three most basic categories of culture systems are *open, semiclosed* and *closed* systems. *Open system* culture generally refers to fish farming in natural water bodies such as oceans, bays, estuaries, coastal lagoons, lakes, or rivers. *Semiclosed* systems are those in which the culture water makes one pass through the system and is discharged. Systems of this nature are referred to as *flow-through* or *once-through* systems. The raceway falls into this category. *Closed* systems are those where the water is reconditioned and recirculated to the culture unit(s). A more recent term for this type system is the *closed recirculating* system. Various combinations of the latter two system types are in everyday use. The combinations are limited only by the imagination of the aquaculturist. Open system culture is the subject of this chapter. Other systems types are discussed in later chapters.

The natural system is the most primal open system and is the one most often used for commercial profit since it provides resources for commercial and recreational fisheries. Its continued popularity is a result of low cost (since natural systems exist without man's influence) and limited management requirements. A step beyond the natural system is what Wheaton (1977) refers to as the *modified open system*. This involves the confinement of fish in enclosures, floating cages, net pens, baskets, trays, etc. In some parts of the world molluscan shellfish are cultured by attaching them to strings or lines suspended from floats or rafts or strung between poles or pilings driven into the substrate. Other techniques involve capturing the spat of benthic shellfish directly onto pilings and harvesting the mature adults at a later time.

The bulk of this chapter discusses cage culture since the technology has evolved using more of an engineering approach. Space limits a lengthy discussion of all of the engineering considerations that go into cage culture. Therefore, this chapter presents an overview of the subject.

CAGE CULTURE

Terminology used in cage culture is often confusing and warrants discussion. The terms *enclosure, pen*, and *cage* appear to be synonymous and are sometimes used interchangeably. In an attempt to reduce confusion, the term *enclosure* as used in this text refers to a natural bay or semisheltered water body where the shoreline forms all but one side of the enclosure. The enclosure bottom is formed by the natural bottom of the water body. In most cases, enclosures are separated from open water by a solid barrier or with netting or plastic mesh material. The mesh size is typically small enough to retain the cultured fish but large enough to allow the ingress and egress of small fish and food organisms. Enclosures commonly range in size from about 0.1 ha to over 1,000 ha in surface area (Beveridge 1987).

Pens differ from enclosures in that they are almost entirely man-made. All sides are constructed from bamboo, wooden poles, or stakes driven into the substrate. The bottom is formed by the natural bottom of the water body. Pens range in size from very small structures of barely over 1 m to several hectares in surface area.

Cages are entirely man-made and generally float off the bottom. They are much smaller than either enclosures or pens, ranging in size from about 1 m^2 to over 1,000 m^2 in surface area (Beveridge 1987). In North America and Europe large cages are often referred to as *net pens*. At what point a cage becomes a net pen is not clear. Therefore, the terms *cage* and *net pen* are often used interchangeably, and some confusion continues to exist.

History of Cage Culture

Early cage types were probably little more than modified fish traps, where the catch was held by the fishermen until a sufficient quantity was accumulated to take to market. This type of fish holding facility has been used for many generations in all parts of the world. True cage culture, where aquatic animals are retained for long periods of time while they increase in biomass, had its origin in southeast Asia approximately a century ago. Modern cage farming originated in Japan where yellowtail were first commercially cage-produced in the 1950s. The growth of modern cage farming in the United States began with tilapia farming in the 1960s and expanded throughout the 1970s and 1980s with channel catfish and rainbow trout culture.

Cage culture of Atlantic and coho salmon began in the 1960s in Norway, Scotland, Canada, and the United States. The fish are first produced in hatcheries and are then moved to the cages as young fingerlings. The time that fish are kept in the cages depends upon the species and the market size desired. For example, salmonids are produced in hatcheries, and the fingerlings are transferred to cages in the spring of the year after which they are fed for approximately 18 months. Harvesting typically runs from September to the following April.

Since many of the sheltered inshore sites were rapidly taken, the industry moved farther offshore as it grew. By the late 1970s the offshore industry was located mostly in Europe since environmental regulations in the United States caused considerable red tape. The larger offshore facilities comprise the bulk of the net pen industry. Over 20 net pen manufacturers are located in Europe, while only one is known to exist in the United States. Cage and net pen permits, costing many thousands of dollars, are required in the United States. For example, in Maine a pen permit costs approximately $100,000 (Riley 1991). The high cost for permits has the effect of driving smaller producers out of business, and they are soon replaced by large companies that can afford the fees.

Cage Culture versus Traditional Aquaculture

Recent reviews of various government publications show that production in freshwater accounts for approximately 80–90% of the world's finfish production, and about 95% of this is in ponds. Cage culture accounts for only 3–4% of freshwater finfish production, but about 40% of marine and brackish finfish production. Cage culture assumes a broader role in certain sectors of the aquaculture industry. For example, over 40% of rainbow trout production in Scotland is in cages. The yellowtail industry in Japan and Western Europe's Atlantic salmon industry are almost exclusively based on cage culture (Beveridge 1987). In the United States over 130 species of finfish are produced in cages. Over a dozen species of prawn, lobster, and crab are also produced in cages, but cage culture of nonfinfish species is in its infancy.

Cage culture offers certain advantages and disadvantages, and trade-offs must be made when weighing the potential of this cultural practice. A significant advantage is that cage culture makes use of existing water bodies, which can be a major benefit in giving nonland-owning sectors of the community access to fish farming. Water bottom can be leased from the local governing authority, eliminating the need to invest large amounts of capital in land. Another advantage is that cage farm management is, in some ways, less complex than land-based systems. Once installed, land-based systems are difficult and expensive to alter. Cage farms can be expanded by simply adding cages as experience grows. Additionally, cages are mobile, meaning that they can be moved to other sites to take advantage of better-quality water, and more abundant food organisms and to escape storms.

Capital and operating costs for cage facilities are highly variable and can be greater than similar-sized land-based systems, depending upon cage sizes, species cultured, and local conditions. However, Shaw and Muir (1986) demonstrated that both capital and operating for intensive cage culture of salmon are significantly lower than traditional land-based rearing methods for this species. It is argued that cage culture is the most economical means of culturing such species as salmonids, yellowtail, and grouper in marine waters, and is a comparatively profitable means of producing other species (Beveridge 1987).

Just as there are advantages, there is also a negative side to cage culture. Since cage facilities are often sited in public or multiuser water bodies, farmers have no control over water quality conditions, and pollution is often responsible for serious damage. Also, cage installations are at the mercy of the weather and may suffer damage from high winds, tides, and waves if they cannot be moved when storms approach. Offshore cage culture is generally not practiced in parts of the world subject to frequent violent storms, such as in many parts of the Orient.

Finfish cultured in cages can suffer damage to skin, scales, and fins through abrasion with the sides and bottom of the cage, which open avenues for infection. However, these problems can be mitigated through careful selection of construction materials, good management, and by proper siting and mooring of the cages.

Cage facilities have an impact on the aquatic environment in that large quantities of uneaten feed and feces are released. This may adversely affect water quality in the general area, and may stimulate primary production which, in highly intensive situations, can lead to eutrophication. The overall impact could be that native species are negatively impacted (Loyacano and Smith 1976; Beveridge 1984). Cages could also introduce or disrupt disease and parasite cycles.

Cages take up space in public waters, which can make navigation hazardous or may deny access to certain areas by commercial and sport fishermen. Additionally, some may consider the sight of a cage facility aesthetically unacceptable, and may be concerned about a reduction in nearby land values.

A final disadvantage is that cages are highly vulnerable to poaching and vandalism. This may preclude their use in some parts of the world.

Cage Types

There is an immense diversity of cage types and designs. They are very versatile, lending themselves to a number of applications. A cage classification system developed by Beveridge (1987) is presented in Figure 5-1. There are four basic types: fixed, floating, submersible, and submerged. Many variations and combinations of these are in use today.

Figure 5-2 illustrates cage designs popular with small-scale cage farmers in the United States for both fresh- and seawater applications. The first (Figure 5-2a) is a cage fabricated from wood or plywood with small wire or plastic mesh openings on the sides for water exchange. Figure 5-2b illustrates a cage constructed with a rigid frame, made from either wood or metal tubing. The latter type is lighter in weight and has more open area at the sides to enhance water

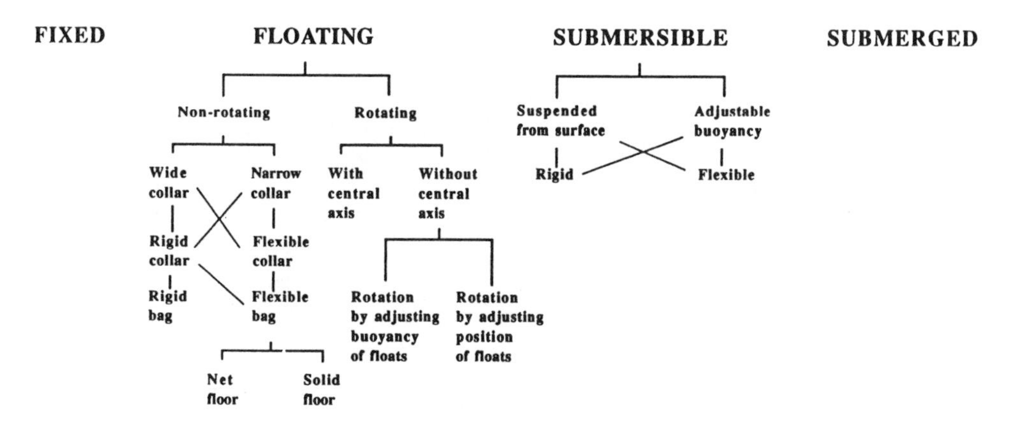

Figure 5-1. A classification system for aquaculture cages (adapted from Beveridge (1987)).

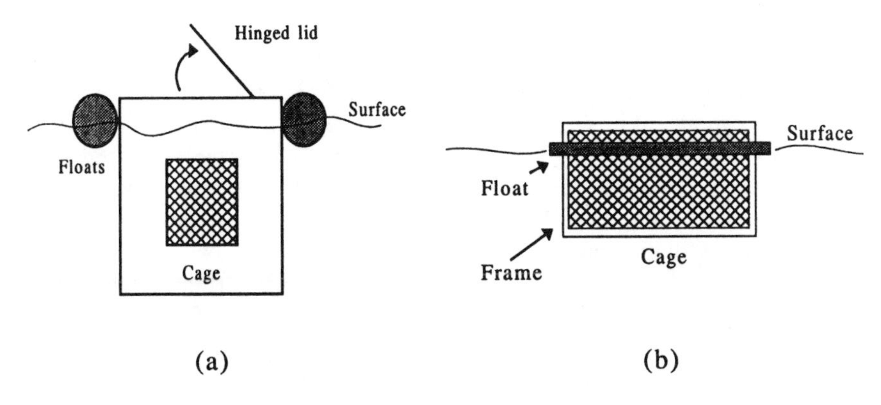

Figure 5-2. Common small aquaculture cages: (a) Wooden sides and bottom; and (b) mesh sides and bottom.

exchange. Both are covered to prevent escape and predation. Hinged lids provide ready access for feeding and harvesting, while locks discourage poaching. The cages are shown with floats as opposed to fixed cages attached to pilings driven into the substrate. Floats eliminate the need for pilings and allow the cages to rise and fall with fluctuating water levels. Larger floating cages and net pens have catwalks and/or working platforms which facilitate feeding, harvesting, and routine activities. Fixed cages normally do not have these structures. They are accessible only by boat. Smaller fixed cages are often attached to piers, which makes them readily accessible from shore.

Both fixed and floating cages can be of either rigid or flexible construction. Rigid cages are those that use a nonflexible framework as well as a stiff mesh material at the sides. The bottom is also sometimes comprised of mesh material. The cages shown in Figure 5-2 are of the rigid type. Rigid cages are used most often in swiftly flowing water currents such as in rivers and streams. Flexible cages are usually of the floating type and have collars that support the catwalks and/or working platform. The collar assembly also supports the cage *bag*, the term given to the flexible netting structure that contains the fish. Large net pens are typically constructed in this fashion. They are large, up to 15 m (50 ft) (square or in diameter), and are often grouped so that a walkway can serve more than one pen. Walkways are constructed with inner and outer handrails as illustrated in Figure 5-3. Modern net pens employ large vertical spars or pontoons for flotation.

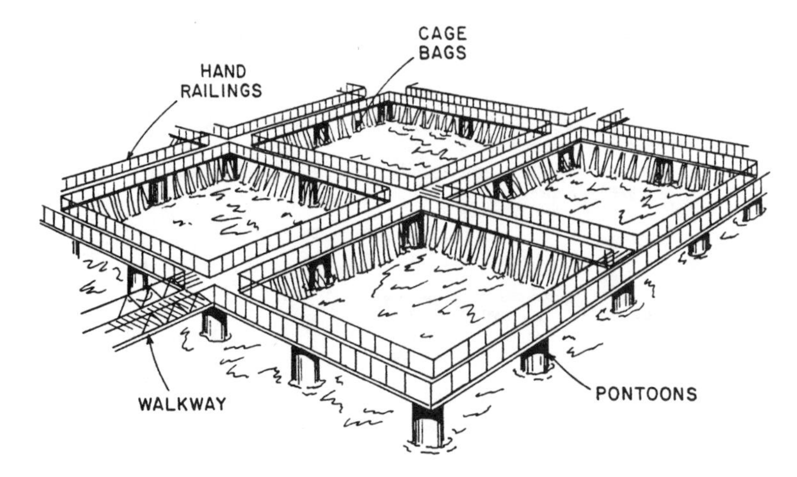

Figure 5-3. Grouped arrangement of floating net pens.

Some cage designs allow the cage to be rotated. They are usually cylindrical in shape and are floated with the cylindrical axis lying horizontal and the cage rotated about the axis. The cages are rotated to control biofouling. Daily rotation allows all sections of the cage to be exposed to the atmosphere, which helps to dry fouling organisms and kill them by exposing them to the sun's ultraviolet rays.

Submersible cage designs have also evolved in recent years and are in use in several countries. They are used primarily as a strategy against stormy weather conditions and to avoid problems with ice. There are both fixed and floating types. A simple, submersible fixed cage, which is of the type used for tilapia culture in the Philippines is illustrated in Figure 5-4. The cages are fitted with net tops, and when a storm threatens, the rigging is untied, and the cages are lowered a 1 m or so beneath the surface.

Submerged cages are used to culture yellowtail and sea-bream in Japan at sites that are exposed to weather extremes. In a typical design the cages are attached to buoys at the surface (Figure 5-5). During normal conditions the cages are fixed at about 2–5 m (6–16 ft) beneath the surface, and the fish are fed via a feeding tube. During storms the feeding tube is tied shut, and the cage is submerged to a depth of about 10–15 m (32–48 ft). Cages are often up to 100 m^2 (3,200 ft^2) in surface area and are congregated in groups of 10–12 (Beveridge 1987).

Submerged cages are used since the water movements produced by waves occur primarily at the surface. As waves advance over the sea, the water particles move in rotating motions with each wave (Figure 5-6). The diameter of the orbits at the surface is equal to H, the height of the wave. The orbit diameter decreases exponentially with depth according to the following relationship from Pond and Pickard (1978):

$$D_z = H \exp\left(\frac{2z}{L}\right) \tag{5-1}$$

where D_z = diameter of the orbit; z = depth; and L = wavelength. At a depth of approximately $L/9$, D_z is reduced by half, and at $L/2$, D_z is reduced to 0.04 H. Wave forces are dramatically reduced by submerging the cages.

Engineering Considerations in Cage Design

In theory, cage design depends upon a number of factors including fish species; environmental conditions; extensive versus intensive culture; properties, cost, and availability of materials; and management skills. Cage designs have largely evolved empirically. Since research of this type is very expensive, and since

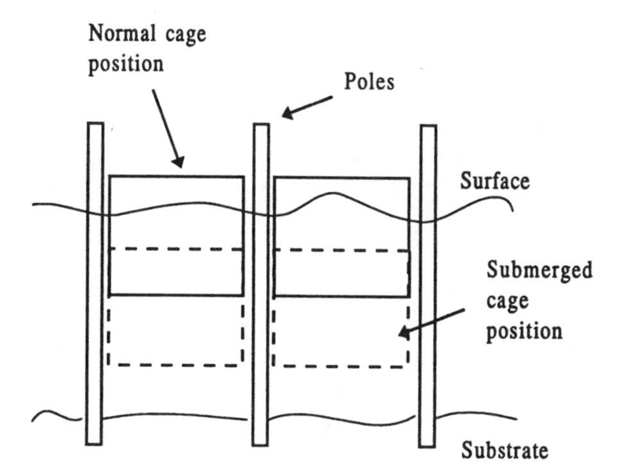

Figure 5-4. Simple small submersible cages.

Figure 5-5. Submersible cage for culturing yellowtail in Japan.

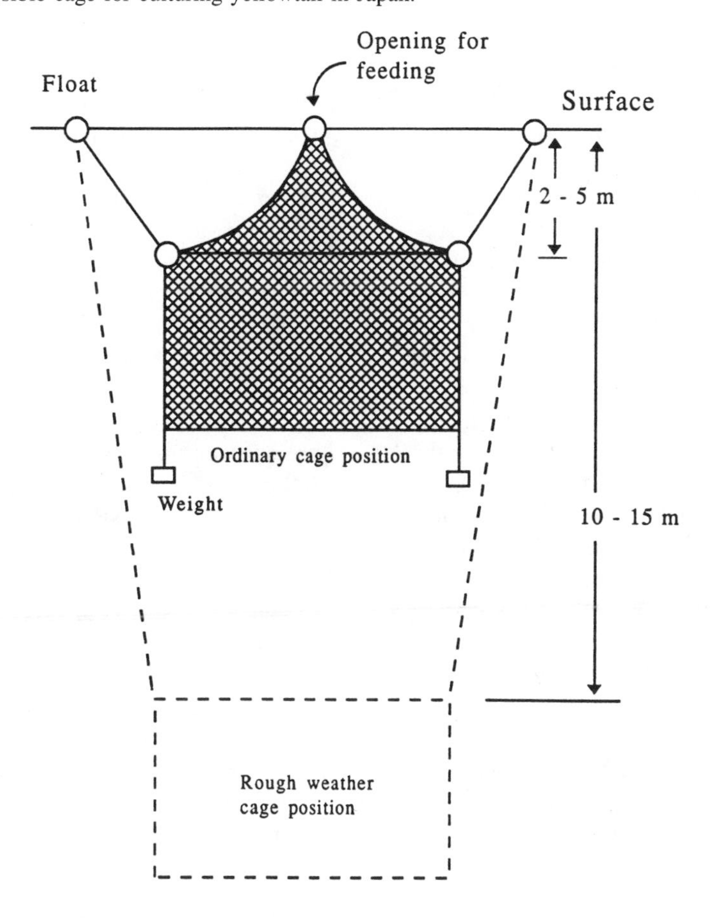

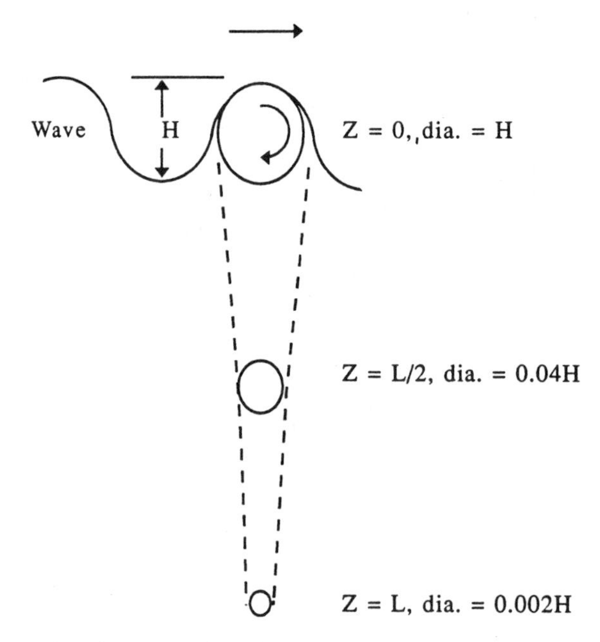

Figure 5-6. Rotational properties of water particles in a wave (adapted from Pond and Pickard (1978)).

the industry is traditionally very small, there has not been much available capital to sink into research and development. This changed somewhat in the 1980s as multinational interests became involved. These large interests, located primarily in the United States, Japan, and Europe, have the capital and are willing to invest in research and development programs. As a result, the industry has expanded rapidly. The fact that cage culture has moved farther offshore has also laid the groundwork for future research. Cages will have to be more stringently designed, tested, and evaluated to satisfy the demands of lending institutions, insurance companies and government agencies. Modern test facilities consist of large wave tanks where scale-model cages and modeling procedures are used to test the forces on cage components as well as new cage designs.

Cage Bag. The cage *bag* is the mesh structure that contains the fish. It can be either round, square, rectangular, six-sided, or eight-sided in shape. Round cage bags are stronger and make the most efficient use of materials at the least cost per unit volume. Also, schooling species, like salmon and milkfish, appear to be less stressed when cultured in round bags since they tend to swim in a circles when enclosed (Sutterlin et al. 1979). However, octagonal and/or square cages probably work just as well. A disadvantage to round cages is that the surface area to volume ratio (*SA*:*V*) is small compared to other shapes, resulting in poor water exchange. Additionally, the supporting structure for round cages is considerably more sophisticated (Beveridge 1987).

The influences of cage surface area and depth are not well understood. Maruyama and Ishida (1976, 1977) showed that cage depth less than 1.2 m (4 ft) retarded growth and affected body shape in carp and tilapia. Flatfish seem to do well in cages with a bag depth between 0.9 and 1.6 m (3 and 5 ft) (Kerr et al. 1980). Maruyama and Ishida (1976, 1977) remarked that cages greater in depth than 10–12 m (30–36 ft) are probably poorly utilized by many fish species.

Beveridge (1987) listed properties the ideal bag material should possess: strength and lightness; rot, corrosion, and weather resistance; fouling resistance; easy reparation; drag force resistance; nonabrasion; and economical cost.

Materials used to construct cage bags are many and varied, ranging from cotton and synthetic fibers to semirigid materials such as extruded plastics and metals. Flexible cages are fabricated from fibers, either natural or synthetic. Natural fibers are used very infrequently today since they are subject to rot and subsequent loss of strength. They also cannot be used in the fabrication of knotless meshes that are recommended for use in cage culture. Modern netting is composed of synthetic fibers such as nylon, polyamide (PA), polyester (PES), polyethylene (PE), and polypropylene (PP). Knotless meshes are generally fabricated from nylon, PE, or PES. The names of these and other synthetic materials vary from one country to another and between manufacturers in a given country. Figure 5-7 shows but a few of the many types of netting material available today. Bag materials (as well as the supporting framework) should be rigid enough to retain shape but flexible enough to minimize the effects of wind and wave forces.

Rigid cages use either plastic or metal mesh fabric. The mash shape is usually square and most are in a range of sizes. Plastics used have a density on the order of 0.35–0.50 kg/m^3 (0.02–0.03 lb/ft^3). Cage bags constructed with plastics are semirigid, while those made from metal are of rigid construction only. Metals have a big disadvantage in that they are subject to corrosion, particularly in marine or brackish water environments. Corrosion effects are mitigated with the use of metal alloys, such as copper-nickel (Cu-Ni), by cathodic protection with sacrificial corrosion or with protective coatings. Metallic coatings are applied by dipping, electroplating, spraying, cladding, or cementation. Many different paints, varnishes, plastics, lacquers, and rubber coatings are also available. Only three types of metal mesh have been widely used in the manufacture of cages: (1) Copper-nickel wire and expanded metal mesh; (2) galvanized steel weld mesh or chain link netting; and (3) plastic-coated galvanized chain link or weld mesh. Rigid mesh materials are available in a limited number of panel sizes or roll widths, which can dictate the size, shape, and design of cages (Beveridge 1987).

Figure 5-7. Examples of cage mesh material (photo by T. Lawson).

It is important that the cage maintain its shape against external forces. Otherwise, the fish would be subject to crowding and severe stress, and high mortalities may result. Two principal types of forces act on cage bags: static and dynamic forces. Static loads act vertically and are composed of the weight of the bag and fouling organisms and the fish load whenever the bag is hoisted from the water. Therefore, the bag material must be strong enough to withstand these forces. The dynamic forces are more important and are caused by water currents and wind-induced surface waves. The bulk of these forces act in a horizontal direction, but waves do exert some degree of vertical dynamic loading. Other loads are exerted by floating objects, such as driftwood and floating plants. Quantification of currents and wave forces and the responsive behavior of the cage materials can help predict how a particular cage design will perform and can aid in cage design. Research is severely lacking in this area.

Frames. The basic frame that supports the cage bag generally follows the shape of the bag. It can be constructed from a variety of materials including bamboo, wood, metal, or synthetic substances. Small, cottage-industry-type cage frames are normally fabricated from wood or plastic, and larger commercial-size cages are fabricated from galvanized steel or 90:10 Cu-Ni pipe (Beveridge 1987). Commercial development of modular systems has spurred the production of specially designed framework constructed from PES and glass composite (fiberglass). These frames are lightweight, strong, resistant to corrosion and weathering, and can be produced in an infinite variety of shapes and sizes (Huguenin and Ansuini 1978). Plastic or nylon fasteners should be used to attach bags to the framework since metal fasteners tend to create galvanic cells and experience rapid corrosion.

Bags and frames are usually designed with a freeboard extending above the water surface to prevent fish from jumping out. The height of freeboard is dependent on fish species as some are better leapers than others.

Collars and Supports. Fixed cages typically do not have collars, catwalks, etc. They are supported by either bamboo poles, wooden posts or columns, or some other supporting structure driven into the substrate. The support structure must be able to withstand the static vertical forces imposed on them by the weight of the cage, working staff, equipment, etc. while at the same time resisting dynamic forces exerted by wind, waves, and currents. Milne (1972) proposed that bamboo support poles be driven into the bottom muds a minimum depth of 1–2 meters (3–6 ft). Unfortunately, there are insufficient experimental data to make recommendations for other materials.

Floating cages present another set of diverse design problems. A collar is normally used to support and buoy the cage bag and to help retain its shape. The collar normally is used to support catwalks, which serve as work platforms. In many cage designs the flotation material is an integral part of the collar. Flotation can be provided by bamboo poles; hard or soft woods; foam-filled drums or tubing; or air-filled drums, tubing, or pontoons.

Both static and dynamic forces work on cage collars. Static forces include the weight of the bag, supporting structure, farmworkers, sacks of fish feed, fuel, equipment, etc. To design a collar to support all of these individual loads, the densities of the different materials must be known, and these loads must be balanced by buoyancy forces. The required buoyant force can be calculated from Equation 5-2 (Beveridge 1987):

$$F_B = V_W Q_W - V_M Q_M \tag{5-2}$$

where F_B = buoyancy force, kg (lb); V_W and V_M = volumes of water and flotation material, respectively, m^3 (ft^3); and Q_W and Q_M = densities of water and flotation material, respectively, kg/m^3 (lb/ft^3).

In theory, it should be a simple matter to design the required buoyancy force based on the known volume and density of the materials. However, in practice it is often difficult. The effect of cages being linked together is typically ignored in design. For design purposes, it should be assumed that each cage is independent and that each side of the cage must be able to support its share of the load exerted by the bag and support framework plus the weight of farmworkers and equipment. This is a design safety factor used to take into consideration those times when vertical forces will be unbalanced.

Many small cages provide flotation only and will not support farm staff, equipment, etc. In these cases the cage must be serviced from a boat, pier or raft. Large net pens collars that support walkways and work platforms must provide a safe base upon which to work and are typically difficult to work from in rough weather. Walkways should be a minimum of 1 m wide. Cages having narrow walkways are often joined together by a central platform several meters wide. Many of the large offshore installations have central platforms large enough to drive small vehicles, forklifts, etc.

Linkages and Groupings. Most small cages are moored individually, however, it is common practice in large cage farms to group several cages together. Figure 5-8 shows an aerial view of an existing cage farm, demonstrating cage grouping. Grouping simplifies and minimizes the cost of mooring systems and makes overall management of the installation easier. The number and arrangement of grouped cages depends upon several factors as discussed by Beveridge (1987): (1) Size of the farm; (2) site conditions; (3) cage design; (4) mooring design; and (5) environmental considerations.

Cages in large farms are often grouped to house fish in various life stages or grouped according to species, if more than one species is being cultured at the same time. The physical characteristics of the site (i.e., size, shape, water depth, etc.) may restrict cage positioning and grouping. Cage shape may influence how cages are grouped. Square and rectangular cages can be grouped in a variety of ways, while circular, hexagonal, and other-shaped cages are somewhat limited.

Cage grouping is extremely important in regard to water exchange within individual cages.

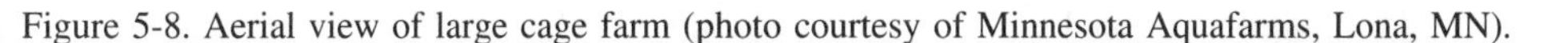

Figure 5-8. Aerial view of large cage farm (photo courtesy of Minnesota Aquafarms, Lona, MN).

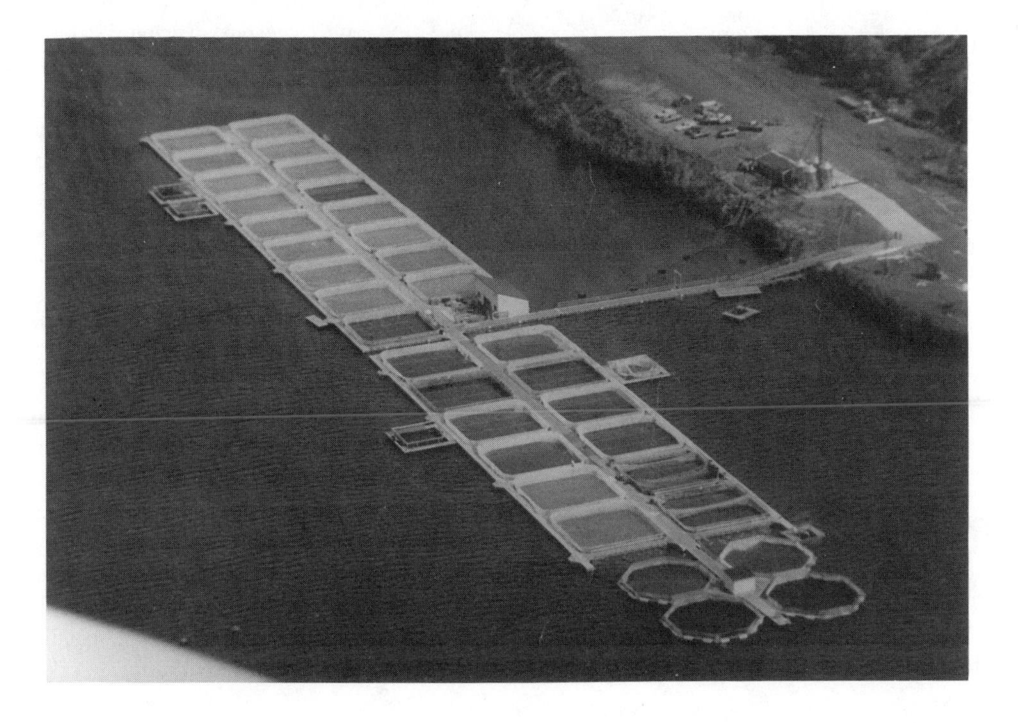

Water exchange may be satisfactory within the cages positioned at the outside of a group, but the innermost cages may not have sufficient water exchange. Experienced cage farmers recommend that cage groupings be no more than two or three units deep in the direction of prevailing water currents. Figure 5-9 illustrates examples of cage grouping.

Cage grouping significantly reduces the forces acting on individual cages and dampens both rectilinear and rotational motions. Cage linkages should dampen pitching motion and minimize yawing and swaying. Ropes or chains form the most simple types of linkages. Old automobile tires may serve as bumpers between cages. More sophisticated linkages are discussed in Beveridge (1987).

Mooring Systems. Mooring systems consist primarily of lines and anchors, and some may also have floats. The purpose of a mooring system is to secure the cages in the desired position. Mooring requirements are determined by cage design and site characteristics.

There are two types of mooring systems: single-point and multi-point systems. Single-point moorings can be used for either individual cages or groups of cages but are primarily used for individual square or octagonal cages of the rigid collar design. They allow the cage to move freely in a circle, and wastes are distributed over a broader area than if the cage were in a fixed position. Examples of single moorings are shown in Figure 5-10. The cage will adopt the position of least resistance to wind and wave forces. Single moorings use less line than multiple mooring systems, however, tremendous forces are often exerted where the line attaches to the cage. Lines are attached to cage collars at flexible attachment points that are capable of resisting the forces.

Figure 5-9. Correct method of grouping cages as viewed from above: (a) Correct; and (b) incorrect.

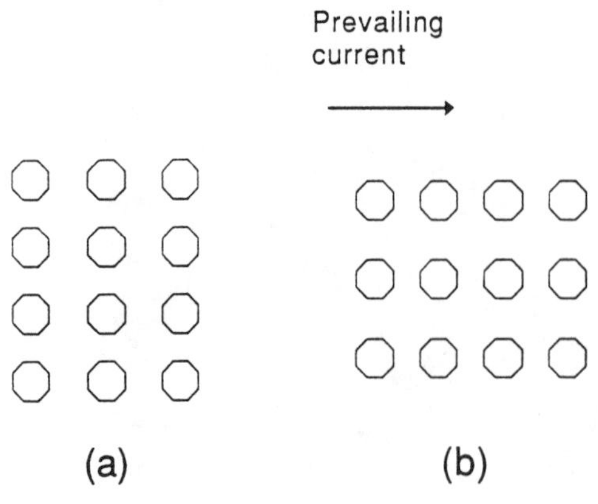

Horizontal and vertical forces acting on cages and supporting structures should be considered when designing mooring systems. Mooring lines must have a high breaking strength and be able to withstand and transmit the forces that influence cages. Lines are be constructed from either braided galvanized steel strands, open-link steel chain, or braided natural or synthetic fibers. The diameter of the lines is determined by the size of the cage group and the forces encountered. The specifications for a seven-strand galvanized steel cable are shown in Table 5-1. The length of mooring lines should be at least three times the maximum water depth anticipated at the site. More detailed discussions of mooring lines are included in Beveridge (1987).

Multi-point moorings secure cages in a fixed position, and forces are distributed among several mooring lines rather than on an individual line. Figure 5-11 illustrates examples of multiple mooring schemes. The choice of orientation of the cages depends on site conditions and the type and group configuration of the cages. If the cages are exposed to strong winds or currents, it is best to secure them in the position offering the least resistance. However, if the cages are sheltered and water exchange is poor, it is often best to position the cages so that water exchange is enhanced.

The most common method of mooring is to connect the cage(s) to anchors with mooring lines and chains. In most installations the mooring line does not connect the cage to the anchor, but rather a short section of chain is used as a buffer between the mooring line and anchor as shown in Figure 5-12. The chain resists abrasive forces exerted by the anchor. Chain lengths of two to three times the maximum water depth are

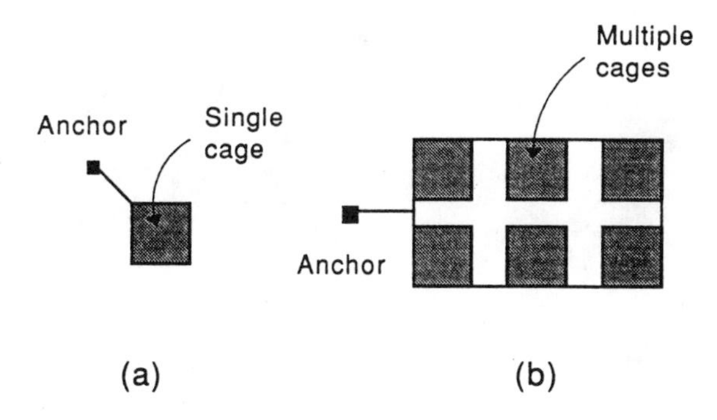

Figure 5-10. Single-point mooring for: (a) One cage; and (b) grouped cages.

Table 5-1. Specifications for seven-strand galvanized steel cable.

Diameter (mm)	Weight per unit length (kg per 100 m)	Safe load[1] (kg)
3.2	4.8	227
4.8	11.2	635
6.4	18.6	1,043
7.9	31.3	1,724
9.5	43.9	2,268
12.7	75.9	3,856
25.4[2]	296.8	14,300[2]

Source: Thomas et al. (1967).

[1] Safe load is approximately 0.25 times breaking load.

[2] Estimated value.

recommended, and they should be no longer than one-third of the total length of the mooring line. Where long chains are used, floats may be required to help support their weight and to minimize the vertical loading on the cages.

Chains are made from a variety of materials including wrought iron and various types of carbon steel and alloys. Stainless steel chains are available for use in marine environments, but they are expensive. A fairly heavy grade of chain is recommended, having a link diameter in the 18–22 mm range (0.71–0.87 in). Specifications for open-link proof steel chain are shown in Table 5-2.

The most simple and least expensive anchor types are deadweight or block anchors. These usually consist of bags of sand or stones, concrete blocks, or large pieces of scrap metal. Deadweight anchors are typically inefficient, however, since they will begin to move across the substrate whenever the horizontal force component exceeds the frictional force between the anchor and the substrate material.

The embedding-type anchor is more efficient. Embedding anchors are constructed so that they can grip the substrate. There are many types and designs. An example is a small boat anchor. A disadvantage of embedding anchors is that they are more expensive than deadweight anchors. Another disadvantage is that they must be dragged a certain distance over the substrate before they take hold. A more detailed discussion of weights, chains, and anchors is included in Beveridge (1987).

An alternative mooring technique in lieu of using chains and anchors is to drive long poles into the substrate and attach the cages directly to them. In theory, the number and dimensions of the poles required and the depth to which they must be driven into the substrate can be determined from estimates of the forces acting on the cages. This mooring method has not been as successful as the chain and anchor method. It obviously has limited application where hard substrates are encountered.

In some cases, at inshore sites where cages are located near the shoreline, land anchors can be used. In this method, wooden stakes are driven into the ground, and the cages are tied to them with ropes or cables. Steel pins can be inserted into rocky ground instead of stakes. Where steel pins are used, the mooring line is attached through an eyelet in one end of the pin.

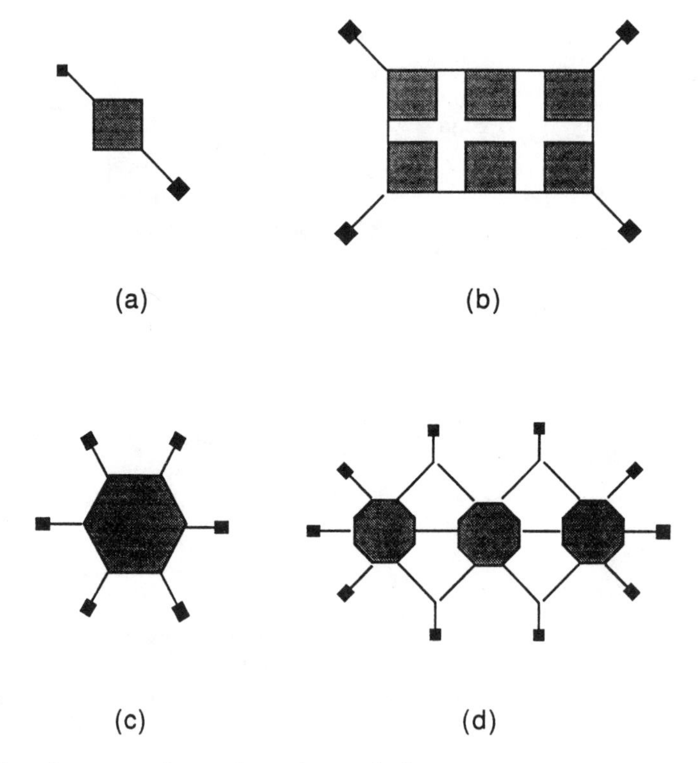

Figure 5-11. Examples of cage groupings and mooring methods.

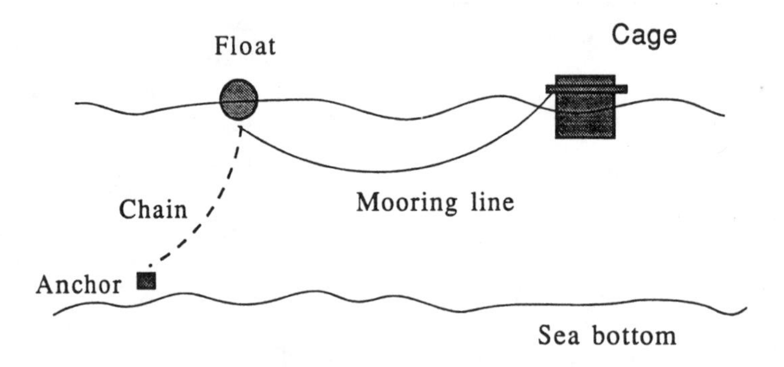

Figure 5-12. Method of mooring a single floating cage.

Cage Aeration. By nature, cage culture is very intensive. Therefore, supplemental aeration must be provided. Aeration can be provided by either compressed air systems, surface agitators, pure oxygen systems, or combinations of these. Air stones can be placed beneath the cages in some cases, but in all likelihood, it would be more advantageous to place air stones and/or surface agitators inside of the cages (Figure 5-13). These devices can be removed during harvesting operations or when changing cage bags. Many types of aeration system options are available to the culturist. Aeration is discussed in Chapter 11.

Factors Determining Site Selection

Cage farm site selection involves many considerations. The correct choice for a site is vitally important since it influences the economic

Table 5-2. Specifications for open link proof steel chain.

Link Diameter (mm)	Weight per meter (kg)	Safe Load[1] (kg)	Link Diameter (mm)	Weight per meter (kg)	Safe Load[1] (kg)
4.8	0.6	263	15.9	6.1	2,268
6.4	1.1	408	19.1	8.6	3,209
7.9	1.6	617	22.2	11.6	4,355
9.5	2.4	844	25.4	14.9	5,625
12.7	4.1	1,497	28.6	18.9	7,076
			30.2	23.2	8,709

Source: Thomas et al. (1967).

[1]Safe load is approximately 0.25 times working load.

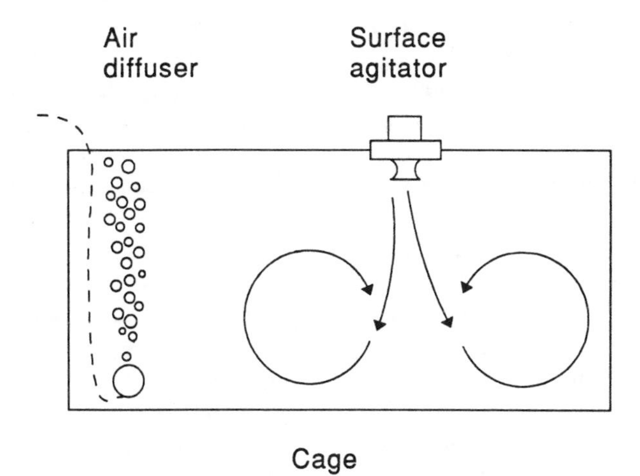

Figure 5-13. Diffused aeration and surface agitation within a cage.

viability of the facility by determining capital outlay, operating costs, transportation costs, storage requirements, rate of production, and mortality. Land-based aquaculture facilities have an advantage over water-based facilities in that the water supply can often be increased by drilling additional wells, and poor water quality can be improved by filtration, sedimentation, and other treatment practices. However, if the site is too exposed, if water exchange is poor, if diseases appear, or if water quality deteriorates, there is little that can be done at a cage farm other than to move the facility to another site.

The criteria for selection of a suitable site for a cage farm can be arranged into three categories as shown in Table 5-3. The first is concerned with the environmental criteria for the cultured organism. The second lists the environmental criteria that affect the cages, and the third reviews those factors that make the farming venture feasible.

Table 5-3. Criteria for selection of a cage farming site.

Category 1	Category 2	Category 3
Water quality	Weather	Legal requirements
Algal blooms	Shelter	Support facilities
Disease organisms	Currents	Security
Water exchange	Water depth	Management strategies
Water currents	Substrate	
Fouling		

Environmental Criteria for the Cultured Organism

WATER QUALITY. As with any aquaculture venture, a cage farm site must exhibit good water quality. This means that the area should be uncontaminated with industrial, municipal, and agricultural pollutants. Other water quality parameters, such as temperature, pH, nitrogenous compounds, dissolved oxygen, etc. should be within the ranges that provide life support and growth for the cultured species. Specific water quality parameters will not be discussed in detail here since they were covered in Chapter 2. This chapter discusses environmental criteria unique to cage farming. Beveridge (1987) and Poxton and Allouse (1982) present comprehensive discussions of water quality criteria relevant to cage fish farming.

Since fish are *poikilothermous*, meaning that their body temperature is governed by the external environment, changes in water temperature are met with changes in metabolism, oxygen consumption rate, and the rate of waste production, among others. Suboptimal conditions can lead to stress, which in turn affects behavior, feeding, feed conversion, and growth. Thus, the thermal environment for a particular species should be carefully evaluated when selecting a site for a cage farm. Cages located in areas receiving thermal effluents from power plants may experience wild temperature fluctuations. Individual cases should be evaluated on their own merits.

Many of the arguments made for water temperature also hold for salinity. Marine cage farms located in areas receiving freshwater discharges will experience rapid salinity changes that can disrupt farming operations and stress the fish.

The importance of adequate dissolved oxygen in any fish farming operation cannot be overstated. The optimal DO range may vary slightly between fish species, but aquaculturists will agree that the DO concentration should not be allowed to drop below 5–6 mg/L. Oxygen availability in a given area is impacted by water temperature, salinity, turbulence, pollutants, and the benthic and algal communities. Oxygen concentration declines with a rise in temperature, salinity, or pollutant concentration, thus, seawater holds less oxygen than freshwater at the same temperature. Organic sediments that accumulate near and beneath cages as a result of feeding activities can greatly increase the oxygen demand of the benthos, causing localized oxygen depletions if water currents are insufficient to flush away the wastes. Cage farms should not be located near areas of sewage discharge or other discharges containing potentially harmful pollutants, organic or otherwise.

Gas supersaturation (O_2 and N_2) is often a problem in waters receiving thermal effluents from power stations where the discharge has been pressurized by pumping, dropping over a weir, or plunging from great heights. The effect is to cause a condition known as *gas bubble disease*, a condition that can be fatal to fish (Marcello and Strawn 1973). Gas bubble disease is discussed in another chapter. In such situations mortalities can be reduced by submerging the cages below the surface to limit exposure. Gas saturation decreases by about 10% for each meter of submergence (Chamberlain and Strawn 1977).

The ideal pH for most fish species is 6.0–8.5. Freshwater can have great variability in pH ranging from less than 3 in acid rain areas to over 11 in areas where the underlying geology is dominated by alkaline sedimentary rock. There are strong diurnal pH fluctuations in poorly buffered inland waters. pH is generally not a problem at marine sites since seawater is strongly buffered. However, rapid pH changes can occur in brackish waters located near freshwater inflows from rivers and streams.

Pollution of various types is responsible for high fish mortalities in numerous cage farming operations. The number and type of common pollutants are many. The principle categories of pollutants often plaguing cage farms are listed in Table 5-4. Expensive and sophisticated laboratory analyses are often required to identify the types and concentrations of many of these pollutants.

Turbidity caused by dissolved and suspended

Table 5-4. Categories of aquatic pollutants.

Acids and alkalis
Anions (sulphide, sulphite, cyanide)
Detergents
Domestic sewage and agricultural wastes
Food processing wastes (including on-farm processes)
Gases (ammonia, chlorine, bromine)
Metals (lead, mercury, cadmium, zinc)
Nutrients (nitrogen, phosphorus)
Oil and derivatives
Toxic organic wastes (formaldehydes, phenols)
Pathogenic organisms
Pesticides
Polychlorinated biphenyls (PCBs)
Radionuclides
Thermal effluents

Source: Beveridge (1987).

organic and inorganic solids, algae, and other suspended substances is often troublesome at cage farm sites. This originates from a variety of sources including industrial discharges, mining operations, soil erosion, agricultural runoff, and sewage discharges. Some of these materials are directly toxic, while others cause problems by altering pH and/or by causing oxygen depletion. Turbidity at sufficiently high concentrations can cause gill damage and may trigger diseases as a result of fish stress. The mortality rate and nature of the damaging material varies with fish species. Turbidity levels below 100 mg/l causes little damage to most fish species. Above this value exposure time is a combined factor that determines damage and mortality levels (Beveridge 1987).

ALGAE BLOOMS. Algae (phytoplankton) blooms are often responsible for sudden oxygen depletions in freshwater and marine water systems. Blooms typically develop during the warm months in areas receiving influxes of nutrients (particularly phosphorus) or sudden upwellings of nutrients from the lower depths. Sewage discharges and agricultural runoff are often important factors contributing to blooms. When environmental conditions affect light levels, water temperature, and nutrient concentrations, heavy algae blooms may die, resulting in oxygen depletions.

The algal community at any time usually consists of a mixture of species, but one species predominates. As environmental conditions change, the dominant species may also change. This process is less pronounced in tropical regions where water conditions are more constant.

Algae may also influx from other areas, driven by wind, tides, or currents. In freshwater the most important groups of algae that cause problems for fish farmers are the diatoms and Cyanobacteria (blue-green algae). Marine algal groups that form dense blooms include the diatoms, Cyanobacteria, prymnesiophytes, and dinoflagellates. The most important of these are the dinoflagellates, which, during bloom conditions, are responsible for the so-called *red tides* which cause serious illness and death in higher organisms further up the food chain, including man. For a complete discussion of algal blooms, species characteristics, and toxins produced, refer to Beveridge (1987).

DISEASES. The worst cage farm sites are those where disease-causing organisms or parasites are likely to thrive under conditions produced by the farm. Pathogenic organisms are typically found in higher numbers in sewage-polluted waters or waters polluted by other materials of organic origin. *Escherichia coli*, a bacterium normally found in the intestinal tract of warm-blooded animals, including man, is an indicator of fecal contamination. Sewage-polluted sites should be avoided completely.

Despite precautions, however, diseases and parasites can influx into the area from other sources. Wild fish populations and other aquatic animals can carry diseases into a seemingly clean cage farm environment. Diseases can also be transmitted by the introduction of infected stock into the cage farm or by birds and other predators attracted to the area. Long-established cage farms are themselves a source of organic nutrients and may consequently attract and support disease organisms. Disease transmission can also function in reverse; there is concern that cage-cultured fish may transmit diseases to wild stocks (Kelley 1992).

WATER EXCHANGE. Good water exchange is essential at cage farm sites to replenish oxygen within the cages and to prevent buildup of wastes beneath the cages. Water exchange is dependent upon currents and is affected by water temperature, salinity, winds, undersea topography, and the geography of the area. Marine cage farms are best located at sites where water is exchanged every few days rather than several weeks. Large-scale intensive culture facilities should not be located in enclosed inlets or bays where water exchanges very slowly. Sites should have good bottom currents as well as surface currents. Locating cages near river outflows can sometimes enhance water exchange, however, the farm should not be located such that the animals are affected by extreme temperature and salinity fluctuations caused by the water flowing out.

Many cage farms, particularly in Northern European countries, are located in fjords or enclosed inlets where the water exchange pattern is more complicated. Fjords were formed by ice sheets moving through river valleys during the Pleistocene era, deepening and widening the river valleys. As a result, rock sills or bars were formed near the mouths, thus forming a stagnant pool at the bottom (Figure 5-14). This water is never exchanged with fresh incoming water. At sites having large tidal fluctuations floating cages may drop near or into this stagnant pool when the tide is out.

The exchange time of inland freshwater ranges from a few seconds, as in the case of swiftly moving rivers, to several years for large lakes. Where cage farming is conducted in inland lakes or reservoirs, the site should be chosen so that water is exchanged at least every few weeks. Less intensive culture methods should be practiced in those cage farms where water exchange rate is low.

Many extensive and semi-intensive cage farms rely on a rich supply of plankton and detritus for supplemental feeding. Primary production and water exchange are inversely related, and as the water exchange rate increases, primary production decreases. Thus, sites with a slower water exchange rate may be preferred in cases where nature is relied upon for supplemental feeding.

FOULING. When cage bags are fouled by algae and other organisms, the specific mesh size of the mesh is decreased, thus increasing drag forces and adding to the weight of the cage. In addition, the decrease in mesh size reduces water exchange through the bag, reducing the rates of oxygen replenishment and waste removal. The increased drag forces cause the bag to deform, crowding the fish.

There are some 200 species of marine fouling plants and animals (Beveridge 1987). The size, diversity, and developmental pattern of the

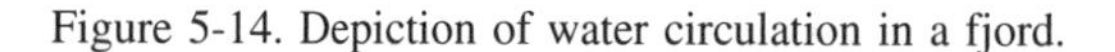

Figure 5-14. Depiction of water circulation in a fjord.

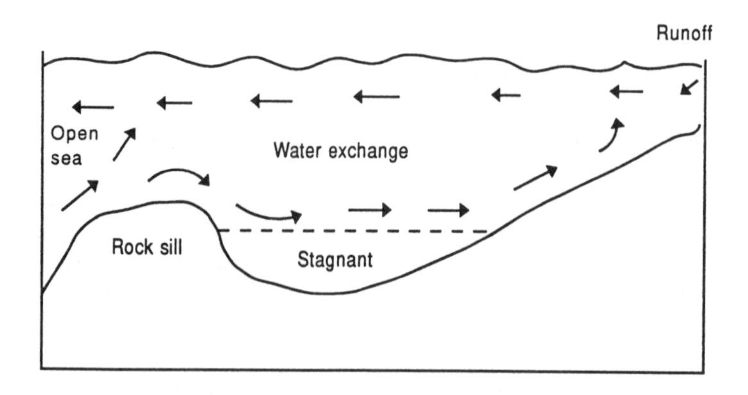

fouling community varies with the materials used and the site environmental conditions. The rate of fouling is high in cages situated near thermal effluents (Chamberlain and Strawn 1977). Fouling rate is most rapid in areas where water currents are slow, and generally becomes less of a problem as salinity decreases. Thus, freshwater cage farms are less affected by fouling than marine farms.

Environmental Criteria Affecting Cages

WEATHER. Weather determines the suitability of a site because of its influence on both cage structure and the contained fish. Violent storms and extreme cold are two weather conditions of particular concern. Storms in tropical latitudes are classified into four types: tropical cyclones, severe tropical storms, moderate tropical storms, and tropical depressions. The wind strengths and speeds corresponding to these classifications are shown in Table 5-5.

The most violent and destructive of these storms are the tropical cyclones. Cyclones are also referred to as hurricanes or typhoons in certain parts of the world. Most occur in the Western Pacific, Western Atlantic, and Indian Oceans. Consequently, the development of marine cage aquaculture has been hampered in these areas. In some cases damage can be minimized with the use of breakwaters or other barriers that dissipate wave forces. Experienced cage farmers have learned to deal with storms and accept occasional losses. There has been recent interest in large oceangoing cage farming systems that can be moved to a different location when storms threaten. One such concept is illustrated in Figure 5-15.

Table 5-5. Storm classifications with corresponding wind strengths and speeds.

Storm classification	Wind strength	Wind speed (m/s)
Tropical cyclone	≥ force 12	≥ 33
Severe tropical storm	10–12	24–32
Moderate tropical storm	8–9	17–23
Tropical depression	< 8	>17

Source: Beveridge (1987).

Icebergs and ice formation on the cages can be damaging. Fish losses by superchilled surface water is also common in such areas. Cage culture is nearly impossible in those areas that are frozen over for long periods or that have prolonged periods of superchilled surface temperatures. Submersible cages have been used with some success in areas that become iced over in winter. The cage is suspended a few meters beneath the water surface and a rigid tube is used to maintain an opening to the surface for feeding. An example of this concept is illustrated in Figure 5-16.

SHELTER. Oil drilling platforms are virtually the only type of structure that can withstand the impact of the open sea for extended periods. Wind blowing for long stretches across the open ocean causes waves to develop, which can damage cages. Most small cages cannot withstand wave heights much over 1–1.5 meters (3–4.5 ft) (Moller 1979). Open areas that have a long, uninterrupted wind run (fetch) should be avoided as potential cage farm sites. Cage farms are often located on the protected leeward side of land masses or islands.

WATER CURRENTS. Rapid water currents can impose additional dynamic loadings on cages, supporting structures, and moorings. In marine systems the predominant source of water currents is tidal fluctuations. Tides are greatly influenced by local topography, surface runoff from land, prevailing winds, and other factors. Tidal current magnitudes are well known throughout most of the world, and are published in tidal charts. Current velocities typically range from 0 m/s during slack water to over 2.5 m/s (8 ft/s) during full flood and ebb tides at some sites. Most researchers have found that tidal currents in the 0.1–0.6 m/s (0.3–2 ft/s) range are satisfactory for most purposes, and velocities greater than 1 m/s are not recommended. The capital costs of constructing cages and moorings escalate as velocities become too high (Chen 1979; Beveridge 1987). Surface current velocities in lakes and reservoirs are usually in

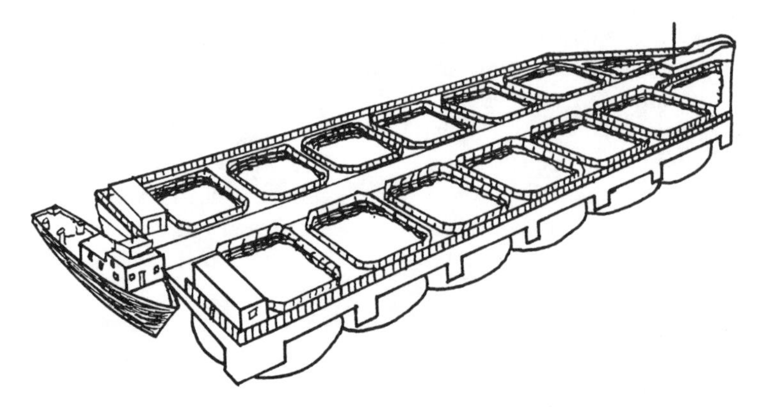

Figure 5-15. Artists example of a mobile, ocean-going cage farm (from cover of *Today's Aquaculturist*, Vol. 2, No. 2, Feb. 1991. Reproduced with permission from Pisces Publishing Group, Devon, CT.).

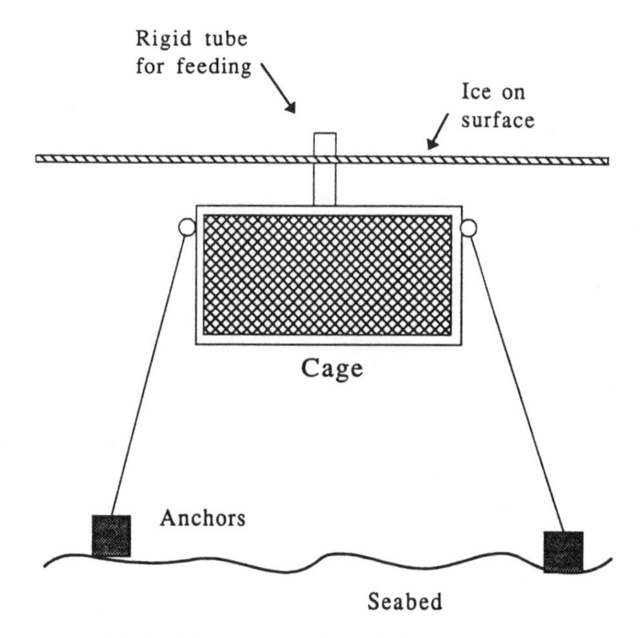

Figure 5-16. A submerged cage with rigid feeding tube through ice.

the range of 0.002–0.02 m/s (0.006–0.06 ft/s), much lower than in rivers or at most marine sites. These currents rarely affect site selection.

The patterns of current velocities at inland sites vary markedly from marine sites. Where marine sites are subjected to four to eight brief periods of swift currents daily associated with flood and ebb conditions, currents in inland rivers, lakes, etc. are relatively constant and vary due to the effects of wind, precipitation, and runoff. Feeding can be temporarily suspended in cages when peak velocities are too high. Feed loss is a problem in cage systems when current velocities are too high, and this is a major reason why cage culture may not prove feasible in swiftly moving rivers and streams.

Water currents impart a load on the cage mesh panel, which, in turn, exerts loads on the supporting frame and mooring system. The load on the panel is dependent upon the type of material, the shape and size of the mesh bag, water velocity and water density as expressed by Equation 5-3 (Kawakami 1964):

$$F_c = C_d V^2 \rho \frac{A}{2} \quad (5\text{-}3)$$

where F_c = drag force applied to the panel, N (lb); C_d = coefficient of drag of the material, (dimensionless); V = current velocity, m/s (ft/s); ρ = density of water, kg/m^3 (lb/ft^3); and A = projected area of the mesh members, m^2 (ft^2).

The projected area of the mesh is calculated from the panel area, number and length of meshes, and the diameter of the mesh members. C_d values for a variety of nonfouled and fouled materials have been experimentally determined and are shown in Table 5-6 for currents flowing perpendicular to the panel. C_d values for materials other than expanded metal meshes can be determined mathematically with Equations 5-4 and 5-5. These values closely approximate those in the table.

$$C_d = 1 + 3.77 \left(\frac{d}{a}\right) \textit{ for knotted materials} \quad (5\text{-}4)$$

$$C_d = 1 + 2.73 \left(\frac{d}{a}\right) \textit{ for knotless materials} \quad (5\text{-}5)$$

where d = yarn diameter, mm (in.); and a = bar length, mm (in.).

By orienting rectangular-shaped net structures with the long axis parallel to the direction of the water current, drag forces on the cage structure can be reduced since the area of the face of the cage is reduced.

DEPTH. Some cages are designed for use in shallow water with their bottoms lying on the substrate. Examples are the rigid wooden and bamboo cages used in such places as Indonesia (Beveridge 1987). The water depth at these sites is not critical so long as sufficient depth is provided. Insufficient depth will crowd the fish and cause water quality to degrade.

Table 5-6. C_d values for perpendicular current forces acting on various knotted netting and rigid mesh materials before and after fouling[1].

Material	Mesh type	Mesh size[2] (mm)	Unfouled year dia. (mm)	C_d	Fouled year dia. (mm)	C_d
Nylon	Diamond	50	2.3	1.42	10.2	3.99
PP (Ulstron)	Diamond	50	2.5	1.47	10.2	3.99
PE (Courlene)	Diamond	50	1.9	1.33	8.9	3.46
PE	Square	50	1.5	1.26	7.6	2.95
PE (cupra-proofed)	Square	50	1.5	1.26	5.1	2.13
Netlon	Square	50	3.3	1.19	7.6	1.48
90:10 Cu-Ni expanded metal	Hexagonal	10	1.3	0.41	-	-
Galvanized steel weld mesh	Square	25	2.5	1.30	3.3	1.41
Galvanized steel chain link	Diamond	25	2.0	1.24	3.8	1.48
Plastabond (PVC-coated chain link)	Diamond	25	2.5	1.30	5.1	1.67
Plastabond (PVC-coated chain link)	Diamond	76	2.5	1.09	6.4	1.25

Source: Beveridge (1987).

[1] Fouling accumulated after two-month immersion in Scottish sea conditions.

[2] Mesh size for netting taken as two times the bar length.

Floating cages should be located at sites where the water depth is sufficient to maximize water exchange and to keep cage bottoms well clear of the substrate at low tide. It is desirable to have a clearance of at least 2–3 m (6–10 ft) at low tide. If the allowable clearance is consistently less, water flow beneath the cages becomes seriously impaired, and wastes buildup beneath the cages (Figure 5-17). Noxious gases, such as H_2S, have been known to develop under such conditions. There is also greater potential for disease buildup in sediments accumulated beneath cages (Rosenthal 1985). Waste accumulation beneath cages can be prevented in some cases with the use of a propeller or water-mixing device as shown in Figure 5-18. Several such devices are commercially available. Floating cages are recommended in those areas having extreme tidal fluctuations. An example is the Bay of Fundy in Newfoundland, where daily tidal fluctuations of over 12 m (40 ft) occur.

SUBSTRATE. The substrate at potential cage culture sites can vary from soft mud to hard rock. The substrate could have significant influence on the choice of a cage site. Floating cages are normally used over a rocky substrate since it

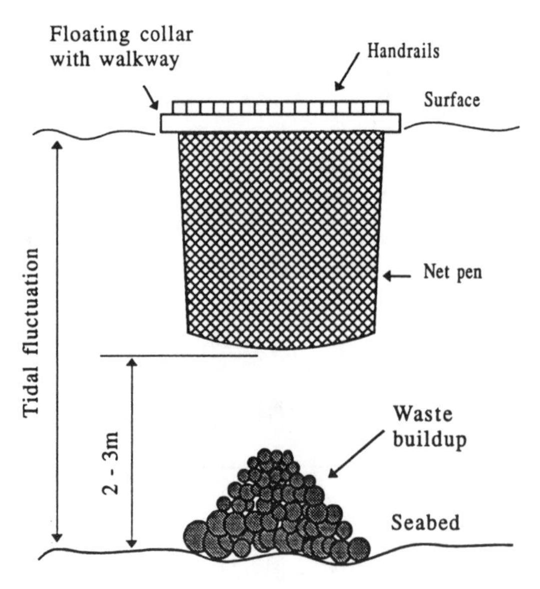

Figure 5-17. Illustration depicting waste buildup beneath cage.

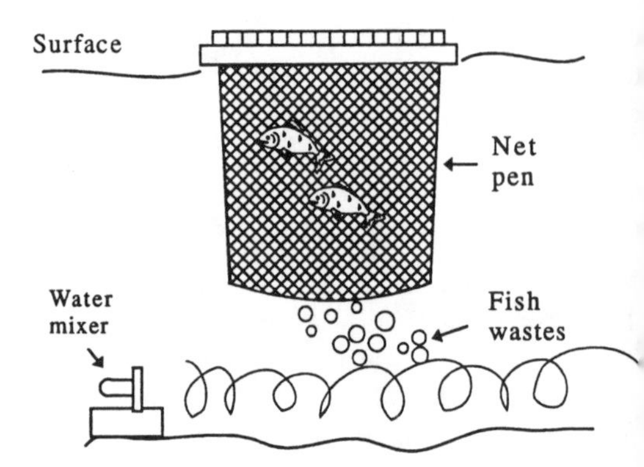

Figure 5-18. Water mixer used to flush wastes from beneath cage.

would be impractical to drill through rock to insert poles or beams to support fixed cages. In many cases a rocky bottom may be more beneficial since it indicates good scour by subsurface currents. Thus, the risk of waste accumulation is reduced. However, mooring becomes more difficult over rocky substrates.

Factors Affecting Cage Farm Feasibility. There are many criteria that determine the success of any aquaculture venture, and cage farms have many unique requirements that must be dealt with. The most common criteria that must be addressed are: legal requirements, support personnel and facilities, security, and management strategies. Many of these criteria were discussed at length in Chapter 3. However, there are many other criteria unique to cage farming that are discussed at length in Beveridge (1987) and other sources.

MOLLUSCAN OFF-BOTTOM CULTURAL PRACTICES

Practices used for off-bottom culture of species other than finfish vary tremendously. While cages are used primarily for finfish culture, other techniques employing floats, rafts, racks, trays, and strings are used to culture mollusks like oysters, clams, mussels, scallops, and abalone.

It is not the purpose of this chapter to discuss

culture methods, but rather to focus on the culture systems themselves. Numerous variations of each technique exist, and it would be impossible to cover all existing variations. Therefore, this chapter focuses on the major techniques used. It bears mention that most, if not all, of the siting considerations previously discussed also apply to off-bottom molluscan culture and will not be repeated.

Probably one of the oldest and more common off-bottom culture techniques is the collection of the young shellfish onto old mollusk shells or some other material (referred to as *cultch*) attached to strings or long lines and suspended from rafts or racks (Figure 5-19). Holes are drilled through the old shells so that they may be strung easily. A variation of this technique is to place cultch into small mesh bags and suspend the bags from strings. Rafts are constructed from wood or styrofoam, or from foam- or air-filled drums. They are typically located in areas of good spatfall until a sufficient density of young shellfish, or spat, are collected on the shells. The young are then either cultured in the same waters where they were collected or the rafts can be towed to other locations where growing conditions are more favorable. The length of strings depends on water depth. Other parameters, such as size of rafts, number of strings, number of collecting shells, etc., are often regulated by local customs or governments. String and raft culture techniques have been used for centuries in some parts of the world, such as Europe and Asia, but they are relatively new to the United States. Governmental regulation in the United States has hindered the development of off-bottom culture to date.

Long-line culture is similar to string culture except that the lines are attached to floats that are attached at intervals along a heavy rope or cable (Figure 5-20). This system is typically used in rougher and deeper water bodies. Sections of chain are normally used between the floats and anchors to help support the weight of the strings as the animals grow. If the floats are not moored and anchored securely they will tend to draw together as the weight of the strings increases.

Rack culture is normally practiced in relatively shallow water. There are many variations of this technique, perhaps the most simple of which consists of wooden beams attached to posts driven into the substrate. Strings are suspended from larger ropes or cables that stretch across the beams as shown in Figure 5-21. This type system is popular for use in shallow, inshore locations where tidal fluctuations alternately flood and expose the racks to enable workers to tend the system from boats during low tide. A modern variation of this method is shown in Figure 5-22. Here young clams are placed into rectangular baskets fabricated from rigid plastic mesh and are laid across and attached to racks constructed from wood or some other material. Advantages of this system are that the animals are protected from predators, and it is a relatively easy matter to lift baskets and move them to other locations if necessary. A disadvantage is that the baskets must be pe-

Figure 5-19. String culture of shellfish.

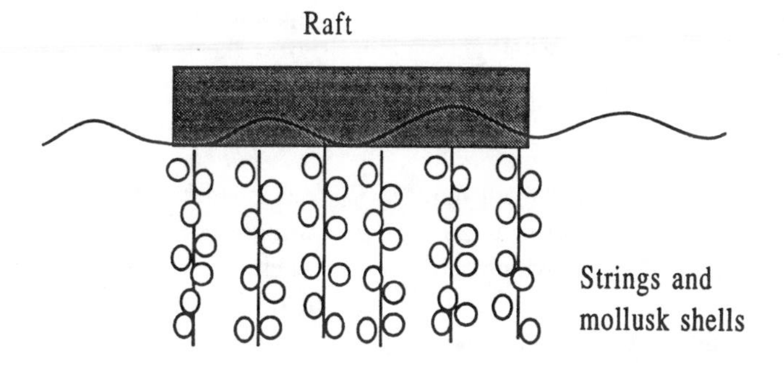

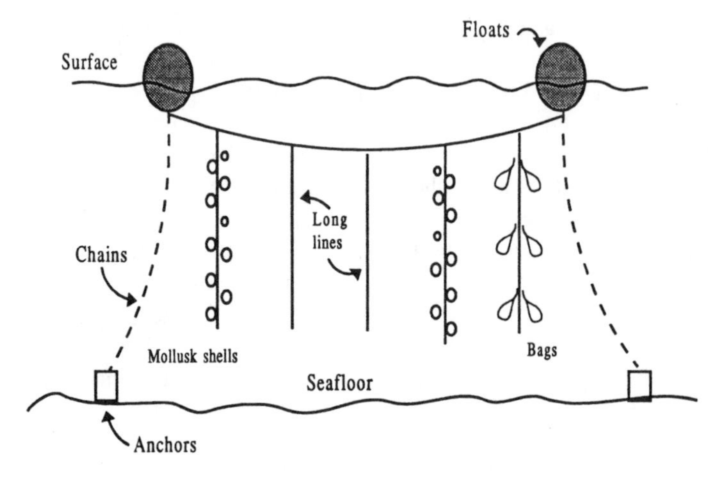

Figure 5-20. Long line culture of shellfish.

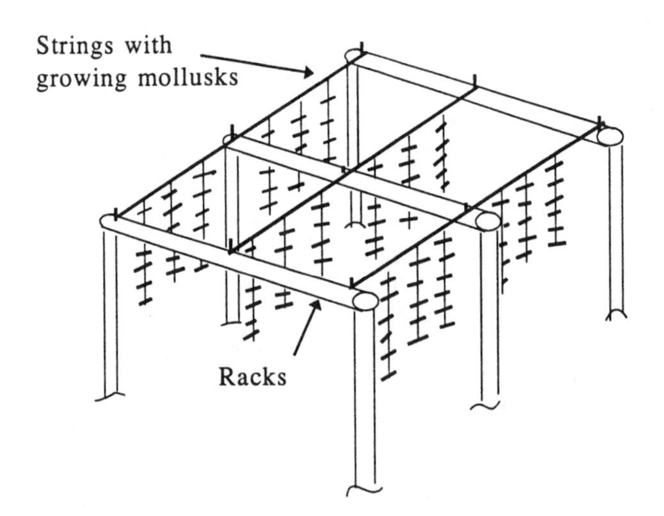

Figure 5-21. Rack culture of shellfish.

riodically cleaned of fouling organisms. The basket mesh size must be large enough to ensure proper water exchange within.

A modern variation of the tray culture method is the *lantern net*, trays constructed so that they resemble Japanese lanterns (Figure 5-23). Multiple trays are used, and the net can have any desired length. The trays are completely enclosed by a protective mesh material. The nets must be completely hoisted from the water for maintenance and handling of stock, or they may be tended by divers, both of which can be expensive operations.

Techniques used to culture mussels in France and certain other parts of Europe are described by Dardignac-Corbeil (1990). Only one technique, culturing mussels on bouchets (posts), was used until the 19th century. Other techniques were developed later. Bouchet culture is by far the most widespread for *Mytilus edulis* and *M. galloprovincialis* culture, the predominate species of mussel.

A *bouchet* is a line of posts driven into the seabed (Figure 5-24). Mussel spat are collected on those bouchet sited farthest from shore (spat bouchet), and are then transferred onto posts sited nearer to shore (rearing bouchet) for grow-out. Bouchet posts are 4–7 m (13–22 ft) in

'igure 5-22. Rack culture of clams in plastic mesh baskets (photo by N. Romanenko, with permission).

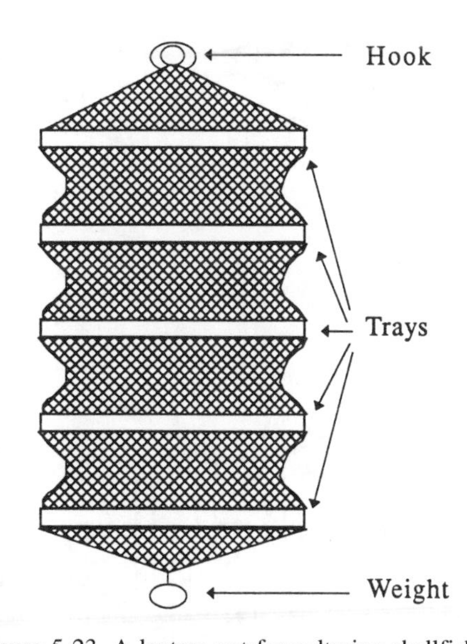

'igure 5-23. A lantern net for culturing shellfish.

ength and 12–25 cm (5–6 in) in diameter. They re made from either oak or pine, but the former s more widespread since oak posts are more esistant to decay and have a useful life of sis to ight years. The posts are embedded to about half their length into the seafloor. Smaller posts can be driven into the substrate with mallets, but pile-driving equipment must be used for the larger posts.

French law dictates the length of the bouchet and the distance between posts, and the rules are not the same for all regions within the country. For that reason bouchet lines vary from 50 to 100 m (160–320 ft) in length and may contain anywhere from 120 to 190 posts. In some regions two rows of posts are allowed per bouchet while only one row is allowed in other regions. The distance between bouchets is variable but must be at least 25 m (80 ft). Posts are usually installed from January to April, and spat set takes place between March and June.

Spat are collected either directly onto the posts or onto coconut fiber ropes, a variation which has been in use since around 1960. Spat collected on ropes are later moved to other posts, located closer to shore, for grow out. The ropes are also regulated, and regulations vary between regions. In some regions the rope bouchet must not exceed 50 m (160 ft) in length or contain more than 42 posts arranged in two

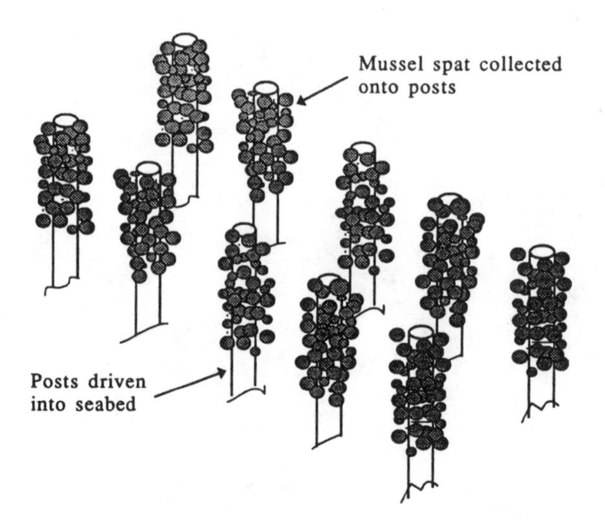

Figure 5-24. A bouchet for capturing mussel spat.

rows. Total rope length cannot exceed 2,400 m per bouchet, and they can be installed in one of two ways: either supported by 3 m (10 ft) long horizontal poles or nailed to vertical posts (Figure 5-25). In the first method the rows of posts must be spaced at least 1 m apart, and the ropes can be arranged in up to three levels. Some regions require that the bouchets all be of the same type, but other regions allow mixed methods for capture.

In order to make posts suitable for spat settlement, the previous year's growth of old mussels and barnacles must be scraped away. In those instances where the posts have not been cleaned, a suitable site for spat set can be made by winding 3–5 m (10–15 ft) lengths of rope around individual posts. Most of the young mussels collected and grown on ropes are used to supply seed to areas where natural settlement is low. The ropes with spat attached can easily be transported from one site to another when the spat are very small. Transfer is usually done between May and July.

The most widespread method of grow out is a process called *boudinage*. In this process young mussels are loaded into long cylindrical tubes called *sausages* (boudin), which are about 12 cm (5 in) in diameter and 3–5 m (10–16 ft) in length. The tubes are constructed from a netting material. These sausages are then wound

Figure 5-25. Bouchets for culturing mussels: (a) Horizontal ropes; and (b) vertical ropes.

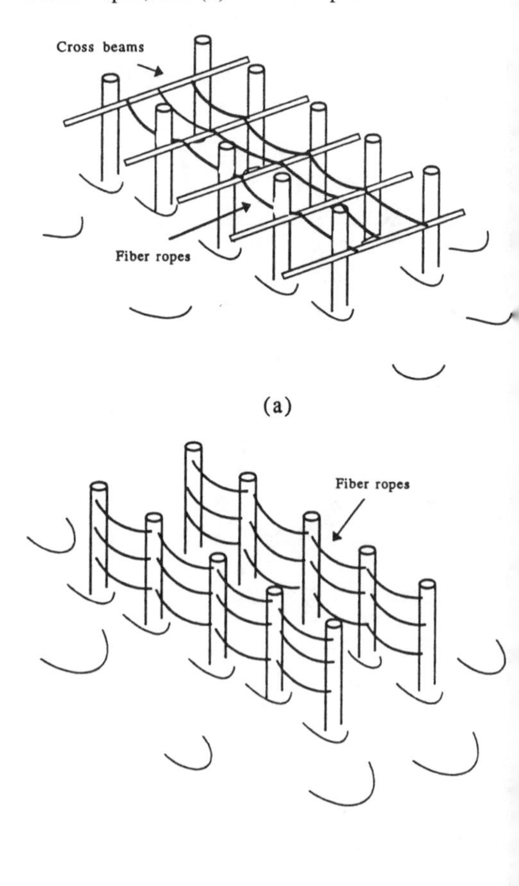

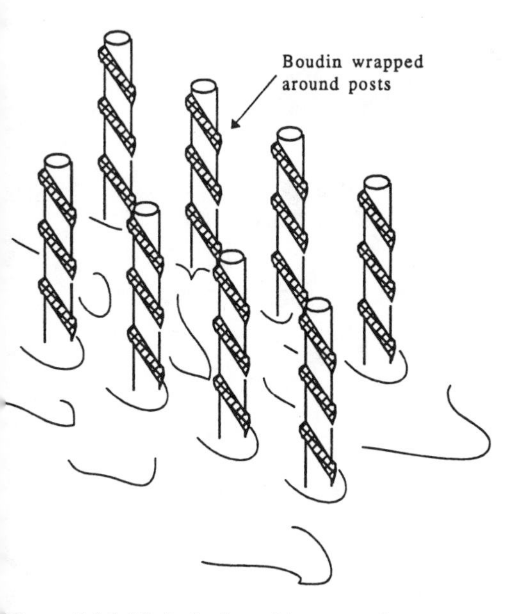

Figure 5-26. Method of attaching mussel sausages to poles.

around posts where they remain until the mussels reach harvestable size (Figure 5-26). The type of netting material varies, depending upon the time of year. Until August, young mussels readily attach to the posts, and a cotton netting material is used. The material rapidly decomposes, leaving the animals unconfined. After August, young mussels do not attach to the posts as readily, and a nylon or plastic netting is used to make the sausages. The process of removing the mussels at harvest thus becomes more labor-intensive. Harvesting typically takes place after about 15 months of culture.

Certain enterprising mussel farmers sought improvement in the boudinage process around 1975. A technique was developed in which a cylindrical tube is slipped over a post, and mussel spat are collected on the outside surface of the tube. Once the outer surfaces of the tubes are covered with spat, the tubes are lifted and transferred to clean posts located in grow-out areas. Another common technique is to raise the cylinders and reverse their position. Reversing the cylinder tends to even out differences in growth since the mussels at the foot of the post always grow fastest.

There are many variations of the off-bottom culture techniques described, and to discus them all would be beyond the limits of this book. For more detailed reading refer to Beveridge (1987) and Barnabé (1990a, 1990b).

CHAPTER 6
FLUID MECHANICS

Fluid mechanics is the science that deals with the forces on fluids and their actions. A *fluid* is defined as a substance consisting of particles that change their position relative to one another or, in other words, a substance that will continuously deform when a shear stress is applied. By contrast, a solid resists shear stresses and will not deform when one is applied providing that the shear stress does not exceed the elastic limit of the material.

Air and water are the two fluids with which the aquaculturist is most concerned. Water is the medium in which fish live. It can exist as liquid, solid, or gas. As a liquid, and for all practical aquaculture applications, water can be considered an incompressible fluid. Air, on the other hand, is a compressible gas.

PROPERTIES OF FLUIDS

Fluids have certain characteristics by which their physical condition is described. These characteristics are called *fluid properties*, a few of which are listed in Tables 6-1 and 6-2 for air and water, respectively. The properties of pure water were discussed at length in Chapter 2. However, the study of fluid mechanics demands a full understanding of the properties of water. Therefore, several important physical properties of water will receive additional discussion.

Specific Weight

The *specific weight* of a fluid is the gravitational force exerted on a unit volume of the fluid, or the weight per unit volume. Specific weight is assigned the symbol γ (gamma) and is defined as:

$$\gamma = \rho g \tag{6-1}$$

where ρ = density, kg/m^3 (slugs/ft^2); and g = acceleration of gravity, 9.81 m/s^2 (32.2 ft/s^2). The specific weight of water at 15°C (50°F) is 910 N/m^3 (N = 9.81 kg) in Standard International (SI) units and 62.4 lb/ft^3 in English units. The specific weight of air at the same temperature is 12 N/m^3 (0.0779 lb/ft^3).

Density

The mass density of a fluid is given by its specific weight at the earth's surface divided by g, the acceleration of gravity. Mass density, often expressed as simply mass per unit volume, has units of kg/m^3 (slugs/ft^2) and is given the symbol ρ (rho). The density of some fluids changes. For example, air can easily be compressed, and its density changes accordingly (Table 6-1). Water, on the other hand, is incom-

Table 6-1. Physical properties of air at standard atmospheric pressure.

Temperature	Density	Specific weight	Dynamic viscosity	Kinematic viscosity
°C	kg/m^3	N/m^3	$N\text{-}s/m^2$ ($\times 10^{-5}$)	m^2/s ($\times 10^{-5}$)
0	1.29	12.7	1.72	1.33
10	1.25	12.2	1.76	1.41
20	1.20	11.8	1.81	1.51
30	1.17	11.4	1.86	1.60
40	1.13	11.1	1.91	1.69
50	1.09	10.7	1.95	1.79
60	1.06	10.4	2.00	1.89
70	1.03	10.1	2.04	1.99
80	1.00	9.81	2.09	2.09
90	0.97	9.54	2.13	2.19
100	0.95	9.28	2.17	2.29
°F	$slugs/ft^3$	lb/ft^3	$lb\text{-}s/ft^2$ ($\times 10^{-7}$)	ft^2/s ($\times 10^{-4}$)
0	0.00269	0.0866	3.39	1.26
20	0.00257	0.0828	3.51	1.37
40	0.00247	0.0794	3.63	1.47
60	0.00237	0.0764	3.74	1.58
80	0.00228	0.0735	3.85	1.69
100	0.00220	0.0709	3.96	1.80
120	0.00213	0.0685	4.07	1.91
150	0.00202	0.0651	4.23	2.09

Source: Roberson and Crowe 1990.

pressible and can be assumed to have a relatively constant density (Table 6-2). A very large pressure is required to change the density of water by even a small amount.

Specific Gravity

The ratio of the specific weight of a fluid to the specific weight of the same volume of water is defined as *specific gravity*, *S*. By definition *S* for water is 1.0. Specific gravity is dimensionless and is therefore independent of the system of units used.

Viscosity

A fluid's resistance to shear or angular deformation is perhaps its most important physical property. The measure of a fluid's resistance to shear is called *absolute* or *dynamic viscosity* and is given the symbol μ (mu). Highly viscous fluids, like molasses or heavy motor oil, flow very slowly when subjected to a shear stress. Viscosity is thus defined as the capacity of a fluid to convert energy of motion into frictional heat energy. This energy either results in warming of the fluid or dissipates into the surrounding atmosphere. The viscosities of air and water are both temperature-dependent (Tables 6-1 and 6-2).

Viscosity is usually measured in *centipoises*. The dynamic viscosity of water at 20°C is one centipoise. This value is used as a relative standard for viscosities of other fluids. By comparison, the dynamic viscosities of air and mercury

Table 6-2. Physical properties of water at atmospheric pressure.

Temperature	Density	Specific weight	Dynamic viscosity	Kinematic viscosity	Vapor pressure
°C	kg/m^3	N/m^3	$N\text{-}s/m^2$	m^2/s	N/m^2 abs.
0	1,000	9,810	1.79×10^{-3}	1.79×10^{-6}	611
10	1,000	9,810	1.31	1.31	1,230
20	998	9,790	1.00	1.00	2,340
30	996	9,771	7.97×10^{-4}	8.00×10^{-7}	4,250
40	992	9,732	6.53	6.58	7,380
50	988	9,693	5.47	5.53	12,300
60	983	9,643	4.66	4.74	20,000
70	978	9,594	4.04	4.13	31,200
80	972	9,535	3.54	3.64	47,400
90	965	9,467	3.15	3.26	70,100
100	958	9,398	2.82	2.94	101,300
°F	$slugs/ft^3$	lb/ft^3	$lb\text{-}s/ft^2$	ft^2/s	psia
40	1.94	62.43	3.23×10^{-5}	1.66×10^{-5}	0.122
50	1.94	62.40	2.73	1.41	0.178
60	1.94	62.37	2.36	1.22	0.256
70	1.94	62.30	2.05	1.06	0.363
80	1.93	62.22	1.80	0.930	0.506
100	1.93	62.00	1.42	0.739	0.949
120	1.92	61.72	1.17	0.609	1.69

Source: Roberson and Crowe 1990.

at the same temperature are 0.17 and 1.7 centipoises, respectively (Simon 1981). A *poise* is 100 centipoises, which is equivalent to 1 g-s/cm. In SI units one poise is equal to a force of 1 $N\text{-}s/m^2$. In the English system dynamic viscosity is expressed in $lb\text{-}s/ft^2$. Dividing the dynamic viscosity by the density of the fluid at the same temperature results in the *kinematic viscosity*, which is given the symbol ν (nu). Thus,

$$\nu = \frac{\mu}{\rho} \tag{6-2}$$

In the SI system ν is expressed in cm^2/s and is called the *Stoke*. Its equivalent in the English system its ft^2/s. One ft^2/s is the equivalent of 929 Stokes (Simon 1981).

Vapor Pressure

Vapor pressure is the pressure exerted by the gaseous phase of a fluid when it is in equilibrium with its liquid phase. Another way of expressing vapor pressure is the pressure at which a liquid just begins to boil and change to a vapor. Water boils at 100°C (212°F) at one atmosphere pressure. Vapor pressure increases with an increase in temperature at sea level. However, at higher elevations the atmospheric pressure is less, and water boils at a temperature lower than 100°C. By increasing water temperature at sea level to 100°C the vapor pressure is increased to the point at which it is equal to atmospheric pressure (760 mm Hg or 14.7 psia), and the water begins to boil. Water can actually boil at temperatures much lower than 100°C if its pressure in the water is reduced to its vapor pressure.

In closed conduits water may change phase from a liquid to a vapor because of a reduction in pressure, even though the temperature remains unchanged. This sometimes occurs on the suction side of pumps. When the water begins to boil, vapor bubbles form in local regions of very low pressure and then collapse when they move into regions of higher pressure downstream. The bubbles collapse in a noisy and violent manner that can cause considerable damage to pipes and pumps. This damaging process is called *cavitation*.

Vapor pressure for pure water is shown in Table 6-2 as a function of temperature and pressure. In the English system of units the vapor pressure of water at standard atmospheric pressure (14.7 psia) and 68°F is 0.34 psia. In metric values it equals 2,335 N/m^2 at 20°C. In SI units vapor pressure is expressed in *Pascals, Pa* (1 *Pa* = 6,895 psia). The addition of salt lowers the vapor pressure of water. The relationship between vapor pressure and salinity is expressed by (Wheaton 1977):

$$VP = (VP)_0\,[1.000016 - 0.000537(S)] \quad (6\text{-}3)$$

where VP = vapor pressure of seawater, *Pa*; VP_o = vapor pressure of pure water at same temperature, *Pa*; and S = salinity (g/L).

FLUID STATICS

Pressure Intensity

Fluid statics is the study of fluids at rest. Fluids at rest produce only normal forces (forces perpendicular to a plane surface) called *pressure forces*. A certain pressure intensity exists at every point in a static fluid. The average pressure intensity (P) at any point is force per unit area and is expressed by

$$P = \frac{F}{A} \quad (6\text{-}4)$$

where P = pressure, *Pa* (lb/ft^3); F = normal forces acting on an area, N (lb); and A = area over which the force is acting, m^2 (ft^2).

Pressure intensity has magnitude only. As the area in Equation 6-4 is reduced to an infinitesimally small size it approaches a point. Thus, the pressure at point 1 and at any other point in a motionless column of liquid acts equally in all directions. This point is illustrated in Figure 6-1.

Equation 6-4 can be used to calculate the pressure on the bottom of a tank filled with a liquid. For example, suppose the tank in Figure 6-1 with a bottom area A is filled to depth h with a fluid having specific weight γ. From Equation 6-4, the pressure at point 2 is the force exerted by the column of fluid above the point divided by the area over which the force occurs. The force exerted over the entire bottom of the tank can be calculated by Equation 6-5 as:

$$F = \gamma V \quad (6\text{-}5)$$

where γ = fluid specific weight, N/m^3 (lb/ft^3); and V = total volume of fluid in tank, m^3 (ft^3).

The volume of the tank is the cross-sectional area times the height of the liquid column, or A x h. Substitution into Equation 6-5 yields

$$F = \gamma A h \quad (6\text{-}6)$$

Combining with Equation 6-4 and simplifying, Equation 6-7 then gives the pressure at the bottom of the tank:

$$P = \gamma h \quad (6\text{-}7)$$

Equation 6-7 defines the pressure at any point in the fluid column. For example, the pressure at point 1 then becomes

$$P = \gamma \frac{h}{3} \quad (6\text{-}8)$$

Figure 6-1. A column of fluid at rest.

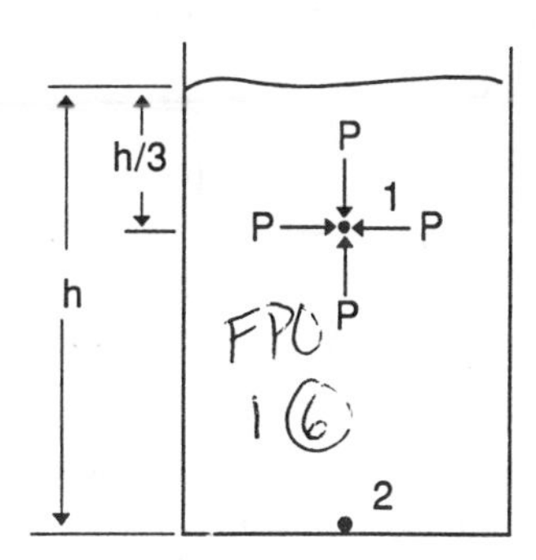

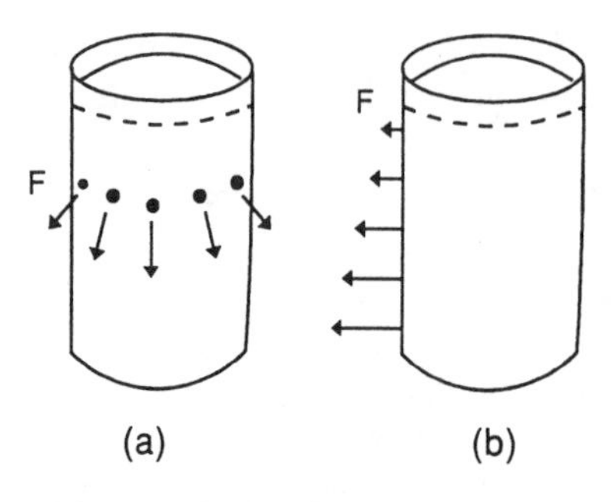

Figure 6-2. Magnitude of pressure forces F in (a) Horizontal plane; and (b) vertical plane.

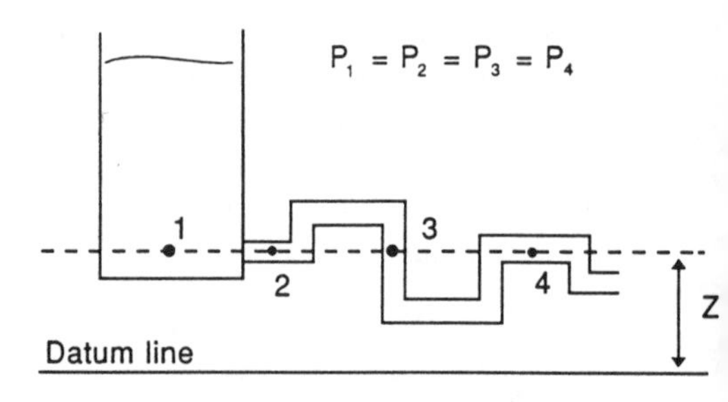

Figure 6-3. Pressure at several points in a fluid at rest.

which shows that the pressure at any point in a fluid is equal to the specific weight of the fluid times the height of the fluid column above the point in question. This principle is illustrated in Figure 6-2a. If holes were drilled through the tank wall in a horizontal line around the tank circumference, the fluid would spurt out an equal distance from each hole. However, as one descends deeper into the column of liquid the pressure increases. If several holes were drilled in a vertical line through the tank wall the length of the fluid jet will increase from top to bottom as the static pressure increases (Figure 6-2b). It may be further illustrated that the pressure is the same at all points lying in a horizontal plane in a fluid system at rest. For example, the pressure is equal at points 1, 2, 3, and 4 in Figure 6-3.

Example problem 6-1. The fluid in the tank in Figure 6-1 is water at 80°F. If h = 12 ft, what is the pressure at points 1 and 2 in: (a) lb/ft^2; and (b) Pascals.

Solution: From Table 6-2 γ = 62.22 lb/ft^3. Therefore,

(a) At point 1: $P = (62.22)(12/3) = 248.88$ lb/ft^2.
At point 2: $P = (62.22)(12) = 746.64$ lb/ft^2.

(b) At point 1: $P = (248.88)(0.048) = 11.95$ kPa.
At point 2: $P = (746.64)(0.048) = 35.84$ kPa.

Pressure Measurement

Two different planes are used to reference pressure, absolute zero and atmospheric pressure. This reference system is illustrated in Figure 6-4. Vacuum pressure is called *absolute zero*. All pressure readings referenced to absolute zero are called *absolute pressures*. Thus, atmospheric pressure at mean sea level (MSL) is given as 101 kN/m^2 in the SI system. Common pressure gauges measure, not absolute pressure, but the difference between the pressure in the fluid into which the gauge is tapped and the pressure of the surrounding atmosphere. The reference in this case is atmospheric pressure at the gauge. This is referred to as *gauge pressure*. Gauge pressure is given by Equation 6-9 as

$$\textit{gauge pressure} = \textit{absolute pressure} - \textit{atmospheric pressure} \qquad (6\text{-}9)$$

In the SI system atmospheric pressure at MSL is 101 kPa (since Pa = N/m^2). In the English system standard atmospheric pressure is 14.7 psia. Pressures in the English system are sometimes reported in lb/ft^2 (psf) or lb/in^2 (psi). Expression of atmospheric pressure at MSL in each system of units is equivalent to a 760-mm (29.9-in.) tall column of mercury (Hg).

Gauge and absolute pressures are usually indicated after the unit. For example, if a pressure of 25 kPa is measured with a gage referenced to the atmosphere, and the atmospheric pressure is 101 kPa, then the pressure can be expressed as either 25 kPa gauge or 126 kPa absolute. Gauge pressure referenced to atmospheric pressure can be either positive or negative. Negative gauge pressures are also referred to as *vacuum pressure*. Thus, if a gage tapped into a tank indicates a vacuum pressure of 10 kPa, this can be re-

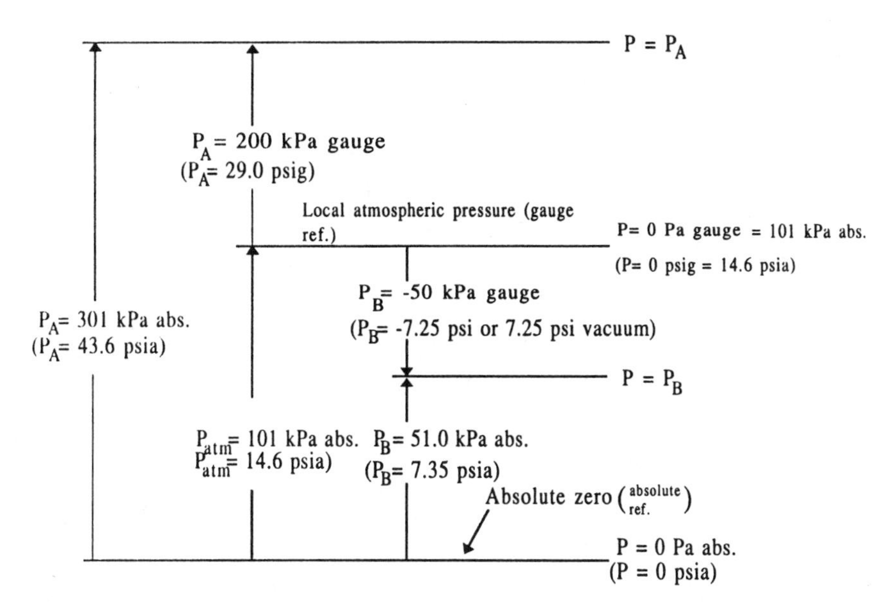

Figure 6-4. Examples of pressure relationships.

ported as either −10 kPa gauge or +91 kPa absolute.

A number of instruments have been developed over the years to indicate pressure. Most of these operate on the principle of either *manometry* (*piezometer* or *differential manometer*) or by measuring the flexing of an elastic member whose deflection is directly proportional to the applied pressure, such as with a *bourbon gauge*.

Manometry. Manometry methods utilize the change in pressure with elevation to determine pressure. The simplest manometer, or *piezometer*, is a straight vertical tube attached to a pipe (Figure 6-5). The gauge pressure at the center of the pipe is determined by $P = \gamma h$ (Eq. 6-7). In its present form this device is impractical for measuring high pressures. Also, it can only be used for measuring liquid pressures.

A U-tube, or *differential manometer*, is used to measure higher pressures (Figure 6-6). A manometer is a liquid-filled tube made from glass, plastic, or other transparent material that is marked with a scale. The most common liquids used are water and mercury. Usually one end of the manometer is connected to the unknown pressure source, and the other end is open to the atmosphere (Figure 6-6). Manometers can also be used to measure the pressure difference between two points in a pipeline (Figure 6-7).

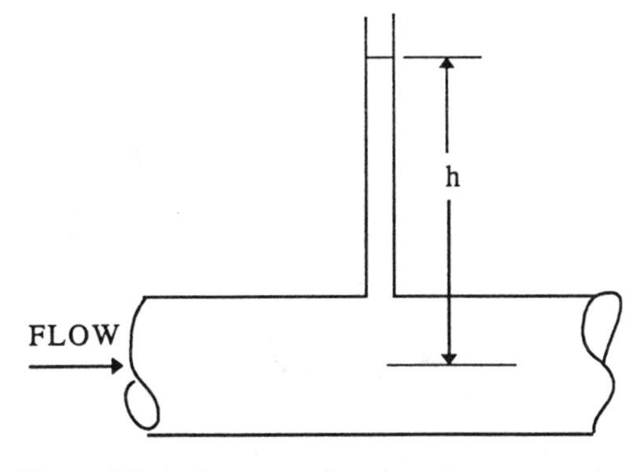

Figure 6-5. A piezometer for measuring pressure in a pipeline.

In the simple manometer in Figure 6-6 pressure forces the liquid in one leg of the U-tube to rise until the head pressure exerted by the column of fluid is in equilibrium with the unknown pressure. The manometer measures the pressure difference between the unknown source and the atmosphere. Hence, the pressure measured is in gauge pressure. A knowledge of the specific weights γ of the fluids involved and the linear dimensions h and Δh is needed to calculate the

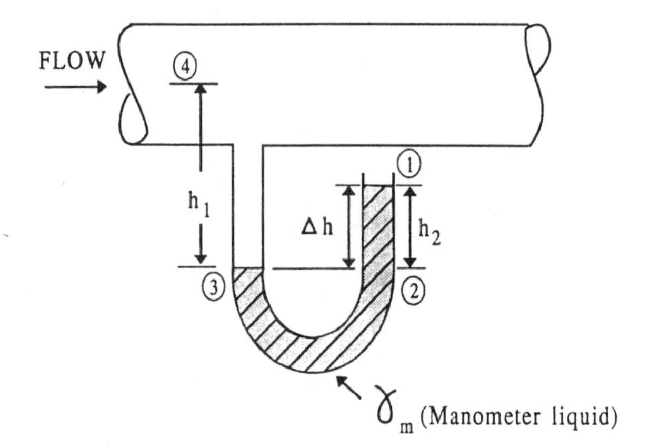

Figure 6-6. A simple U-tube manometer.

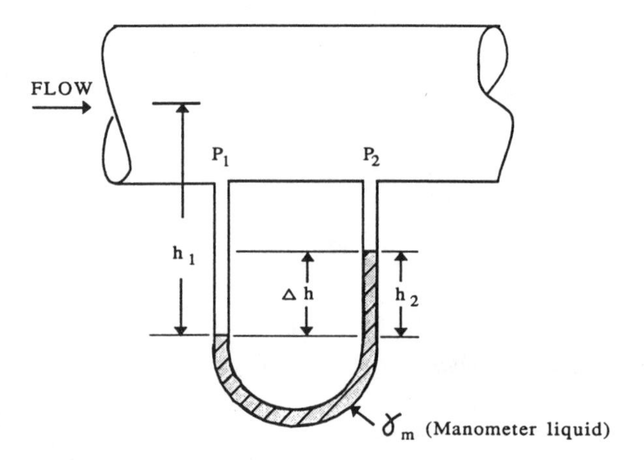

Figure 6-7. A manometer used to measure pressure drop in a pipeline.

pressure in the pipe. The procedure used to evaluate pressure is to calculate the step-by-step pressure changes from one level to the next in each fluid.

Bourbon-Tube Gauge. A Bourbon-tube gauge uses a tube having an elliptical cross section to indicate pressure (Figure 6-8). The tube is bent into the shape of a circular arc, and when the pressure in the gauge is atmospheric (zero gauge pressure) the tube is not deflected. When a pressure is applied to the gauge the tube tends to straighten, and a pointer is actuated that directly corresponds to pressure. The face of the gauge has a printed scale that is read in pressure units (i.e., N/m^2, psi, psf, etc.).

FLUIDS IN MOTION

The science of fluid dynamics deals with fluids in motion. Fluid dynamics is of primary concern to aquacultural engineers since virtually all productive aquaculture systems are dynamic hydraulic systems.

Conservation of Mass

The law of *conservation of mass* is one of three basic laws of physics that states that mass cannot be created or destroyed. This concept results in the *continuity equation* which states that within any hydraulic system,the system influent, storage volume, and effluent must be in balance. All hydraulic quantities must be ac-

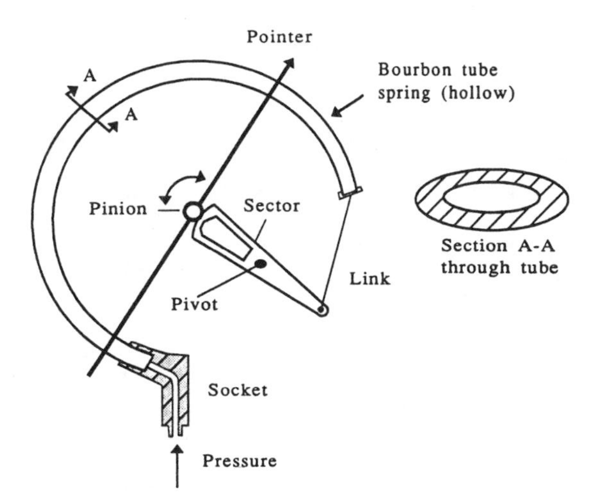

Figure 6-8. Schematic of the internal mechanism of a Bourbon-tube pressure gauge.

counted for since they have nowhere else to go. Mathematically this can be expressed as

$$Q_{in} - Q_{out} = \Delta \, storage \qquad (6\text{-}10)$$

where Q_{in} quantity flowing into the system; and Q_{out} = quantity flowing out of the system. Equation 6-10 is often used to determine the change in storage volume of fish ponds and tanks. Referring to Figure 6-9, Equation 6-10 can be rewritten as

$$Q_{in} - Q_{out} = \frac{\Delta storage}{time\ interval} \qquad (6\text{-}11)$$

If the influent rate is greater than the effluent rate, the net change in storage will be positive, and the pond/tank will fill over time. If the effluent (discharge) rate is greater than the influent rate, the net change in storage will be negative, and the pond/tank will empty over time.

Example problem 6-2. A 10 m^3 water storage tank in a fish hatchery is continuously replenished from a well by a pump delivering 2 L/min. Assuming that the tank was initially full, how long will it tank the tank to empty if 30 L/min is used to supply the needs of the hatchery?
Solution: From Equation 6-10
Δ storage = 2–30 = −28 L/min

Thus, there is a net flow of 28 L/min out of the tank. Then,

$$\frac{\text{volume}}{\text{flow rate}} = \frac{10{,}000 \text{ L}}{28 \text{ L/min}} = 357 \text{ minutes} \cong 6 \text{ hours}$$

In a situation where no change in storage volume is possible, the right side of Equation 6-10 is zero. An example is flow through a closed pipeline. According to the continuity equation, the quantity of fluid flowing through one section of the pipe must equal the flow through all other sections. To put the concept into practice, consider Figure 6-10. A given quantity of fluid is flowing through section 1 with velocity V_1 and through section 2 with velocity V_2. Since the mass of fluid flowing through both sections must be equal, we say that *mass is conserved*, and a mass balance equation can be written for the system as follows:

$$Q_1 = Q_2 \qquad (6\text{-}12)$$

or

$$\rho_1 V_1 A_1 = \rho_2 V_2 A_2 \qquad (6\text{-}13)$$

where V = velocity, m/s (f/s); A = cross-sectional area of conduit, m^2 (ft^2); and ρ = fluid density, kg/m^3 (slugs/ft^2).

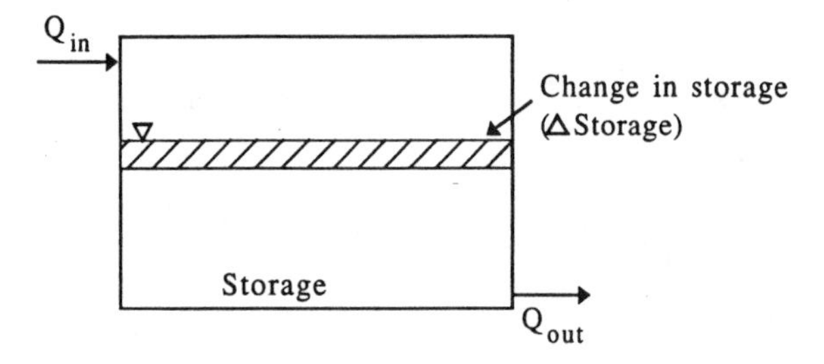

Figure 6-9. Change in storage volume in a tank.

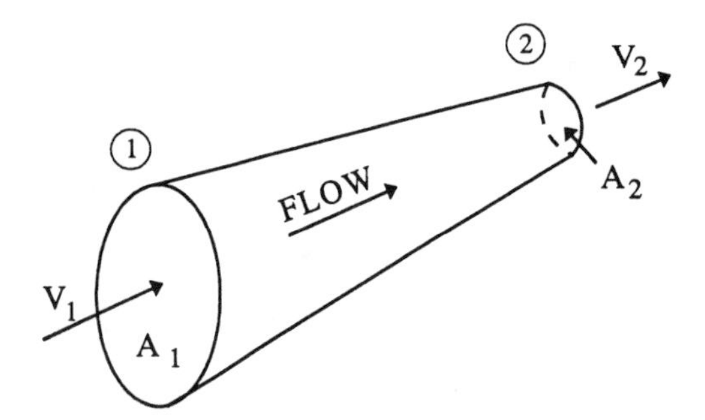

Figure 6-10. Flow through a closed conduit of variable cross sectional area.

For fluids that are essentially incompressible, like water

$$\rho_1 = \rho_2 \tag{6-14}$$

and Equation 6-13 reduces to

$$V_1 A_1 = V_2 A_2 \tag{6-15}$$

Equation 6-15 is one form of the continuity equation, which is one of the most basic relationships in fluid mechanics. It is used in numerous fluid flow applications where the flow is considered one-dimensional and incompressible. The example problem that follows may better illustrate this principle.

Example problem 6-3. What is the velocity V_2 in Figure 6-10 if $A_1 = 0.1\ \text{m}^2$, $A_2 = 0.01\ \text{m}^2$ and $V_1 = 0.05$ m/s?

Soultion: From the continuity equation (Eq. 6-13)

$$(0.05)(0.1) = V_2(0.01)$$
$$V_2 = 0.5 \text{ m/s}$$

Conservation of Energy

Another important concept is the law of *conservation of energy*. The total energy at any point in a fluid system consists of three components:

1. *Potential energy due to elevation above a point of reference*. As an example, a rock held in one's hand a given distance above the floor has potential energy in relation to the floor. The *potential energy* $(PE)_e$ is equal to the product of the vertical distance from the floor to the rock and the weight of the rock. This is expressed mathematically as

$$(PE)_e = ZW \tag{6-16}$$

where Z represents the vertical distance from the reference plane to the object in question, and W is the weight of the object.

2. *Potential energy due to pressure*: This is expressed as the product of the weight of the fluid, and the height of the column of fluid above the reference point, or

$$(PE)_p = hW \tag{6-17}$$

but, from Equation 6-7, $h = P/\gamma$. Therefore,

$$(PE)_p = \frac{P}{\gamma} W \tag{6-18}$$

3. *Energy due to motion of the fluid*: energy of motion is called *kinetic energy (KE)*, and its magnitude is expressed as

$$KE = \frac{1}{2} MV^2 \tag{6-19}$$

where M = mass, kg/m^3 (slugs/ft^2); and V = velocity, m/s (ft/s).

From Newton's laws we can write

$$M = \frac{W}{g} \qquad (6\text{-}20)$$

Substituting into Equation 6-19 produces

$$KE = W\frac{V^2}{2g} \qquad (6\text{-}21)$$

Therefore, Equation 6-21 gives the kinetic energy of an element of fluid with weight W and velocity V.

The total energy E_T of an element of fluid in motion is the sum of its potential and kinetic energy and can be calculated by summing Equations 6-16, 6-18, and 6-21

$$E_T = ZW + \frac{P}{\gamma}W + \frac{V^2}{2g}W \qquad (6\text{-}22)$$

By the law of conservation of energy, the total energy at any point in a fluid system must be the same. Referring to Figure 6-11, the energy at point 1 must equal the energy at point 2,

$$E_1 = E_2 \qquad (6\text{-}23)$$

or

$$Z_1W + \frac{P_1}{\gamma}W + \frac{V_1^2}{2g}W = Z_2W + \frac{P_2}{\gamma}W + \frac{V_2^2}{2g}W \qquad (6\text{-}24)$$

Dividing through by the weight W yields

$$Z_1 + \frac{P_1}{\gamma} + \frac{V_1^2}{2g} = Z_2 + \frac{P_2}{\gamma} + \frac{V_2^2}{2g} \qquad (6\text{-}25)$$

Equation 6-25 is known as *Bernoulli's equation* and is applicable to flow of an *ideal fluid* having no energy losses between points 1 and 2. The term Z on both sides of the equation is often referred to as the *pressure head*, and the term $V^2/2g$ is called the *velocity head*.

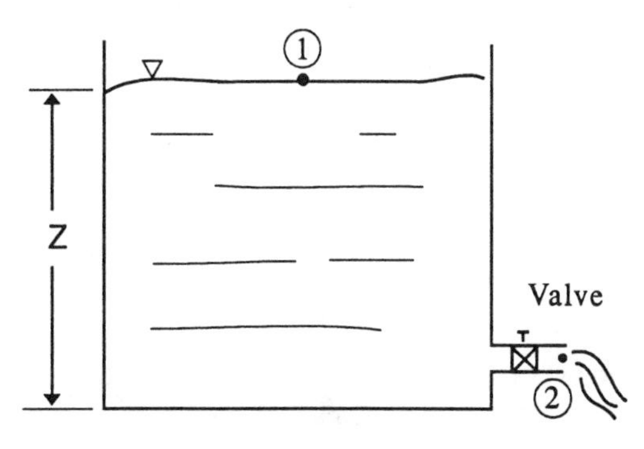

Figure 6-11. Energy at two points in a fluid system.

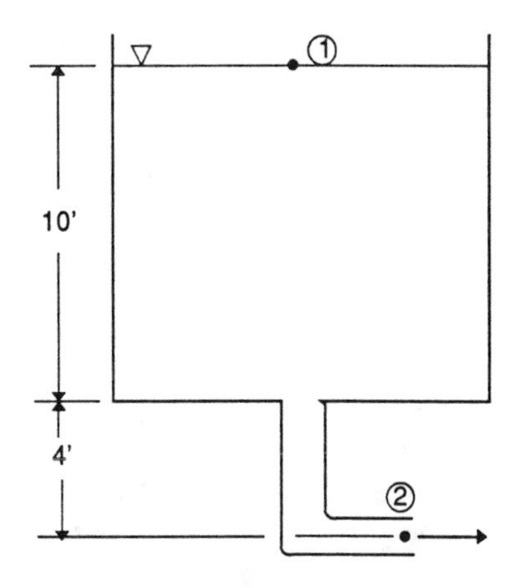

Figure 6-12. Figure for example problem 6-4.

Example problem 6-4. Calculate the velocity of the water discharging from the overhead storage tank shown in Figure 6-12.

Solution: According to Equations 6-23 and 6-24, the energy at points 1 and 2 must be the same. Also, since points 1 and 2 are both open to the atmosphere, the gauge pressure is zero at both points. Using the elevation at point 2 as the point of reference, Equation 8–25 then becomes

$$14 + 0 + \frac{V_1^2}{2g} = 0 + 0 + \frac{V_2^2}{2g}$$

If the tank in question is very large, we can assume that the velocity at the water's surface as the tank drains is very small and can be ignored. Thus,

$$14 = \frac{V_2^2}{2g}$$

Thus, $V_2^{\,2} = (14)(2)(32.2)$

From which $V_2 \cong 30$ ft/s

In actual practice fluids have losses due to friction in the pipes and minor losses associated with elbows, tees, valves, etc. In addition, systems often have an external energy source, such as a pump. Thus, Equation 6-25 can be rewritten to include energy inputs and losses. External energy inputs are normally included on the left side of the equation with losses on the right side as in Equation 6-26.

$$Z_1 + \frac{P_1}{\gamma} + \frac{V_1^2}{2g} + \begin{matrix}\textit{external}\\ \textit{energy}\\ \textit{input}\end{matrix} = Z_2 + \frac{P_2}{\gamma} + \frac{V_2^2}{2g} + \begin{matrix}\textit{minor}\\ \textit{losses}\end{matrix} + \begin{matrix}\textit{friction}\\ \textit{losses}\end{matrix} \qquad (6\text{-}26)$$

Energy from a pump is designated as E_p, and minor and pipe friction losses are represented as *head losses* h_m and h_f, respectively

$$Z_1 + \frac{P_1}{\gamma} + \frac{V_1^2}{2g} + E_p = Z_2 + \frac{P_2}{\gamma} + \frac{V_2^2}{2g} + h_m + h_f \qquad (6\text{-}27)$$

In systems having no pump, $E_p = 0$, and the external energy term drops out of Equation 6-27.

Minor Losses. Minor pipe losses are those associated with friction occurring when the fluid encounters restrictions in the system (i.e., valves), changes in direction (elbows, bends, tees), changes in pipe size (reducers, expanders), and losses caused by the fluid entering or leaving a conduit. Other system components can have energy losses associated with them, such as foot valves and pipe intake screens. A loss coefficient K is associated with each component causing a loss, and total minor losses are obtained from the product of the sum of the K-values and the velocity head (Eq. 6-28).

$$h_m = \Sigma K \frac{V^2}{2g} \qquad (6\text{-}28)$$

Loss coefficients (K) must be developed experimentally for each pipe system component. K-values for valves, fittings, and other other common pipe components are shown in Tables 6-3 and 6-4.

Example problem 6-5. Calculate the total minor losses through the pipe system shown in Figure 6-13 when the gate valve is three-quarters open, $D = 6$ in., $d = 3$ in. and $V = 2$ ft/s.

Solution: From Table 6-3 the K values for the elbow and gate valve are 0.9 and 1.15, respectively. From Table 6-4, and interpolating for $(d/D)^2 = 0.25$, $K = 0.40$. Thus, from Equation 6-28

$$h_m = \frac{(0.9 + 1.15 + 0.40)(2)^2}{(2)(32.2)}$$

$$h_m = 0.15 \text{ ft}$$

Pipe Friction Losses. Pipe friction losses are caused by the friction generated by the movement of the fluid against the walls of the pipe, fittings, etc. The magnitude of the loss depends on several factors: the internal pipe diameter, fluid velocity, roughness of the internal pipe surfaces, and certain physical properties of the fluid, namely, density, and viscosity. The relationship of these parameters is expressed mathematically by

$$f = \textit{function}\ (D, V, \varepsilon, \rho, \mu) \qquad (6\text{-}29)$$

Figure 6-13. Figure for example problem 6-5.

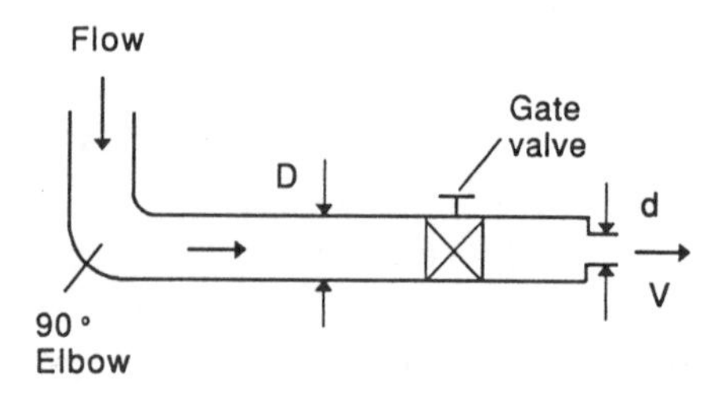

Table 6-3. Minor loss coefficient K for common fittings.

Fitting	K	Fitting	K
Globe valve (fully open)	10.0	Strainer bucket:	
Angle valve (fully open)	5.0	With foot valve	10.0
Check valve (fully open):		Without foot valve	5.5
Swing type	2.5	Close return bend	2.2
Ball type	70.0	Standard tee (entrance to minor line)	1.8
Lift type	12.0	Standard 90° elbow	0.9
Gate valve:		Standard 45° elbow	0.26
Fully open	0.19	Medium sweep elbow	0.75
3/4 open	1.15	Long sweep elbow	0.6
1/2 open	5.6	Pipe entrance: Square edge	0.50
1/4 open	24.0	Pipe entrance: Slightly rounded	0.23
Diaphragm valve:		Pipe entrance: Inward projecting	0.78
Fully open	2.3	Pipe entrance: Well rounded	0.04
3/4 open	2.6	Pipe exit (all)	1.00
1/2 open	4.3		
1/4 open	21.0		

Sources: Wheaton (1977), Roberson and Crowe (1990), Simon (1981), Eschbach (1952), and Kövári (1984).

where f = friction factor; D = inside pipe diameter; V = fluid velocity; ϵ = absolute roughness; ρ = fluid density; and μ = absolute viscosity. The terms in parentheses can be grouped into two dimensionless terms as in Equation 6-30.

$$f = function\left(\frac{\rho VD}{\mu}, \frac{\varepsilon}{D}\right) \tag{6-30}$$

The first term on the right side of Equation 6-30, $\rho VD/\mu$, is the *Reynold's number*, RN, which can also be written as VD/ν, where ν is the fluid kinematic viscosity. RN is the ratio of the internal forces to viscous forces, and is often used to determine whether the flow is laminar or turbulent. RN below 2,000 denotes laminar flow. The friction factor f for laminar flow is a linear function of RN and may be defined as

$$f = \frac{64}{RN} \tag{6-31}$$

Turbulent flow is characterized by RN greater than 4,000. Between 2,000 and 4,000 is the *transition zone* where flow is very unstable. Pipe designers should avoid using RN values in the transition zone.

The second term in Equation 6-30, ϵ/D, is called the *relative roughness* and is the ratio of the *absolute roughness* and the inside pipe diameter. The absolute roughness is a measure of the height of the irregularities on the inner surfaces of pipes, which vary according to construction material, service use and degree of fouling. Values for ϵ are presented in Table 6-5 for different pipe materials. These values are not specifically for fresh- or saltwater use and do not allow for significant biofouling. Plastic or some other tubing material that is heavily fouled is likely to have an ϵ in the range of steel or concrete even after rigorous cleaning to remove deposits. Pipes with significant biofouling in effect have reduced diameters, and losses can be calculated using an estimated effective diameter or by using the actual pipe diameter and a high ϵ-value in the range of about 0.003 m (0.01 ft) (Huguenin and Colt 1989). It is better to be on the conservative side rather than to underdesign the system.

The relationship between RN, ϵ/D and f is illustrated in Figure 6-14. This figure is com-

Table 6-4. Minor loss coefficients. (Note: Use Equation $h_m = Kv^2/2g$ unless otherwise indicated).

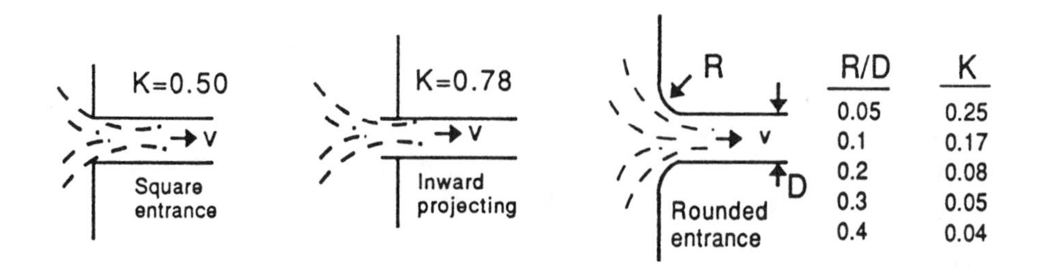

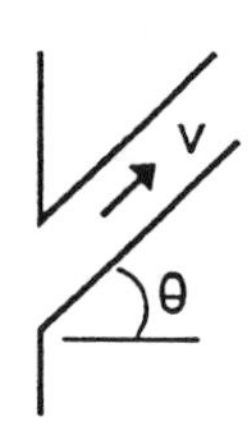

Loss due to skewed entrance

$$K = 0.505 + 0.303 \sin\theta + 0.226 \sin^2\theta$$

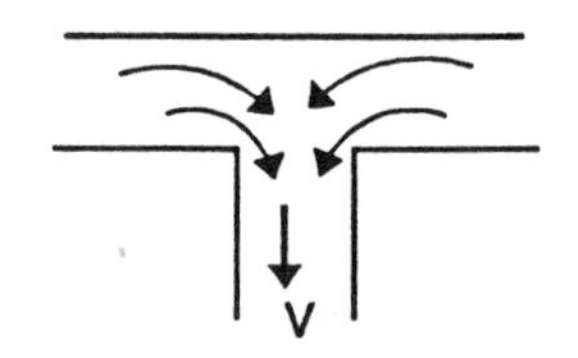

Standard tee, entrance to minor line

K = 1.8

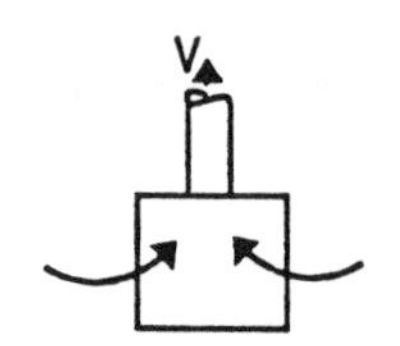

Screened intake

K = 10 with foot valve

K = 5.5 without foot valve

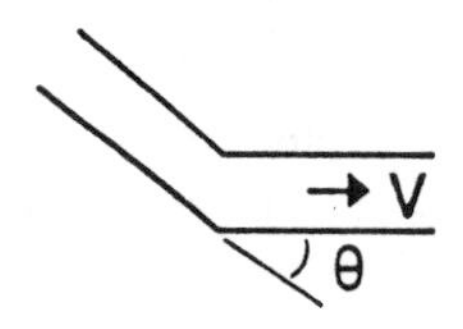

Sharp elbow

$$K = 67.6 \times 10^{-6} \ (\theta \text{ in degrees})^{2.17}$$

Table 6-4. Continued

Sudden expansion

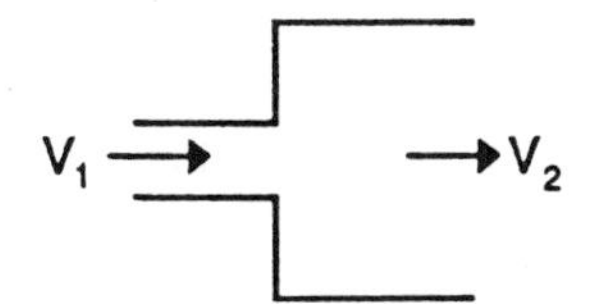

$$h_m = (1 - v_2/v_1)^2 \times v_1^2/2g$$

or

$$h_m = (v_1/v_2 - 1)^2 \times v_2^2/2g$$

Sudden contraction

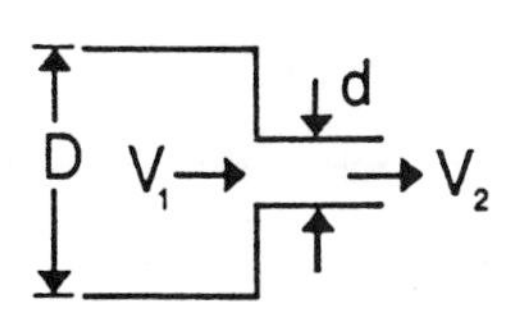

$(d/D)^2 =$	0.01	0.1	0.2	0.4	0.6	0.8
K =	0.5	0.5	0.42	0.33	0.25	0.15

use v_2 to compute h_m

Confusor

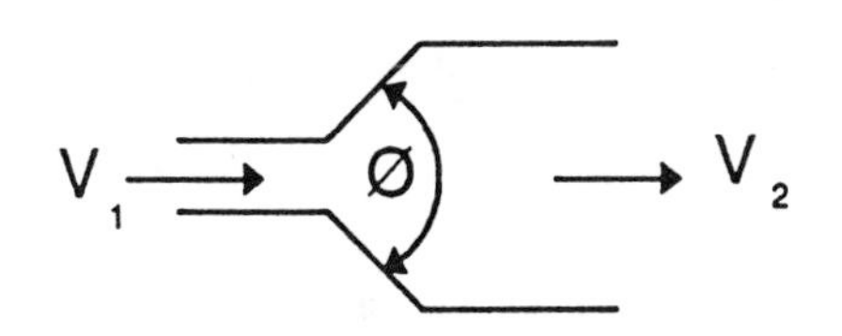

$$h_m = K(v_1^2 - v_2^2)/2g$$

Ø° =	20	40	60	80
K =	0.20	0.28	0.32	0.35

Diffusor

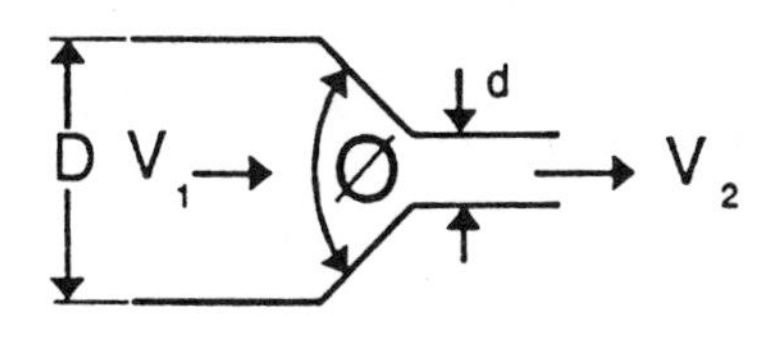

$$h_m = K(v_2^2 - v_1^2)/2g$$

Ø°	6	10	20	30	40	60	80	120	140
K for D=3d	0.12	0.16	0.39	0.80	1.0	1.06	1.04	1.04	1.04
D=1.5d	0.12	0.16	0.39	0.96	1.22	1.16	1.10	1.06	1.04

Close return bend

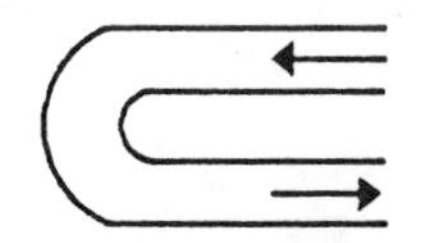

K = 2.2

Table 6-4. Continued

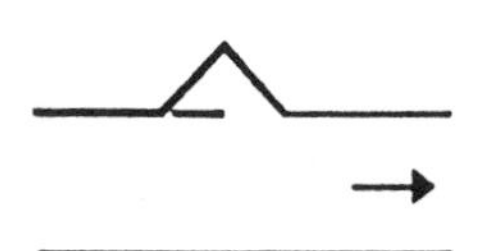

Check valves

Swing type K = 2.5 when fully open
Ball type K = 70.0
Lift type K = 12.0

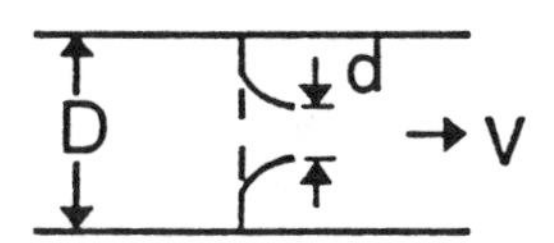

Measuring nozzle

$h_m = 0.3\Delta p$ for $d = 0.8D$
$h_m = 0.95\,\Delta p$ for $d = 0.2D$

where Δp is the measured pressure drop.

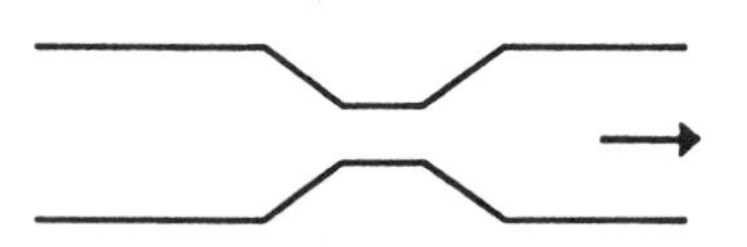

Venturi meter

$h_m = 0.1\Delta p$ to $0.2\Delta p$
where Δp is the measured pressure drop.

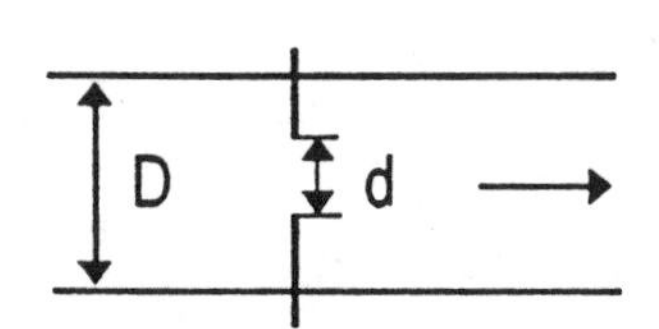

Measuring orifice

$$h = \Delta p\left(1 - \left(\frac{d}{D}\right)^2\right)$$

where Δp is the measured pressure drop.

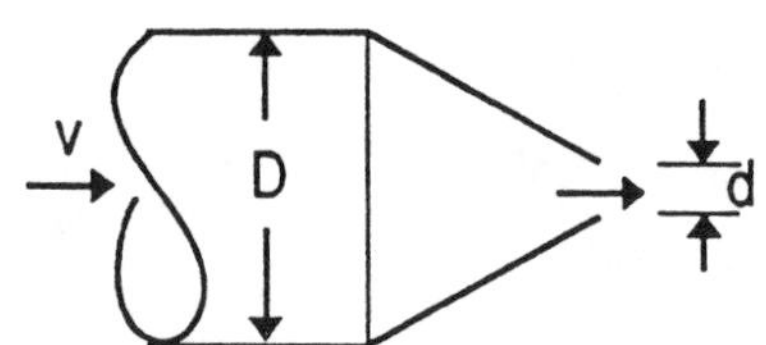

Confusor outlet

d/D =	0.5	0.6	0.8	0.9
K =	5.5	4	2.55	1.1

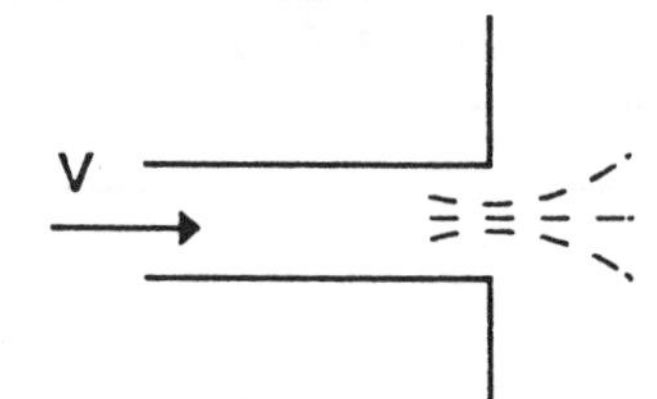

Exit from pipe into pond

K = 1.0

Table 6-4. Continued

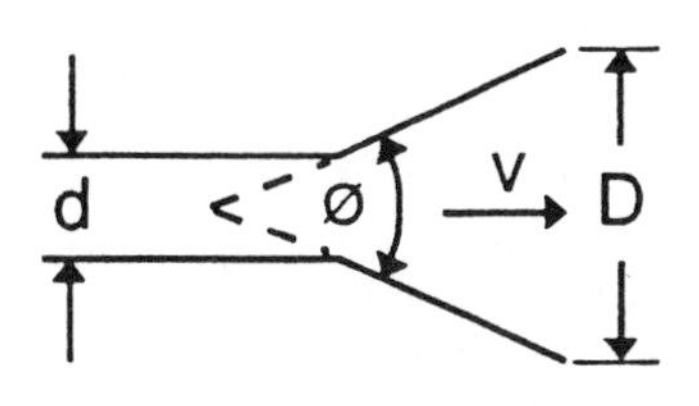

Diffuser outlet for D/d > 2

Ø° =	8	15	30	45
K =	0.05	0.18	0.5	0.6

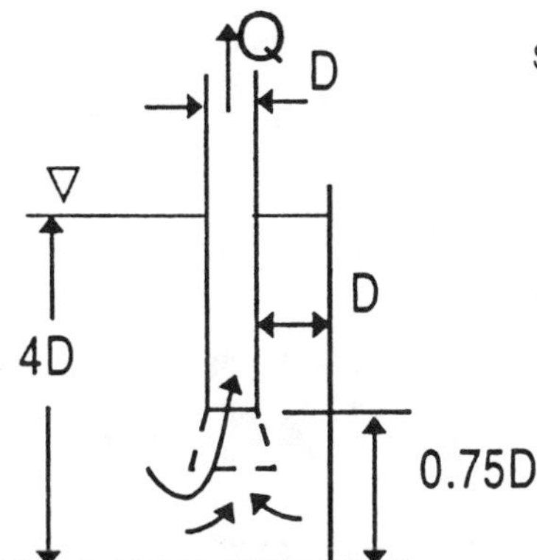

Suction pipe in sump with conical mouthpiece

$$h_m = D + \frac{5.6Q}{(2g)^{0.5} D^{1.5}} - \frac{v^2}{2g}$$

Without conical mouthpiece

$$h_m = 0.53D + \frac{4Q}{(2g)^{0.5} D^{1.5}} - \frac{v^2}{2g}$$

Width of sump shown = 3.5D

Sources: Kovári 1984, Simon 1981.

monly known as *Moody's diagram.* The horizontal axis is Reynold's number, the vertical axis is the friction factor *f*, and separate curves are shown for values of the relative roughness ϵ/D. For a given situation the friction factor can be found by first calculating RN and ϵ/D. By following the RN line vertically upward until it intersects the ϵ/D curve, the magnitude of *f* is found by extending a horizontal line to the *f*-axis. Most practical applications will fall into the right half of the diagram and toward the top if significant biofouling is present in the system (Huguenin and Colt 1989).

Once *f* is known, the head loss due to pipe friction can be calculated from the *Darcy-Weisbach equation*:

$$h_f = f \frac{L}{D} \frac{V^2}{2g} \tag{6-32}$$

where h_f = pipe friction head loss m (ft); *f* = friction factor, dimensionless; *L* = total straight length of pipe, m (ft); *D* = inside pipe diameter, m (ft); *V* = fluid velocity, m/s (ft/s); and g = gravitational constant, m/s^2 (ft/s^2).

A major advantage of the Darcy-Weisbach equation is that it is readily usable in both the SI and English systems. It should be noted that, if the fluid travels through several different pipe diameters, or if the pipe materials or flow conditions change along the line, calculations must be done for each pipe section and the results added to get the total pipe losses.

Once system minor and friction losses are determined they can be entered into Bernoulli's equation (Eq. 6-27) to calculate the external energy E_p required to drive the system.

Example problem 6-6. Water at 20°C is flowing through a 500-m section of 10-cm diameter old cast iron pipe at a velocity of 1.5 m/s. Calculate the total friction losses h_f using the Darcy-Weisbach equation.

Table 6-5. Absolute roughness coefficient ϵ for different pipe materials.

Pipe material	Absolute roughness (cm)	(in.)	Source
Riveted steel	0.091–0.91	0.036–0.358	Wheaton (1977)
Concrete	0.03–0.31	0.012–0.122	Wheaton (1977)
Wood stave	0.018–0.091	0.007–0.036	Wheaton (1977)
Cast iron	0.026	0.010	Wheaton (1977)
Galvanized iron	0.015	0.0059	Wheaton (1977)
Asphalted cast iron	0.012	0.0047	Wheaton (1977)
Commercial steel or wrought iron	0.0046	0.0018	Wheaton (1977)
Drawn tubing	0.00015	0.000059	Wheaton (1977)
PVC	0.000005	0.00000197	Wheaton (1977)
	(m)	(ft)	Source
Glass, plastic, fiberglass, copper, brass, drawn tubing	0.000013	0.000042	Huguenin and Colt (1989)
Steel, wrought iron	0.000046	0.000147	Huguenin and Colt (1989)
Asphalted cast iron	0.00012	0.00038	Huguenin and Colt (1989)
Cast iron	0.00026	0.00083	Huguenin and Colt (1989)
Riveted steel	0.009–0.0009	0.028–0.0029	Huguenin and Colt (1989)
Concrete	0.003–0.0003	0.0096–0.00096	Huguenin and Colt (1989)

Solution: The Reynold's number RN is calculated using the equation

$$\mathrm{RN} = \frac{VD}{\upsilon}$$

The kinematic viscosity $\upsilon = 10^{-6}$ m^2/s (from Table 6-2). Thus,

$$RN = \frac{(1.5)(0.1)}{0.000001} = 150{,}000$$

The absolute roughness coefficient ϵ for cast iron pipe = 0.026 cm (from Table 6-5), and the relative roughness ϵ/D = 0.0026. Then, using Moody's diagram (Figure 6-12) and projecting a straight line upward from RN = 150,000 to where it intersects the curve for $\epsilon/D = 0.0026$, f is approximately 0.027. Then, from Equation 6-32

$$h_f = \frac{(0.027)(500)(1.5)^2}{(0.1)(2)(9.81)}$$

$$h_f = 15.5 \text{ m}$$

An alternate method can be used to estimate friction losses involving the use of the *Hazen-Williams equation* and coefficient *C*. Two forms are presented, in SI and English units.

$$(SI) \qquad h_f = \frac{10.7\, LQ^{1.852}}{C^{1.852}\, D^{4.87}} \qquad (6\text{-}33)$$

$$(English) \qquad h_f \frac{4.73\, LQ^{1.852}}{C^{1.852}\, D^{4.87}} \qquad (6\text{-}34)$$

where h_f = pipe friction loss, m (ft); L = length, m (ft); Q = flow rate, m^3/s (ft^3/s); C = Hazen-Williams coefficient, dimensionless; and D = pipe diameter, m (ft).

Commonly used values of the Hazen-

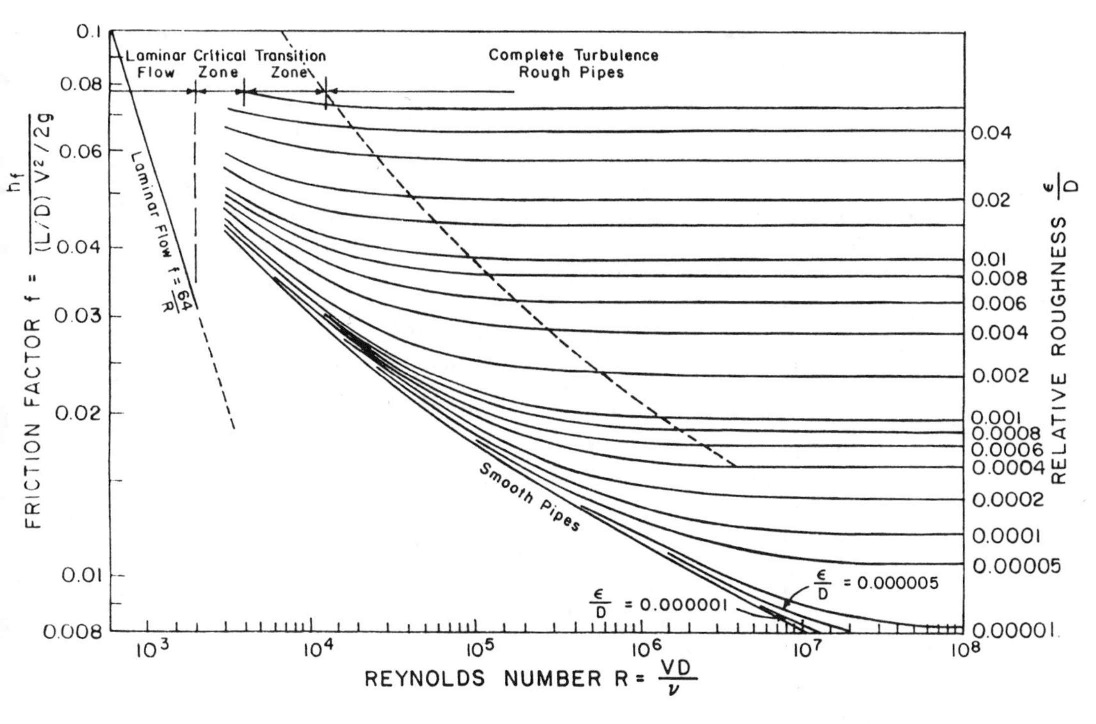

Figure 6-14. Moody's diagram (Moody 1944) with permission from the American Society of Mechanical Engineers.

Williams coefficient C for various pipe materials are shown in Table 6-6. C-values may range from as low as 50 for badly corroded, severely fouled, or old concrete pipe to as high as 150 for very smooth PVC. The choice is a matter of judgment on the part of the engineer. Basically, C-values of 140–150 are used for new pipe or for temporary installations. For permanent installations, C-values of 100 or less should be used to indicate average conditions. Again, it is best to be conservative when selecting C-values.

Example problem 6-7. Calculate the friction losses for the pipe in the previous example in both SI and English units using the Hazen-Williams equation.

Solution: Using the information given in the previous example, the flow rate Q is calculated: $Q = 0.01$ m^3/s (0.35 ft^3/sec). A Hazen-Williams coefficient $C = 100$ is assumed (from Table 6-6). Thus,

(a) SI units (using Eq. 6-33)

$$h_f = \frac{(10.7)(500)(0.01)^{1.852}}{(100)^{1.852}(0.1)^{4.87}} = \frac{1.06}{0.068} = 15.6\text{m}$$

The answer obtained in SI units is very close to the previous answer.

(b). English units (from Eq. 6-34)

$$h_f = \frac{(4.73)(1{,}640)(0.35)^{1.852}}{(100)^{1.852}(0.33)^{4.87}} = 49.5 \text{ ft}$$

Many engineering handbooks contain tables that can save a lot of time performing calculations by listing friction head losses per 100 ft of straight pipe for certain materials. Friction losses for schedule 80 PVC pipe are shown in the appendix. Schedule 80 pipe information is given since it is smaller in diameter than schedule 40 and will give a more conservative estimate of friction losses.

Minor losses for pipe fittings can be calculated by using an equivalent length of straight pipe. Equivalent lengths for plumbing fixtures

Table 6-6. Hazen-Williams coefficient C for selected pipe materials.

Pipe material	C	Source
Asbestos cement	140	Viessman and Hammer (1985)
Brick	100	Metcalf and Eddy (1972)
Concrete		
Average	130	Viessman and Hammer (1985)
Old	60–80	Viessman and Hammer (1985)
Copper	130–140	Simon (1981)
Fire hose (rubber-lined)	135	Simon (1981)
Glass	140	Simon (1981)
	150	Viessman and Hammer (1985)
Iron		
Cast		
New, unlined	130	Viessman and Hammer (1985)
Old, unlined	40–120	Simon (1981)
Old, bad condition	60–80	Metcalf and Eddy (1972)
Cement lined	130–150	Metcalf and Eddy (1972)
Tar-coated	115–135	Metcalf and Eddy (1972)
5-year-old	120	Viessman and Hammer (1985)
20-year-old	100	Viessman and Hammer (1985)
Galvanized	120	Simon (1981)
Lead	130–140	Simon (1981)
Pipes		
Very smooth	130	Metcalf and Eddy (1972)
Extremely smooth	140	Metcalf and Eddy (1972)
Plastic	140–150	Simon (1981)
	150	Viessman and Hammer (1985)
PVC	150	Viessman and Hammer (1985)
Smooth wood, smooth masonry	120	Metcalf and Eddy (1972)
Steel		
New, welded	120	Viessman and Hammer (1985)
New, riveted	110	Metcalf and Eddy (1972)
Old, riveted	95	Metcalf and Eddy (1972)

have been determined experimentally and are presented in Table 6-7 as equivalent length in pipe diameters (L/D). Alternatively, values for equivalent length of straight pipe can be obtained from the nomograph in Figure 6-15. The equivalent length is found by laying a straight edge connecting the point where the fitting in question intersects the left vertical line to the inside diameter line on the right. The equivalent length value is read from the center vertical line. The reader should be aware that the centerline scale in Figure 6-15 is in meters.

Example problem 6-8. Estimate the friction losses in a 6-in. diameter piping system containing 200 ft of straight pipe, a gate valve in the half-closed position, two close return bends, and four standard 90° elbows. The water velocity in the pipe is 2.5 ft/s. Use the Hazen-Williams equation to solve for minor losses: (a) using Table 6-7; and (b) using Figure 6-15.

Solution: (a) From the information given, Q is calculated to be 0.49 cfs. The equivalent lengths of the plumbing fixtures are as follows:

Table 6-7. Representative equivalent length[1] in pipe diameters (*L/D*) for selected valves and fittings.

Description of fitting		Equivalent length in pipe diameters (*L/D*)
Globe valve	Conventional (fully open)	340
	Y pattern (fully open)	160
Angle valve	Conventional (fully open)	145
Gate valve	Conventional (fully open)	13
	(3/4 open)	35
	(1/2 open)	160
	(1/4 open)	900
Check valve	Conventional swing (fully open)	135
	Clearway swing (fully open)	50
	Globe lift or stop (fully open)	same as globe
	Angle lift or stop (fully open)	same as angle
	In-line ball (fully open)	150
Foot valve with strainer	With poppet lift-type disc	420
	With leather-hinged disc	75
Butterfly valve	6-in. and larger (fully open)	20
Cock	Straight-through (fully open)	18
	Three-way (flow straight through)	44
	(flow through branch)	140
Elbow	90° square corner	57
	90° standard	30
	90° long radius	20
	45° standard	16
	90° street	
	50	
	45° street	26
Tee	Standard (flow through run)	20
	Standard (flow through branch)	60
Close return bend		50

Sources: Creswell (1993) and Crane Company, Catalog No. 60, Chicago, IL, 1960.

[1] Total equivalent length = *L/D* × *d*/12 × number of fittings for *d* in inches.

Total equivalent length = *L/D* × *d*/100 × number of fittings for *d* in centimeters.

Total equivalent length (ft) = *L/D* × 1/12 × Number of fittings

One gate valve = 160 × 6/12 = 80 ft

Two close return bends = 50 × 6/12 × 2 = 50 ft

Four elbows = 30 × 6/12 × 4 = 60 ft

Total equivalent length = 80 + 50 + 60 = 190 ft

The total length of straight pipe = 190 + 200 = 390 ft. Using Equation 6-34 with C = 120, the total friction loss is estimated by

$$h_f = \frac{(4.73)(390)(0.49)^{1.852}}{(120)^{1.852}(0.5)^{4.87}}$$

$$h_f = \frac{492.2}{241.1} \cong 2 \text{ ft}$$

(b) From Figure 6-13

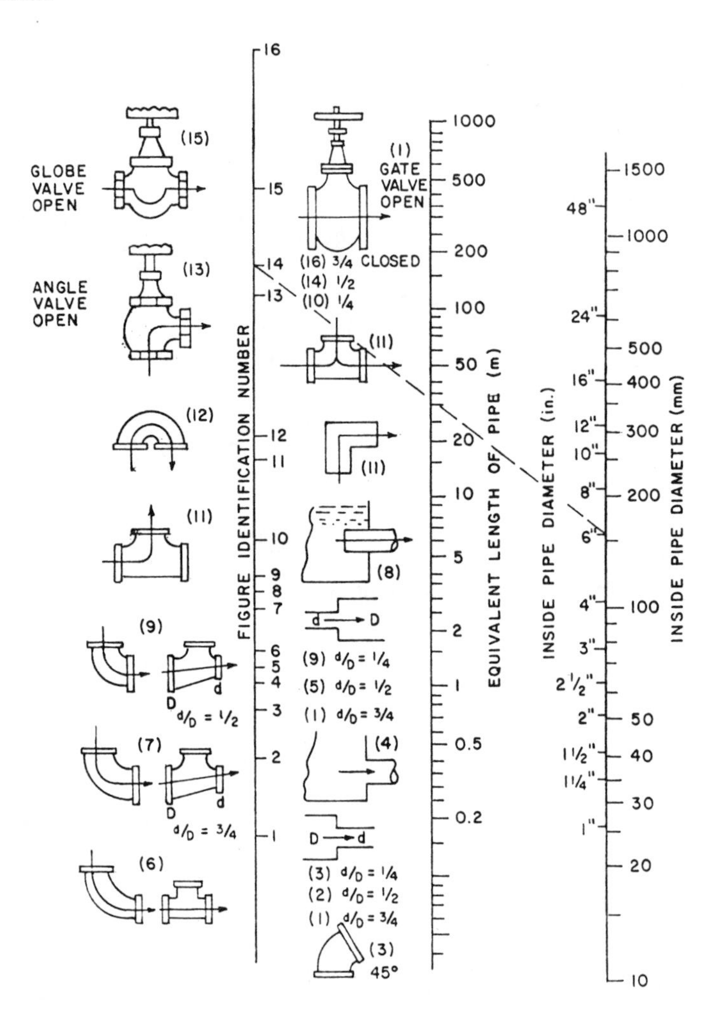

Figure 6-15. Nomograph used to calculate resistance of pipe fittings in equivalent lengths of straight pipe (Finkel 1982) with permission from CRC Press.

Fixture	units	Equivalent length per unit, m	subtotal
gate valve	1	32	32
close returns	2	12	24
elbows	4	5	20

Total equivalent length = 84 m

The total length of straight pipe is 84 + 200/3.28 = 145 m. The total friction loss (from Eq. 6-33) with Q = 0.014 m^3/s is therefore

$$h_f = \frac{(10.7)(145)(0.014)^{1.852}}{(120)^{1.852}(0.152)^{4.87}}$$

$$h_f = \frac{0.573}{0.737} = 0.78 \text{ m } (\cong 2.6 \text{ ft}),$$

close to previous answer.

OPEN CHANNEL FLOW

In addition to pressurized closed conduits water may also be conveyed in open channels, either for delivery to or drainage from a site. Open channels are conduits in which the flowing water has a free surface open to the atmosphere. Pressure is always zero gauge pressure at the liquid surface, and the energy for moving the water is primarily derived from gravitational forces. In some cases pumping systems may be employed. An advantage of open channels over closed conduits is ease of cleaning. Open channels used for draining ponds or tanks should be oversized since the maximum flow will occur during this time. Huguenin and Colt (1989) recommend that open channels used for drainage be designed to carry at least 10 times the normal average flow for the system.

Flow Velocity. The flow characteristics in open channels are determined by the physical parameters of the channel (i.e., shape, slope, cross section, and conditions of the channel bottom and side slopes). Different flow regimes occur in open channels, however, for aquaculture purposes, the treatment of open channel hydraulics will be limited to the most simple, steady-state system where flow does not change with time, and the channel conditions are constant down the line.

Flow velocity in open channels is found from

$$V = \frac{1}{n} R^{2/3} S^{1/2} \quad (SI) \qquad (6\text{-}35)$$

$$V = \frac{1.486}{n} R^{2/3} S^{1/2} \quad (English) \qquad (6\text{-}36)$$

where V = velocity, m/s (ft/s); R = hydraulic radius, m (ft); S = slope, m/m (ft/ft); and n = roughness coefficient, dimensionless.

The constant 1.486 has the dimensions $ft^{0.333}/s$, making Equation 6-36 dimensionally correct (Wheaton 1977). Equations 6-35 and 6-36 are both forms of *Manning's equation*, and the coefficient, n, is known as *Manning's roughness coefficient* (Chow 1959). Values of n vary with the conditions of the channel bottom and side slopes and must be determined experimentally. Experimentally determined values of n for common channel materials are listed in Table 6-8. Stones, rubble, vegetation, or similar materials increase the value of n. Therefore, for older, established channels a higher design

Table 6-8. Typical values of the roughness coefficient n used in Manning's equation.

Lined channels	n
Cement plaster	0.011
Untreated gunite	0.016
Wood: Planed (flumes)	0.012
Unplaned (flumes)	0.013
Concrete: Smooth	0.015
Rough	0.023
Rubble in cement	0.020
Asphalt: Smooth	0.013
Rough	0.016
Metal: Corrugated	0.024
Smooth (flumes)	0.013
Plastic	0.013
Unlined channels	
Earth: Straight and uniform	0.023
Winding and weedy banks	0.035
Dredged	0.027
Cut in rock: Straight and uniform	0.030
Jagged and irregular	0.045
Natural channels	
Gravel beds: Straight	0.025
Straight, large boulders	0.040
Earth: Straight with some grass	0.026
Winding, no vegetation	0.030
Winding, some pools	0.035
Winding, weedy banks	0.050
Very weedy and overgrown	0.080

Sources: Roberson and Crowe (1990); Schwab et al. (1966); and Metcalf and Eddy (1972).

n-value should be used to be on the conservative side.

The design flow velocity of open channels should be low enough to prevent erosion but high enough to limit deposition of solid material suspended in the flow. Usually, a flow velocity greater than 0.6 m/s (2 ft/s) will minimize deposition (James 1988). The maximum allowable velocity that does not cause excessive erosion depends on the material of construction. Obviously, severe erosion is undesirable since the channel sidewalls and bottom may fail resulting in more damage and sedimentation downstream. Too low a velocity will cause sedimentation and reduction in the effective channel cross section and a restriction in flow. Usually, an average velocity of 0.6–1.0 m/s (2–3 ft/s) is sufficient to prevent sedimentation in shallow channels (Schwab et al. 1966). Both erosion and sedimentation can result in channel failure and subsequent property damage. Local experience is often the most effective way to determine maximum allowable velocities for particular channel materials. When local experience is unavailable, Tables 6-9 and 6-10 can be used as a guide to selecting maximum velocities for unlined and vegetation-lined channels, respectively. Lower design velocities may be used for clear water than for water transporting solids. Where a powerful abrasive material is transported with the flow, the velocities shown should be reduced by about 0.16 m/s (0.5 ft/s). For water depths over 1 m, velocities can be increased by 0.16 m/s. If the channel is winding or curved, limiting velocities should be reduced by about 25% (Schwab et al. 1966).

Once the velocity of flow is calculated the flow rate through the channel can be estimated using the continuity equation

$$(SI) \qquad Q = AV = A\frac{1}{n}R^{2/3}S^{1/2} \qquad (6\text{-}37)$$

$$(English) \qquad Q = A\frac{1.486}{n}R^{2/3}S^{1/2} \qquad (6\text{-}38)$$

Manning's equation is convenient for solving for different unknowns, depending upon what information is available. For example, if the channel dimensions and slope are known, then a value can be assumed for n, depending on channel conditions, and flow velocity can be estimated. From there, an estimate of the channel flow rate can be obtained.

Bottom Slope. The slope S in Equations 6-35 through 6-38 is the slope of the channel bottom along its length and is found by dividing the elevation difference between two points (points A and B in Figure 6-16) by the horizontal distance between the two points.

Cross Section. The size of cross section a channel requires is determined by quantity of water to be moved, water velocity, economics, and other site parameters. The most hydraulically efficient cross-sectional shape for an open channel is a semicircle. But, this shape is not practical for unlined earth channels since the sides would be nearly vertical. Thus, this shape is generally limited to precast concrete channels and preshaped metal or plastic flumes (James 1988). Most open channels used for aquaculture purposes will be of earthen construction and are trapezoidal in cross section. Over time, the side slopes of earth channels slump somewhat, and the channel assumes a U-shape. Where funds permit, rectangular or square channels can be constructed from concrete or some other rigid material.

Side Slopes. Channel side slopes are determined mainly by soil texture and stability. Collapse of the channel sides can occur after a rapid drop in water level if the slopes are too steep. For a given side slope, the deeper the channel, the more likely it is to collapse. Side slopes should be designed to suit soil and cover conditions. Suggested channel side slopes are

Figure 6-16. Side view of open channel illustrating method of calculating bottom slope.

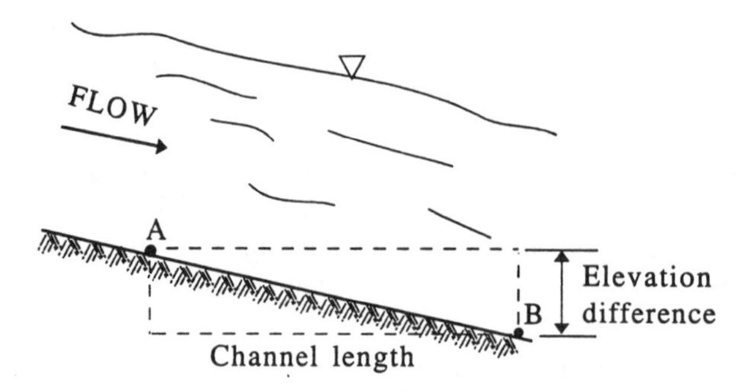

shown in Table 6-11. Very narrow channels should have flatter side slopes than wide channels because of greater reduction in cross section should the channel collapse (Schwab et al. 1966).

Hydraulic Radius. The *hydraulic radius*, R, is the ratio of the channel cross-sectional area and the wetted perimeter. The *wetted perimeter*, P, is the length of the channel perimeter in contact with the water. For example, the wetted perimeter of the rectangular channel shown in Figure 6-17 is the linear dimension $2h + b$. Thus, R for the channel is

$$R = \frac{A}{P} + \frac{bh}{2h + b} \tag{6-39}$$

The ideal cross section is one which minimizes the wetted perimeter P while maximizing the cross-sectional area of flow A. By minimizing P, the resistance to flow is minimized, and the hydraulic radius R is maximized. Therefore, open channels should be designed with the *best hydraulic cross section*. For a rectangular channel, the best hydraulic cross section occurs when the base width is equal to two times the water depth, or

$$\mathrm{b} = 2\mathrm{h} \tag{6-40}$$

for a trapezoidal channel, the best hydraulic cross section occurs when b is given by Equation 6-41

Figure 6-17. Cross section of a rectangular open channel.

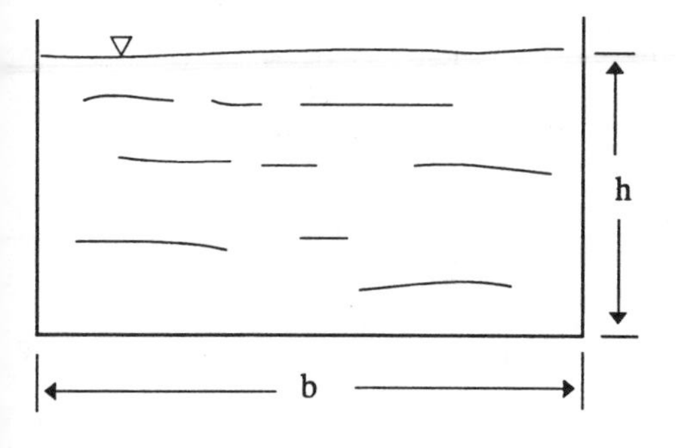

$$b = 2h \tan\frac{\theta}{2} \tag{6-41}$$

where θ = the angle the channel side slope makes with the horizontal.

Figure 6-18 shows design relationships for open channels of various cross sectional shapes. The figure also shows how to determine channel freeboard. *Freeboard* is the vertical dimension from the water's surface when the channel is flowing to the physical top of the channel. The depth of the constructed channel should have a minimum freeboard (D-d) of 15–20% of the total depth for lined channels and possibly as large as 30–35% of the total depth for unlined channels (Schwab et al. 1966).

Example problem 6-9. An open channel is used to transport 2,000 gpm from a lake to a series of fish ponds. The triangular-shaped channel is constructed in loose sandy soil and has a slope of 0.5%. The channel is designed to limit deposition of colloidal material. Determine the channel dimensions.

Solution: The maximum permissible velocity that prevents deposition but limits erosion is 2.5 ft/s (from Table 6-9). Other chosen parameters are: $n = 0.03$ (from Table 6-8) and $Z = 2$ (from Table 6-11). From Figure 6-16 for a triangular-shaped channel,

$$A = Zd^2 = 2d^2$$

$$R = 0.5d \text{ (approximately)}$$

Then

$$Q = AV = 2d^2 \times \frac{1.486}{n} \times R^{2/3} \times S^{1/2}$$

$$2{,}000 \text{ gpm} = 4.5 \text{ cfs}$$

$$= \frac{2d^2 \times 1.486 \times (0.5d)^{2/3} \times (0.005)^{1/2}}{0.03}$$

$$0.64 = d^2 \times (0.5d)^{2/3}$$

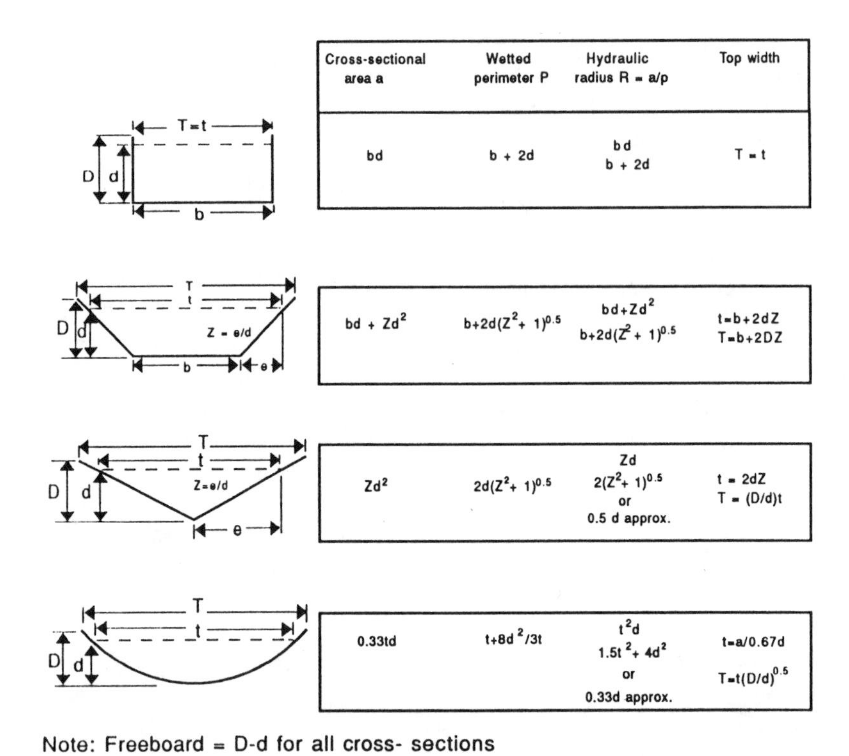

Figure 6-18. Open channel cross sections, dimensions and equations.

Table 6-9. Limiting velocities for essentially straight channels after aging.

	Velocity			
	Clear water		Water transporting colloidal solids	
Material	(m/s)	(ft/s)	(m/s)	(ft/s)
Fine sand, colloidal	0.46	1.50	0.76	2.50
Sandy loam, noncolloidal	0.53	1.75	0.76	2.50
Silt loam, noncolloidal	0.61	2.00	0.91	3.00
Alluvial silts, noncolloidal	0.61	2.00	1.07	3.50
Ordinary firm loam	0.76	2.50	1.07	3.50
Volcanic ash	0.76	2.50	1.07	3.50
Stiff clay, very colloidal	1.14	3.75	1.52	5.00
Alluvial silts, colloidal	1.14	3.75	1.52	5.00
Shales and hardpans	1.83	6.00	1.83	6.00
Fine gravel	0.76	2.50	1.52	5.00
Graded loam to cobbles, noncolloidal	1.14	3.75	1.52	5.00
Graded silts to cobbles, colloidal	1.22	4.00	1.68	5.50
Coarse gravel, noncolloidal	1.22	4.00	1.83	6.00
Cobbles and shingles	1.52	5.00	1.68	5.50

Source: *Principles Of Farm Irrigation System Design* by L.G. James ©1988. Reprinted by permission of John Wiley and Sons, Inc., New York.

Table 6-10. Permissible velocities (m/s) for vegetation-lined channels.

		Soil Type	
Cover	Slope (%)	Erosion Resistant	Easily Eroded
Bermuda grass	0–5	2.50	1.85
	5–10	2.15	1.50
	>10	1.85	1.25
Buffalo grass	0–5	2.15	1.50
Kentucky bluegrass	5–10	1.85	1.25
Smooth brome	>10	1.50	0.90
Blue grama			
Centipede grass	0–5	2.75	2.15
Bermuda grass (mowed)	5–10	2.50	1.85
	>10	2.15	1.50
Grass mixture	0–5	1.50	1.25
	5–10	1.25	0.90
	>10	—Not recommended—	
Lespedeza sericea	0–5	1.0	0.75
Weeping lovegrass	5–10	—Not recommended—	
Ischaemum (yellow bluestem)	>10	—Not recommended—	
Kudzu			
Alfalfa			
Crabgrass			
Annuals used for temporary	0–5	1.0	0.75
cover (common lespedeza	5–10	—Not recommended—	
and Sudan grass)	>10	—Not recommended—	

Source: *Aquacultural Engineering* by F.W. Wheaton ©1977. Reprinted by permission of John Wiley and Sons, Inc., New York.

Table 6-11. Recommended side slopes for unlined open channels.

Soil	Shallow channels (depths up to 1.2 m or 4 ft)	Deep channels (depths over 1.2 m or 4 ft)
Peat and muck	vertical	0.25:1
Stiff, heavy clay	0.5:1	1:1
Clay or silt loam	1:1	1.5:1
Sandy loam	1.5:1	2:1
Loose sandy	2:1	3:1

Source: *Soil And Water Conservation Engineering* by G.O. Schwab, R.K. Frevert, T.W. Edminster and K.K. Barnes © 1966. Reprinted by permission of John Wiley and Sons, Inc., New York.

$$0.64 = d^{6/3} \times 0.63d^{2/3} = 0.63d^{8/3}$$

$$d = (1)^{3/8} = 1 \text{ and } A = 2(1)^2 = 2$$

Checking stream velocity,

$$V = Q/A = 4.5/2 = 2.25$$

The calculated stream velocity is acceptable since it is less than the maximum permissible design velocity (i.e., $2.25 \leq 2.50$).

Channel Liners. Channel liners are classified as hard surface linings, covered membranes, and soil sealant liners. The selection depends largely on economics, but other factors are availability of materials, soil conditions, and size of channel. Advantages are that linings reduce seepage losses through the channel bottom and sides, consequently reducing potential drainage problems. They also reduce water losses caused by burrowing animals. Smooth linings reduce friction, and therefore, increase carrying capacity. Channels can have steeper bottom and side slopes when lined because they are more resistant to erosion. Because lined channels have steeper banks and smaller cross-sectional areas, they occupy less land area. Generally, overall maintenance costs are less for lined channels (James 1988).

The design of open channels can be somewhat complicated and is best left to the competent engineer. The design process normally involves some trial and error. For a more comprehensive treatment of open channel flow the reader is referred to texts on hydraulics (Chow 1959; Simon 1981; Roberson and Crowe 1991).

CHAPTER 7

PUMPS

'umps are mechanical devices that add energy ɔ fluids by changing mechanical energy pro-uced by gasoline, diesel, liquid petroleum gas, nd natural gas engines or electric motors to otential and/or kinetic energy. Pumps are used ı aquacultural systems usually to increase sys-:m pressure and thereby force the water to ıove against an energy gradient. In most quaculture situations pumps are used to lift wa-:r from one point of elevation to some higher levation or to convey water from one location) another. Water will flow only when energy is vailable to cause flow. Pumps are fairly effi-ient mechanisms for transferring energy to wa-:r provided that the correct pump is selected to o the job. The key requirement in pump selec-on is that there be a close correlation between ıe aquaculture system requirements and the ıaximum operating efficiency of the pump. oor pump selection can result in significantly ıcreased operating and maintenance costs nd/or result in system failure.

UMP CLASSIFICATION

umps are available in many combinations of ressures and discharges. In pump design, pres-ıre and discharge are inversely related. Therefore pumps producing high pressures normally have a small discharge, and pumps having large discharges normally operate under low pressure (Cuenca 1989).

There are many types of pumps. Pumps are classified as centrifugal, rotary, reciprocating, or air-lift (Figure 7-1). Each is designed for a specific application or for a variety of applications. Centrifugal and air-lift pumps are the types most often used for aquaculture applications. Other types are used to a lesser degree. No attempt is made here to discuss the operation of all types of pumps. Emphasis will be placed on those types applicable to aquaculture.

Centrifugal Pumps

Centrifugal pumps are *roto dynamic*, meaning that they use the centrifugal force imparted to water by one or more rotating vanes, called *impellers*, to increase the kinetic and pressure energy of the water. The impellers are enclosed within a rigid stationary housing called a *casing*. Water is forced into the center (eye) of the impeller by atmospheric or other pressure and caused to rotate by the impeller vanes. Centrifugal forces cause the water to accelerate outward and impart a high velocity to the water, which is converted into a pressure head, and the

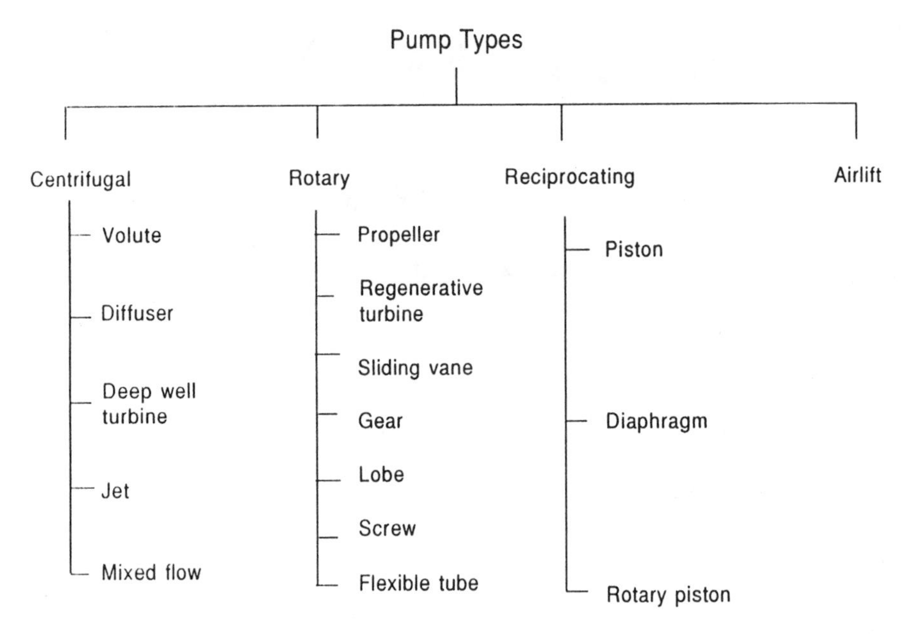

Figure 7-1. Classification of pumps used for aquaculture.

water is directed to the pump outlet. Centrifugal pumps are used in applications requiring high head but limited discharge (Cuenca 1989).

Centrifugal pumps are classified as either single or multi staged and as either horizontal or vertical, depending upon the orientation of the rotating shaft. In multi staged pumps the outflow from the first impeller is directed into the eye of the second, and so forth. Thus, the head delivered by a multi stage pump can be increased as desired, limited only by the number of stages. Horizontal pumps are further classified according to the location of the suction nozzle (inlet) as end-suction, side-suction, bottom-suction, or top-suction (James 1988).

Pump Casings. Centrifugal pumps are also classified according to casing design as either the *volute* or *diffuser* type. Both types are diagrammed in Figure 7-2. Volute casings have spiral-shaped passages for the water that increase in cross-sectional area as the water progresses toward the outlet. The rate of area increase within the volute casing is usually sufficient to reduce the fluid velocity as it approaches the outlet, resulting in an increase in fluid pressure. Most single-stage pumps have

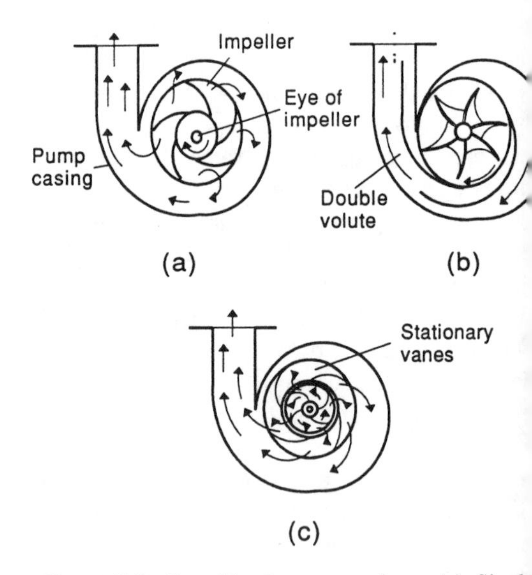

Figure 7-2. Centrifugal pump casings: (a) Singl volute casing; (b) double-volute casing; and (c) di fuser turbine casing.

volute-type casings. Casings like the one in Fi ure 7-2a are called single-volute casings and a normally used when pumps are operated at near peak efficiency. Double-volute casin (Figure 7-2b) are used when the pump is ope ted at part capacity for long periods of time because radial forces are more balanced (James 1988).

In diffuser-type pumps (Figure 7-2c) the impeller is surrounded by diffuser vanes that are fixed to the pump housing. The vanes are designed to streamline the transition of velocity head to static head. By reducing turbulence and other internal losses the diffuser vanes are able to develop higher pressure heads and high pump efficiencies. These pumps are used where very high heads are required, but where the fluid is nearly free of solids, because they clog easily. Diffuser casings are typically used in multistaged pumps. Large diffuser pumps often have efficiencies approaching 90% (Wheaton 1977).

Impellers. The impeller design in pumps determines pump characteristics to a large degree. Impellers are classified as either radial, axial, or mixed flow, depending on the direction of flow of water through the impeller relative to the axis of rotation (James 1988). In pumps having radial flow impellers, the water enters the impellers in a plane parallel to the axis of rotation and is discharged at a right angle to it (Figure 7-3a). Axial flow (propeller) impellers receive and discharge water in a plane parallel to the axis of rotation (Figure 7-3b). Mixed flow impellers receive water in a plane parallel to the axis of rotation and discharge it at an angle between 0° and 90° to the axis of rotation (Figure 7-3c).

Impellers are also classified as totally en-

Figure 7-3. Centrifugal pump impellers: (a) Radial flow; (b) axial flow; and (c) mixed flow.

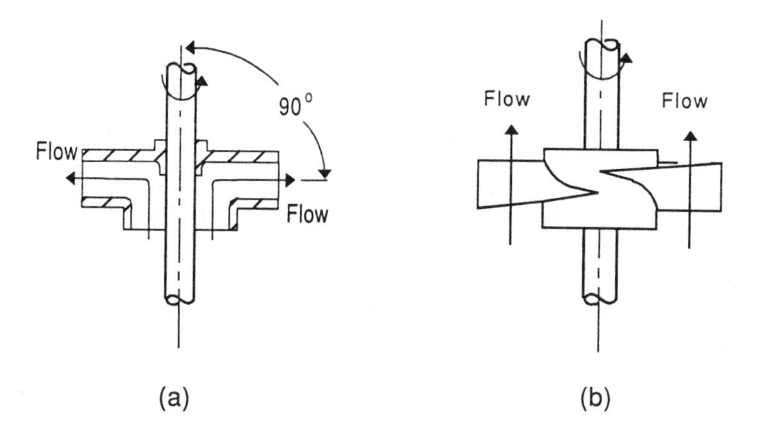

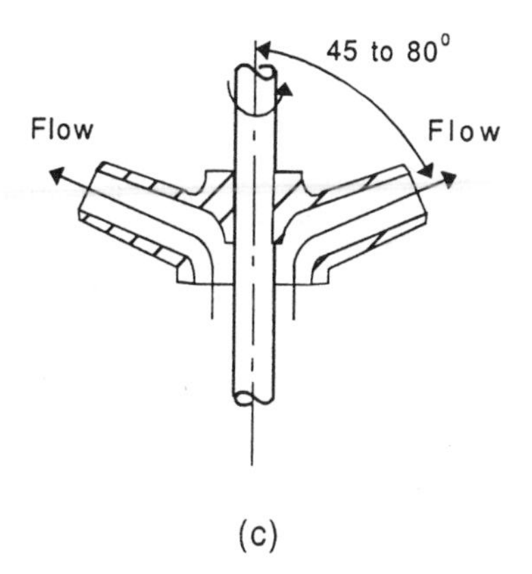

closed, semienclosed, or open impellers. Totally enclosed impellers have shrouds (face plates) that completely enclose the waterways between the vanes on both sides (Figure 7-4a). Semienclosed (also called semiopen) impellers have one shroud or backwall (Figure 7-4b), while open impellers consist only of vanes attached to a partial backwall or hub that is as small as possible (Figure 7-4c). In open and semienclosed impeller pumps the clearances between and around the vanes are large, allowing large solids to pass through the pump. These are used where high solids content liquid is being pumped, but efficiencies are low. The efficiency of semienclosed pumps are higher than open pumps. Totally enclosed impeller pumps have the highest efficiencies, but are not well suited for pumping solids or suspended material due to the close tolerances between and around the vanes. The impellers will wear quickly if used to pump water containing solids.

A third classification of impellers is that they

Figure 7-4. Centrifugal pump impellers: (a) Closed; (b) semiclosed; and (c) open.

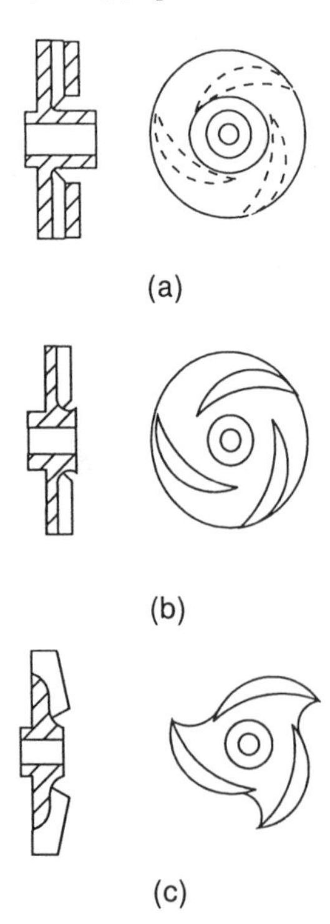

are either single- or double-suction. Water enters single-suction impellers from one side only Water flows symmetrically into both sides of a double-suction impeller. A double-suction impeller is essentially two single-suction impellers placed back-to-back in a single casing. They are usually preferred over single-suction impellers for pumping water containing suspended materials, for multi staged pumps, and for small pumps (James 1988).

Performance Parameters. A pump's performance is described by the following parameters: capacity, head, power, efficiency, required net positive suction head, and specific speed.

PUMP CAPACITY. Capacity, Q, is the volume of water delivered per unit of time by the pump In SI units Q is usually expressed in liters per minute (L/min) or cubic meters per second (m^3/s). In the English system the corresponding units are gallons per minute (gpm) and cubic feet per second (ft^3/s or cfs).

HEAD. System *head* is the net work done on a unit weight of water by the pump and is given by the following equation from James (1988):

$$H_2 = SL + DL + DD + h_m + h_f + h_o + h_v \quad (7\text{-}1$$

where H_s = system head, m (ft); SL = suction side lift, m (ft); DD = water source drawdown m (ft); h_m = system minor losses, m (ft); h_f = pipe friction losses, m (ft); h_o = operating head (pressure), m (ft); and h_v = velocity head (V^2/2g), m (ft). The suction- and discharge-side static lifts are measured when the system is not operating and are independent of system water flow. Drawdown (DD) is the decline of the water surface elevation of the source water due to pumping (such as well drawdown). Drawdown is extremely small in large bodies of water and can usually be neglected. However, DD can be considerable for wells and where water is pumped from a tank. DD, h_m, h_f, h_o and h_v all increase with increasing Q. Friction losses (h are independent of the pump except for losse occurring in the column pipe of vertical diffuser-type pumps.

POWER. The power required to operate a pump is directly proportional to discharge, head, and specific gravity of the fluid and is inversely proportional to the pump efficiency. The power imparted to the water by the pump is called *water horsepower* and is denoted by *WHP*. Equation 7-2 is used to compute water horsepower.

$$WHP = \frac{QHS}{K} \qquad (7\text{-}2)$$

where *WHP* = water horsepower, kW (hp); Q = pump capacity or discharge, m^3/s (cfs) or L/min (gpm); H = head, m (ft); S = fluid specific gravity, dimensionless; and K = unit constant (K = 6,116 for *WHP* in kW and Q in L/min; K = 0.102 for *WHP* in kW and Q in m^3/s; K = 3,960 for *WHP* in hp and Q in gpm; and K = 8.81 for *WHP* in hp and Q in cfs).

The specific gravity (S) of water is 1.0. Therefore, this term is usually ignored in the power equation. However, it must be remembered that the specific gravity of water is 1.0 at 20°C (68°F) and changes slightly for other temperatures. Therefore, in order to be more accurate the specific gravity term should be included in Equation 7-2 for water temperatures other than 20°C. S should be included when fluids other than water are being considered since its value will be some value other than 1.0. This is particularly true for seawater systems.

Another form of Equation 7-2 is

$$WHP = \frac{Q(TDH)}{3{,}960} \qquad (7\text{-}3)$$

where Q = pump discharge, gpm; and *TDH* = total dynamic head or the sum of all losses while the system is operating. *TDH* is normally reported in meters or feet.

Efficiency. Brake horsepower (*BHP*) is the power that must be applied to the shaft of the pump by a driving device such as an electric motor or an engine in order to turn the impeller and impart power to the water. *BHP* and *WHP* are related by an efficiency term:

$$E_p = 100\,\frac{WHP}{BHP} = \frac{output}{input} \qquad (7\text{-}4)$$

where E_p = pump efficiency, %; *WHP* = water horsepower, kW (hp); and *BHP* = brake horsepower, kW (hp).

Pumps are never 100% efficient. Energy losses that contribute to reduced efficiency are the friction in the bearings around the pump shaft, friction between the shaft and packing in the stuffing box, leakage, friction of the moving water against the metal pump housing, and pitting on the impeller and casing. For a centrifugal pump, efficiencies range from 25% to 85%. Older, worn pumps have lower efficiencies. Also, efficiency is lower in situations where a pump is not sized right for the job.

Example problem 7-1. What brake power is required to drive the system given if the pump efficiency is 85%?

Given: 1. Total vertical lift = 50 m
2. $Q = 0.025\ m^3/s$
3. $h_o = 0.52$ m
4. $h_v = 1.5$ m
5. $h_m + h_f = 5$ m

Solution: The total system head (Eq. 7-1) is

$$H_s = (SL + DL) + DD + (h_m + h_f) + h_o + h_v$$
$$= 50 + 0 + 5 + 0.52 + 1.5 = 57.02 \text{ m}$$

From Eq. 7-2

$$WHP = \frac{(0.025)(57.02)(1)}{6{,}166} \cong 14 \text{ kW}$$

Then, from Eq. 7-4 the power requirement is

$$\text{BHP} = \text{WHP/e} = \frac{14}{0.85} = 16.5 \text{ kW}$$

Suction Head. Certain general limitations are placed on pumps based on its suction-side flow conditions. An important point to remember when determining pump location is the elevation of the pump relative to the water source. The vertical distance from the static water surface to the centerline of the pump intake (for horizontal pumps) is the *static suction lift*, *SL*, as illustrated in Figure 7-5. This dimension is determined when the pump is not operating, that is, when there is no water flow. For vertical pumps *SL* is measured from the static water surface to the top of the pump's dicharge pipe rather than its centerline. Suction lift is positive if the pump is positioned above the water surface and negative if positioned below the water surface, as illustrated in Figure 7-6. The total suction head (H_s) is the sum of the static lift and the friction losses, minor losses and velocity head on the suction side of the pump. The relationship is expressed as

$$H_s = SL + (h_m + h_f) + \frac{V_s^2}{2g} \tag{7-5}$$

All components in the equation are in meters or feet. The subscript *s* denotes the suction side of the pump. Friction and minor losses are usually small in relation to suction lift and can often be ignored.

The water in the suction line of a pump is in tension. If the total suction head corresponds to a pressure reduction in the pump that is equal to or greater than the vapor pressure of the water,

Figure 7-5. Illustration of static lift with a centrifugal pump.

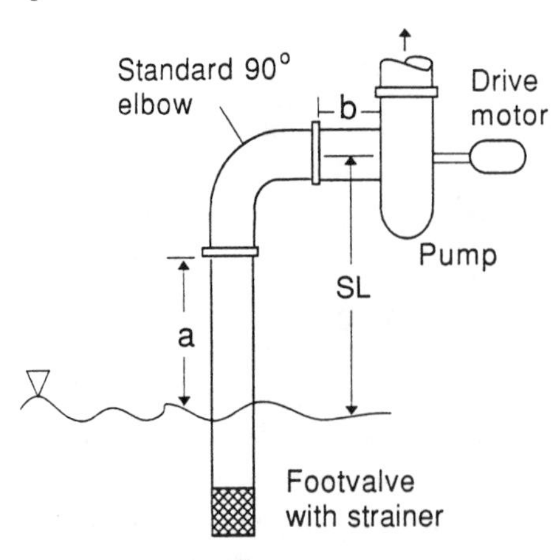

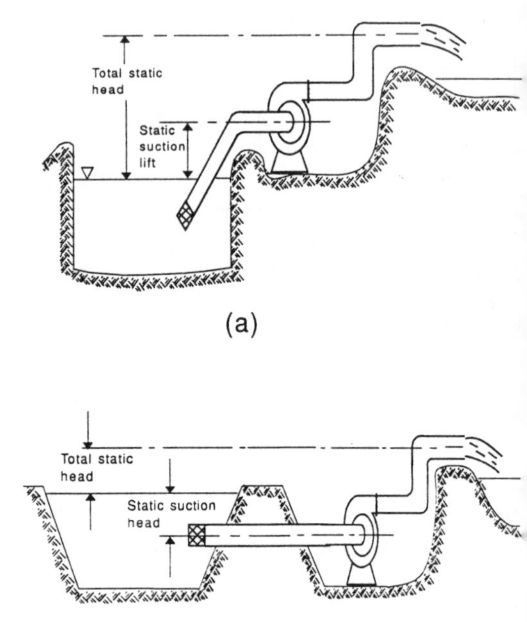

Figure 7-6. (a) Positive; and (b) negative static suction heads.

the water will turn to vapor (Simon 1981). If the water vaporizes small bubbles form inside the pump. The bubbles collapse when they reach regions of high pressure. Often the collapses occur with such violent force that stresses are induced in metal pump housings and impellers, and pitting occurs on their surfaces. This phenomena is called *cavitation*. Cavitation in a pump can be easily recognized since it sounds as though the pump were full of gravel. Cavitation causes losses in pump efficiency and fatigue damage if the condition is not corrected.

At mean sea level, the greatest theoretical suction lift possible is 10.6 m (34 ft) at a water temperature of 10°C (50°F) or lower. Pump manufacturers usually recommend that suction lifts be limited to 705 of the theoretical value, or about 7.4 m (23.8 ft) at MSL. At altitudes above MSL, and at higher water temperatures, the theoretical maximum suction lift is reduced. Table 7-1 lists maximum practical suction lifts at different elevations. For example, at 5,000 ft altitude and a water temperature of 80°F, the safe

Table 7-1. Practical static suction lift in meters (feet) [1].

Altitude		Temperature, °C (°F)					
m	ft	Under 10(50)[2]	15.6(60)	21.1(70)	26.7(80)	32.2(90)	37.8(100)
0	0	10.4 (34)	7.1 (23.4)	7.1 (23.2)	7 (23)	6.9 (22.6)	6.8 (22.2)
152.4	500	10.2 (33.4)	7.0 (23)	7 (22.8)	6.9 (22.5)	6.8 (22.2)	6.6 (21.8)
304.9	1,000	10 (32.7)	6.8 (22.4)	6.8 (22.3)	6.7 (22)	6.6 (21.8)	6.5 (21.4)
457.3	1,500	9.9 (32.1)	6.7 (22)	6.7 (21.9)	6.6 (21.6)	6.5 (21.4)	6.4 (20.9)
609.7	2,000	9.6 (31.5)	6.6 (21.6)	6.6 (21.5)	6.5 (21.2)	6.4 (20.9)	6.2 (20.5)
914.6	3,000	9.2 (30.3)	6.3 (20.8)	6.3 (20.6)	6.2 (20.4)	6.1 (20.1)	6.0 (19.7)
1,219.5	4,000	8.9 (29.2)	6.1 (20)	6.1 (19.9)	6 (19.6)	5.9 (19.3)	5.8 (18.9)
1,524.4	5,000	8.6 (28.1)	5.8 (19.2)	5.8 (19)	5.7 (18.8)	5.7 (18.6)	5.5 (18.1)
1,829.3	6,000	8.2 (27)	5.6 (18.5)	5.6 (18.3)	5.5 (18.1)	5.4 (17.8)	5.3 (17.4)
2,134.1	7,000	7.9 (26)	5.4 (17.8)	5.4 (17.6)	5.3 (17.4)	5.2 (17.1)	5.1 (16.7)
2,439.0	8,000	7.6 (25)	5.2 (17.1)	5.2 (16.9)	5.1 (16.7)	5 (16.4)	4.9 (16)

[1] 70% of maximum theoretical suction lift.

[2] Maximum theoretical suction lift at 10°C (50°F) and lower.

suction lift is 18.1 ft. (This is 70% of the maximum theoretical value).

Example problem 7-2. Given the pump system shown in Figure 7-5. Will the pump function properly under the conditions given?

Given: 1. Suction pipe length $(a + b)$ = 15 m

2. Suction pipe (smooth new iron) diameter = 20 cm

3. System altitude = 900 m above MSL

4. Maximum summertime water temperature = 37.8°C (100°F)

5. Desired pump discharge = 3,500 L/min

6. Static suction lift = 5 m

Solution: Flow velocity

$$V = Q/A = \frac{3.5}{(3.14)(0.1)^2\,(60)} = 1.86 \text{ m/s}(60)$$

Minor losses for fittings (from Table 6-3 and Eq. 6-28)

$$h_m = \frac{(10 + 0.9)(1.86)^2}{(2)(9.81)} = 1.92 \text{ m}$$

Friction losses for pipe (from Table 6-6 and Eq. 6-33)

$$h_f = \frac{(160.5)(0.00513)}{(8223)(0.00039)} = 0.26\text{m}$$

Suction head (from Eq. 7-5)

$$H_s = 5 + 1.92 + 0.26 + \frac{(1.85)^2}{(2)(9.81)} = 7.4 \text{ m}$$

Reference to Table 7-1 shows the maximum practical suction lift at 900 m above MSL to be about 6 m. Therefore, the system will not function as planned. The situation can be corrected by either decreasing the static lift or by using a larger-diameter suction pipe.

REQUIRED NET POSITIVE SUCTION HEAD. The net positive suction head relates the atmospheric pressure on the water on the suction side of the pump to the vapor pressure of the water. Water must enter the eye of single- or first-stage impellers under pressure in order for the pump to work. This pressure is the *required net positive suction head, $(NPSH)_r$*, and deals only with the suction side of the pump. If the pressure at the eye falls below the vapor pressure of the water the pump will cavitate. The net positive suction head required to prevent cavitation is strictly a function of pump design and is determined experimentally for each pump. To ensure that their pumps perform as specified manufacturers conduct laboratory tests to determine the $(NPSH)_r$ for each pump model they produce.

Once a pump is installed the engineer must make on-site calculations to verify that the net *available positive suction head*, $(NPSH)_a$, is greater than the net positive suction head required by the manufacturer. Thus,

$$(NPSH)_a \geq (NPSH)_r \qquad (7\text{-}6)$$

where $(NPSH)_a$ = net available positive suction head, m (ft) and $NPSH_r$ = required net positive suction head, m (ft). The $NPSH_a$ is determined by Equation 7-7 (Cuenca 1989):

$$(NPSH)_a = H_a - H_s - H_f - H_m - H_v \qquad (7\text{-}7)$$

where H_a = atmospheric pressure; H_s = static lift on suction side of pump; H_f = sum of pipe friction losses on suction side of pump; H_m = sum of minor pipe losses on suction side of pump; and H_v = fluid vapor pressure at the operating temperature. All components in Equation 7-7 have units of meters or feet.

Atmospheric pressure is commonly based on the Standard U. S. Atmosphere, which quantifies atmospheric pressure as a decreasing function of altitude (Cuenca 1989). Values for atmospheric pressure at different altitudes are shown in Table 7-2 or can be computed with Equation 7-8 (James 1988):

$$AP = K_1 - 0.00117h + K_2 h^2 \qquad (7\text{-}8)$$

where AP = atmospheric pressure, m (ft) of water; h = elevation above MSL, m (ft); and K_1, K_2 = unit constants (K_1 = 10.33 and $K_2 = 5.55 \times 10^{-8}$ for AP in meters of water and h in meters; K_1 = 33.89 and $K_2 = 1.69 \times 10^{-8}$ for AP in feet of water and h in feet).

Example problem 7-3. Calculate the $(NPSH)_a$ for the pump in example problem 7-2. The XYZ Pump Company offers a model ABC pump that requires a *NPSH* of 1.9 m. Will this pump do the job?

Solution: Atmospheric pressure H_a (from Table 7-2) = 9.3 m H_2O

Vapor pressure H_v (from Table 6-2 at 37.8°C) = 6,691 N/m^2 = 0.68 m H_2O

Thus, from Eq. 7-7

$$(NPSH)_a = 10.3 - 5 - 0.26 - 1.92 - 0.68 = 2.44 \text{ m } H_2O$$

which is greater than $(NPSH)_r$, therefore, this pump will do the job.

Table 7-2. Standard U.S. atmospheric pressure at different altitudes.

Altitude		Atmospheric Pressure				
m	ft	mm Hg[1]	kPa abs.	psia	m H_2O[2]	ft H_2O[2]
−200	−656.2	778	103.7	15.0	10.6	34.8
0	0	760	101.3	14.7	10.3	33.8
200	656.2	742	98.9	14.3	10.1	33.1
400	1,312.3	725	96.6	14.0	9.9	32.5
600	1,968.5	707	94.3	13.7	9.6	31.5
800	2,624.7	690	92.0	13.3	9.4	30.8
1,000	3,280.8	674	89.8	13.0	9.2	30.2
1,200	3,937.0	657	87.6	12.7	8.9	29.2
1,400	4,593.2	641	85.5	12.4	8.7	28.5
1,600	5,249.3	626	83.4	12.1	8.5	27.9
1,800	5,905.5	611	81.4	11.8	8.3	27.2
2,000	6,561.7	596	79.4	11.5	8.1	26.6
3,000	9,842.5	526	70.1	10.2	7.1	23.3
4,000	13,123.4	462	61.6	8.9	6.3	20.7

[1] Mercury at 0°C.

[2] Water at 4°C.

If the water source is not exposed to the atmosphere, the atmospheric pressure term in Equation 7-7 must be replaced by the absolute pressure at the water surface. This would be the case, for example, when water is pumped from a pressurized tank. The vapor pressure of the water varies with temperature. Vapor pressures for water at different temperatures can be obtained from Table 6-2.

SPECIFIC SPEED. Specific speed (N_s) consolidates a pump's speed, design capacity, and head into one term. Specific speed is found from Equation 7-9 (Cuenca 1989):

$$N_s = 0.2108 \frac{NQ^{0.5}}{H^{0.75}} \tag{7-9}$$

where N_s = specific speed, dimensionless; N = pump speed (rpm); Q = pump discharge (L/min); and H = design head (m).

Geometrically similar pumps have similar performance characteristics. They also have identical specific speeds, regardless of size (James 1988). Specific speed ranges from about 500 for a centrifugal pump to 10,000 for a propeller pump.

Pump Performance Curves. Performance characteristics are data, usually presented in graphical form that relate the major pump characteristics of head, efficiency, power requirements, and *(NPSH)*$_r$ to capacity. This set of curves is known as *characteristic curves*. Each pump has its own unique set of characteristic curves established by the pump's geometry and dimensions of the impeller and casing. These characteristics change as the pump ages and parts wear. Characteristic curves for a typical single-stage pump are illustrated in Figure 7-7. Pump manufacturers normally publish a set of characteristic curves for each model pump they offer. Data for the curves are developed by testing several pumps of a specific model. Some curves represent the average performance of all pumps of a particular model tested. Others are prepared for the pump having the poorest performance (James 1988).

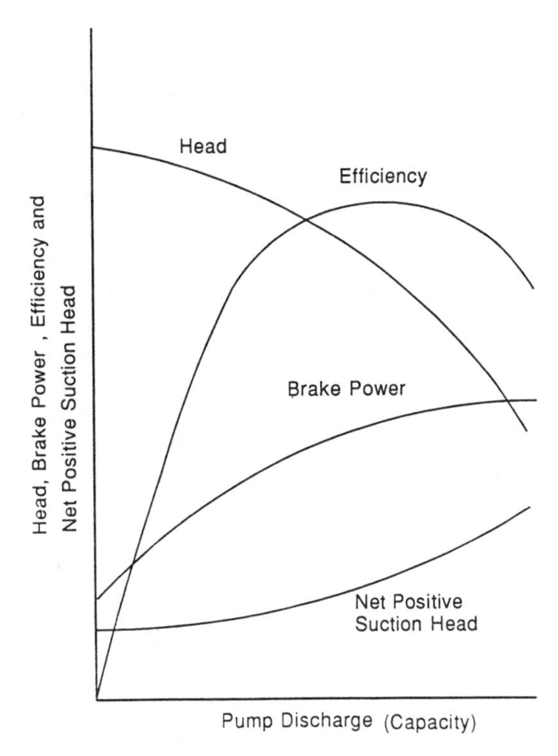

Figure 7-7. Characteristic curves for a single-stage centrifugal pump.

HEAD VERSUS CAPACITY. The head versus capacity (*H-Q*) curve relates the pressure head produced by the pump to the volume of water being pumped per unit of time. The head steadily decreases as the volume of water being pumped increases. The shape of the *H-Q* curve varies with specific speed (N_s). As Figure 7-7 illustrates, the pump produces the maximum head at zero discharge. Manufacturers determine the maximum head of a pump by closing down a valve on the discharge side of the pump until no water flows. The maximum head-zero discharge point is called the *dead head* or *shut-off head* (Cuenca 1989). This value must be known in order to design piping systems on the discharge side of the pump. Piping on the discharge side must be able to withstand the shut-off head when the valve is closed. Note that pump efficiency is zero at the shutoff head.

EFFICIENCY VERSUS CAPACITY. The efficiency versus capacity (E_p-Q) curve for a typical pump is also illustrated in Figure 7-7. E_p steadily increases with an increase in Q, reaches a maximum at a particular combination of discharge and head, and then declines below the maximum as Q continues to increase. The shape and magnitude of the curve vary depending on the pump design and the impeller diameter. The E_p-Q relationship is often shown as a family of curves for different impeller diameters (Figure 7-8).

E_p-Q curves are usually drawn for a specific number of stages. If multistages are needed for a particular application, efficiencies must be adjusted upward or downward depending on the number of stages. Manufacturers usually provide guidelines for making these adjustments.

BRAKE POWER VERSUS CAPACITY. The *BHP*-Q curve is derived from the pump's *H*-Q and E_p-Q curves. Brake horsepower (*BHP*) can be computed by solving Equation 7-4 for *BHP* and substituting Equation 7-2 for *WHP*, resulting in Equation 7-10:

$$BHP = \frac{100(WHP)}{E_p} = \frac{100QHS}{E_pK} \tag{7-10}$$

where K is the unit constant from Equation 7-2.

The shape of the *BHP*-Q curve depends on the pump's specific speed and impeller design. For radial flow impellers, *BHP* generally increases from some non zero value and reaches a maximum, after which it declines with increasing Q. For axial flow impellers *BHP* is maximum when Q is zero and steadily declines as Q increases. For mixed flow impellers *BHP* increases steadily from some non-zero value as Q

Figure 7-8. Example pump characteristic curves from a pump manufacturer's sales literature.

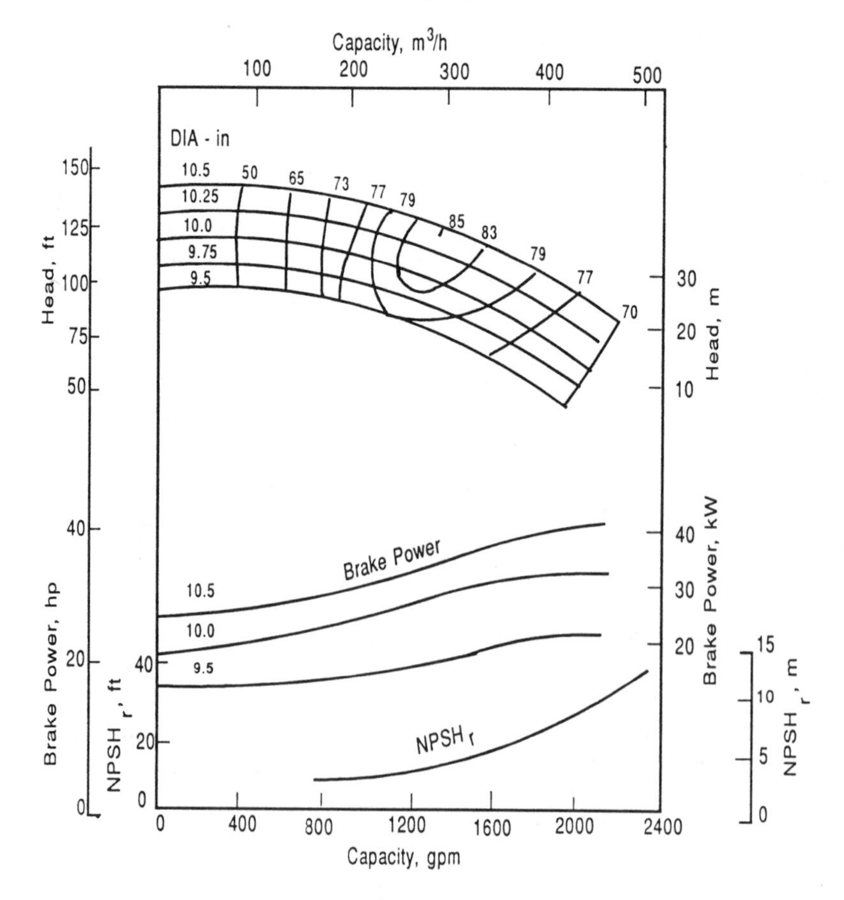

increases. To minimize start-up load on axial flow pumps, the discharge side of the pump should be open to the atmosphere. Similarly, the discharge side valve should be closed when starting radial and mixed flow pumps (James 1988).

$(NPSH)_r$ VERSUS CAPACITY. Figure 7-7 illustrates that $(NPSH)_r$ steadily increases as Q increases for a typical radial flow pump.

Affinity Laws. A pump's characteristic curves are altered if the impeller diameter and/or speed is changed. Thus, a single-impeller design may be used for a variety of head and discharge conditions. It also allows pump owners to alter pump performance to better match changes in their pumping systems.

CHANGES IN PUMP SPEED. Changes in a pump's performance resulting from changes in speed can be computed using the following equations:

Discharge $$Q_2 = Q_1\left(\frac{N_2}{N_1}\right) \quad (7\text{-}11)$$

Head $$H_2 = H_1\left(\frac{N_2}{N_1}\right)^2 \quad (7\text{-}12)$$

Brake power $$BHP_2 = BHP_1\left(\frac{N_2}{N_1}\right)^3 \quad (7\text{-}13)$$

$(NPSH)_r$ $$(NPSH)_{r,2} = (NPSH)_{r,1}\left(\frac{N_2}{N_1}\right)^2 \quad (7\text{-}14)$$

where the subscripts 1 and 2 refer to initial conditions and new conditions, respectively.

Example problem 7-4. A pump operating at 1,750 rpm discharges 1,000 gpm against a head of 300 ft. The original $(NPSH)_r$ is 12 ft and the original BHP is 100 hp. Calculate the new Q, H, BHP, and $(NPSH)_r$ for a new pump speed of 2,000 rpm.
Solution: Using Equations 7-10 through 7-13

Q_2 = (1000)(2000/1750) = 1,143 gpm
H_2 = (300)(2000/1750)^2 = 392 ft
BHP_2 = (100)(2000/1750)^3 = 149 hp
$(NPSH)_{r,2}$ = (12)(2000/1750)^2 = 15.7 ft

CHANGES IN IMPELLER DIAMETER

Discharge $$Q_2 = Q_1\left(\frac{D_2}{D_1}\right) \quad (7\text{-}15)$$

Head $$H_2 = H_1\left(\frac{D_2}{D_1}\right)^2 \quad (7\text{-}16)$$

Brake power $$BHP_2 = BHP_1\left(\frac{D_2}{D_1}\right)^3 \quad (7\text{-}17)$$

$(NPSH)_r$ $$(NPSH)_{r,2} = (NPSH)_{r,1}\left(\frac{D_2}{D_1}\right)^2 \quad (7\text{-}18)$$

Since changing the diameter of an impeller may alter its geometry, Equations 7-15 through 7-18 give approximate changes in pump performance. These equations are most reliable when diameter changes are less than 20% (James 1988).

Example problem 7-5. Given the pump in the previous example, having an original impeller diameter of 8 in., calculate the new Q, H, BHP, and $(NPSH)_r$ for an impeller diameter of 7.5 in.
Solution: Using Eqs. 7-14 through 7-17

Q_2 = (1,000)(7.5/8) = 937 gpm
H_2 = (300)(7.5/8)^2 = 264 ft
BHP_2 = (100)(7.5/8)^3 = 82 hp
$(NPSH)_{r,2}$ = (12)(7.5/8)^2 = 10.5 ft

Changes in Pump System Configuration. Water flow requirements of aquacultural systems at times necessitate the combination of pumps to achieve certain discharges or pressures. Pumps can be operated in series or in parallel to produce the design requirements. When pumps are located close together, such as in a common pumping station, only the $(NPSH)_r$ needs of the first pump upstream must be provided. However, if pumps are widely separated, the $(NPSH)_r$ of each individual pump should be provided (James 1988).

PUMPS OPERATING IN SERIES. Two or more pumps can be used in series by linking them so that the upstream pump discharges directly into the intake of the following pump, as shown in Figure 7-9. Series hookups are used when the discharge flow rate from a single pump is adequate but a larger system pressure head is desired than can be produced by a single pump. Figure 7-10 illustrates the individual H-Q curves for pumps A and B and the combined H-Q curve for the pumps operating in series. The combined curve is obtained by adding the individual heads of pumps A and B at a constant flow rate.

Considering the two pumps, A and B, with discharge Q, output heads H_A, and H_B, and power requirements P_A and P_B. The following equations govern the series system (Cuenca 1989):

$$Q_{series} = Q_A = Q_B \tag{7-19}$$

$$H_{series} = H_A + H_B \tag{7-20}$$

$$P_{series} = P_A + P_B \tag{7-21}$$

Equations 7-19 and 7-20 show that the discharge of the two pumps is the same, and the head and power of the two is additive. The combined pump efficiency is evaluated considering the H_{series} and P_{series} in Equation 7-22:

$$E_{series} = \frac{Q_{series}\,(H_A + H_B)}{K(P_A + P_B)} \tag{7-22}$$

where Q_{series} = discharge, m^3/s (ft^3/s); H_A and H_B = discharge pressure head, m (ft); and P_A and P_B = power requirement, kW (hp).

Figure 7-9. Two identical centrifugal pumps operating in series.

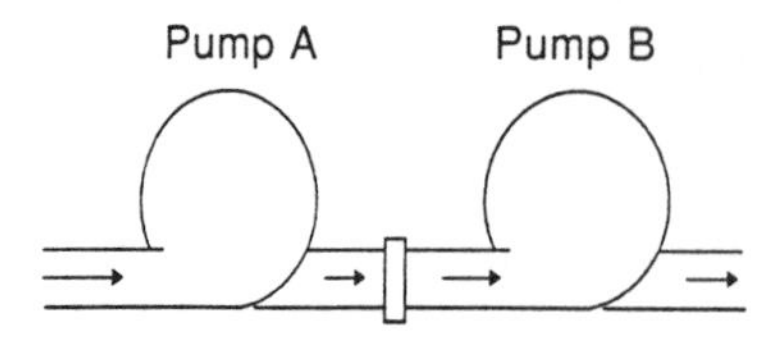

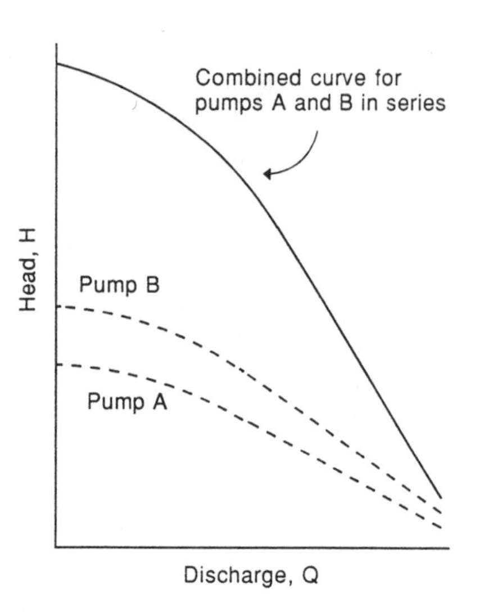

Figure 7-10. Individual H-Q curves for pumps A and B and combined H-Q curve for pumps A and B operating in series.

Example problem 7-6. Calculate the discharge pressure head, total power required and overall pump system efficiency for two pumps (Model 123A and Model 456B) when the pumps are linked in series. The pumps are operated at 1,750 rpm, and the system design discharge is 0.0473 m^3/s (750 gpm). The impeller diameters are 25.4 cm (10 in.) for the Model 123A and 30.5 cm (12 in.) for the Model 456B.

Solution:

Model 123A	Model 456B
H_A = 26.8 m (88 ft)	H_B = 33.5 m (110 ft)
E_A = 83 %	E_B = 78 %
$P_A = \dfrac{(0.0473)(26.8)}{(0.0102)(0.83)}$	$P_B = \dfrac{(0.0473)(33.5)}{(0.0102)(0.78)}$
P_A = 15 kW	P_B = 19.9 kW

Then, for the pumps installed in series,

$$Q_{series} = Q_A = Q_B = 0.0473 \text{ m}^3/\text{s}$$

$$H_{series} = H_A + H_B = 26.8 + 33.5 = 60.3 \text{ m}$$

$$P_{series} = P_A + P_B = 15 + 19.9 = 34.9 \text{ kW}$$

$$E_{series} = \frac{(0.0473)(60.3)}{(0.102)(34.9)} = 0.80 = 80\%$$

PUMPS OPERATING IN PARALLEL. Pumps can be operated in parallel to increase the total discharge while maintaining a constant head. The discharge of identical pumps operating in parallel is additive, but the head remains unchanged. Pumps are used in parallel in situations where the pumps remove water from a common source and discharge it to a single outlet, as shown in Figure 7-11. This is a common method of meeting variable discharge requirements, since P and energy use are minimized by operating only those pumps needed to meet a given demand. Figure 7-12 illustrates the individual and combined H-Q curves for pumps A and B operating in parallel. The combined curve is obtained by adding the individual pump discharges. A design system discharge pressure head H is selected, and the discharge for each pump in parallel is read off of its respective pump performance curve at the given head. The overall system power requirement is the sum of the power required for each pump at the same head.

The governing equations for two pumps A and B operating in parallel are given by the following equations (Cuenca 1989):

$$H_{para} = H_A = H_B \tag{7-23}$$

$$Q_{para} = Q_A + Q_B \tag{7-24}$$

$$P_{para} = P_A + P_B \tag{7-25}$$

Figure 7-11. Two identical centrifugal pumps operating in parallel.

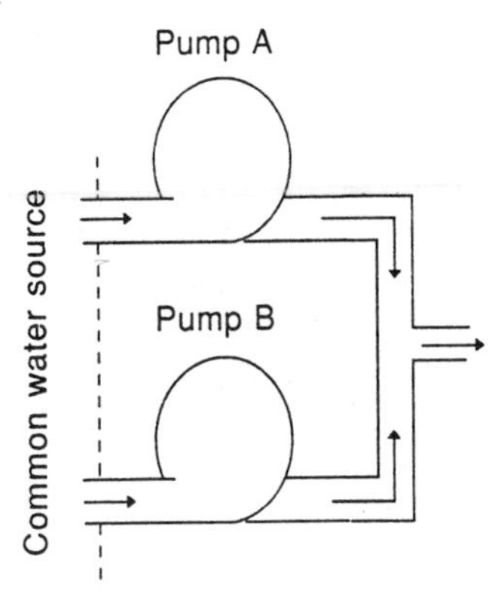

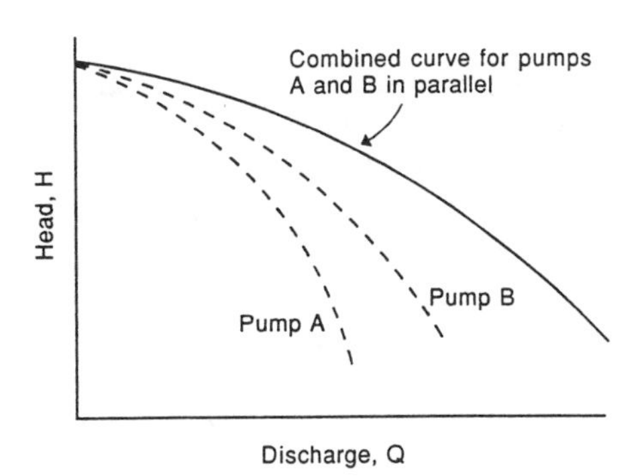

Figure 7-12. Individual H-Q curves for pumps A and B and combined H-Q curve for pumps A and B operating in parallel.

The overall system efficiency is then computed considering that the individual pumps operate at the same head:

$$E_{para} = \frac{(Q_A + Q_B)\, H_{para}}{K(P_A + P_B)} \tag{7-26}$$

Example problem 7-7. Compute the system discharge, power requirements and efficiency if the pumps in the previous example are operated in parallel against a head of 27.4 m (90 ft).

Solution:

Model 123A	Model 456B
$Q_A = 0.0473\ m^3/s$	$Q_B = 0.052\ m^3/s$
$E_A = 83\ \%$	$E_B = 73\ \%$
$P_A = \frac{(0.0473)(27.4)}{(0.0102)(0.83)}$	$P_B = \frac{(0.0473)(27.4)}{(0.0102)(0.73)}$
$P_A = 15.3$ kW	$P_B = 19.1$ kW

Then, for the pumps installed in parallel,

$$Q_{para} = Q_A + Q_B = 0.0473 + 0.052 = 0.0993\ m^3/s$$

$$H_{para} = H_A = H_B = 27.4\ m$$

$$P_{para} = P_A + P_B = 15.3 + 19.1 = 34.4\ kW$$

$$E_{para} = \frac{(0.0993)(27.4)}{(0.102)(34.4)} = 0.78 = 78\%$$

System Operating Point. The system operating point, where the pump and the system function at their greatest efficiency, is defined by the pump system performance curves. Where these two curves intersect is the system operating point. This is the point at which the *H-Q* requirements of the fluid system balance the *H-Q* requirements of the pump. The pump performance curve is established as discussed previously. The system operating curve, which describes the *H-Q* requirements of the system, is constructed by computing heads required by the piping system to deliver different volumes of water per unit time. It is the responsibility of the engineer to match the system and pump performance curves so that the system functions at its designed capacity and pressure.

The system head (H_{sys}) is composed of two parts: (1) The fixed system head, H_{fix}; and (2) the variable system head, H_{var}. These are summed to give the system head according to the equation

$$H_{sys} = H_{fix} + H_{var} \tag{7-27}$$

The fixed system head does not vary with Q. It is made up of the difference in elevation between the static water surface and the discharge point. The fixed system head is given by

$$H_{fix} = H_s + H_d \tag{7-28}$$

where H_s = static suction lift, m (ft); and H_d = discharge-side lift, m (ft). This relationship is illustrated graphically in Figure 7-13.

Lift on the discharge side of the pump, H_d, is found using Equation 7-29 (James 1988):

$$H_d = Elev_u - Elev_p \tag{7-29}$$

where $Elev_u$ = elevation of water surface at point of delivery, m (ft); and $Elev_p$ = elevation of pump discharge as defined for horizontal and vertical pumps, m (ft).

The variable system head, H_{var}, is given by

$$H_{var} = H_{dd} + H_f + H_m + H_p + H_{vel} \tag{7-30}$$

where H_{dd} = water source drawdown, m (ft); H_f = pipe frictional losses, m (ft); H_m = minor losses through pipe fittings, m (ft); H_p = operating pressure, m (ft); H_{vel} = velocity head = $V^2/2g$; V = fluid velocity, m/s (ft/s); and g = acceleration of gravity, 9.81 m/s^2 (32.2 ft/s^2).

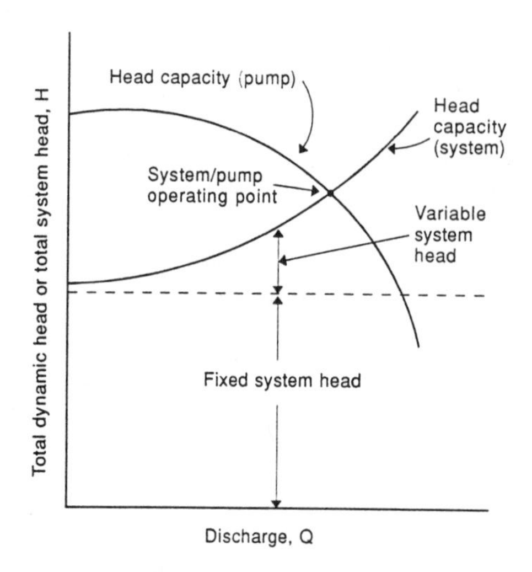

Figure 7-13. Matching pump and system operating curves by observing point of intersection.

The variable head increases with increases in Q. It is made up of the well drawdown (if the water source is a well), friction and minor losses in the pipeline, pressure at the outlet, and the velocity head. If the water discharges into the atmosphere, the outlet pressure is zero. If the water discharges to a closed tank, the pressure is not zero but is equal to the absolute pressure inside the tank.

Water surface drawdown is the difference in elevation between the static water surface when the pump is not operating and the water surface during pumping. Calculations for determining well drawdown and yield are considered beyond the scope of this text.

The sum of the fixed and variable system operating heads is the total dynamic head, *TDH*, given by

$$TDH = H_{fix} + H_{var} \tag{7-31}$$

Deep Well Turbine Pumps. Deep well turbine pumps are multi-staged centrifugal pumps that are specially manufactured so that one pump bowl (the housing around the impeller) bolts directly onto the other (Figure 7-14). They are either of the volute type or diffuser type. This is a special case of pumps operating in series. Theoretically, single-stage centrifugal pumps cannot lift water a greater distance than

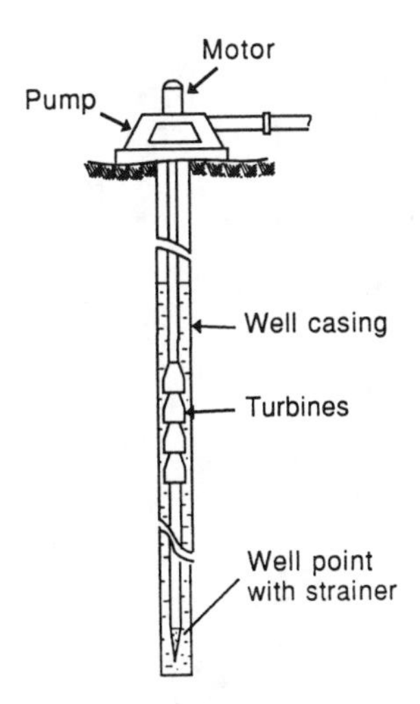

Figure 7-14. A deep well turbine pump with the motor located at the surface.

10.5 m (34 ft). By multistaging the impellers, the discharge of the first impeller becomes the intake of the next, and so on, and the pressure head is increased with each successive stage. Therefore, water can be lifted from any depth, depending upon the number of stages used. The number of stages used is determined by the discharge and head requirements of the system.

There are two types of deep well pumps: (1) Pump power source at the surface; and (2) power source submerged. In the case of an electric motor or engine at ground level, the pump impeller(s) are located at the bottom of a well shaft and are connected to the power source with a long shaft. The shaft must be straight and not wobble or efficiencies will be low and energy requirements high. The weight of the shaft usually limits the depth of these wells to about 305 m (1,000 ft) (Wheaton 1977).

In the case of submersible deep well turbine pumps, the pump and drive motor are both located at the bottom of the well. Only a cable connects the pump to available electric service. These are used for wells 10 cm (4 in.) in diameter or greater. Well depth is usually not a problem. Some wells of this type exceed 2,500 m (8,000 ft) in depth (Wheaton 1977).

Low-Capacity Centrifugal Pumps. An array of small centrifugal pumps (less than 3.7 kW or 5 hp) are available to perform a variety small aquaculture jobs, as shown in Figure 7-15. They are available in both submersible and nonsubmersible models. Most small submersible pumps are designed to operate submerged only since the surrounding water functions to prevent the pump motors from overheating. Others may be operated either submerged or nonsubmerged. These pumps can be used for aquaria, small aquaculture systems, outdoor fish pools, fountains, etc. Small-model off-the-shelf pumps cannot be modified. That is, different-sized impellers cannot be interchanged, and the impeller rotational speeds are fixed. Many small pumps have sealed housings so that it is difficult to repair them or change performance characteristics. They are usually inexpensive enough so that they can simply be discarded and replaced when they malfunction. Since the performance characteristics of these small pumps cannot be changed, one must have a detailed analysis of pump requirements in terms of system head and discharge requirements in order to select a pump.

Usually only *H-Q* curves are available for fractional horsepower pumps. However, dealer's catalogs normally contain simple graphs and tables, such as those illustrated in Figure 7-16 and Table 7-3, for selecting pumps. One must be careful to check whether a particular pump model is designed for continuous or intermittent duty. Also, some models are designed specifically for freshwater use while others can operate in fresh- or seawater.

Power Sources For Pumps. A pumping plant consists of a pump and a power source for the pump. The overall efficiency of a pumping plant depends upon the efficiency of the pump, the power source efficiency, and the efficiency of the drive unit connecting the two.

POWER SOURCE. Electric motors are probably the most frequently used power source for pumps. They are more efficient than internal combustion engines, they are small and lightweight, they operate cleanly and quietly, and they can be mounted either in a horizontal or vertical position. Electric motors are relatively

Figure 7-15. Fractional horsepower centrifugal pumps (photo: T. Lawson).

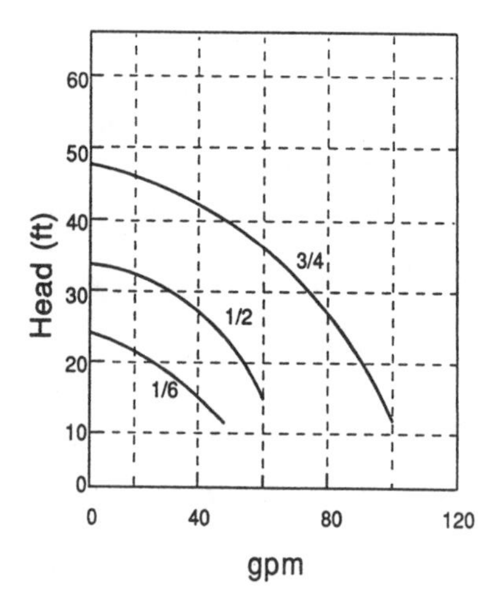

Figure 7-16. *H-Q* curves for small centrifugal pumps from a manufacturer's catalog.

Table 7-3. Catalog information for fractional horsepower centrifugal pump.

Part Number	Description	hp	Shipping weight (lb)	Price
X1	pump, centrifugal	1/6	18	$175.00
X2	pump, centrifugal	1/2	18	$185.00
X3	pump, centrifugal	3/4	20	$199.00

maintenance-free and can be loaded up to 100% of their rated output so long as they are rated for continuous duty. If the estimated motor size falls between standard motor sizes, the next larger size should be selected. For example, if a 17.7-hp motor is required, then a 20- or 25-hp motor should be chosen. Motors typically are not available in fractional horsepowers over 5.6 kW (7.5 hp). Vertical, squirrel-cage induction-type or synchronous-type constant speed motors are the most common types used for pump drives. Propeller pumps require high starting torque. Therefore, the available line voltage must be checked with the local utility company. If the line voltage is sustained under 90% or over 110% of the voltage on the pump motor nameplate, then a special motor is required (ASAE 1985).

Single-phase motors are usually limited to loads up to 7.5 kW. Above 7.5 kW three-phase power is necessary. Three-phase motors are inherently more efficient than single-phase motors. Three-phase motors larger than 22 kW (30 hp) will operate above 90% efficiency. However, three-phase power is expensive to install and is not available in many rural areas. In some cases, three-phase power converters can be used to provide three-phase power with balanced current from single-phase power lines.

Electric motors vary in efficiency ranging about 85–92%. When sizing new systems, an efficiency of about 90% can be assumed. Older, used motors lose efficiency slightly, therefore a lower efficiency of about 85% can be assumed. Efficiency decreases as bearings and packing wears and shafts become misaligned, and through faulty wiring or improper installation. Electric motor efficiency should be entered into the brake horsepower equation (Eq. 7-10) as a second efficiency term:

$$BHP = \frac{100(Q)(H)(S)}{K(E_p)(E_m)} \qquad (7\text{-}32)$$

Unfortunately, electric power is not available in many rural areas. Therefore, an alternative power source is the internal combustion engine. The most commonly used engine type is the diesel. Diesel engines operate more efficiently than other internal combustion engines, and diesel fuel is generally less expensive than other fuels. Also, diesel fuel is usually more readily available than natural gas or liquid petroleum (LP) gas in many areas. However, the purchase price of a diesel engine is usually higher than other engines. Therefore, a diesel must be used a minimum of 800 to 1,000 hours annually to justify the additional cost. Table 7-4 lists operational efficiencies of internal combustion engines. This efficiency factor should be used in place of electric motor efficiency in Equation 7-32.

Power sources operate under a continuous load while the pump is in operation, but an engine cannot be operated continuously at its full-rated power without suffering damage. Therefore, internal combustion power sources must

Table 7-4. Relative efficiencies of electric motors and internal combustion engines.

Power unit	Efficiency, %
Electric motor	85–92
Gasoline engine	20–26
Natural and LP gas engine	20–26
Diesel engine	25–37

Source: Baker and Bankston (1988).

be selected based on their continuous service rating rather than their maximum brake horsepower rating. The continuous *BHP* available depends upon the engine type, and engines are derated accordingly: 40% for air-cooled gasoline engines, 30% for water-cooled gasoline engines, and 20% for diesels (Wheaton 1977). A conservative assumption for natural gas and LP gas engines would be a 40% derating of their maximum *BHP*. Thus, the *BHP* calculated using Equation 7-32 is the engine's derated power and not the full-rated power.

When the full-rated power for an engine is given, it should be derated for continuous duty using the engine derating factors in Table 7-5. The continuous power can then be compared to the required continuous BHP calculated with Equation 7-32 to determine if the engine has enough power to perform the job intended.

DRIVE UNIT. Drive units or drive heads are required between the power source and pump. Loss of efficiency through drive units varies from none for direct drives to about 5% for gear connections and up to 10% for multiple V-belt and flat-belt connections (ASAE 1985). Efficiencies for various drive units are shown in Table 7-6. The drive unit adds a third efficiency term to the power equation:

$$BHP = \frac{100(Q)(H)(S)}{K(E)_p(E_m)(E_d)} \qquad (7\text{-}33)$$

Example problem 7-8. Given the pump in example problem 7-1. (a) What size electric motor is required to drive the pump (direct drive)? (b) What size diesel engine if a flat-belt drive is

Table 7-5. Derated engine for pumping.

Type of service	Deduct from maximum brake horsepower (%)
Continuous load	20
For each 1,000 ft elevation above sea level	3
For each 10°F rise in ambient temperature above 60°F	1
Accessories (generator, air cleaner, water pump-heat exchanger, etc)	5
Fan and radiator are used	5
Right-angle drive (if not used to calculate water horsepower)	3

Note: The derating indicated is for diesel engines only. Derate other internal combustion engines as follows: air-cooled gasoline engine, 40%; water-cooled gasoline engine, 30%; natural gas and LP gas engines, 40%.

Source: Baker and Bankston (1988).

Table 7-6. Efficiency of drive units.

Drive unit	Efficiency, %
Direct drive	almost 100
Gear head (right-angled)	90–95
V-belt drive	90–95
Flat-belt drive	80–95

Source: Baker and Bankston (1988).

used? Assume the engine is water-cooled and is operated at an ambient air temperature of 32°C (90°F).

Solution:

(a) The pump *BHP* requirement is 16.5 kW (from example problem 7-1). Assuming an electric motor efficiency of 90% (Table 7-4) and a drive efficiency of 100% (Table 7-6), the electric motor power requirement is

$$BHP = \frac{16.5 \text{ kW}}{(0.90)(1.00)} = 18.3 \text{ kW (24.6hp)}$$

Therefore, select an 18.6 kW (25-hp) electric motor.

(b) Assume a new diesel engine efficiency of 30% (Table 7-4) and a belt drive efficiency of 90% (Table 7-6). The engine should be derated from its maximum BHP rating by 20% for continuous operation (Table 7-5). The engine should also be derated 5% for a fan and radiator and 1% for each degree rise in ambient air temperature above 60°F (Table 7-5). Thus, the diesel engine maximum *BHP* is

$$BHP = \frac{16.5 \text{ kW}}{(0.30)(0.90)(0.72)} = 85 \text{ kw (114 hp)}$$

Pumping Costs. Pumping costs can be determined knowing the type of power source, type of drive, fuel cost, and energy factor. Operating costs for internal combustion engines can be estimated using Equation 7-34 and efficiencies and thermal energy factors obtained from Tables 7-4 through 7-7. If a component is not to be used (e.g., a generator or an electric motor), then that component should be removed from Equation 7-34. Equation 7-35 can be used to estimate electric power costs.

$$\text{\$/hr to operate internal combustion engine} = \frac{(\textit{energy factor})(\textit{BHP})(\textit{fuel cost in \$/unit of fuel})}{(\textit{power source eff.})(\textit{drive eff.})(\textit{generator eff.})(\textit{electric motor eff.})} \quad (7\text{-}34)$$

$$\textit{\$/hr to operate from electric power} = \frac{0.746(\textit{BHP})(\textit{energy cost in \$/kW-hr})}{(\textit{electric motor efficiency})(\textit{drive efficiency})} \quad (7\text{-}35)$$

Example Problem 7-9. The *BHP* requirement for a water pump is calculated to be 28.65 hp. Calculate the cost to operate: (a) A used diesel engine sized properly (80% of maximum rated *BHP*, 22% efficiency), right angle gear drive (90% efficiency), diesel cost = \$0.90 per gallon; (b) used diesel engine oversized (10% efficiency), right angle gear drive; and (c) electric motor (90% eff.), direct couple drive (100% efficiency), cost of electricity = \$0.08 per kWhr.

Solution:

(a) (From Eq. 7-34)

Table 7-7. Typical thermal efficiencies for electrical and internal combustion power units.

Power type	Efficiency, %	Energy factor	Unit of fuel
Diesel engine:		0.01887	gallon
New, matched to load	25–37		
Used	18–25		
Oversized	7–15		
Gasoline engine	7–28	0.02054	gallon
Natural gas engine	7–25	0.02547	CCF[1]
LP gas (propane) engine	7–25	0.02768	gallon
Electric motor	85–92		kW-hr[2]
Generator conversion efficiency	75–85		

Source: Baker and Bankston (1988).

[1] 100 ft^3.

[2] Kilowatt-hour.

$$\text{Cost} = \frac{(0.01887)(28.65)(0.90)}{(0.22)(0.90)} = \$2.46/\text{hr}$$

(b) $$\text{Cost} = \frac{(0.01877)(28.65)(0.90)}{(0.10)(0.90)} = \$4.84/\text{hr}$$

(c) (From Eq. 7-35)

$$\text{Cost} = \frac{(0.746)(28.65)(0.08)}{(0.90)(1.00)} = \$1.90/\text{hr}$$

Rotary Pumps

Rotary pumps consist of a casing and a rotating member or members that force the fluid to move. There are several types of rotary pumps, however, the only type that will receive mention here is the propeller pump, since, next to the centrifugal pump, it is the most common type of mechanical pump used in aquaculture. For a discussion of other types of rotary pumps, the reader is referred to Wheaton (1977).

The most simple propeller pump has a rotating element similar to a boat propeller. The propeller is located inside of a pipe or shaft that serves as a conduit to convey the fluid, and a power source turns the propeller. Figure 7-17 illustrates propellers shaped like fan blades. The rotating shaft may be mounted either in the horizontal (Figure 7-17a) or vertical (Figure 7-17b) position. Propeller pumps are best suited for conditions of low head and high discharge. The *H-Q* curves tend to be flatter for propeller pumps than they are for centrifugal pumps. Power requirements increase as discharge decreases and head increases. Propeller pumps may be staged to increase head requirements.

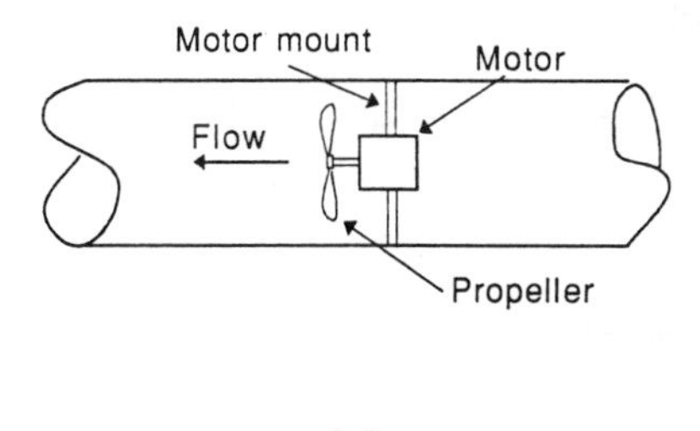

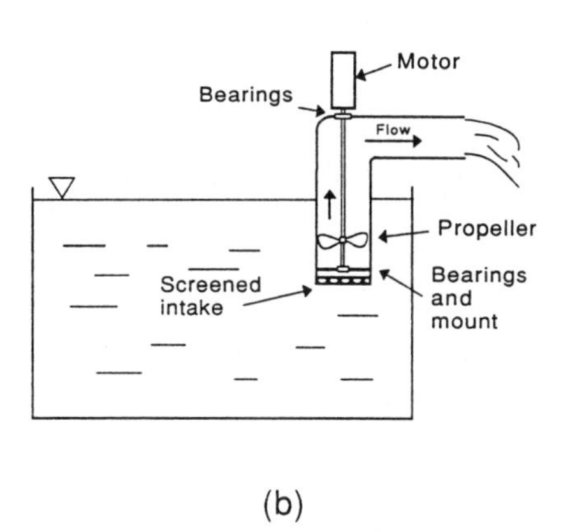

Figure 7-17. Propeller pumps: (a) Horizontal; and (b) vertical position.

Large propeller pumps commonly operate at slow speeds (100–300 rpm) to prevent cavitation (Wheaton 1977).

Air Lift Pumps

Next to centrifugal pumps, air lift pumps are probably the most common type of pump used in the aquaculture industry. Air lift pumps are described by Wheaton (1977) and Spotte (1979, 1992). An air lift pump uses a rising column of air to generate flow in a liquid system. The most common type air lift consists of an open-ended tube or pipe that is partially submerged in fluid into which air is injected (Figure 7-18). Air lift pumps operate due to the difference in specific gravity between the fluid on the outside and the air-fluid mixture on the inside of the tube. If no air is injected the pressure of the surrounding liquid will cause the fluid to rise to level *S*. Air injection into the tube causes the specific gravity of the fluid mixture in the tube to be lowered. Thus, Equation 7-36 describes the column of fluid:

$$(S + L)\gamma_m = S\gamma_o \tag{7-36}$$

where S = submergence depth of liquid-air mixture in tube; L = lift or height of liquid-air mixture above water surface; γ_m = specific gravity of liquid-air mixture; and γ_o = specific gravity of liquid outside tube.

Since air is less dense than water, the specific gravity of the liquid-air mixture will be less than that of the liquid alone, and $(S + L)$ must be larger than S for Equation 7-36 to remain in equality. If sufficient air is continuously injected into the tube, the specific gravities will be sufficiently unbalanced such that $(S + L)$—S will exceed L, which is the pump lift, and the mixture will discharge from the top of the tube. So long as sufficient air is flowing, pumping will continue.

One of the main factors affecting the efficiency of an air lift is the *submergence* of the lift tube. Submergence is the percentage of the overall length of the lift tube beneath the surface of the liquid, expressed as a decimal value. As the submergence increases, and more of the lift

Figure 7-18. Operating characteristics of an air lift pump.

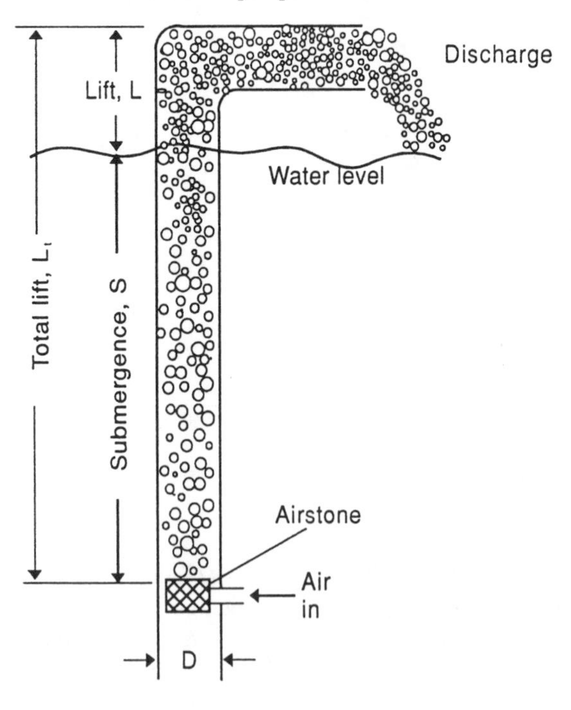

Table 7-8. Flow rate, Q (l/min), for airlift pumps as a function of length and diameter (using Eq. 7-37) at varying submergence ratio values.

Length, L_t (cm)	Submergence ratio	Lift pipe diameter, D (cm)					
		1.0	2.0	3.0	4.0	6.0	8.0
30	0.8	1.7	7.7	18.7	35.3	86.1	162.1
	0.9	2.0	9.2	22.4	42.2	102.9	193.7
	1.0	2.4	10.9	26.5	50.0	122.0	230.0
50	0.8	2.0	9.1	22.2	41.9	102.2	192.4
	0.9	2.4	10.9	26.6	40.0	122.1	229.8
	1.0	2.8	12.9	31.4	59.2	144.0	272.0
75	0.8	2.3	10.4	25.5	48.0	117.0	220.4
	0.9	2.7	12.5	30.4	57.3	139.8	263.2
	1.0	3.2	14.7	36.0	67.7	165.0	311.0
100	0.8	2.5	11.5	28.1	52.8	128.9	242.7
	0.9	3.0	13.7	33.5	63.1	153.9	289.8
	1.0	3.5	16.2	39.6	74.5	182.0	342.0
150	0.8	2.9	13.2	32.1	60.5	147.6	277.9
	0.9	3.4	15.7	38.4	72.2	176.2	331.9
	1.0	4.0	18.6	45.3	85.3	208.0	392.0
200	0.8	3.2	14.5	35.4	66.6	162.5	306.0
	0.9	3.8	17.3	42.2	79.5	194.0	365.0
	1.0	4.4	20.4	49.8	93.8	229.0	431.0
300	0.8	3.6	16.6	40.5	76.3	186.1	350.4
	0.9	4.3	19.8	48.4	91.0	222.2	418.3
	1.0	5.1	23.4	57.0	107.4	262.0	493.0

Source: Spotte, S.H. 1979. *Seawater Aquariums: The Captive Environment*. New York: John Wiley and Sons, Table 4-1, p. 68 (reproduced with permission).

tube is beneath the liquid surface, the efficiency increases. The *submergence ratio* is the ratio of the length of the tube beneath the surface to the total tube length. Spotte (1979) presented the following equation for estimating water flow rate with air lift pumps based on the work of Castro et al. (1975):

$$Q = (0.758\, S^{1.5} L_t^{0.333} + 0.01196) D^{2.2} \quad (7\text{-}37)$$

where Q = maximum flow rate, L/min; S = submergence; L_t = total lift tube length, cm; and D = diameter of tube, cm. The minimum acceptable value for submergence ratio for the operation of aquarium air lifts is S = 0.8. Spotte (1979) presented data by which to estimate air lift pump flow rate knowing L_t, S and D. These data are presented in Table 7-8.

Figure 7-19. Typical air lift pump used in ponds.

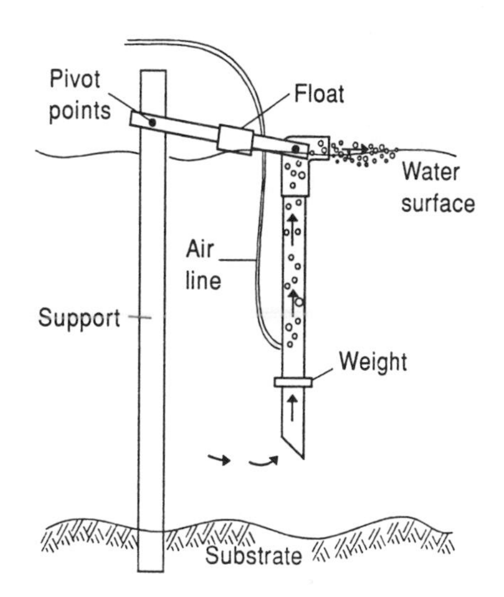

Larger-scale air lift pumps can be used in fish ponds or tanks for aeration and/or water circulation. In addition to air, liquid oxygen can be injected to enhance aeration. Figure 7-19 illustrates air lifts that are typical of those used in ponds and fish tanks. Submergence ratios for systems outside of the range of values presented in Table 7-8 should be determined experimentally.

CHAPTER 8
FLOW ESTIMATION AND MEASUREMENT

Very often it becomes necessary to maintain a given water level in a tank or monitor water flow from a tank, through pipelines or in open channels. Flow control means having a knowledge of what the water requirements are to individual parts of the system and being able to adjust flow rates as required. This chapter explores various methods commonly used in the aquaculture industry for sensing and maintaining water levels in tanks or ponds, and for sensing and measuring water flow through conduits and channels.

Instruments used to measure fluid flow vary considerably. Some measure volumetric flow rate directly while others are used to measure velocity.

CLASSIFICATION OF FLOW MEASUREMENT

Methods of flow measurement can be classified as either direct or indirect. *Direct* methods involve the actual measurement of the quantity of flow for a given time interval. Quantity of flow can be measured either by volume or by weight. An example of direct measurement would be the collection of water flowing from a tank into a bucket or tub over a specific time interval. A flowmeter that converts the rotations of an impeller into flow rate, which is then read from a dial, is another example.

Indirect flow measurements involve measurement of a pressure change or some other variable related to the rate of flow. Examples are instruments that measure pressure changes, such as venturi meters, orifices, and nozzle meters. Weirs and flumes are devices that indirectly indicate flow rates in open channels. Another example of an indirect measuring device is an electromagnetic flowmeter, which uses the voltage generated when a conductor passes through a magnetic field to indicate flow rate. These methods and others will be discussed in sections that follow.

MEASURING FLOW IN PIPELINES

There are numerous devices available for measuring flow rate in closed conduits or pipelines. These devices are classified as differential pressure, rotating mechanical, bypass, electromagnetic, ultrasonic, insertion, or variable area flowmeters. Another device is the anemometer.

Differential Pressure Flowmeters

Differential pressure flowmeters create a pressure difference that is proportional to the square of the volumetric flow rate. The pressure difference is usually created by passing the flow through a contraction in the conduit. The most common differential flowmeters are venturi tubes, orifice plates, nozzle meters, and elbow meters. The pressure difference can be measured with a manometer, Bourbon gauge, or pressure transducer (James 1988).

Venturi Tube. Venturi tube meters consist of converging (upstream), throat, and diverging (downstream) pipe sections (Figure 8-1). A pressure drop is created between the converging and throat sections as fluid passes through the throat. The lower pressure in the throat section is produced by the higher fluid velocity. In the downstream section a gradual increase in cross-sectional area of the conduit causes both the velocity and pressure to increase. The pressure drop ($\Delta P = P_1 - P_2$) between the upstream and throat sections is related to the volumetric flow rate by Equation 8-1 as follows (James 1988):

$$Q = \frac{Cd^2K\sqrt{P_1 - P_2}}{\sqrt{1 - \left(\frac{d}{D}\right)^2}} \qquad (8\text{-}1)$$

where Q = discharge, L/min (gpm); C = flow coefficient (dimensionless); d = diameter of throat section, cm (in.); D = diameter of upstream section, cm (in.); P_1 = pressure in upstream section, kPa (psi); P_2 = pressure in throat section, kPa (psi); and K = unit constant (K = 6.66 for SI units; K = 29.86 for English units). The flow coefficient C varies with Reynold's Number and is usually determined by calibration. C is normally supplied by the manufacturer for commercially produced venturi tubes. C is equal to 0.98 for Reynold's Numbers exceeding 200,000. The coefficient may approach unity for extremely smooth venturi tubes, such as those fabricated from glass or PVC.

Orifice Plate Meter. An *orifice* consists of a thin plate with a circular, square-edged hole in the center that is clamped between two flanges in a pipe (Figure 8-2). In principle, orifice plates function similarly to venturi tubes. Increased velocity in the orifice creates a pressure drop between the upstream and downstream sections of the pipe. Orifices normally have a greater pressure drop than venturi tubes having the same d to D ratio. On the plus side, orifices are not as costly to manufacture as venturi tubes. Equation 8-1 can also be used to calculate the flow rate through an orifice plate, however, the flow coefficient C varies considerably as the ratio d/D varies. Values of C for an orifice plate can be obtained from Figure 8-3 for the conditions shown and for Reynold's numbers ≥ 100,000. The orifice plate must maintain a sharp upstream edge. If the upstream edge becomes rounded because of chemical abrasion or other wear, the orifice will no longer be a reliable flow measuring device because the flow coefficient will change.

Figure 8-1. A typical venturi meter.

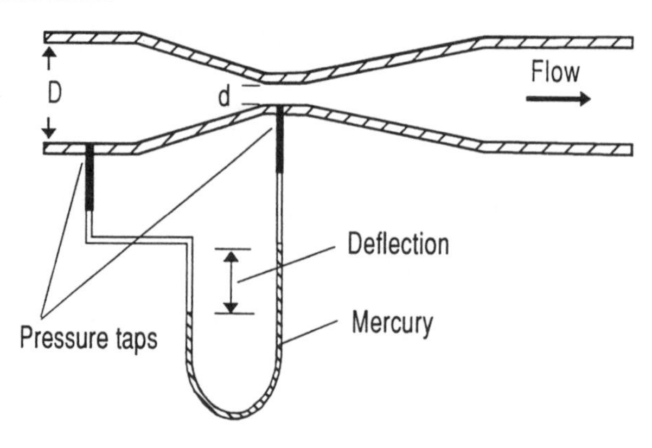

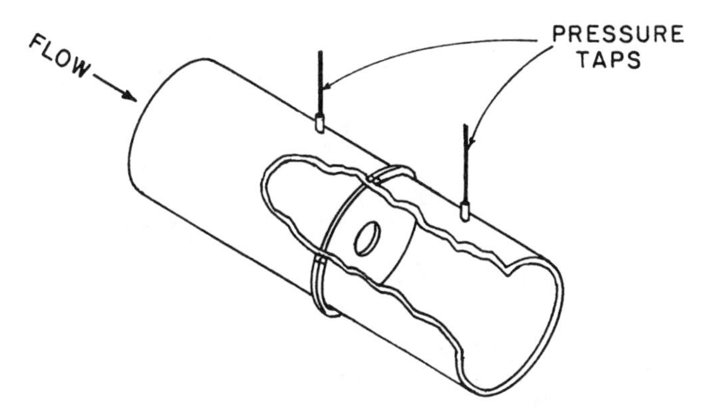

Figure 8-2. An orifice meter.

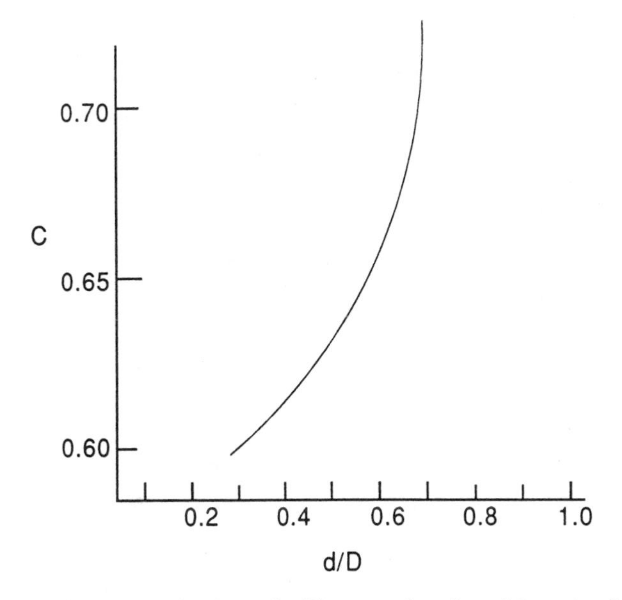

Figure 8-3. Flow coefficient C for square-edged round orifices as a function of the ratio of orifice diameter to inside pipe diameter for Reynolds numbers greater than 100,000 (Data estimated from James (1988)).

Nozzle Meter. The nozzle meter has an advantage over the orifice meter in that it is less susceptible to wear. If the pressure taps are installed as shown in Figure 8-4, the nozzle meter has approximately the same flow coefficients as a venturi meter and Equation 8-1 can be used to calculate flow. The head loss through a nozzle meter is much the same as that for an orifice meter (Roberson and Crowe 1990).

Elbow Meter. Another differential pressure flow measuring device is the elbow meter. As fluid flows through a bend in a pipe, the direction of the velocity changes, and a greater force acts upon the outer curvature of the bend. By measuring the pressure difference between the inner and outer curvature of a 90° pipe bend, as shown in Figure 8-5, the effect of the force is determined and can be correlated to flow rate by the following equation (James 1988):

$$Q = CKA\sqrt{P_0 - P_i} \tag{8-2}$$

where Q = discharge, L/min (gpm); C = elbow meter flow coefficient; A = elbow cross-sec-

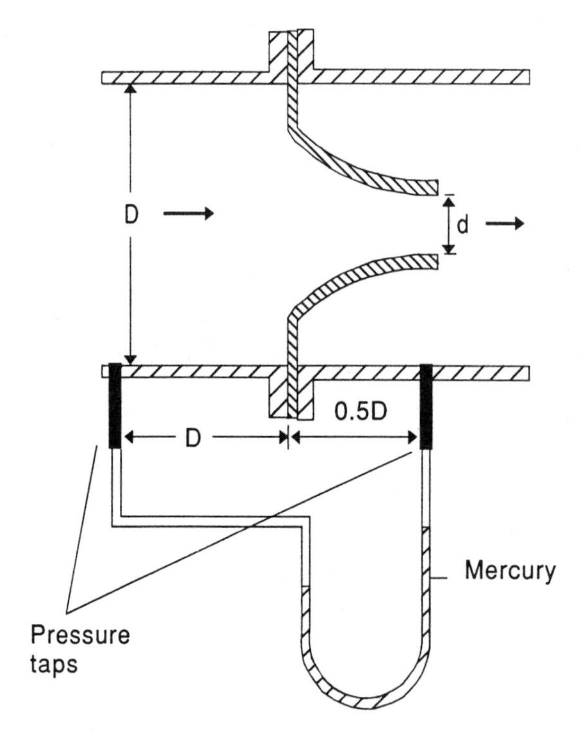

Figure 8-4. A typical flow nozzle.

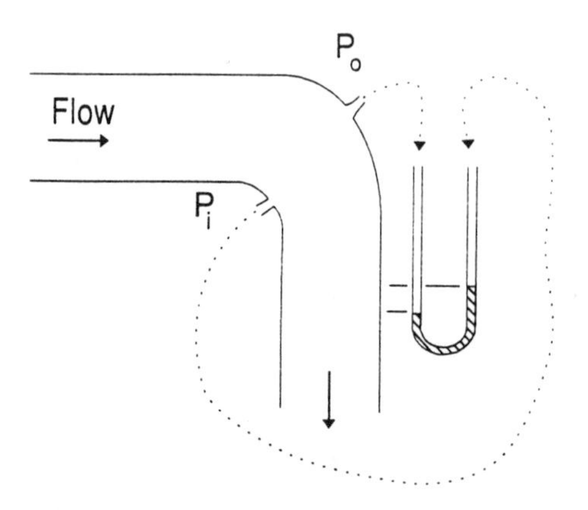

Figure 8-5. Pressure distribution in a pipe elbow.

tional area, cm^2 ($in.^2$); P_o = pressure outside of elbow, kPa (psi); P_i = pressure inside of elbow, kPa (psi); and K = unit constant (K = 8.49 for SI units; K = 38.02 for English units). The flow coefficient C is obtained from Figure 8-6.

Rotating Mechanical Flowmeters

There are many types of rotating mechanical flowmeters used to measure flow in pipes.

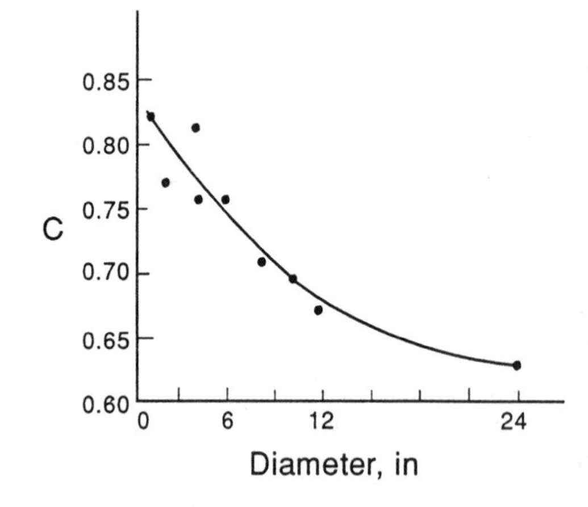

Figure 8-6. Flow coefficient C for elbow meters for Reynolds Numbers greater than 100,000 (Data estimated from James (1988)).

These meters use rotating propellers, impellers, rotors, turbines, vanes, etc., that revolve at a speed proportional to flow rate. They are usually manufactured with a device for counting and displaying the number of revolutions made by the impeller per unit of time. The number of

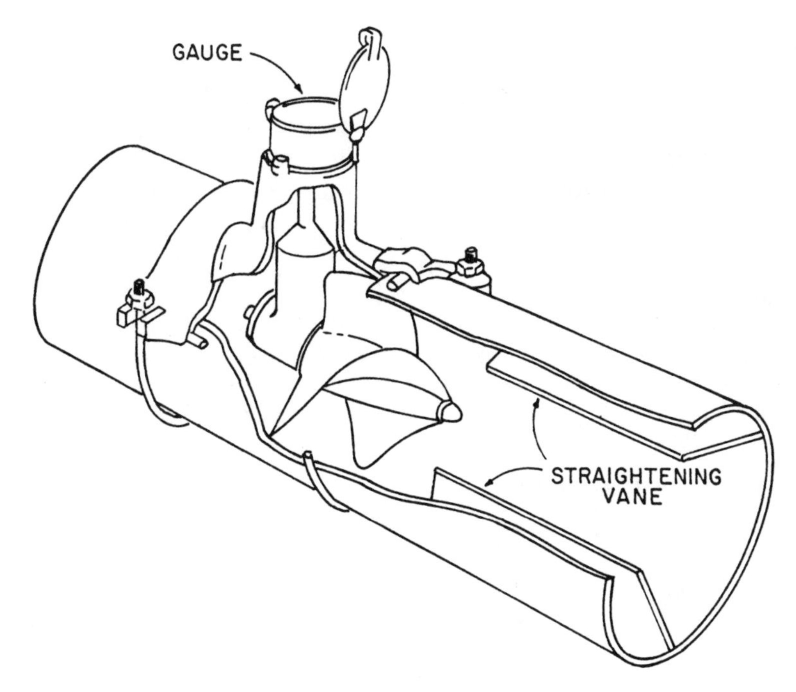

Figure 8-7. A propeller meter installed in a pipe section (Source: SCS (1969)).

revolutions is then converted to the total volume of flow and/or volumetric flow rate. Calibration tests are usually done to accurately relate revolutions to flow rate. Figure 8-7 is an illustration of a propeller meter installed in a pipe.

Rotating mechanical flowmeters are widely used to measure flow in large diameter closed conduits. Some are used in open channels, streams, rivers, lakes and for measuring ocean currents. They are used extensively for measurements in the field because they are reasonably accurate, rugged, and simple to operate. In addition, they are relatively inexpensive in comparison to other types of flowmeters. Head losses are usually low.

Bypass Flowmeters

A *bypass* flowmeter, sometimes referred to as a *shunt* meter, is simply an orifice or other differential pressure device with a small mechanical flowmeter across the pressure taps as illustrated in Figure 8-8. The small flowmeter is calibrated to indicate flow rate in the larger pipe. Bypass meters are manufactured in several

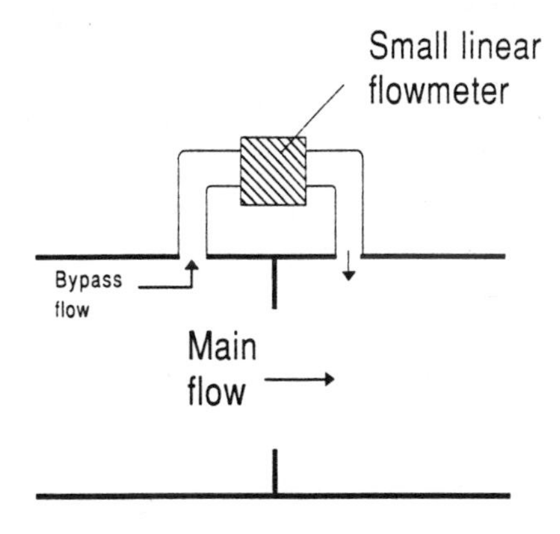

Figure 8-8. A bypass or shunt flowmeter.

different pipe diameters, and calibration data is normally supplied by the manufacturer.

Electromagnetic Flowmeters

The basic principle of the electromagnetic flowmeter is that liquids having a certain degree of conductivity generate a voltage proportional to velocity. The voltage can be indicated by two electrodes placed directly across the pipe from one another as shown in Figure 8-9. The output signal varies linearly with the flow rate. Advantages of this type flowmeter are that it causes no resistance to flow, and pressure taps are not required. Pressure taps are subject to clogging. Disadvantages are its high cost and its unsuitability to measuring gas flow (Roberson and Crowe 1990).

Ultrasonic Flowmeters

The ultrasonic flowmeter is another form of nonintrusive flowmeter that has wide application from measuring blood flow to flow in open channels. These meters use ultrasound to measure flow velocity and hence volumetric flow rate. Ultrasonic meters have no moving parts, do not obstruct flow, and do not cause a pressure drop in the conduit. Since ultrasonic beams will travel through the wall of a pipe, the meters can be either portable or built-in. Built-in units are more accurate. There are several types of ultrasonic flowmeters. One type is the *cross-correlation meter*. This meter uses two transverse beams of ultrasound, one located a short distance upstream from the other, as illustrated in Figure 8-10. The volumetric flow rate is cal-

Figure 8-9. An electromagnetic flowmeter.

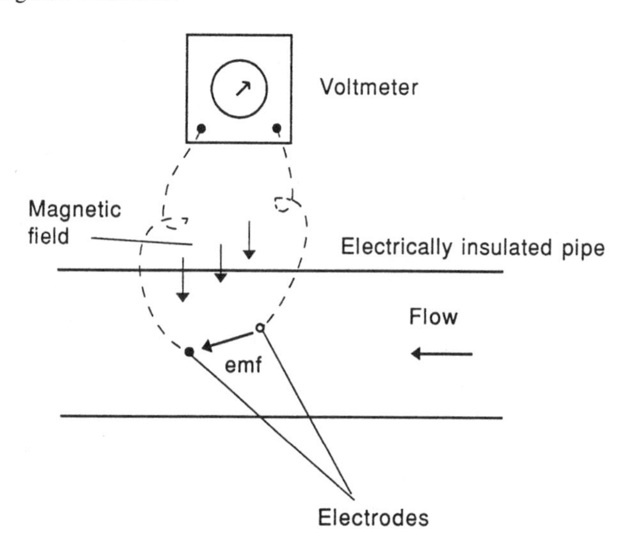

Figure 8-10. A cross-correction ultrasonic flowmeter.

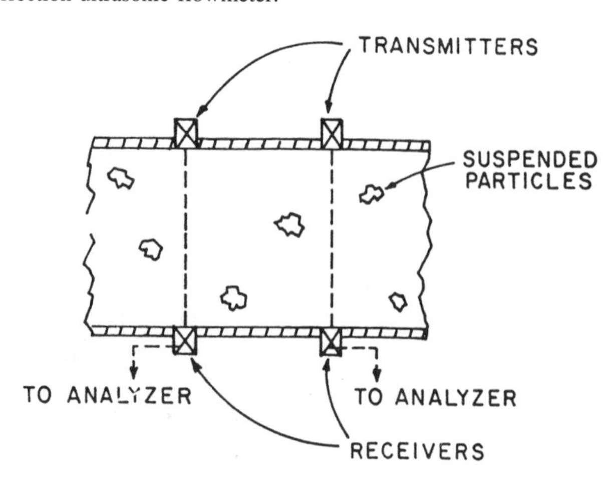

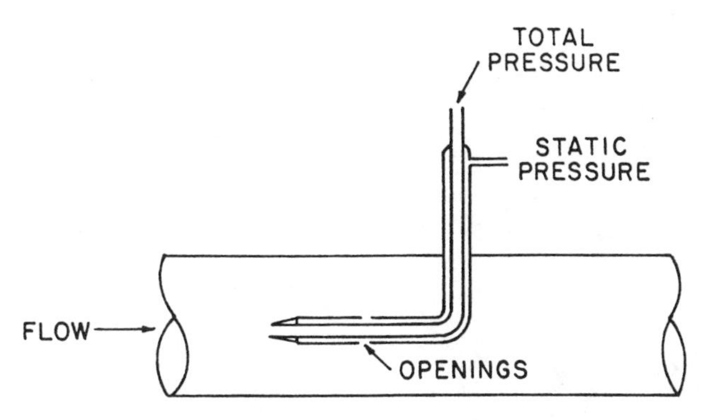

Figure 8-11. A pitot tube.

culated from the time it takes fluid particles, air bubbles, or fluid turbulence (eddies) to pass from the upstream to the downstream beam. Other types of ultrasonic flowmeters exist and are discussed in detail in James (1988).

Insertion Flowmeters

The major type of insertion meter is the *pitot tube*. These are used primarily for measuring gas flow, but they can also be used for liquids. A pitot tube consists of two small-diameter concentric tubes that face uptream against the direction of flow (Figure 8-11). The inner tube opens facing upstream and measures the total pressure P_t (kinetic plus static), while the outer tube opens to the outside and measures only static pressure P_s. The difference in pressure (head difference) between the tubes is the kinetic energy of flow and is related to fluid velocity by the following equations (James 1988):

$$V = CK\sqrt{P_t - P_s} \quad (8\text{-}3)$$

$$V = CKH^{0.5} \quad (8\text{-}4)$$

where V = flow velocity, m/s (ft/s); C = flow coefficient (C = 1.0 for well-designed tubes); H = head difference, cm (in.), and K = unit constant (K = 0.443 for SI units; K = 2.315 for English units). Once the velocity V is known, Q can be determined from the continuity equation (Q = AV) given the cross-sectional area of the pipe.

Variable Area Flowmeters

Variable area flowmeters are also referred to as *rotameters*. A rotameter consists of a vertical tapered tube fabricated from glass or plastic in order that the position of a float or rotating impeller located inside the tube is visible (Figs. 8-12 and 8-13). The liquid or gas flows upward through the tube, and the float rides on the fluid. The tube is tapered so that the fluid velocity is

Figure 8-12. An illustration of a rotameter.

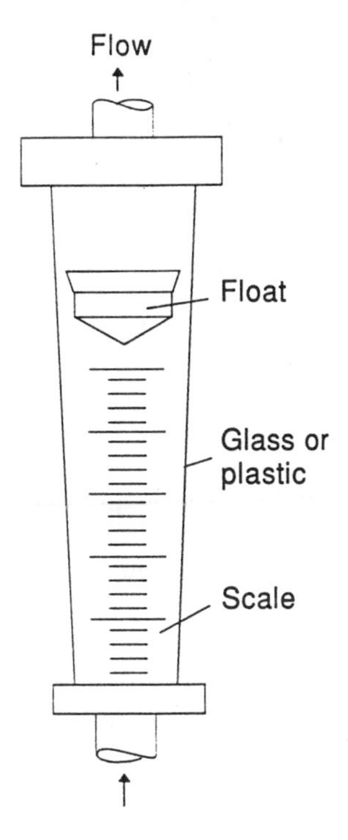

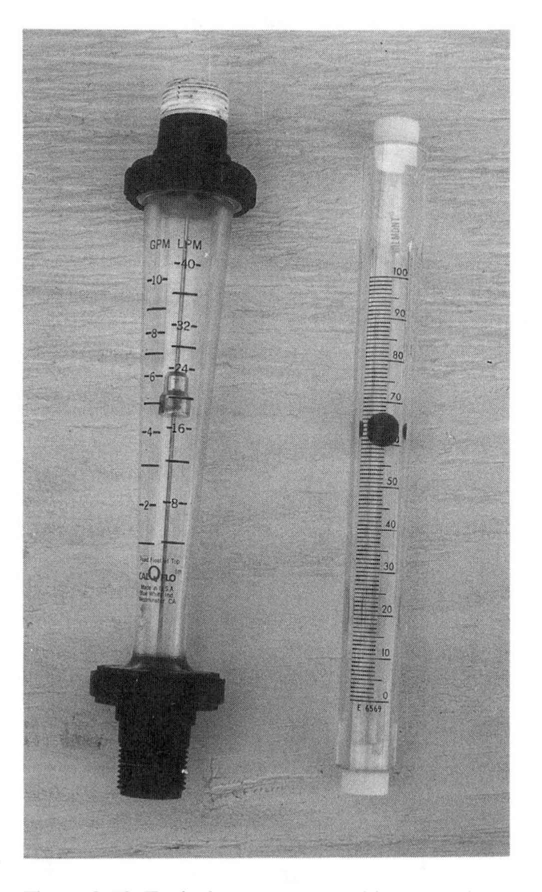

Figure 8-13. Typical rotameters used in aquaculture.

less the farther the float travels up the tube. Thus, the float seeks a neutral position where the fluid drag force just balances its weight. The float will rest higher or lower in the tube depending on fluid flow rate, fluid density and viscosity, fluid temperature, and weight density. A calibrated scale on the side of the tube indicates the rate of flow. The rotameter is calibrated depending on the properties of the fluid to be measured. Floats are usually interchangeable, and floats made from different materials of various densities are available so that the instrument can be calibrated for use with a variety of fluids.

Anemometers

Anemometers are instruments used for measuring wind velocity and flow in wind tunnels, closed conduits, or open channels. There are two basic types: vane or propeller anemometers and hot-wire anemometers. Vane anemometers have been used for a numbers of years in a variety of applications to measure velocity of either gases or liquids. The type commonly used for airflows consists of vanes or blades attached to a rotor that drives a gear train, which, in turn, drives a pointer on a dial (Figure 8-14). The dial indicates total distance that the air travels per unit of time. Thus, if the device is held in an airstream for one minute and the pointer indicates a 100-m change on the scale, then the average air speed is 100 m/min. Another type of vane anemometer is used for measuring the velocity of flowing water in an open channel.

The response of vane or propeller anemometers is too slow to measure rapid changes in velocity, and the device is too large to measure velocity in tight areas or in boundary layers. An instrument that can meet these challenges is the *hot-wire anemometer*. Hot-wire anemometers are also very sensitive to low-velocity flows. Drawbacks are that they are delicate and expensive. Because of the extreme fragility of the wire element, they are used almost exclusively for gas flow measurements (Wheaton 1977). Figure 8-15 illustrates how a hot-wire anemometer is used to measure airflow in a closed conduit.

Hot-wire anemometers consist of an electrical power source and a very small-diameter wire-the sensing element of the instrument. The wire is heated by electrical current passing

Figure 8-14. A vane anemometer for measuring air velocity.

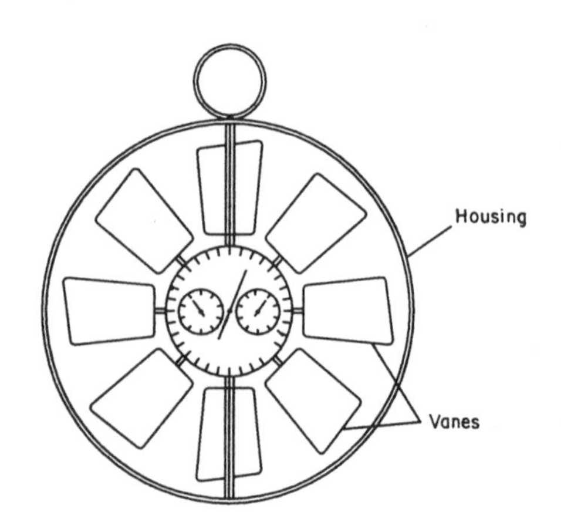

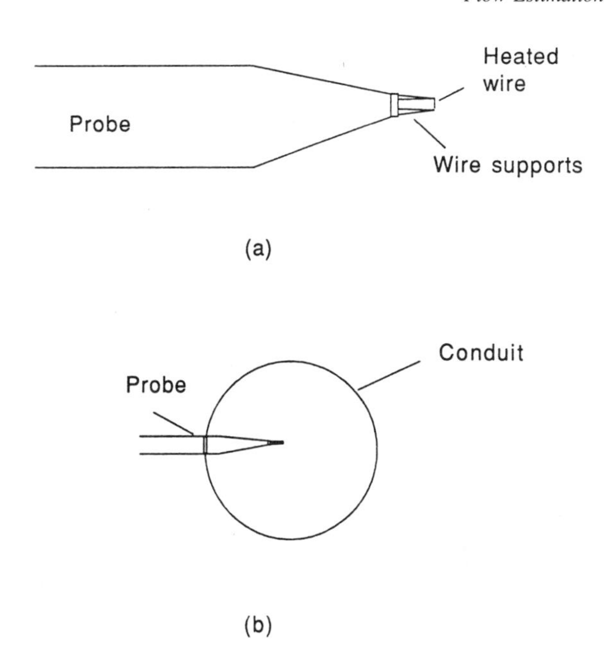

Figure 8-15. A hot-wire anemometer (a) Probe; and (b) cross section of probe in a conduit.

through it (normally to 150°C), and fluid flowing past the wire causes it to cool because of convective heat transfer. The wire is typically 1–2 mm in length and 10 μm or less in diameter (Roberson and Crowe 1990).

There are two types of hot-wire anemometers, constant current and constant temperature. Constant current anemometers provide a constant current flow through the heated wire. Changing fluid velocity causes the temperature of the wire to change, which, in turn, changes the wire's electrical resistance. The resistance change is proportional to fluid velocity. The constant-temperature hot-wire anemometer is the more popular of the two. This device operates by varying the electrical current in such a manner that the wire temperature (and resistance) remain constant. Current flow correlates to fluid velocity.

MEASURING DISCHARGES FROM OPEN PIPES

Water discharging from the open end of a pipe flowing freely into the air can be estimated by a variety of methods. The simplest method, which is commonly used for measuring flow from small-diameter pipes and small flows from large diameter pipes, is the *volumetric* method, which involves measuring the time it takes for the flow to fill a vessel of known volume. Other methods, the so-called *trajectory* methods, are also simple and relatively easy to use. Basically, they consist of measuring the horizontal and vertical coordinates of a point in the water jet discharging from the end of a open pipe. The pipe can be horizontal, vertical or at an angle. The difficulty comes in getting an accurate estimate of the horizontal and vertical coordinates of the water jet.

The calculations involving trajectory methods for estimating discharges from open pipes were obtained from English references dating back many years before the use of metric or SI units became standard. English units were used throughout the calculations and are left as such for simplicity purposes. Once velocity and/or discharge is obtained in English units, conversion to the SI system is relatively simple.

Volumetric Method

This is perhaps the most simple method used for measuring flows from small conduits or

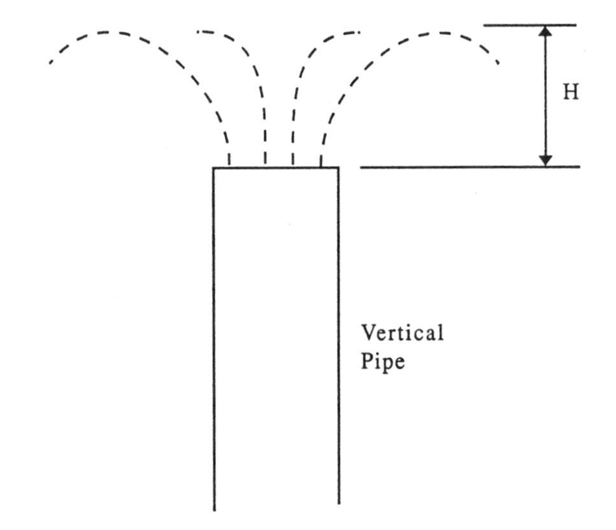

Figure 8-16. Measuring flow from a vertical pipe.

tanks, but it is also one of the most inaccurate. With this method, the time it takes to fill a vessel of known volume is timed with a stopwatch. The flow rate is then determined by dividing the volume collected by the time. To increase accuracy, the procedure should be repeated several times and a mean flow rate calculated.

Trajectory Methods

Vertical Pipe. Lawrence and Braunworth (1906) developed a method for estimating flows from vertical pipes. Referring to Figure 8-16, when the height of the water jet above the lip of the pipe exceeds 1.4D, the flow may be estimated by Equation 8-5:

$$Q = 5.01\,D^{1.99}\,H^{0.53} \qquad (8\text{-}5)$$

where Q = discharge, gpm; D = inside pipe diameter, in., and H = height of water jet, in. When the jet height is less than 0.37D, the discharge is estimated by Equation 8-6:

$$Q = 6.17\,D^{1.25}\,H^{1.35} \qquad (8\text{-}6)$$

For jet heights between 0.37 and 1.4D, the flow is less than that given by either of these equations. Figure 8-17 can be used to estimate discharges for standard pipes 2–12 in. in diameter and for jet heights from 1.5–60 in. (Bellport and Burnett 1967).

Example problem 8-1. Estimate the discharge from a vertical 4-in. diameter pipe when the jet height H = 8 in.
Solution: From Figure 8-17 for H = 8 in. and D = 4 in., the discharge is approximately 220 gpm.

Horizontal Pipe. There are two methods by which one can determine flow from horizontal pipes flowing full or partially full: (1) The California pipe method; and (2) the Purdue method.

CALIFORNIA PIPE METHOD. The *California pipe method* was developed from experimental data for pipes 3–10 in. (7.6–25.4 cm) in diameter, which are not flowing full. The method has four essential requirements that must be met to obtain reasonably accurate results: (1) The pipe must be horizontal; (2) it must be flowing partially full; (3) it must discharge freely into the air; and (4) the discharge end of the pipe must be at least six pipe diameters downstream from a restriction or change in direction (such as a valve or elbow). The method of complying with these requirements is illustrated in Figure 8-18. Other designs are possible. The only measurements necessary are the vertical distance from

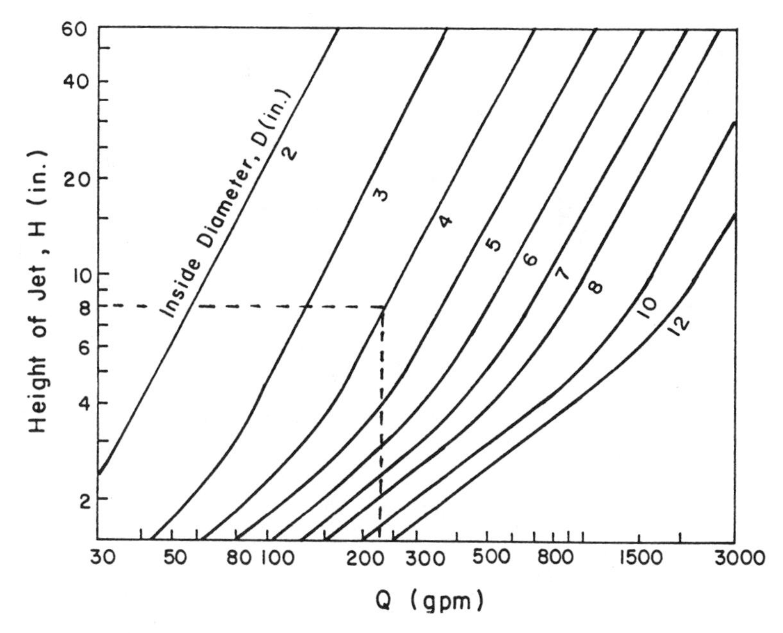

Figure 8-17. Discharge curves for measuring flow from vertical standard pipes (Source: U.S. Department of the Interior (1967)).

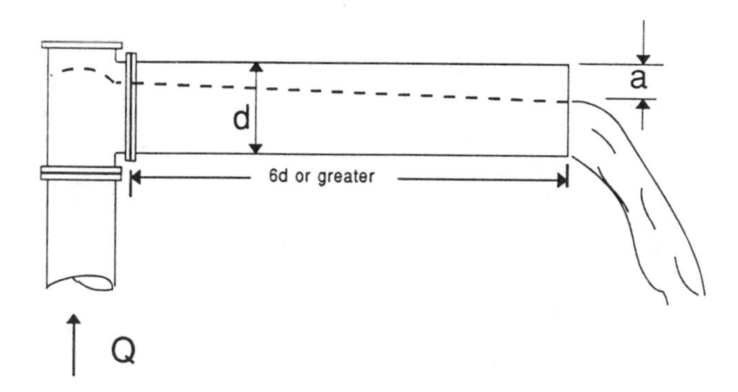

Figure 8-18. Typical arrangement for measuring flows by the California pipe method (Source: U.S. Department of the Interior (1967)).

the upper inside surface of the pipe to the water surface and the inside pipe diameter. With this information the discharge can be computed using Equation 8-7 (Bellport and Burnett 1967):

$$Q = 8.69 \left(1 - \frac{a}{d}\right)^{1.88} d^{2.48} \qquad (8\text{-}7)$$

where Q = discharge, ft^3/s; a = distance from upper inner pipe surface to water surface, ft; and d = inside pipe diameter, ft. Equation 8-7 is only valid for a/d greater than 0.5 (water depth less than 0.5d), therefore, one should be cautious when making these computations.

PURDUE METHOD. The *Purdue* or *yardstick method* is another trajectory method that is used to estimate discharge from horizontal or inclined pipes that are flowing either partially or completely full (Grieve 1928). This method consists of measuring the coordinates of a point on the upper surface of a water jet as shown in

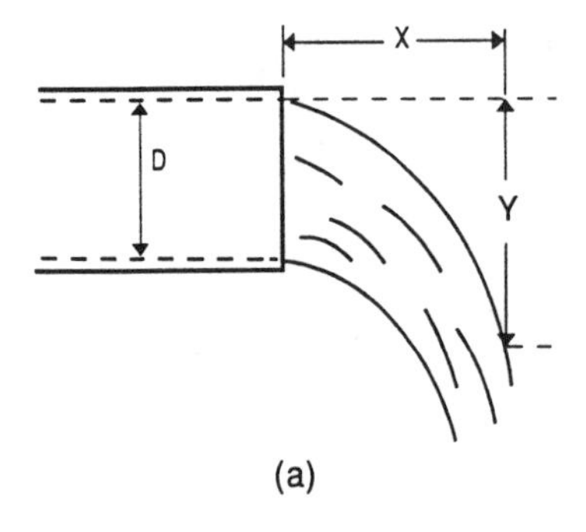

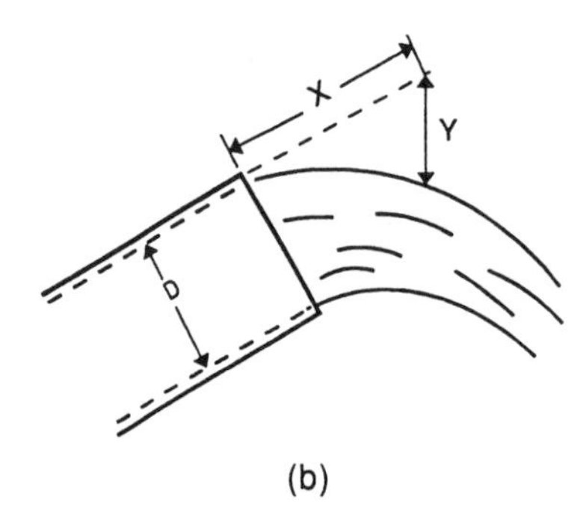

Figure 8-19. Measuring flows from (a) Horizontal; and (b) inclined standard pipes.

Figure 8-19. If the pipe is flowing with a water depth of less than 0.8D at the outlet, the vertical distance Y should be measured in the pipe opening where X = 0. For pipes flowing full, Y can be measured at horizontal distances X from the end of the pipe, for example at 6, 12 and 18 in. Discharges in gpm for 2–6 in. diameter standard pipes when Y is between 1 and 6 in can be obtained from Figure 8-20.

Table 8-1 lists flow rates for 2–12 in. diameter horizontal or inclined pipes for distances X when Y is 12 in. For Y = 6 in., multiply the value from the graph by 1.4 to obtain the new discharge.

Example problem 8-2. Estimate the discharge from the horizontal pipe shown in Figure 8-19a for (a) Y = 6 inches; and (b) Y = 12 inches.
Solution: (a) From Table 8-1, Q = 1,017 gpm.
(b) From Table 8-1 footnote, Q = 1,017 x 1.4 = 1,424 gpm.

Figure 8-21 and Table 8-2 can be used to estimate the discharge from pipes flowing partially full. Multiply the flow rate for a pipe of diameter D flowing full (from Table 8-1) by the factor in Table 8-2 to obtain the new discharge.

Example problem 8-3. Estimate the discharge from the 8 inch diameter pipe in Figure 8-21 for Y = 12 in., X = 18 in., and E = 2 in.
Solution: From Table 8-2 for E = 2 in, the ratio E/D is 0.25. The factor F is then 0.81. The discharge for a pipe flowing partially full is the full-flowing discharge times the factor F. From Table 8-1, the full-flow discharge is 915 gpm. The partial discharge is thus Q = 915 x 0.81 = 741 gpm.

MEASURING FLOW IN OPEN CHANNELS

There are many methods by which one may estimate water flow in furrows, gullies, channels, or small streams. These include the volumetric and velocity-area methods as well as the use of velocity meters, floats, salts, and dyes. Flow measurement in large streams or rivers is beyond the scope of this text and will not be dealt with herein.

Volumetric Method. Flows in furrows, gullies, and small channels can be estimated by collecting the flow in a container for a specified time interval. The time is measured with a stopwatch, and the flow rate is determined by dividing the volume of liquid collected in the vessel by the time indicated on the stopwatch. An installation for measuring such flows is shown in Figure 8-22. The accuracy of the measurement depends on the size of the collecting vessel relative to the flow rate and the accuracy of the timepiece. The procedure should be repeated several times to determine the mean discharge.

Velocity-Area Method. This approach involves measuring the flow velocity and cross

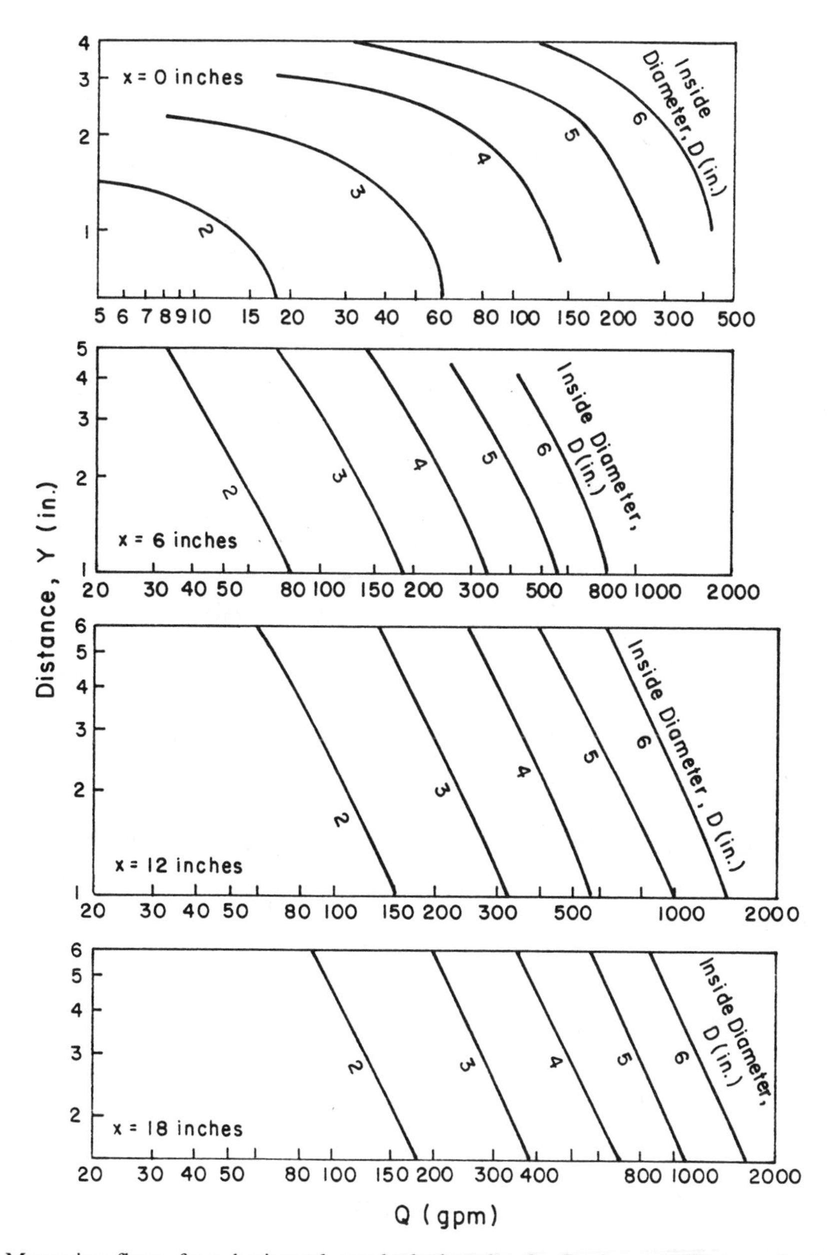

gure 8-20. Measuring flows from horizontal standard pipes by the Purdue coordinate method (Source: U.S. epartment of the Interior (1967)).

ctional area of the channel and then using quation 8-8 to compute the flow rate.

$$Q = K\,V\,A \qquad (8\text{-}8)$$

here Q = flow rate, L/s (gpm); V = average flow velocity, m/s (ft/s); A = channel cross-sectional area, m^2 (ft^2); and K = unit constant (K = 1000 for SI units; K = 448.8 for English units).

The cross-sectional area normal to the water flow can be estimated by dividing the channel

Table 8-1. Flow rate (gpm) for pipes flowing full with pipe diameter D, horizontal distance X and Y = 12 in.[1]

Pipe diameter D (in.)	Horizontal distance X (in.)									
	12	14	16	18	20	22	24	26	28	30
2	41	48	55	61	68	75	82	89	96	102
3	90	105	120	135	150	165	180	195	210	225
4	150	181	207	232	258	284	310	336	361	387
6	352	410	470	528	587	645	705	762	821	880
8	610	712	813	915	1,017	1,119	1,221	1,322	1,425	1,527
10	960	1,120	1,280	1,440	1,600	1,760	1,920	2,080	2,240	2,400
12	1,378	1,607	1,835	2,032	2,300	2,521	2,760	2,980	3,210	3,430

[1] For Y = 6 in. multiply gpm by 1.4.

Source: Berkely Pump Company, Berkely, CA.

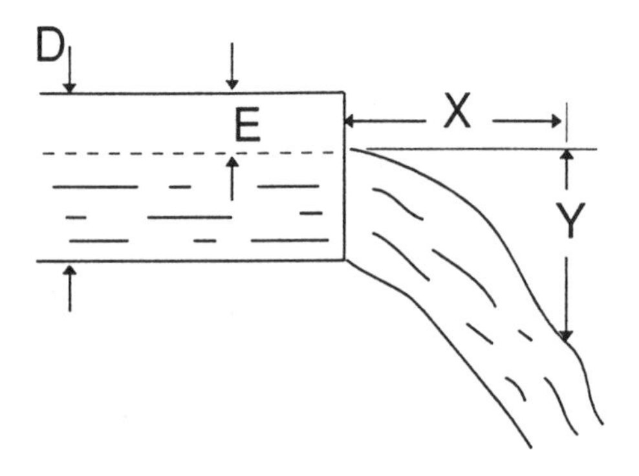

Figure 8-21. Measuring flow from horizontal pipe flowing partially full.

Table 8-2. Factor *F* to calculate flow rates (gpm) for pipes flowing partially full.

E/D	Factor F	E/D	Factor F
10	0.95	50	0.50
20	0.86	60	0.38
25	0.81	65	0.31
30	0.75	70	0.25
35	0.69	80	0.14
40	0.63	90	0.05
45	0.56	100	0.00

Source: Berkeley Pump Company, Berkeley, CA.

into segments as shown in Figure 8-23. The area of each segment is the product of its width times its average depth. The total cross-sectional area is obtained by adding the individual sums. I channels where the water depth is too deep fo a person to walk across the channel, the depth may be obtained from a boat by suspending weight on a string that has been marked at pre determined intervals.

The flow velocity is more difficult to deter mine since the velocity may vary significantl among channel segments. Several methods ar available for determining velocity.

Floats. The float method can be used in si uations where a high degree of accuracy is nc required. Velocity is determined by measurin and marking off two points along the length c the channel and measuring the time required fc a float to traverse the distance between the tw points. The distance divided by the time the gives the estimated velocity. The procedur should be repeated several times to calculate mean. The type of float used in this procedure i very important. If the majority of the float bod is visible above the surface of the water, th speed of the float can be severely affected b wind. Oranges make good floats because the have about the same density as water, they flo so that only a small part of the orange is visib above the water's surface, and they are round i shape so that wind effects are minimal.

Since the float velocity is representative c the surface velocity of the channel, the me sured velocities should be reduced by a veloci

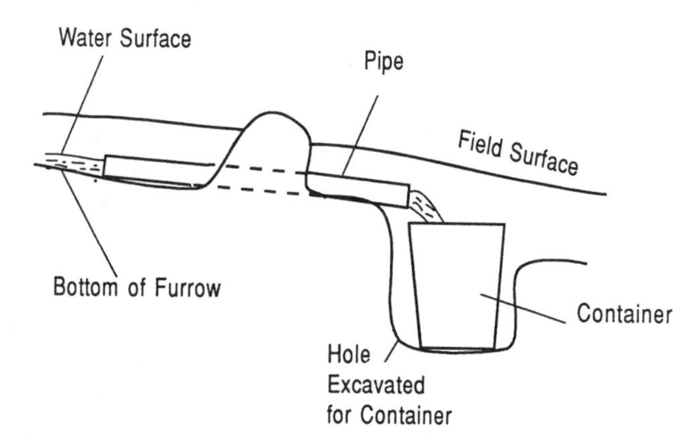

Figure 8-22. Volumetric measurement of outflow from an irrigation furrow.

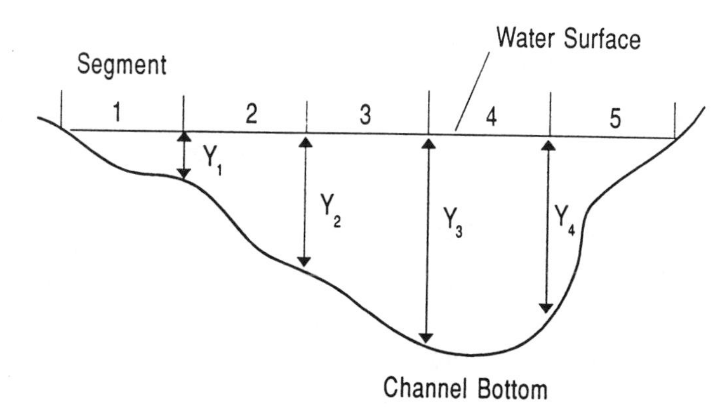

Figure 8-23. Cross section of an open channel that has been divided into five equal sections. Y_1 through Y_4 indicate the water depths at the respective points shown.

orrection factor C. Velocity correction factors re shown in Table 8-3. The average velocity of he channel can then be computed with Equaion 8-9 (James 1988):

$$V = C\,V_f \tag{8-9}$$

vhere V = average channel velocity; C = veocity correction factor, and V_f = observed float elocity.

For more accurate measurements with floats more sophisticated approach can be used. *Subnerging* floats can be fabricated from long, lender aluminum or PVC tubes sealed at both nds and containing enough weight to cause hem to float upright (Figure 8-24). The float hould have a length selected so that it submerges to a depth of about 25–38 cm (10–15 in.) above the bottom and protrudes above the water's surface by about 5–10 cm (2–4 in.). The observed velocity, V_f, of the submerging float then can be used in Equation 8-10 to calculate average channel velocity (Simon 1981).

$$V_{ave} = V_f\left(0.9 - 0.116\sqrt{1 - \frac{h_f}{y}}\right) \tag{8-10}$$

where h_f = the length of the float; and y = the depth of the stream.

Velocity Meters. Velocity or current meters can be used to measure flow velocity in open channels, ditches, furrows, streams, or rivers, and they can also be used to measure water

Table 8-3. Correction factor for Equation 8-9.

Average Flow Depth (m)	(ft)	C
0.3	1.0	0.66
0.6	2.0	0.68
0.9	3.0	0.70
1.2	4.0	0.72
1.5	5.0	0.74
1.8	6.0	0.76
2.7	9.0	0.77
3.7	12.0	0.78
4.6	15.0	0.79
≥ 6.1	≥ 20.0	0.80

Source: James (1988).

velocity in open water bodies such as lakes and oceans. This is the simplest and most straightforward approach, however, good velocity meters cost upwards of $2,000. Two types of velocity meters are shown in Figure 8-25. These meters use either a rotating propeller or impeller, which revolves at a speed proportional to water velocity. The rotating element is connected to a counter, and water velocity is determined by noting the time required for so many revolutions and using a meter calibration curve that relates the speed of rotation to water velocity.

Another type of velocity meter uses an electromagnetic sensor to measure water velocity directly (Figure 8-26). When the sensor is immersed in flowing water, its magnetic field creates a voltage that is sensed by electrodes embedded in the probe. The voltage amplitude represents the water flow rate around the probe. The meter readout is directly in m/s, ft/s, or knots. An advantage of this meter is that there are no revolving elements and no parts to wear out. This meter is also rugged and well-adapted for field use.

Velocity meters can also be used inside large closed conduits, such as sewer pipes, to measure flow velocity. The flow rate through the pipe can then be calculated knowing the cross-sectional area of the wetted area and by using the continuity equation.

Control Sections. This method uses known stage-discharge relationships of certain types of structures that are installed across a body of flowing water. Obviously, this method cannot be used for very large streams or rivers since installation usually involves blocking the water flow for a limited period of time. Control sections consist primarily of weirs and flumes. Other hydraulic control structures, such as sluice gates and culverts, can also be used to accurately measure flow in small channels, but the design and operation of these structures is considered beyond the scope of this book.

Weirs. Thin-plate weirs, also called *sharp-crested weirs*, consist of a smooth, vertical, flat plate installed across a channel perpendicular to the flow. The weir restricts water flow, causing water to back up a distance behind the weir plate and to flow over the crest as shown in

Figure 8-24. Measuring stream velocity with a submerged float.

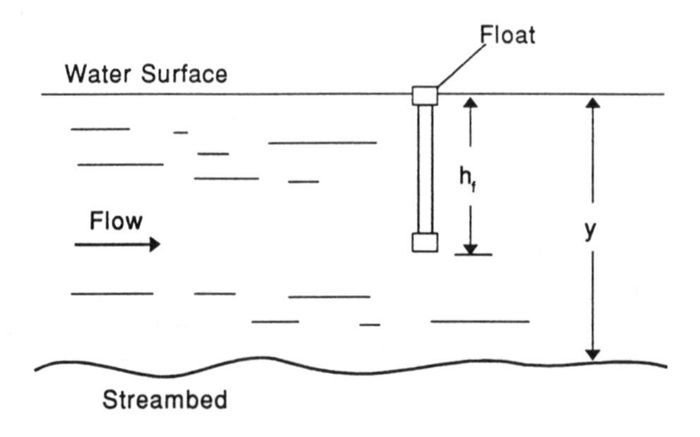

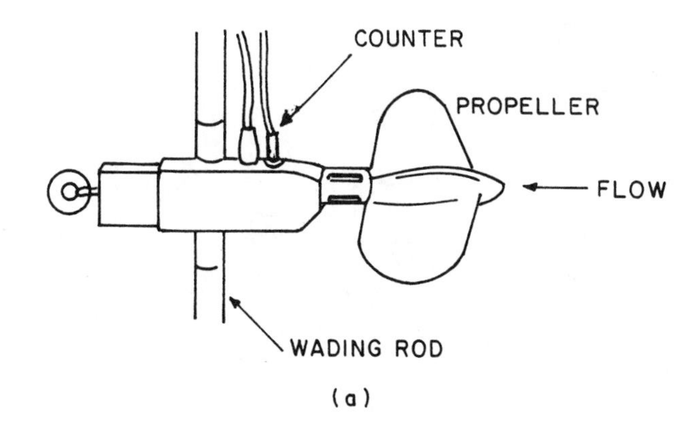

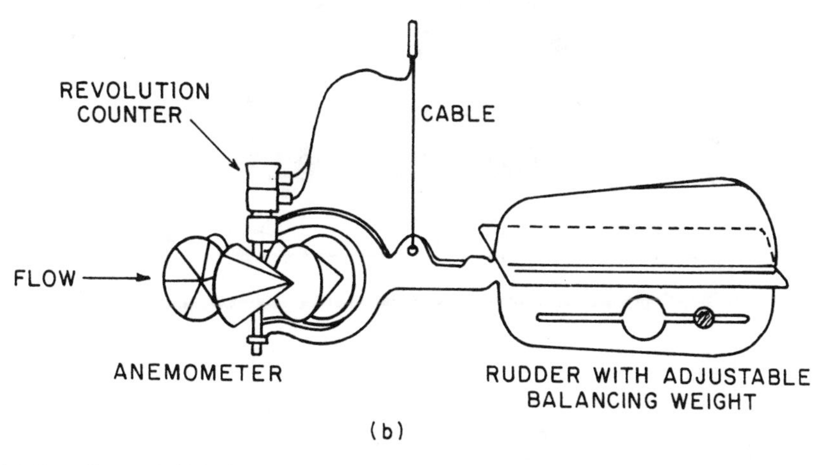

igure 8-25. (a) Propeller and (b) Price-type current meters. Redrawn from Principles of Farm Irrigation System esign by L.G. James. © 1988 Reprinted by permission of John Wiley and Sons, Inc. New York.

igure 8-27. The distance from the channel botom to the weir crest is the *crest height*. The epth of the water over the weir crest, measured t a specified distance upstream, is called the *ead*. The water overflowing the weir is called e *nappe*. Thin-plate weirs are most accurate hen the nappe completely springs free of the eir crest and air circulates freely around the appe, as shown in the figure. A head of at least cm (2 in.) and a crest thickness of no more an 1–2 mm (0.03–0.08 in.) are required for the ater to spring free from the weir crest. Very arp knife edges are not recommended because ey are easily damaged by debris, sediment, nd by rust pitting. The discussion of weirs that llows was extrapolated from Bellport and Burnett (1967) and James (1988) unless otherwise noted.

The position of the weir crest relative to the bottom of the approach channel can affect the performance of thin-crested weirs. The crest height should be at least twice the head on the weir (Figure 8-27). If not, standard tables and equations relating head and discharge will not be valid. Thin-plate weirs may not always be suitable. They are not recommended when the flow contains sediment or debris because deposition upstream of the weir will change the upstream hydraulic characteristics.

The weir crest may extend across the full width of the channel, or it may be notched. When the weir crest extends the full width

Figure 8-26. Measuring stream velocity with an electromagnetic velocity meter.

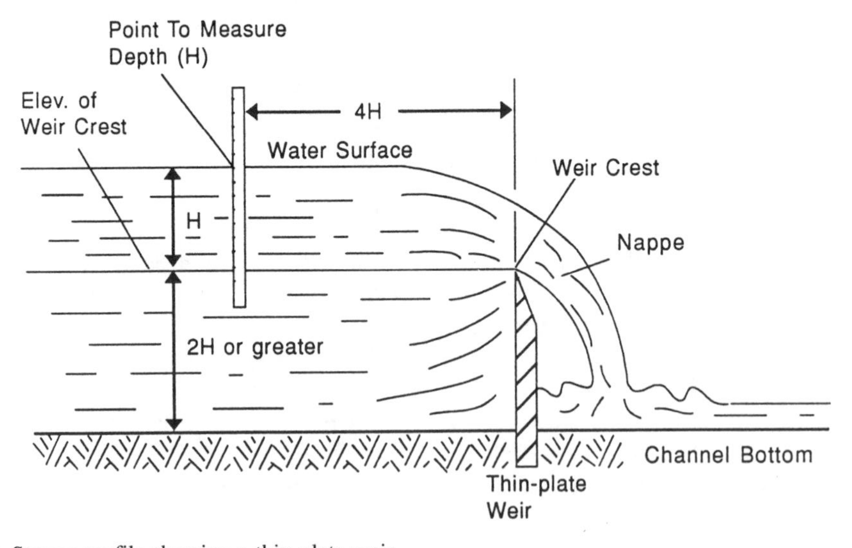

Figure 8-27. Stream profile showing a thin-plate weir.

of the channel, the weir is referred to as a suppressed rectangular weir because its sides are coincident with the sides of the approach channel (Figure 8-28a). The discharge for this type of weir can be computed with Equation 8-11.

$$Q = K\,L\,H^{1.5} \qquad (8\text{-}11)$$

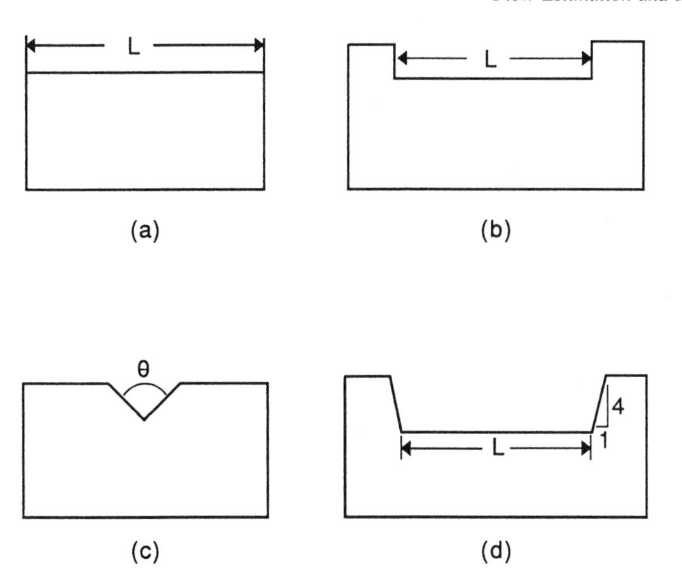

Figure 8-28. Weir types: (a) Rectangular suppressed; (b) rectangular contracted; (c) V-notch; and (d) Cipolletti.

where Q = discharge, L/s (gpm, cfs); L = crest length, m (ft); H = difference between crest and water surface at a point upstream from the weir a distance of 4H, m (ft); and K = unit constant (K = 1,838 for Q in L/s, L and H in in.; K = 1,495 for Q in gpm, b and H in ft; K = 3.33 for Q in cfs, b and H in ft).

Equation 8-11 is valid for H < P/2, H > 6 cm (2 in.), and P > 0.3 m (1 ft) where P is the crest height in m or ft. The equation applies only when the velocity of the water approaching the weir is negligible. If the approach velocity is appreciable and cannot be neglected, then Equation 8-12 applies

$$Q = K\,L\,((H + h)^{1.5} - h^{1.5}) \tag{8-12}$$

where $h = V^2/2g$ = head due to velocity of approach, m (ft); and V = velocity of approach, m/s (ft/s).

The most common notch shapes for notched weirs are rectangular, triangular, or trapezoidal. Weirs constructed with a rectangular notch are called *contracted rectangular weirs* (Figure 8-28b), and the discharge is obtained with Equation 8-13.

$$Q = K\,(L - 0.2H)\,H^{1.5} \tag{8-13}$$

Weirs having a triangular-shaped notch are referred to as *V-notch weirs*. These are accurate flow measuring devices. The V-notch weir is more accurate than other weir types for flows less than 30 L/s and at least as accurate as other types for flows of 30–300 L/s (450–4500 gpm). A V-notch weir is diagrammed in Figure 8-28c. For proper operation V-notch weirs should be installed so that the minimum distance from the channel side to the point of the weir notch and the distance from the bottom of the approach channel to the point of the weir notch is at least twice the head on the weir. The head-discharge relationship for the V-notch weir is given by Equation 8-14.

$$Q = K_1\,C\,\tan\left(\frac{\theta}{2}\right)h_e^{2.5} \tag{8-14}$$

where Q = discharge, L/s (gpm); C = coefficient of discharge from Figure 8-29; θ = notch angle (degrees); h_e = effective head, m (ft) = $H + K_2 k_h$; H = difference between notch point and water surface at a point upstream from the weir a minimum distance of 4H, m (ft); k_h = constant from Figure 8-30, m (ft); K_1 = unit constant (K_1 = 2,362 for SI units; K_1 = 1,920 for English units); and K_2 = unit constant (K_2 = 3.28 for SI units; K_2 = 1.00 for English units).

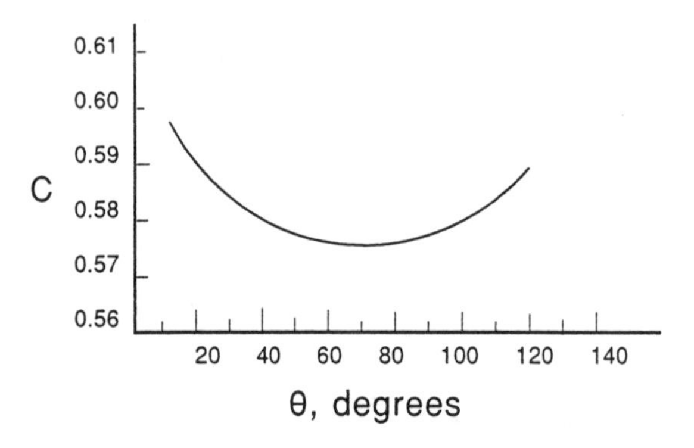

Figure 8-29. Coefficient of discharge C as a function of the notch angle of a V-notch thin-plate weir.

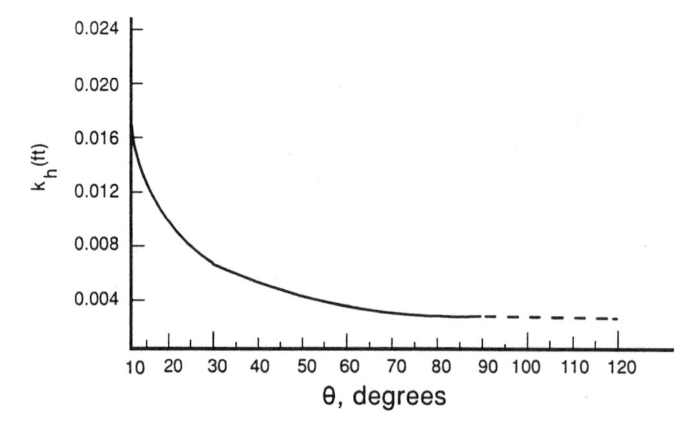

Figure 8-30. Constant k_h used for solving for effective head h_e in Equation 8-14 as a function of the notch angle for a V-notch thin-plate weir.

A final weir type that will be discussed is the trapezoidal or *Cipolletti weir*. The side slopes of the notch are constructed at a slope of 1 horizontal to 4 vertical (Figure 8-28d). The distance from the sides of the notch to the sides of the channel should be at least twice the head, and other requirements are the same as for other weir types. Cipolletti weirs should not be used for heads less than about 6 cm (2 in.), nor for heads greater than one-third the crest length. The head-discharge relationship is described in Equation 8-15.

$$Q = K\,L\,H^{1.5} \qquad (8\text{-}15)$$

where Q, L and H are defined as before; and K = unit constant (K = 1.859 for SI units; K = 3.367 for English units).

Flumes. A *flume* is a specially-shaped channel section that can be installed into small channels to obtain a head-discharge relationship. Like weirs, there are many types of flumes, each with its own unique qualities. Flumes have a converging inlet section that directs flow into a level constricted, or *throat* section. The inlet serves as a transition between the channel and the throat. A diverging section downstream of the throat returns flow to the channel. The throat acts as a control, and head measurements are made in the converging and diverging sections by means of water stilling wells. Flumes are useful for measuring channel flow because they are quite accurate and eliminate many of the problems associated with weirs. Flumes normally do not present as much of a restriction to flow as weirs. Velocity in the throat section is increased, there-

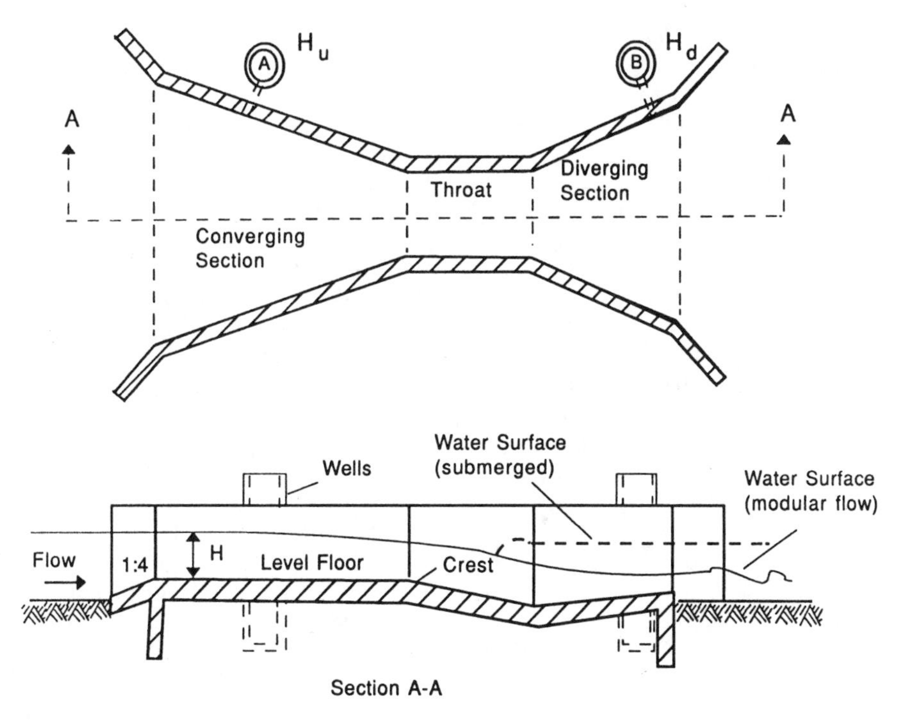

Figure 8-31. Geometry of a Parshall flume.

fore sediment deposition is usually not a problem unless the flows encountered are too small to keep sediment in suspension. Probably the most widely used flume type is the *Parshall flume*, and this is the only type that will be discussed since an extensive discussion of flumes is considered beyond the scope of this text.

Figure 8-31 illustrates a Parshall flume. A Parshall flume has a depressed bottom, and measurements are made at two stilling wells, indicated at A and B in the figure. Accurate measurements over a wide range of flows can be obtained with properly sized flumes and with proper construction and installation. Flumes are constructed from a variety of materials, including sheet metal, wood, and concrete. Large flumes are normally constructed on-site, while small flumes can be prefabricated and installed in small open channels.

The water flow rate through a Parshall flume can be affected by the depth of flow downstream from the throat section. If the tailwater depth is not great enough to affect flow, free-flow conditions are said to exist. If the tailwater depth affects flow, and the submergence ratio, S, exceeds the transition submergence, S_t, the flume is considered to be submerged. The submergence ratio, S, can be computed with Equation 8-16 (James 1988). Values for S_t for different Parshall flume designs are shown in Table 8-4.

$$S = \frac{H_d}{H_u} \tag{8-16}$$

where S = submergence ratio (dimensionless); H_d = head downstream from throat, m (ft); and H_u = head upstream from throat, m (ft).

When the tailwater does not affect flow through a Parshall flume, Equation 8-17 can be used to calculate the discharge (adapted from James (1988)):

$$Q = C_f\,(KH_u)^a \; for \; S \leq S_t \tag{8-17}$$

where Q = discharge, L/s (cfs); C_f = coefficient from Table 8-4; a = flow exponent (dimensionless); K = unit constant (K = 3.28 for SI units;

Table 8-4. Free-flow and submerged-flow coefficients and exponents for Parshall flumes.

	C_f for Q in		C_s for Q in				
Width	(L/s)	(cfs)	(L/s)	(cfs)	a	b	S_t
1 in.	9.57	0.338	8.47	0.299	1.55	1.000	0.56
2	19.14	0.676	17.33	0.612	1.55	1.000	0.61
3	28.09	0.992	25.91	0.915	1.55	1.000	0.64
6	58.34	2.06	47.01	1.66	1.58	1.080	0.55
9	86.94	3.07	71.08	2.51	1.53	1.060	0.63
12	113.28	4.00	88.08	3.11	1.52	1.080	0.62
18	169.92	6.00	125.17	4.42	1.54	1.115	0.64
24	226.56	8.00	168.22	5.94	1.55	1.140	0.66
30	283.20	10.00	204.47	7.22	1.55	1.150	0.67
3 ft.	339.84	12.00	243.55	8.60	1.56	1.160	0.68
4	453.12	16.00	314.35	11.10	1.57	1.185	0.70
5	566.40	20.00	383.74	13.55	1.58	1.205	0.72
6	679.68	24.00	448.87	15.85	1.59	1.230	0.74
7	792.96	28.00	514.01	18.15	1.60	1.250	0.76
8	906.24	32.00	577.73	20.40	1.60	1.260	0.78
10	1,136.48	40.13	702.05	24.79	1.59	1.275	0.80
12	1,345.20	47.50	830.91	29.34	1.59	1.275	0.80
15	1,658.42	58.56	1,024.33	36.17	1.59	1.275	0.80
20	2,180.86	77.00	1,346.90	47.56	1.59	1.275	0.80
25	2,702.86	95.44	1,669.46	58.95	1.59	1.275	0.80
30	3,225.08	113.88	1,992.03	70.34	1.59	1.275	0.80
40	4,269.24	150.75	2,636.88	93.11	1.59	1.275	0.80
50	5,313.68	187.63	3,282.00	115.89	1.59	1.275	0.80

Source: Skogerboe et al. (1967).

K = 1.0 for English units); and S_t = transition submergence (dimensionless).

When the flume is submerged, Equation 8-18 applies (adapted from James (1988)):

$$Q = \frac{C_s(K\,(H_u - H_d))^a}{(-(\log S + C))} \; for\; S > S_t \qquad (8\text{-}18)$$

where C_s = coefficient from Table 8-4 (dimensionless); b = coefficient from Table 8-4 (dimensionless); C = 0.0044 for Parshall flumes; and K = unit constant from Equation 8-17.

Chemical Discharge Measurements. *Chemical discharge measurements* are most valuable for use in small streams and open channels of irregular shape or cross-sectional area or for use where other measurement methods would be too expensive or impractical. This method involves the introduction of a chemical of known concentration into the flowing water, into which it completely diffuses after traveling a certain distance. The concentration in which the chemical is found after complete mixing is proportional to the stream discharge. Thus, by determining the degree of dilution of the chemical, the stream discharge can be computed. For complete mixing the chemical must travel a given distance L downstream. The equation for computing L in SI units is given by Simon (1981) as:

$$L = \frac{0.13\,C\,(0.7C - 6)B^2}{g\,H} \qquad (8\text{-}19)$$

where L = distance downstream; C = Chezy's coefficient for the stream given by Equations

8-20 and 8-21; B = stream surface width; H = stream depth and g = gravity term.

The following equations are used to calculate Chezy's coefficient:

$$(SI) \qquad C = \frac{R^{1/6}}{n^{1/2}} \tag{8-20}$$

$$(English) \qquad C = \frac{1.49\, R^{1/6}}{n^{1/2}} \tag{8-21}$$

where C = Chezy's coefficient; n = channel roughness coefficient (from Table 6-8); and R = channel hydraulic radius (Eq. 6-38).

The concentration of the chemical downstream from the point of entry is determined by sampling and/or with a probe. A variety of analytical procedures can be used to determine the tracer concentration, depending on which chemical is used. The easiest chemicals to use are rhodamine dyes and common salt (sodium chloride). The concentration of these chemicals can be determined in the field with the proper instrumentation. Dye concentrations can be determined either with a field fluorometer, or samples can be returned to the laboratory for analysis with a lab model fluorometer. Salt concentrations can be determined with an electric conductivity meter.

Two methods of chemical tracer introduction into a stream are discussed by Simon (1981). The first is the continuous introduction of a constant tracer discharge Q_t of known concentration C_t into the stream. Samples are collected at some point D ≥ L downstream from the entry point, and the tracer concentration C determined. The stream discharge is then determined by Equation 8-22.

$$Q = \frac{C_t}{C} Q_t \tag{8-22}$$

A second method of determining stream discharge is with the use of a *tracer wave*. This method involves dissolving a known amount of a tracer material (dye or salt) in a container of water. The mixture is then dumped into the stream, and samples are collected at a predetermined location downstream at timed intervals (15 or 30 s) until the concentration of the passing wave of the tracer substance is very small. The wave of the passing tracer at the sampling location is then plotted as shown in Figure 8-32, and the area under the curve is calculated. The area under the curve, A, has the units s-mg/L and defines the discharge of the stream as (Simon 1981):

$$Q = \frac{E}{A} \tag{8-23}$$

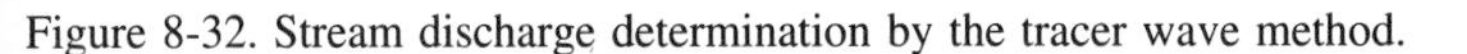

Figure 8-32. Stream discharge determination by the tracer wave method.

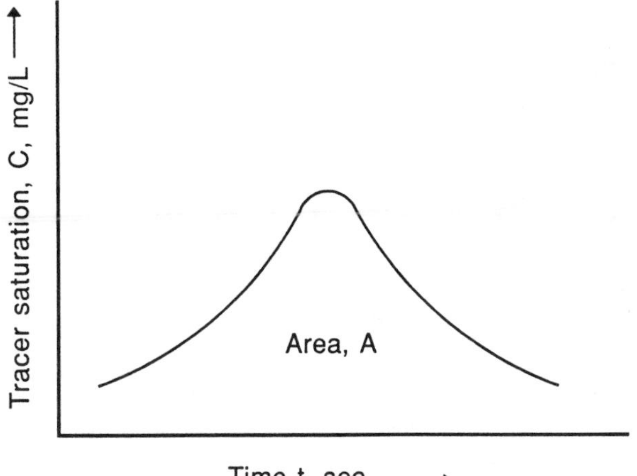

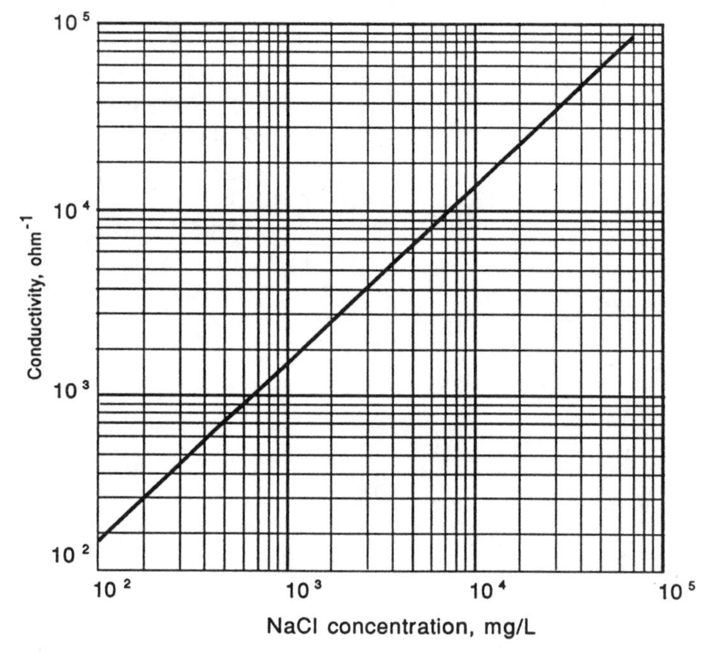

Figure 8-33. Electric conductivity versus salt concentration in water.

where Q = stream discharge, L/s; and E = tracer input, mg.

Sodium chloride (common salt) is perhaps the easiest of all materials to use when employing this method for stream flow determination. The maximum degree of saturation of salt is 240 g/L, which is the equivalent of 240 parts per thousand (ppt). Therefore, an 18.9-L (5-gal) container of water will dissolve 4.5 kg (10 lb) of salt. The electrical conductivity of the mixture can then be determined after it is dumped into the stream. Figure 8-33 shows a plot of conductivity versus salt concentration. The electrical conductivity of the salt solution can be measured at the downstream sampling point and plotted versus concentration as shown in the figure. Once the salt concentrations over time are determined, a tracer wave can be plotted as before, and the area under the tracer curve is the stream discharge. Since salt is often found to exist naturally in the environment, a background salt concentration should first be determined and this value subtracted from the tracer readings.

Simple instrumentation is available that directly indicates electrical conductivity, such as with the Yellow Springs Instrument Co. (YSI) Model 33 conductivity meter, which retails for under $1,000.

CHAPTER 9

AQUACULTURE IN PONDS, RACEWAYS, AND TANKS

An earlier chapter discussed aquaculture in natural and modified systems, those systems requiring little or no input by man to achieve a desired production level. This chapter deals with more advanced culture systems, that is, those requiring significant inputs by man in terms of resources, energy, and investment. Confined fish culture may be carried out in ponds, raceways, or tanks. Many considerations determine which type system to use. These considerations are the subject of this chapter.

POND CULTURE SYSTEMS

Historically, the practice of fish husbandry, or aquaculture, in ponds dates back several thousand years. Many types of finfish and crustaceans are routinely produced in ponds. A recent worldwide overview of aquaculture by Nash (1988) points out that pond aquaculture accounts for about 41% of worldwide aquaculture production. On a worldwide scale it is estimated that about 7.9×10^9 kg (1.74×10^{10} lb) of aquatic biomass is produced in 6×10^6 ha (1.48×10^7 ac) of fish ponds, or roughly 1,317 kg/ha (1,176 lb/ac). The species considered important by Nash (1988) break down as follows: 46% carp and tilapia, 37% oysters and mussels, 4% salmon and trout, 2.2% shrimp and prawns, and 2.2% catfish.

Commercial pond fish production is profitable only when a combination of resources is available at a reasonable cost: (1) Pond culture usually requires the purchase of large tracts of land with the proper soils, slope, and topography; (2) pond culture is water-intensive, hence, large volumes of good-quality water must be available; (3) since water temperature cannot be controlled in ponds, commercial production is feasible only in regions where the growing season is long enough to economically produce a market-sized animal; and (4) there must be a ready market for the product (Tucker and Robinson 1990). Pond production costs are generally lower than in other types of man-made production systems, largely because the natural pond processes maintain a healthy environment for fish growth. For example, oxygen in ponds is supplied by photosynthesis and diffusion, and fish wastes are decomposed by naturally occurring microorganisms. Relatively high yields are possible with limited technological input. Water quality management in other types of culture systems, such as raceways and tanks, requires more energy to provide oxygen and water flow to the system and to process fish wastes. Production in these systems is economical only when a high value crop is being cultured or when other circumstances occur to depress production costs.

Fish ponds provide either a *static* or *dynamic* environment. Static water refers to ponds where there is no flow of water. The ponds are filled by rainfall and the water is regulated primarily by relying on local precipitation, but it may be occasionally supplemented by pumped water. In either case, the water remains static or nonflowing as opposed to dynamic ponds where a certain percentage of the pond water volume is exchanged daily or water continuously flows through the pond, much like a raceway. Dynamic pond systems are more expensive in terms of startup and operating costs, but production per unit volume is much higher than in static systems, which helps to offset the additional costs.

Pond Site Selection

Aquaculture site selection was discussed in detail in Chapter 3. However, since farm pond design is so site-specific, a few details related to pond culture warrant attention here. One must keep in mind that the *ideal* site is not always available. A universal pond design may not be adequate in many situations, therefore one design is not practical. Deficiencies of the site are often mitigated by making a series of trade-offs. However, many of the major design features can be defined and used in a variety of situations. The discussion that follows is primarily aimed at land-based pond farms, but many of the same basic principals and design procedures are applicable to coastal brackish water or saltwater pond farms. Differences will be pointed out as the discussion progresses.

The two basic types of ponds used for aquaculture are *watershed* and *levee* ponds. The choice of which of these pond types depends on many factors, such as topography, land slope, capital investment required, and level of production desired. These factors and others will be addressed as the discussion progresses.

Watershed Ponds

Watershed ponds are created by constructing a dam or dike across a small stream or watercourse. These ponds are sometimes referred to as *hill* or *embankment* ponds. The pond is filled entirely by precipitation and runoff from the surrounding watershed. Watershed ponds are typically used for recreational fishing, fee fishing, or stock watering since they are not ideally suited for commercial fish production. Water depth at the bottom of the dam is usually too deep to harvest by seining. Also, the pond bottoms in watershed ponds are usually uneven, allowing fish to escape beneath a seine. They must therefore be harvested by draining, which results in lost production time. Water quality in deep ponds is also harder to manage because of stratification. Foul deep, bottom waters may mix with surface waters during seasonal overturns.

Watershed ponds can be constructed in a wide range of topographic conditions. Ponds for commercial fish farming preferably should be constructed in gently sloping, shallow valleys rather than in deep valleys with steep slopes. Ponds built in shallow valleys can be cut and filled so that the finished basin has a uniform and shallow depth (Figure 9-1) that will facilitate harvest without draining. Construction costs are determined by the amount of cut and fill necessary. Steep slopes make higher dams, resulting in higher construction costs. If there is enough soil remaining from the cut sections after the pond is shaped it can be used to construct the dam. Normally, it is necessary to haul additional soil in from another site for dam construction.

Pond Capacity and Water Source. Watershed pond sizes vary considerably. The size is determined primarily by topography and personal preference. Ponds should be large enough to supply sufficient water to satisfy the intended use requirements (i.e., stock watering or irrigation in addition to fish culture). Shapes vary, but are usually irregular since they tend to follow land contours.

Water runoff from the surrounding water-

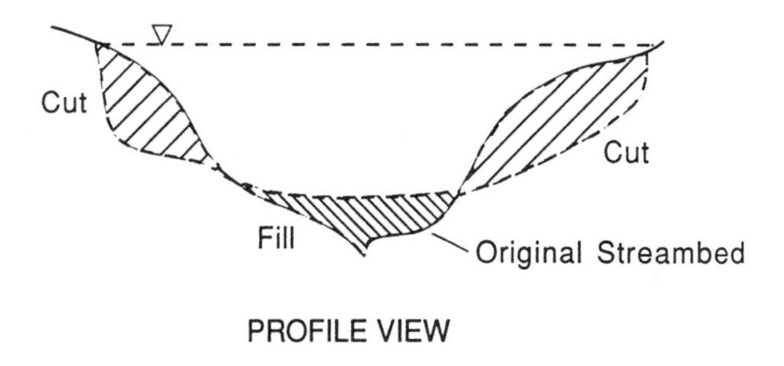

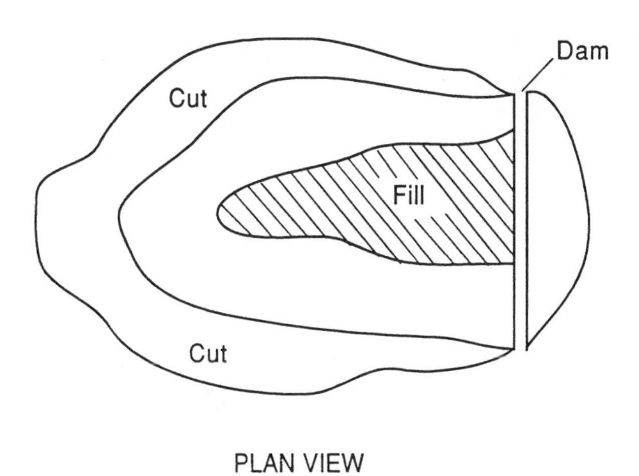

Figure 9-1. Illustration of how cutting and filling areas in a watershed pond can make the pond bottom more uniform.

shed is the main source of water. The pond must therefore be large enough to maintain sufficient water during dry periods. The amount of watershed required per hectare of pond varies with runoff characteristics and with rainfall patterns. In the Southeastern United States, usually 0.8–2 ha of watershed is adequate for each hectare (2–5 ac per ac) of pond surface (SCS 1982). The pond should be located in an area where it will not receive runoff from feedlots, orchards, row crops, or other sites where pesticides and/or manure are applied. The watershed should have sufficient grass cover to prevent runoff of soil into the pond. The site should be fenced to prevent livestock from trampling and damaging the pond side slopes and the dam.

A supplemental water source, such as a small well, is desirable even in areas of high rainfall. A well delivering 0.01–0.02 m^3/s/ha (10-15gpm/ac) is normally adequate during dry weather periods in the Southern United States (Tucker and Robinson 1990). Well water can also be used to supplement runoff for refilling ponds that were drained for harvest, thus reducing lost production time.

Soils. If additional soil is required for constructing the dam, potential borrow sites should be within an economical hauling distance, and the site should be studied to determine the nature of the soil to be used. Cohesive soils having a plasticity index of at least 15 are suitable for dam construction (Pillay 1990). The plasticity index is a measure of the interaction between the water and the cohesive plastic components in the soil. Soils that contain too much sand or gravel should not be used since they will not hold water. Clays and silty clays are best. Soils containing at least 20% clay by volume are suitable for fish ponds (Boyd 1982).

The soil forming the foundation where the dam is to be located must be able to support the weight of the dam. Swampy, muddy, or plastic soils should be removed and the dam constructed on the underlying consolidated soil material. In lieu of this procedure, suitable foundation soil can be hauled in from another location. Highly organic soils should not be used as foundation material since they decompose and cause the dam to settle and eventually leak (and possibly fail altogether). Foundation material should have a low permeability to prevent excessive seepage losses. A highly permeable soil can be used for foundation material provided that seepage can be prevented.

Dams. Detailed dam design is best left to a professional engineer, especially in the event that human life or valuable property may be lost due to dam failure. In the United States, the Soil Conservation Service (SCS) provides assistance for pond and dam design, the procedures of which are included in Soil Conservation Service (1982). The equations and calculations that follow can be used as a guide. For simplicity, all equations are left in their original English form.

Only dams constructed from earth are considered in this chapter. Dam construction using materials other than earth are beyond the scope of this book.

DAM HEIGHT. Fish pond dams are rarely higher than about 8 m (25 ft). The usual depth at the toe of the dam ranges from about 5–8 m (16–26 ft) (SCS 1982). Dam height is calculated as follows:

$$H = h_d + h_{st} + h_s + h_f \qquad (9\text{-}1)$$

where H = total dam height; h_d = normal water depth; h_{st} = storage depth; h_s = settlement factor; and h_f = freeboard. The cross section of a typical small, earthen dam is illustrated in Figure 9-2. The units for all of the above components are in meters or feet.

The *normal water depth* is the depth at the bottom of the dam. It is determined largely by topography. Ponds shouldn't be extremely deep because dam construction will be costly. In addition, water depths greater than about 1.5–2 m (5–6 ft) are difficult to harvest by seining. Deep ponds must be partially or completely drained to harvest the fish.

Storage depth is the amount of storage volume required to retain any additional water due to storms and floods to prevent damage to or complete failure of the dam. Storage height is the difference in elevation from the crest of the mechanical spillway to the emergency spillway. Storage may be obtained from the following relationship (SCS 1982):

$$h_{st} = \frac{\textit{drainage area, acres}}{(\textit{pond surface area, acres}) \times f} \qquad (9\text{-}2)$$

Figure 9-2. Cross section of dam showing placement of impermeable soil key for seepage control.

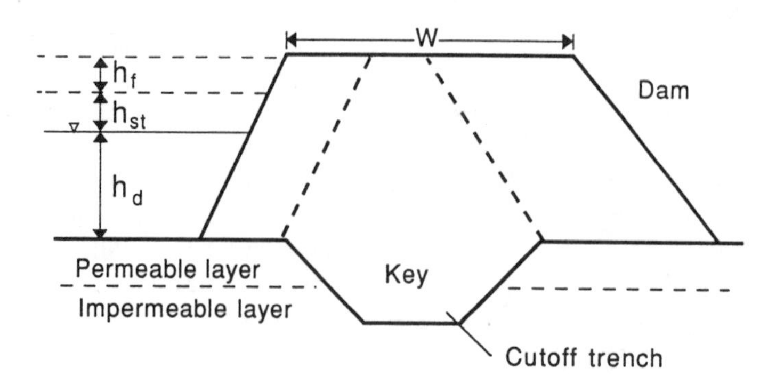

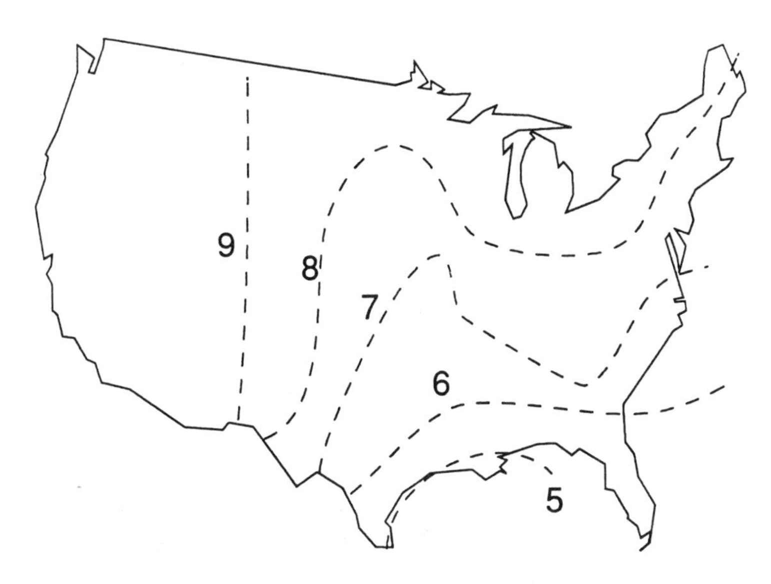

Figure 9-3. Rainfall amounts from a 24-hour storm based on a 50-year frequency (Source: SCS (1982)).

where h_{st} = water storage depth (ft); and f is obtained from Figure 9-3.

Example problem 9-1. Determine the storage depth for a 10-ac pond near Baton Rouge, LA, if the area that it drains is 30 ac.
Solution: From Figure 9-3, $f = 5$. Then,

$$h_{st} = \frac{30}{10 \times 5} = 0.6 \text{ ft}$$

The dam should be well compacted during construction to minimize settlement. Earth dams are normally constructed a little higher than the design dimension to allow for settlement. A soil of low permeability should be used for the dam foundation and applied in layers, each layer no greater than 0.2 m thick (Piedrahita 1991). Each layer should be well compacted with heavy construction equipment before the next layer is applied. This is called *roll-filled* construction. *Settlement* allowance for roll-filled dams averages about 1/12 of the water depth d (SCS 1982). Before the material is compacted, all tree roots, rocks, etc. should be removed.

Freeboard is an added dam height provided as a safety factor to prevent overtopping by wind and other causes. For ponds up to 200 m (660 ft) long, a freeboard no less than 0.3 m (1.0 ft) should be provided (SCS 1982). For ponds 200–400 m in length (660–1,320 ft) about 0.5 m (1.5 ft) of freeboard should be added.

Example problem 9-2. Calculate the total height of a dam for the 10-ac pond in example problem 9-1 if the water depth at the toe of the dam is to be 15 ft and the desired freeboard is 0.5 ft.
Solution: From Equation 9-1,

$$H = 15 \text{ ft} + 0.5 \text{ ft} + 1/12(15) \text{ ft} + 0.6 \text{ ft}$$
$$H = 17.25 \text{ ft}$$

CUTOFF TRENCH. Foundation *cutoff trenches* are used to deal with high seepage losses under the dam by joining the impervious strata in the foundation with the base of the dam. These are also called *core* trenches. The most common type is constructed from compacted clay material. A trench is cut parallel to the centerline of the dam to a depth that cuts well into the impervious strata beneath the surface. A cross section of a dam with a cutoff is shown in Figure

9-2. The bottom of the trench should be no less than 2.4 m (8 ft) wide and the side slopes no steeper than 1:1 (SCS 1982). The trench may be wider if construction equipment is required to work the trench. A layer of impermeable soil 15–20 cm (6–8 in.) thick is placed into the trench and compacted. Successive layers are added until the trench or *key* is complete. The impermeable key is then sealed to the underlying impermeable layer, effectively preventing leakage. Once the key is in place, the dam is completed in a similar manner. In some instances the key may extend to the top of the dam, as shown in the figure.

TOP WIDTH. The dam top width will vary according to the total dam height. The U.S. Soil Conservation Service (1982) recommends a minimum top width of 1.8 m (6 ft) for dams under 3 m (10 ft) tall (Table 9-1). The width is increased as the height of the dam increases. If the dam is to be used for a roadway, the minimum recommended top width is 4.9 m (16 ft). This provides for a shoulder on each side to prevent raveling.

SIDE SLOPES. The steepness of the side slopes on a dam depends on the stability of the fill and foundation materials. Unstable materials require flatter slopes; the more stable the material, the steeper the slopes allowed. Recommended slopes are shown in Table 9-2. For stability, the slopes should not be steeper than those recommended in the table, but they can be flatter. Side slopes can be contoured to blend with the surrounding landform.

Table 9-1. Dam height and minimum recommended top width.

Height of dam		Minimum top width	
(m)	(ft)	(m)	(ft)
under 3	under 10	1.8	6
3.4–4.3	11–14	2.4	8
4.6–5.8	15–19	3.0	10
6.1–7.3	20–24	3.7	12
7.6–10.4	25–34	4.3	14

Source: SCS (1982).

Table 9-2. Recommended side slopes for earth dams.

	Side slope	
Fill material	Upstream	Downstream
Clayey sand, clayey gravel, sandy clay, silty clay, silty gravel	3:1	2:1
Silty clay, clayey silt	3:1	3:1

Source: SCS (1982).

ESTIMATING THE VOLUME OF EARTH FILL. Once the dam height is determined using Equation 9-1, the volume of fill for the dam must then be estimated. The most simple method for estimating the fill required is with the sum-of-end-area method described in SCS (1982). A centerline profile should be used to establish the ground surface elevation at all points along the dam centerline where the side slopes change significantly. Once the settled top elevation of the dam is established, the fill height can be obtained by subtracting the ground surface elevation from the dam settled top elevation. Once the fill height, side slope, and top width dimensions are established, the end areas at each point along the centerline can be estimated using Table 9-3. The volume for each dam section can then be estimated by summing the two end areas of the section, dividing by 2, and multiplying the result by the length of the section. The total volume is then the sum of the sections.

POND OUTLETS. Ponds should be constructed with two types of outlets: a mechanical and an emergency spillway. A *mechanical spillway* carries the normal expected water flow from the pond and can be designed to either partially or completely drain the pond to facilitate harvest. Two types of mechanical spillways commonly used for fish ponds are illustrated in Figure 9-4a. The first, called a *drop-inlet trickle tube*, consists of a common interior riser pipe with a drainpipe. The crest of the riser maintains the desired water depth in the pond. A hood is sometimes used to cover the riser inlet to prevent trash from clog-

Table 9-3. End areas (square feet) of embankment sections for various side slopes and top widths[1].

Fill Height (ft)	Side slopes					Top width (ft)				
	2.5:1 2.5:1 2:1 3:1	2.5:1 3:1 2:1 3.5:1	3:1 3:1 2.5:1 3.5:1	3.5:1 3.5:1 3:1 4:1	4:1 4:1 3:1 5:1	8	10	12	14	16
1.0	3	3	3	4	4	8	10	12	14	16
1.2	4	4	4	5	6	10	12	14	17	19
1.4	5	5	6	7	8	11	14	17	20	22
1.6	6	7	8	9	10	13	16	19	22	26
1.8	8	9	10	11	13	14	18	22	25	29
2.0	10	11	12	14	16	16	20	24	28	32
2.2	12	13	15	17	19	18	22	27	31	35
2.4	14	16	17	20	23	19	24	29	34	39
2.6	17	19	20	24	27	21	26	31	36	42
2.8	20	22	23	27	31	22	28	34	39	45
3.0	22	25	27	32	36	24	30	36	42	48
3.2	26	28	31	36	41	26	32	38	45	51
3.4	29	32	35	40	46	27	34	41	47	55
3.6	32	36	39	45	52	29	36	43	50	58
3.8	36	40	43	50	58	30	38	46	53	61
4.0	40	44	48	56	64	32	40	48	56	64
4.2	44	49	53	62	71	34	42	50	59	67
4.4	48	53	58	68	77	35	44	53	61	71
4.6	53	58	63	74	85	37	46	55	64	74
4.8	57	63	69	81	92	38	48	57	67	77
5.0	62	69	75	87	100	40	50	60	70	80
5.2	67	74	81	94	108	42	52	62	73	83
5.4	73	80	87	102	117	43	54	65	75	87
5.6	78	86	94	110	125	45	56	67	78	90
5.8	84	93	101	118	135	46	58	69	81	93
6.0	90	99	108	126	144	48	60	72	84	96
6.2	96	106	115	135	154	50	62	74	87	99
6.4	102	113	123	143	164	51	64	77	89	103
6.6	109	120	131	152	174	53	66	79	92	106
6.8	116	128	139	162	185	54	68	81	95	109
7.0	123	135	147	172	196	56	70	84	98	112
7.2	130	143	156	172	207	58	72	86	101	115
7.4	138	152	165	193	219	59	74	89	103	119
7.6	145	159	174	203	231	61	76	91	106	122
7.8	153	168	183	214	243	62	78	93	109	125
8.0	160	176	192	224	256	64	80	96	112	128
8.2	169	185	202	235	269	66	82	98	115	131
8.4	177	194	212	247	282	67	84	101	117	135
8.6	186	204	222	259	296	69	86	103	120	138
8.8	194	213	232	271	310	70	88	105	123	141
9.0	203	223	243	283	324	72	90	108	126	144
9.2	212	233	254	296	339	74	92	110	129	147

(continued)

Table 9-3. End areas (square feet) of embankment sections for various side slopes and top widths[1] (continued)

Fill Height (ft)	Side slopes					Top width (ft)				
	2.5:1 2.5:1 2:1 3:1	2.5:1 3:1 2:1 3.5:1	3:1 3:1 2.5:1 3.5:1	3.5:1 3.5:1 3:1 4:1	4:1 4:1 3:1 5:1	8	10	12	14	16
9.4	222	244	266	310	353	75	94	113	131	151
9.6	231	254	277	323	369	77	96	115	134	154
9.8	241	265	289	337	384	78	98	117	137	157
10.0	250	275	300	350	400	80	100	120	140	160
11.0	302	333	363	424	484	-	110	132	154	176
12.0	360	396	432	504	576	-	120	144	168	192
13.0	422	465	507	592	676	-	130	156	182	208
14.0	490	539	588	686	784	-	140	168	196	224
15.0	563	619	675	788	900	-	150	180	210	240
16.0	640	704	768	896	1024	-	160	192	224	256
17.0	723	795	867	1,012	1,156	-	-	204	238	272
18.0	810	891	972	1,134	1,296	-	-	216	252	288
19.0	903	993	1083	1,264	1,444	-	-	228	266	304
20.0	1,000	1,100	1,200	1,400	1,600	-	-	240	280	320
21.0	1,103	1,213	1,323	1,544	1,764	-	-	252	294	336
22.0	1,210	1,331	1,452	1,694	1,936	-	-	264	308	252
23.0	1,323	1,455	1,587	1,852	2,116	-	-	276	322	268

[1] To find end area for any height, add square feet given under side slopes to that under the top width for total section. Example: 6.4-ft fill with 3:1 front and back slopes, 14-ft top width (123 ft^2 + 89 ft^2 for a total of 212 ft^2 for the section.

Source: SCS (1982).

ging the riser. A valve on the drainpipe exterior to the pond can be opened for draining. The drainpipe outlet should be located above the high-water line of the receiving channel to prevent unwanted fish from entering through the drainpipe during high-water periods. Riser pipes are fabricated from a variety of materials, including concrete, smooth steel, PVC, and corrugated metal. When the required discharge capacity is known, Tables 9-4 and 9-5 can be used to select adequate pipe sizes for the riser and drainpipe. The diameter of the riser must be larger than that of the drainpipe so the drainpipe will flow full. SCS (1982) recommends a minimum riser pipe diameter of 15.2 cm (6 in.). A second common type of mechanical spillway is illustrated in Figure 9-4b. This spillway consists of an external riser pipe and valve. Outside riser pipes and valves are easier to use and maintain (Tucker and Robinson 1990). For both types, SCS (1982) recommends that the drainpipe extend a minimum of 2.4–3.0 m (8–10 ft) beyond the toe of the dam to prevent erosion problems. Erosion protection in the form of rip-rap, concrete, gravel, etc., should be placed around the outlet at the toe of the dam as shown in Figure 9-4. The drain lines should have at least a 1% slope for good flow.

Another alternative is illustrated in Figure 9-5. This system is called a *monk* and is used primarily for larger ponds (Huet 1970). Horizontally laid boards are used to control water level. The pond can be drained completely by removing all of the boards. This outlet is less favorable than the others because it drains only surface water. When draining ponds, it is most

Table 9-4. Discharge values for trickle tubes of smooth pipe.[1]

Total head		Ratio of barrel diameter to riser diameter in inches					
(m)	(ft)	6:8	8:10	10:12	12:15	15:24	18:36
		ft³/s					
1.8	6	1.54	3.1	5.3	8.1	13.6	20.6
2.4	8	1.66	3.3	5.7	8.9	14.8	22.5
3.0	10	1.76	3.5	6.1	9.6	15.8	24.3
3.7	12	1.86	3.7	6.5	10.7	16.8	26.1
4.3	14	1.94	3.9	6.8	10.7	17.8	27.8
4.9	16	2.00	4.0	7.0	11.1	18.6	29.2
5.5	18	2.06	4.1	7.2	11.5	19.3	30.4
6.1	20	2.10	4.2	7.4	11.8	19.9	31.3
6.7	22	2.14	4.3	7.6	12.1	20.5	32.2
7.3	24	2.18	4.4	7.8	12.4	21.0	33.0
7.9	26	2.21	4.5	8.0	12.6	21.5	33.8

[1] Length of pipe barrel used in calculations is based on a dam with a 12-ft top width and 2.5:1 side slopes. Discharge values are based on a minimum head on the riser crest of 12 in. Pipe flow based on Manning's $n = 0.012$.

Source: SCS (1982).

Table 9-5. Discharge values for trickle tubes of corrugated metal pipe.[1]

Total head		Ratio of barrel diameter to riser diameter in inches					
(m)	(ft)	6:8	8:10	10:12	12:15	15:24	18:36
		ft³/s					
1.8	6	0.85	1.73	3.1	5.1	8.8	14.1
2.4	8	0.90	1.85	3.3	5.4	9.4	15.0
3.0	10	0.94	1.96	3.5	5.7	9.9	15.9
3.7	12	0.98	2.07	3.7	6.0	10.4	16.7
4.3	14	1.02	2.15	3.8	6.2	10.8	17.5
4.9	16	1.05	2.21	3.9	6.4	11.1	18.1
5.5	18	1.07	2.26	4.0	6.6	11.4	18.6
6.1	20	1.09	2.30	4.1	6.7	11.7	18.9
6.7	22	1.11	2.34	4.2	6.8	11.9	19.3
7.3	24	1.12	2.37	4.2	6.9	12.1	19.6
7.9	26	1.13	2.40	4.3	7.0	12.3	19.9

[1] Length of pipe barrel used in calculations is based on a dam with a 12-ft top width and 2.5:1 side slopes. Discharge values are based on a minimum head on the riser crest of 12 in. Pipe flow based on Manning's $n = 0.025$.

Source: SCS (1982).

often advantageous to drain off the bottom to remove the stale lower water layers first.

To retard seepage through the dam along the drainpipe, the fill around the pipe should be compacted. In addition, *antiseep collars* should be constructed around the drainpipe to increase the flow path that water must travel, as shown in Figures 9-4 and 9-5. Antiseep collars are com-

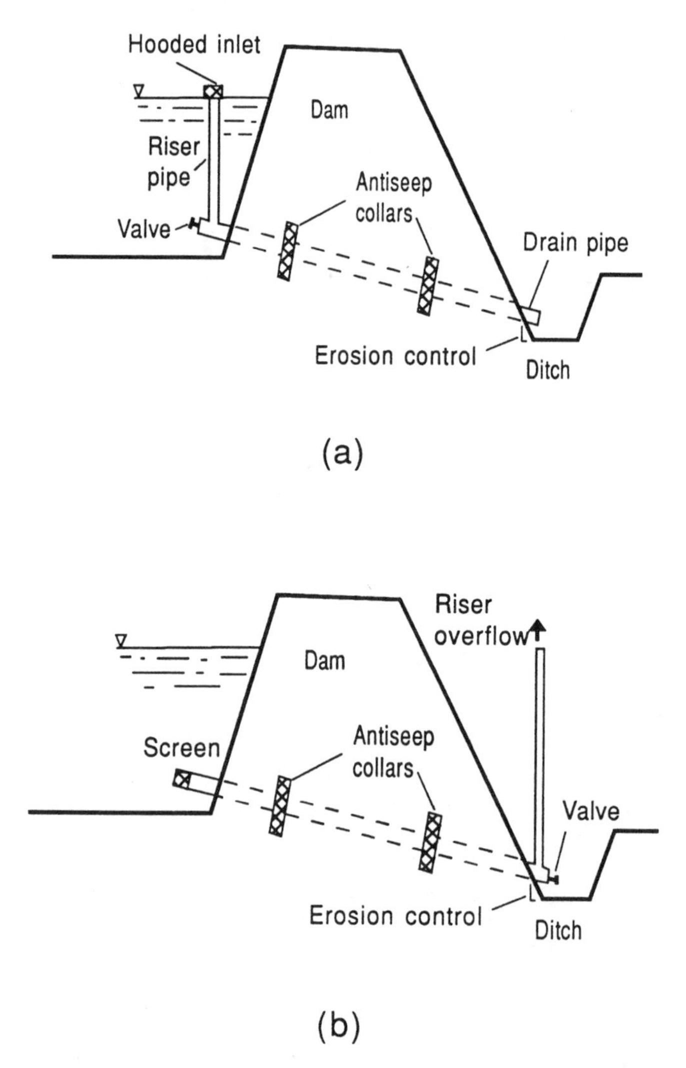

Figure 9-4. Cross section of dam showing two types of drains commonly used: (a) Hooded inlet pipe; and (b) outside drain with valve. Also note use of antiseep collars.

monly constructed from concrete or metal. The collars should extend a minimum of 61 cm (24 in.) into the fill. For dams less than 4.6 m (15 ft) high one antiseep collar at the dam centerline is sufficient. For taller dams two or more collars equally spaced between the fill centerline and the upstream end of the drainpipe should be used (SCS 1982).

The second type of pond outlet is called an *emergency spillway*. The emergency spillway takes care of peak storm peak flows that exceed the capacity of the mechanical spillway. Small farm ponds may not have an emergency spillway. The emergency spillway is usually constructed into the undisturbed soil at either end of the dam. Soil borings should be taken to locate an adequate site for an emergency spillway. Loose, sandy soils and other highly erodible soils should be avoided. No matter how well the dam is constructed, it will probably be severely damaged during the first heavy storm after construction if the capacity of the spillway is exceeded. Emergency spillway design is discussed in detail in SCS (1982).

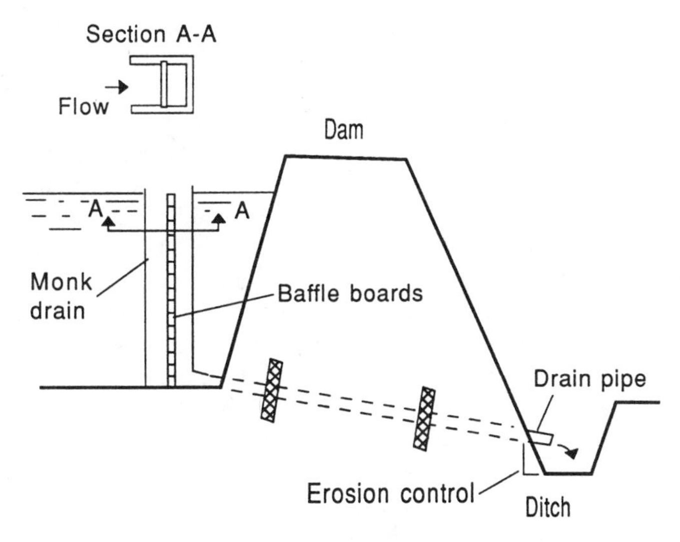

Figure 9-5. Mechanical spillway with variable level monk drain.

Levee Ponds. *Levee ponds* are the principal type used for commercial finfish, shrimp, and crawfish production. They are constructed either by digging holes in the ground and building levees around them or by constructing above-grade dikes or levees to impound water. Ponds constructed by digging holes in the ground are sometimes referred to as *excavated* ponds.

WATER SUPPLY. Water flow rates for fish ponds should be sufficient to fill a pond in a reasonable amount of time. However, what is reasonable at one site may not be adequate at another. Tucker and Robinson (1990) recommend that the time required to fill ponds should be a maximum of two weeks. The flow rate required to fill a pond in a two-week period can be estimated using Equation 9-3:

$$Q = V/336 \tag{9-3}$$

where Q = flow rate (m^3/hr); and V = pond volume (m^3). This equation assumes that infiltration, rainfall, and evaporation are negligible.

The water exchange rate in most fish ponds is relatively small, with residence times of 20–30 days being common (Piedrahita 1991). Exceptions are shrimp and crawfish ponds where water exchange rates of 25–33% per day, are not unusual (Brune and Drapcho 1991; Lawson et al. 1992).

Levee ponds cannot be filled by surface runoff. Therefore, water must be pumped from a well or a surface source. A well is usually drilled after ponds are constructed so that it can be strategically located to maximize efficiency of water delivery. Well water must be of the proper salinity, total alkalinity, and hardness, as discussed in Chapter 2. It should also be free of pollutants, and usually must be aerated before use. Aeration is discussed in Chapter 11.

LAND USE. Land may have certain restrictions that will limit its use for aquaculture purposes. Land use restrictions and zoning laws should be investigated before land is purchased or it should be included as part of a site evaluation. The U.S. Army Corps of Engineers requires a permit before clearing and/or building on land in the United States and/or territories that are classified as *wetlands*. Other permits may be required for aquaculture land use purposes. The entrepreneur should contact federal, state, and local agencies concerning permits before proceeding with pond construction.

SIZE, SHAPE AND LAYOUT. The ideal pond size, shape, and layout are a compromise among land topography, property lines, parcel shape, construction and operating costs, available resources, and management ease. The farm should be affordable and economical to construct and operate. Pond sizes should be compatible with production goals, expected income, and marketing strategy. Sizes vary depending upon their use (i.e., either for food fish production, fingerling production, holding brood fish, spawning or other use), and they should match the water supply and desired filling time. Pond sizes typically range from 0.05–2.0 ha (0.12–5 ac) for nursery ponds and 0.25–10 ha (0.5–25 ac) for brood stock or grow-out ponds. Ponds for intensive culture range from about 1–5 ha (2.5–12 ac) (Pillay 1990). Spawning ponds average about 0.01–0.05 ha (0.03–0.15 ac).

Small ponds are generally easier to manage, and fish production per unit surface area is greater (Tucker and Robinson 1990). However, construction costs are greater as the pond size decreases, due largely to additional levee requirements. Additionally, small ponds have less water area on a given amount of land because a larger land area is covered by levees, supply channels and drainage ditches. In the channel catfish industry in the Southern United States most ponds are about 6.9 ha (17 ac) in size, occupying 8.1 ha (20 ac) of land area (Boyd 1985a; Tucker and Robinson 1990). About 1.2 ha (3 ac) is taken up by levees and canals. Many farmers prefer to have smaller ponds (2–4 ha or 5–10 ac) for brood fish or fingerlings and larger ponds (6–8 ha or 15–20 ac) for food fish production.

Pond water depth largely depends on climatic conditions and topography, but some guidelines are helpful. For catfish production, Tucker and Robinson (1990) recommend a minimum depth of 0.9 m (3 ft) and a maximum depth of about 1.5 m (5 ft) if the pond is to be harvested by seining. Water depth in trout ponds averages about 1.0 m (3.3 ft) at the inflow and slope to 1.5–2.0 m (5.0–6.6 ft) at the outflow (Pillay 1990). Shrimp ponds are rarely deeper than about 0.6–0.9 m (2–3 ft). Ponds used for crawfish production are fairly shallow, averaging 0.5–0.6 m (1.5–2.0 ft).

Although levee ponds can be built to almost any desired shape, it appears that there is a preference for rectangular-shaped ponds in freshwater aquaculture since they can be constructed adjacent to one another, making water supply and management practices easier. Regular features and flat bottoms also make harvesting easier (Pillay 1990). Square ponds are more economical to construct than rectangular ponds, but on steep slopes, rectangular ponds are easier to drain. Regardless of the shape, ponds should be no more than 213 m (700 ft) wide to accommodate harvesting by seining (Tucker and Robinson 1990).

The natural land slope should be utilized to lay out ponds and drainage canals. Flatland ponds should be constructed with a gently sloping (0.05–0.1% slope) bottom to facilitate drainage. In hilly areas the deep end of the pond should be located at the lowest elevation. Catch basins (10% of the area of the pond; 2–3 ft deep) are often located at the drain end of fish ponds to facilitate harvesting and draining.

Ponds constructed adjacent to one another conserve land area by using a common levee. Figure 9-6 illustrates a typical layout for four levee ponds. Notice that the well is located in the center of the group to minimize the distances that water must be pumped to each pond. This reduces the cost for pipe fittings, etc. The well should be located at the highest point of elevation to take advantage of gravity flow. It is also a good idea to locate the well off-center as shown so that it will not block vehicular traffic.

Ponds should be laid out to take advantage of prevailing winds. If rectangular ponds are laid out with the long axis parallel to prevailing winds, the windward levee may be subject to erosion due to wave action. The fetch (straight-line distance the wind travels) is reduced and potential erosion is reduced if the pond is laid out with long axis vertical to prevailing winds. Pond mixing is also accomplished as the wind travels across the pond.

ESTIMATING VOLUME OF EXCAVATION. Once the dimensions and side slopes of ponds are selected, the volume of excavation required

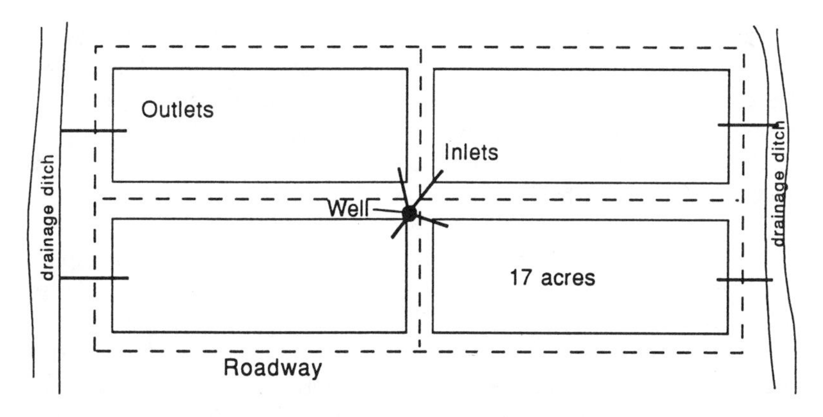

Figure 9-6. Typical layout for four 17-ac catfish ponds showing location of well.

can be estimated. Knowing the volume of excavation will help determine the cost of the excavation, it will give an accurate estimate of the volume of water the pond will hold and it will enable the designer to determine if additional soil is required to construct the levees. The volume of excavation required can be estimated with good accuracy using the *prismoidal formula*. Equation 9-4 is used for the SI system of units:

$$(SI) \qquad V = \frac{(A + 4B + C)}{6} \times D \qquad (9\text{-}4)$$

where V = volume of excavation, m^3; A = area of excavation at the ground surface, m^2; B = area of excavation at mid-depth, m^2; C = area of excavation at the pond bottom, m^2; and D = average depth of the pond, m. Equation 9-5 is used for the English system:

$$(English) \qquad V = \frac{(A + 4B + C)}{6} \times \frac{D}{27} \qquad (9\text{-}5)$$

where V is in yd^3; A, B and C are in ft^2; and D is in ft.

LEVEE CONSTRUCTION. Many of the same considerations used for constructing earthen dams are also utilized for levee construction. Soil that is removed when excavating ponds is usually used to construct the levees. If insufficient soil volume is removed during excavation, suitable soil must be hauled in from some other location. The most common side slopes for levees are 4:1 (four units horizontal to one unit vertical) on the water side and 3:1 on the outside. Steeper side slopes (3:1) can be used throughout if the soil is well compacted and contains enough clay to prevent slumping. Levees with flatter slopes are more expensive to build, and flatter slopes encourage aquatic weed growth. Levee side slopes should have a good sod cover to minimize erosion.

Levee width is determined by the pond size, types of equipment used for construction, and the types of vehicles that will be using the levees. Levee size should not be skimped on to save money since erosion can often further reduce the size. Levees should be built to last and must be large enough and compact enough to support large trucks or heavy equipment. Tucker and Robinson (1990) recommend a minimum levee top width of 4.9 m (16 ft) to accommodate feeding and harvesting equipment. Wider levees are more convenient and more resistant to wave damage but are more expensive to build. Good pond management dictates that motorized vehicles should have access to ponds on at least two sides. Also, levee tops on at least one side of the pond should be covered with gravel for all-weather access.

Levees should be adequately compacted as they are constructed. This is done by putting down a layer of soil and then compacting it with a sheepsfoot roller, bulldozer, or other heavy equipment before putting another layer on top.

Soil should be compacted after each successive layer is put down. In some cases a core trench may be required to key the levee into underlying subsoils to prevent leakage at the toe and to prevent shearing of the levee from the ground surface due to hydrostatic pressure behind the levee.

The volume of fill required for levee construction can be estimated using the end-area method that was demonstrated for dams. Knowing the total linear dimension around the pond along the centerline of the levee, the volume of fill (m^3 per meter of length) for levees can be estimated using Table 9-6.

Freeboard for levee ponds should be a minimum of 0.3 m (1.0 ft) and a maximum of 0.6 m (2 ft) (Tucker and Robinson 1990). This allows the pond to hold rainwater without flooding and also prevents overtopping of the levee by wave action during windy periods. The volume of water required to fill a pond to a certain depth can be calculated using Equations 9-4 and 9-5 using the actual water depth for *D*.

DRAINAGE FROM PONDS. Good pond drainage is essential. Drainage ditches should be lower in elevation than pond bottoms to permit complete drainage of all ponds. Ponds should be laid out and constructed to permit complete drainage of each by gravity flow. The elevation of the pond bottom at the drain should be 0.3–0.6 m (1.0–2.0 ft) above the water level in the receiving ditch to prevent fish from entering the pond through the drainpipe during floods. Two types of drain structures are the most common: inside turn-down drains and outside valve drains (Figure 9-7). The turn-down drain (Figure 9-7a) is the most popular. It serves as both an overflow and a drainpipe. The water level can be adjusted by pivoting the pipe. A special antiseep collar should be constructed around the pipe to prevent seepage. Pipes may have to be weighted in some instances to prevent them from floating. One disadvantage of the turn-down drain is that it removes water from the surface. To compensate for this, a second type of turn-down has a double sleeve that allows draining off the bottom (Figure 9-7b).

Another type of drain structure is the drainpipe with the valve located outside the pond (Figure 9-7c). The drain inlet should be

Table 9-6. Approximate volume of dirt to fill a 1-ft section of pond levee of various heights, slopes, and top widths.

Leeve height		Slopes total 6:1[1]					Slopes total 7:1[2]				
		Top widths					Top widths				
(m)	(ft)	12	14	16	18	20	12	14	16	18	20
		ft³									
1.52	5.0	135	145	155	165	175	147	157	167	177	187
1.59	5.2	143	154	164	175	185	157	167	177	188	198
1.65	5.4	152	163	174	185	195	167	177	188	199	210
1.71	5.6	161	172	184	195	206	176	188	199	211	222
1.77	5.8	170	182	193	205	217	187	199	211	223	234
1.83	6.0	170	182	204	216	228	198	211	223	234	246
1.89	6.2	190	202	214	226	239	209	223	234	246	259
1.95	6.4	200	212	225	238	251	220	233	245	259	217
2.01	6.6	210	223	236	250	263	232	245	258	271	285
2.07	6.8	220	234	248	261	275	244	257	270	285	298

[1] 6:1 total slope equals levee with inside and outside slopes of 3:1.

[2] 7:1 total slope equals levee with 3:1 slope on one side and 4:1 slope on the other side.

Source: SCS (1982).

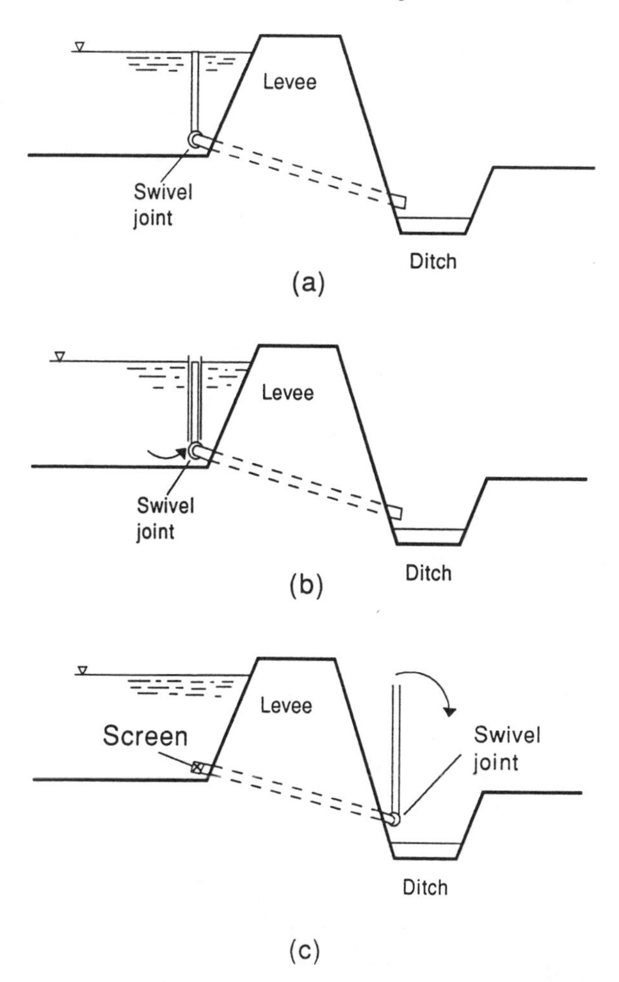

Figure 9-7. Drainage methods: (a) Turn-down drain; (b) double-sleeve turndown drain; and (c) swivel joint on the outside of the pond.

positioned as near to the pond bottom as possible for complete drainage. Drainpipes extend through the levee and should have a slope of about 1% for good drainage. Table 9-7 shows estimated discharge rates for short drainpipes with low head pressure. Tucker and Robinson (1990) recommend a minimum 30.5-cm (12in.) diameter drainpipe for ponds larger than 4 ha (10 ac). Smaller drainpipes (11-cm or 4 in. in diameter) are adequate for smaller ponds. The drain size should be adequate to drain the pond in two to three days. Multiple drainpipes can be used for larger ponds.

Table 9-7. Estimated average discharge rates for short drainpipes under low head pressure[1].

Pipe diameter		Approximate discharge	
(mm)	(in.)	(L/min)	(gpm)
102	4	454	120
152	6	1,325	350
203	8	2,271	600
254	10	3,785	1,000
305	12	6,056	1,600
356	14	9,084	2,400

[1] To calculate the time to drain a pond in days:

$$\frac{ac\text{-}ft \times 325{,}851}{discharge\ gpm \times 1{,}440} = drain\ time\ in\ days.$$

Sealing Leaking Ponds

Excessive seepage will occur in ponds that are built on sites where the soil is too permeable to hold water. Poor site selection is often the result of an inadequate site investigation. In some instances there is no choice but to use soils of an inferior quality for pond construction. If so, the original pond designs must have a provision for reducing seepage by sealing the pond in some manner. Seepage can be reduced by decreasing the soil's permeability, either by compaction, by incorporating clay into the soil, or by chemical addition. Often the only recourse is to use impermeable plastic liners.

The least expensive method of sealing leaky ponds is to compact the soil. If the soil contains a wide range of sand, clay, small gravel, etc. The pond area should be cleared of all trees, other vegetation, rocks, and roots. Stump holes should be filled and compacted. The soil should then be loosened by rototilling, and it should then be compacted with four to six passes of compacting equipment. The compacted seal should be no less than 20 cm (8 in.) thick where 3 m (10 ft) or less of water is to be impounded. The thickness of the seal should be increased where the water depth is greater.

A wide variety of compactors are available such as the sheepsfoot roller (Figure 9-8), rubber-tired rollers, and steel drum rollers. They are difficult to use in mucky soil conditions, however. In such conditions the dragline is more useful, and dikes can be constructed in layers with compacting by several passes of the dragline between layers. Bulldozers can also be used to compact soils.

Soils low in clay content can be sealed by adding additional clay particles. Bentonite is a

Figure 9-8. A sheepsfoot roller used to compact pond levees (photo by T. Lawson).

ine-textured colloidal clay. When wet it absorbs several times its own weight in water and wells to 8–20 times its original volume. When nixed into the soil in the proper proportions and ompacted, the bentonite swells on contact with vater so that the pores between large soil granles are filled, effectively sealing the soil. A soil nixture of at least 20% clay should be used to ine the surface to seal the soil at the bottom and ides of the pond. The blanket should be a minmum of 30 cm (12 in) thick for a 3 m (10 ft) vater depth. It's thickness should be increased y 5 cm (2 in.) for every 0.3 m (1 ft) of water lepth over 3 m (10 ft). The clay is applied by preading a thick (15–20 cm) layer over the area nd compacting by several passes with comacting equipment (SCS 1982).

Good pond soils should contain 15–50% lay, otherwise they are too clumpy and hard to vork. The structure of soils containing more han 50% clay can be improved with soil addiives called *dispersing agents*. The most comnonly used dispersing agents are the sodium olyphosphate and sodium chloride (common able salt), although others are used. Tetrasolium pyrophosphate and sodium tripolyphoshate are the two most effective of the sodium olyphosphates. Technical grade soda ash, 99–00 percent sodium carbonate, can also be used. Sodium polyphosphates are applied at the rate f 0.25–0.5 kg/m^2, sodium chloride at 1.0–1.6 g/m^2, and soda ash at 0.5–1.0 kg/m^2 (SCS 1982). In order to determine which dispersing gent and what application rate will be most ffective, a laboratory analysis of the soil beorehand is essential. The dispersing agent hould be applied as a blanket in a single layer r several layers, as previously described for entonite.

Probably the most expensive method of sealng ponds is with impermeable plastic liners. Pond liners are fabricated in a number of sizes nd thicknesses by several manufacturers. Prices typically range from about $2.16–3.78 er m^2 ($0.20–0.35 per ft^2) in the United States. They are made from polyethylene, polybutyene, vinyl, rubber, and asphalt-sealed fabric. Some liners are not made for fish contact and may contain toxic substances. The aquaculturist should check with the manufacturer concerning toxicity and use only aquaculture-grade liners. In addition, UV-resistant liners should be used since the sun's rays can rapidly decompose plastics which aren't resistant. When using liners, the pond bottom and sides should be carefully prepared before the liner is installed, and all sticks, rocks, gravel, and other sharp objects that can puncture the liner should be removed.

Special Types of Ponds

There are several special types of aquaculture ponds that do not fall into general aquaculture pond categories, and they warrant a discussion of their own. One special pond type, the crawfish pond, is similar to other types of earthen ponds in construction except for differences in water depth and levee arrangement. Other pond types are not really ponds in the general sense of the word, but are tanks. Two factors distinguishing a pond from a tank are size and construction materials. Ponds are usually much larger than tanks and are of typically of earthen construction whereas tanks are smaller and are fabricated from man-made materials.

Crawfish Ponds. Crawfish ponds warrant a section of their own because of the size of the industry in the United States and worldwide, and because they require special design features not seen in ordinary finfish ponds. Two species of crawfish are of commercial importance in the United States: the red swamp crawfish (*Procambarus clarkii*) and the white river crawfish (*P. zonangulus*). Other species are also pond-cultured to a lesser degree. The industry is focused in Louisiana where over 48,000 ha (120,000 ac) of ponds produce over 22,000 MT (50 million lb) of crawfish annually.

Crawfish ponds are constructed in a similar manner to other types of earthen fish ponds with two exceptions: they are usually shallow (0.3–0.6 m deep) and they are laid out with a network of small internal baffle levees inside the main perimeter levees for water control. Two water management techniques dominate the industry:

(1) A nonrecirculating technique where water exchange is practiced; and (2) a recirculating technique where the same culture water is used repeatedly.

Figure 9-9 illustrates the non-recirculating crawfish pond type. The perimeter levees are normally about 0.9–1.2 m (3–4 ft) high after compaction and settling. Levee freeboard is about 0.3–0.7 m (1.0–2.0 ft). The perimeter levees should have a minimum top width of 1.8–2.4 m (6–8 ft), which makes the levee wide enough to prevent crawfish from burrowing through and causing leaks. The levees on at least two sides of the pond should be wide enough to support vehicular traffic (de la Bretonne and Romaire 1989).

Baffle levees are used to direct water flow through the pond in a serpentine fashion (Figure 9-9a). This is to eliminate dead areas in the pond (areas of low dissolved oxygen concentration). They are usually somewhat smaller than the perimeter levees, averaging about 0.6–0.9 m (2–3 ft) in height. Baffle levees are usually no compacted like the perimeter levees, but the should be built to a sufficient height so that the extend above the water surface during a ful flood. Baffle levee spacing is normally abou 30–60 m (100–200 ft) (de la Bretonne and Ro maire 1989). A cross-sectional view of perime ter and baffle levees is illustrated in Figur 9-9b. Special crawfish pond designs incorporat ing paddle wheel aerators for water circulatio are described in Chapter 11.

Foster-Lucas Tank. A variation from th traditional aquaculture pond is the oval-shape Foster-Lucas pond. The original Foster-Luca tank had the dimensions shown in Figure 9-10 Larger systems are referred to as ponds. Thi type culture unit may be used in the recirculat ing mode with external water purification, or i the flow-through mode. Burrows and Che

Figure 9-9. Layout of levees in a nonrecirculating or flow- through crawfish pond.

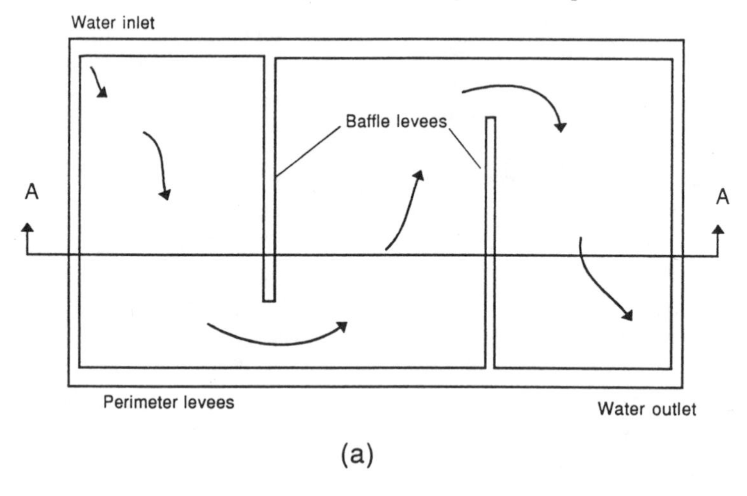

(a)

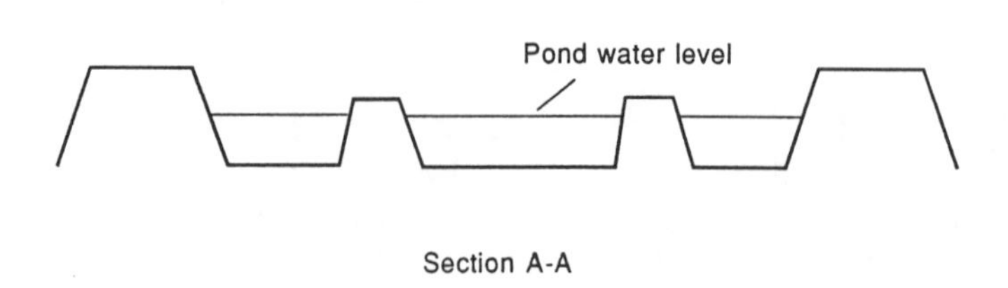

Section A-A

(b)

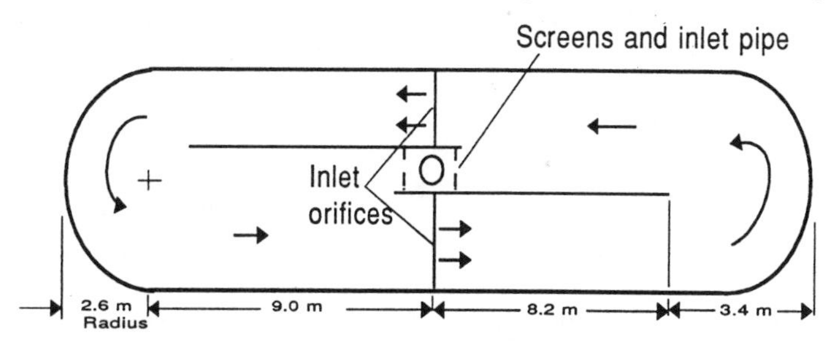

igure 9-10. A Foster-Lucas tank.

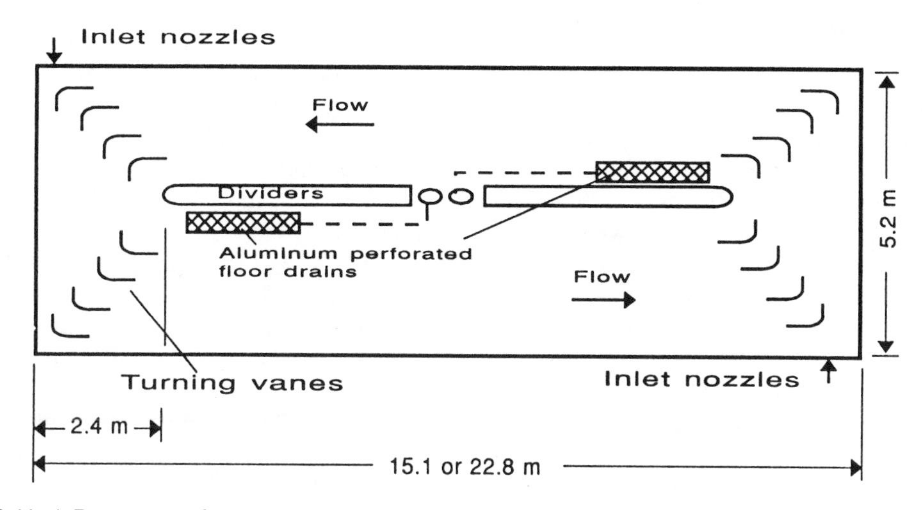

igure 9-11. A Burrows pond.

oweth (1955) compared water flow patterns nd short-circuiting characteristics of a Foster-ucas pond to those of a circular pond and a iceway. The dimensions of the tank and flow atterns observed are shown in the figure. Two alls offset from one another form dividers. Vater circulation is created by nozzles located oove the tank at the midpoint, which forces ater to flow in opposite directions around the viders. The discharge is centrally located be-veen the dividers. This type of pond was once idely used for salmonid culture.

Burrows Pond. Burrows and Chenoweth 970) modified the Foster-Lucas pond into hat they called a "recirculating rectangular nk," which later became known as the *Bur-ws pond* (Figure 9-11). The primary modifi-tions consisted of squaring the corners and using vertical turning vanes to force water to flow around the corners. The turning vanes prevented the deposition of waste in the corners and, hence, the development of dead zones. However, their addition made fish harvest difficult. A central wall divides the pond into two sections of equal width. Water under pressure and relatively high velocity is introduced through two inflow pipes located at opposite ends of the pond. The water flows parallel to the outside walls, gradually moves toward the center, and discharges through perforated plates located in the pond bottom at opposite ends of the center wall (not shown). Like the Foster-Lucas tank, the Burrows pond may be operated as a component in a recirculating system, or it may stand alone as a raceway-type unit.

Design dimensions and flow rates of the Burrows pond are very specific. Pond length is either 15 or 22 m (48 or 70 ft), and the width is 5 m (16 ft). Water depth is maintained by a removable standpipe located in the drain line. The pond functions well at a depth of 76 or 91 cm (30 or 36 in.) and is relatively self-cleaning at water inflows of 1.5 m^3/min (400 gpm) or greater. Fish are well distributed throughout the pond as water currents carry food to them (Piper et al. 1982).

RACEWAY CULTURE SYSTEMS

Raceways are culture units in which water flows continuously, making a single pass through the unit before being discharged. These systems are also referred to as *flow-through systems*. The residence time of the water in a raceway is very short, usually on the order of a few minutes instead of hours or days as in ponds. In the United States, raceways have traditionally been used for salmonid culture, but other species (i.e., catfish, oysters, and clams) are also cultured in raceways to a lesser degree.

Raceway culture has been traditionally practiced for hatchery production of fish for decades, and there are several good reasons for this. Environmental parameters (i.e., water quality, temperature, etc.) and water quantity are easier to manage in raceways than in pond systems. Flowing water flushes wastes from the culture units. Flowing water also forces the fish to exercise. Studies have shown that exercised fish have better survival rates when stocked into the wild (Wheaton 1977). The shallow water in raceways allows visual observation of the fish so that diet and/or disease problems can be promptly corrected. Finally, feeding and harvesting are generally easier in raceway systems. Feeding and disease treatment are more easily managed in raceway systems than in open systems or ponds. On the negative side, fish are normally cultured in very high densities in raceways, leading to increased risk of disease due to the stress caused by confinement and crowding. Better management skills are required for raceway culture.

Site Selection and Layout

Site selection for earthen raceways must be done with care. Raceway culture is water-intensive, therefore, probably the most important consideration is the water supply. Most raceway culture in the United States is in mountainous regions where gravity flow conditions can be taken advantage of to supply the needed water to the raceway systems. In the trout culture industries in the Northeast and Northwest United States, freshwater springs are the main water sources due to their relatively low and constant water temperatures. The water quantity and temperature required depends upon the species cultured and the size of the operation.

Figure 9-12 illustrates the most simple raceway design. In actuality, incoming water usually requires some degree of pretreatment before it can be used. Pretreatment may consist of nothing more than sedimentation, or it may consist of a combination of a number of processe including aeration, heating/cooling, degassing or filtration. Depending on the effluent water quality, discharge water may also require posttreatment before it can be released into the environment.

Multiple raceway facilities are arranged either in series (Figure 9-13) or in parallel (Figure 9-14). Series raceway systems can be used in regions where there is sufficient land slope so that the outlet for one raceway serves as the inlet for the next in the series. A 1–2% slope is considered adequate for good water flow (Pillay 1990). Waste buildup can be a serious problem in the series systems since the wastes from the upstream unit enter the next unit downstream, etc. Wastes tend to increase as the water traverses through the system. For this reaso there is a practical limit to the length of raceways and the number of units in series. Thes limitations will be demonstrated in example problems that follow in later sections.

The waste problem is somewhat alleviated by arranging the raceway units in parallel. There are no culture units downstream receiving wastes from upstream units, therefore, waste buildup and aeration problems are minimized

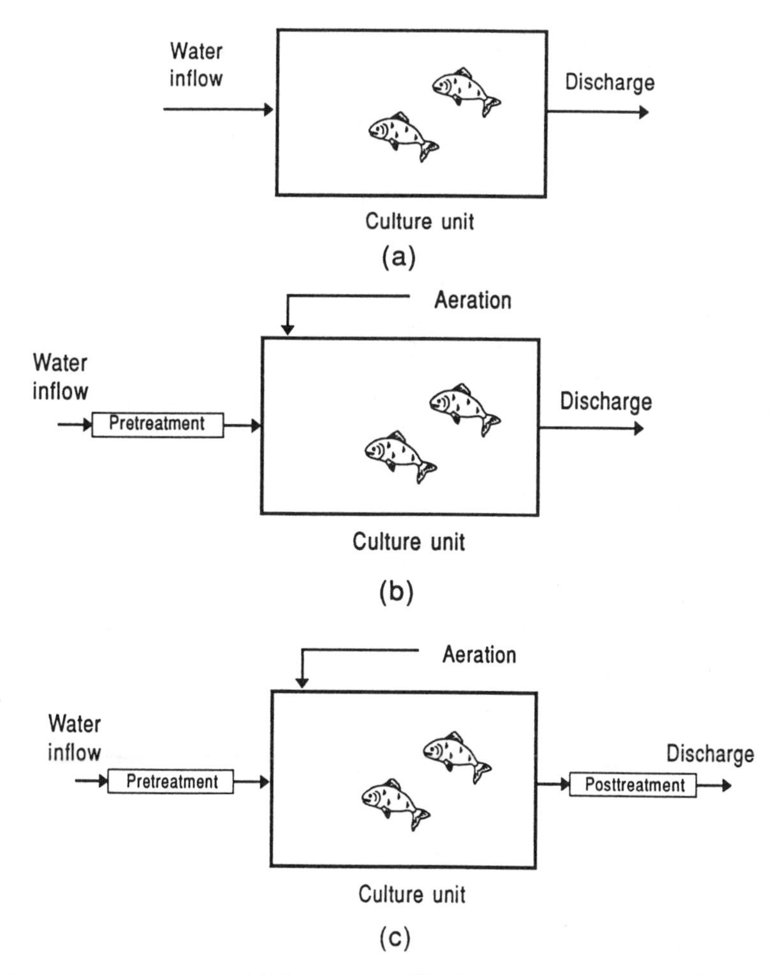

'igure 9-12. Single-pass raceway systems: (a) No treatment; (b) with pretreatment and aeration; and (c) with pre- and osttreatment.

Iowever, the quantity of water required inreases in direct proportion to the number of aceways. Thus, if pumping is required, operatng costs are higher in the parallel arrangement.

Many large raceway farms use a combinaon of the series and parallel configurations. 'igure 9-15 illustrates a recirculation technique sing a large pond for waste deposition and war recycling. Thus, this configuration relies eavily on water conservation.

Figure 9-16 shows a commercial raceway ystem in Southern Arkansas for producing hannel catfish. The facility consists of 108 oncrete raceways, each 11 m (36 ft) long by 5 (16 ft) wide with a water depth of 1.2 m (4 ft). The raceway sections are arranged in three descending tiers, creating three oxygen-producing waterfalls. Supplemental oxygen is added at the inflow to each section with a liquid oxygen system. Water flows by gravity from a header tank, and water retention time in each section is about 6–7 minutes. The effluent from the raceway at the bottom of each tier discharges into a pond and then circulates through a series of about 57 ha (140 ac) of fish ponds to settle solids, oxidize nitrogenous wastes, and reoxygenate the water. The water is eventually pumped back up into the header tanks at the rate of about 1.9 m^3/s (30,000 gpm).

In the Arkansas raceway facility, catfish

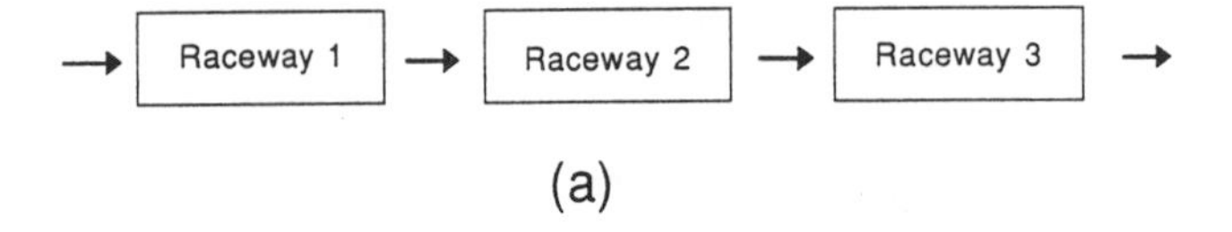

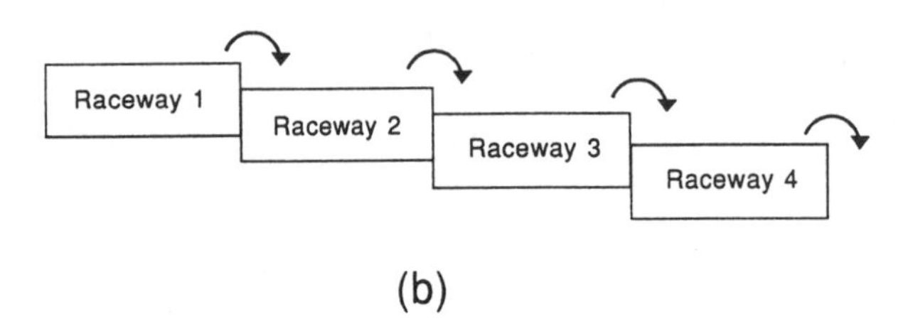

Figure 9-13. Raceway units in series: (a) On flat ground; and (b) on sloping ground.

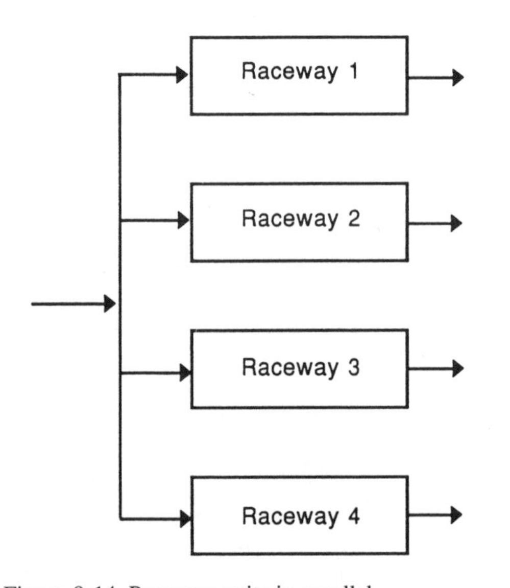

Figure 9-14. Raceway units in parallel.

weighing about 113 g (0.25 lb) or more are stocked at 108 fish/m^3 (10 fish/ft^3) or about 15,000 fish in each raceway section. Large fish are stocked so that two crops of fish can be reared during a 200-day growing season. The facility produces about 907,000 kg (2 million lb) of catfish annually.

Construction Materials

Although earthen raceways are sometimes used, the majority are constructed from concrete or cement blocks, like the Arkansas raceway facility previously described. Earthen raceways are sometimes lined with waterproof liners to reduce water loss through leakage. Many small experimental raceways are fabricated from wood, metal, fiberglass, plastic, or other materials.

Water Supply

Raceways are designed to provide flow-through for culturing very dense populations of fish. An abundant flow of good-quality water is essential to provide for the health of the animals and to flush wastes from the system. Water quality in a raceway is maintained by manipulation of the water flow rate and by adjusting other water treatment processes, aeration for example, in response to the demands of the cultured animals. Water quality tends to vary along the long axis of a raceway, and a distinct degradation of conditions is evident between the inlet and outlet. The specific flow rate required to meet the oxygen demands of the fish and flushing of metabolites is determined by the influent water temperature and dissolved oxygen concentration and by the oxygen consumption and ammonia excretion of the fish in the raceway.

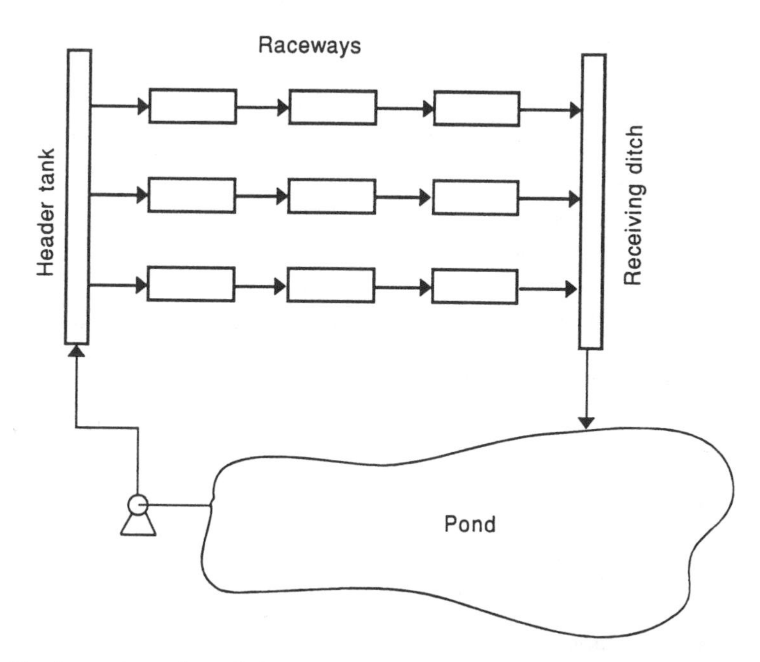

Figure 9-15. Combination series and parallel raceway units with water recirculation.

Figure 9-16. A commercial raceway culture system producing channel catfish (photo by T. Lawson).

Engineering Aspects of Raceway Design

The amount of aquatic animals that a flow-through system will be capable of supporting is dependent on water quality considerations, management skills, specific conditions, and species biology, including the ability of a particular species to tolerate crowded conditions. Carrying capacity may also be governed by other factors of which researchers and producers are not yet aware (Meade 1991). Some generally valid rules-of-thumb can be used for aquaculture planning purposes.

Definitions. The following raceway terminology is encouraged in Huguenin and Colt (1989) and Westers (1991):

$$\textit{Volumetric density} = \frac{\textit{Mass of animals}}{\textit{Volume of rearing unit}} \qquad (9\text{-}6)$$

$$\textit{Areal density} = \frac{\textit{Mass of animals}}{\textit{Area of rearing unit}} \qquad (9\text{-}7)$$

$$\textit{Loading} = \frac{\textit{Mass of animals}}{\textit{Flow through rearing unit}} \qquad (9\text{-}8)$$

$$\textit{Exchange rate} = \frac{\textit{Flow through rearing unit}}{\textit{Volume of rearing unit}} \qquad (9\text{-}9)$$

Loading, density, and exchange rate are related by Equations 9-10 through 9-15 for both SI and English units as follows:

LOADING

$$(SI) \qquad L = \frac{D \times 0.06}{R} \qquad (9\text{-}10)$$

where L = loading rate, kg/L/min; D = fish density, kg/m^3; and R = number of water exchanges/hr, and

$$(English) \qquad L = \frac{D \times 8.02}{R} \qquad (9\text{-}11)$$

where L has units of lb/gal/min; and D has units of lb/ft^3.

DENSITY

$$(SI) \qquad D = \frac{L \times R}{0.06} \qquad (9\text{-}12)$$

$$(English) \qquad D = \frac{L \times R}{8.02} \qquad (9\text{-}13)$$

EXCHANGE RATE

$$(SI) \qquad R = \frac{D \times 0.06}{L} \qquad (9\text{-}14)$$

$$(English) \qquad R = \frac{D \times 8.02}{L} \qquad (9\text{-}15)$$

Flow-Space Relationship. According to Westers (1991) there are two aspects to fish-carrying capacity in flow-through systems. One is based on flow and the *loading* term is used. The other is space related, and the density term is used. Loading capacity depends on water quality, primarily dissolved oxygen, temperature, and pH, but also on fish species and size. Density is generally a function of fish species and size, but it is also governed by type of rearing container and hydraulics of the rearing container. Density is the most difficult parameter to define and is still highly controversial.

Westers (1991) reported that salmonids are routinely maintained at densities ranging from 60 to 120 kg/m^3 in Michigan hatcheries. Poston (1983) reported a density of 234 kg/m^3. Figures are also available for catfish culture. Avault (1989) reported that catfish were cultured at densities of 32–64 kg/m^3 (2–4 lb/ft^3) in raceways in Idaho. Channel catfish were stocked at 40 kg/m^3 (2.5 lb/ft^3) or more in concrete raceways in Southern Arkansas (Tucker and Robinson 1990). Piper at al. (1982) recommended that the density (kg/m^3) of trout should not exceed 3.16 times the length of the fish in centimeters. The equivalent in English units is density (lb/ft^3) equals one-half of the fish length in inches.

For example, density equals 48.2 kg/m^3 for 15.2-cm trout or 3 lb/ft^3 for 6-in. fish.

The allowable loading based on oxygen as the limiting parameter is expressed by Equation 11-78 (see Chapter 11):

$$L_{OXY} = \frac{DO_{avail}}{2.0 \times \%BW}$$

where L_{OXY} = loading level based on oxygen control (kg fish/L/min—multiply by 8.33 to convert to lb fish/gpm); DO_{avail} = dissolved oxygen available to the fish (oxygen differential between inlet and outlet, mg/L); and $\%BW$ = the daily feeding rate based on body weight of fish in the system expressed as a whole number.

Equation 11-77 was derived based on a 16.7-hour feeding day. If a 24-hour day were used the equation would change to:

$$L_{OXY} = \frac{1.44 \times DO_{avail}}{2.0 \times \%BW} \tag{9-16}$$

This relationship holds since 1.0 mg/L of dissolved oxygen per 1.0 L/min per 24 hours amounts to 1,440 mg (1 × 1 × 60 min/hr × 24 hr/day) or 1.44 g versus 1 × 1 × 60 × 16.7 = 1,002 mg (≈ 1 g) for the 16.7-hour day. The result is a more conservative, simpler equation (Westers 1991).

The allowable loading based on un-ionized ammonia criteria is from Timmons and Youngs (1991):

$$L_{NH_3} = \frac{UA_{allow} \times 1000}{DO_{avail} \times \%UA} \tag{9-17}$$

where L_{NH3} = allowable fish loading based on un-ionized ammonia control (kg/L/min); UA_{allow} = allowable or upper design limit for un-ionized ammonia level (mg/L); and $\%UA$ = the percentage of the total ammonia nitrogen that is in the un-ionized form (temperature- and pH-dependent). Some aquaculturists use a design criteria of 0.01 mg/L NH_3-N for salmonids. Others, such as Westers (1991) use a more lenient value of 0.02 mg/L NH_3-N. If we use 0.02 mg/L for UA_{allow} in Equation 9-17, we are left with

$$L_{NH_3} = \frac{20}{DO_{avail} \times \%UA} \tag{9-18}$$

which is used by Westers (1991) for expressing the loading rate based on ammonia control.

Required Flow Rate. After calculating the allowable loading based on oxygen and un-ionized ammonia control, the smaller of the two values should be used to determine the water flow rate required to support the weight of fish it is desired to hold in the system, as shown in Equation 9-19:

$$Q = \frac{Fish\ weight}{(L_{OXY})\ (L_{NH_3})} \tag{9-19}$$

where Q = required water flow rate, L/min (multiply by 3.785 to convert to gpm).

Number Of Uses. The number of times that the water can be reused is found by the ratio of Equation 9-18 and 11–77:

$$Number\ of\ uses = \frac{40 \times \%BW}{(DO_{avail})^2 \times \%UA} \tag{9-20}$$

Hydraulic Retention Time. The hydraulic retention time, or the time that a given volume of water is in the raceway, is calculated as follows:

$$T = V/Q \tag{9-21}$$

where T = retention time (min); V = raceway volume (m^3); and Q = flow rate (m^3/min).

Raceway Length. The raceway, or rearing unit, length is calculated based on the water exchange rate R and the *minimum required average velocity* V_{min} for the raceway:

$$L_r = \frac{36 \times V_{\min}}{R} \tag{9-22}$$

where L_r = raceway length in meters (multiply by 3.28 to convert to feet); and $Vmin$ = minimum required average velocity (cm/s). Minimum required average velocity will be discussed in a later section.

An allowable V_{min} of 3 cm/s (0.1 ft/s) and a low exchange rate of one per hour (R = 1) will result in an excessively long rearing unit. For example, in Equation 9-22 if we use V_{min} = 3 and R = 1, a raceway length of 108 meters results, which is too long. Standard raceway lengths are typically 19–31 m (60–100 ft) (Westers 1991).

Raceway Velocity and Geometry. The minimum raceway velocity required for cleaning can be calculated using Equation 9-23 from Timmons and Youngs (1991):

$$V_{\min} = V_{clean} = 0.5\, d^{\,0.444}\, (G - 1)^{0.5} \quad (9\text{-}23)$$

where V_{clean} = cleaning velocity, ft/sec (multiply by 0.305 to change to m/s); G = specific gravity of material; and d = particle diameter (mm). This expression was developed for materials having a specific gravity between 1.83 and 2.64. Timmons and Youngs (1991) made the assumption that the equation is valid as a first approximation to predict cleaning velocities for fish feces and uneaten food.

Specific gravity for fish feces from fish fed salmonid diets was determined by Chen (1991) to be 1.19. Minimum velocities for cleaning for several materials are presented in Table 9-8. Westers (1991) considers a V_{clean} of 3 cm/s to be acceptable. This velocity is adequately low to settle solids, yet fast enough to provide good raceway hydraulics. Good raceway hydraulics is necessary for scouring settled wastes from the unit and for preventing stagnant (low DO) spots from developing, resulting in a healthier rearing environment.

Table 9-8. Minimum required velocity for channel self-cleaning[1].

		V_{clean}	
Channel material	Particle size	(cm/s)	(ft/s)
Feed/feces (G = 1.19)	100 micron	2.4	0.08
Feed/feces	1/16 in.	3.7	0.12
Silt	0.002 mm	0.9	0.03
Fine sand	0.05 mm	4.0	0.13

[1] To convert to m/s multiply ft/s by 0.305.
Source: Timmons and Youngs (1991).

Average velocity of flow in a raceway can be calculated using the equations for open channel flow from Chapter 6. The average velocity, V_{ave}, can be calculated using Manning's equation (Eq. 6-37):

$$V_{ave} = \frac{1}{n} R^{2/3} S^{1/2}$$

where V_{ave} = average velocity (m/s); R = hydraulic radius (m); S = channel bottom slope (m/m); and n = Manning's roughness coefficient. Values for the roughness coefficient for different channel materials can be obtained from Table 6-8 (see Chapter 6). Having calculated the average velocity, the channel cross sectional area can be calculated from the continuity equation since the water flow rate (m^3/s) is determined by the fish requirements:

$$Q = A \times V$$

The average velocity , V_{ave}, must be compared to the minimum raceway velocity required for cleaning V_{min}. If these values are too far apart, use 3 cm/s for V_{min} and calculate a new raceway cross-sectional area that will satisfy the given conditions of velocity and flow rate. It becomes immediately obvious that open channel design procedures cannot always be used to design fish production raceways. Often the design velocity may be sufficiently fast enough for cleaning and slow enough to not fatigue the fish but will require an unusually large cross sectional area. For this reason, some scientists go so far as to say that raceways do not work properly for fish production, and many use round tanks since the hydraulics are much simpler.

The majority of raceways for fish production are constructed from concrete. Thus, the raceway cross section can be either square or rectngular in shape with vertical side slopes. For aceways with vertical side slopes, the best hydraulic cross section occurs when the base vidth is twice the water depth (Equation 6-40 rom Chapter 6):

$$b = 2h$$

vhere b = raceway bottom width; and h = water lepth.

Earthen raceways typically have a trapezoidal cross section. The bottom width providing he best hydraulic cross section is given by Equation 6-41:

$$b = 2h \tan\left(\frac{\theta}{2}\right)$$

vhere θ = side slope angle. The side slope angle for vertical channels is 90°. The tangent of 90 ÷ 2 = 45° = 1.00. For raceways cut into the earth, allowable side slopes are given in Table 9-9.

Waste Removal. Fish wastes and uneaten food are normally flushed from raceways if the water velocity is appropriate. Figure 9-17 illustrates the use of dam boards for maintaining water level and waste removal. When installed as shown, the raceway becomes self-cleaning by removing wastes from the bottom of the unit. A screen should be placed upstream from the dam boards to prevent the fish from exiting the culture unit. Boards and screen are held in place by slots formed into the sides of the raceway.

Harvesting Raceways. One of the major advantages of raceway culture systems is ease of harvest. Harvesting is usually accomplished by pulling a seine from one end of the raceways to the other, crowding the fish into a small area where they can be netted. Some facilities have automated harvesting screens or grader bars that move on tracks located on either side of the raceway. Grader bars allow smaller fish to escape from the crowding area and are left in the raceway to grow larger. Either automated or hand-labor systems work well, harvesting fish with a minimum of labor and effort and minimal damage to the fish.

Table 9-9. Allowable side slope for soil constructed channels up to 1.2 m (4 ft) in width.

Soil type	Side slope
Peat, muck	0.5:1
Stiff heavy clay	1:1
Sandy loam	1.5:1
Loose sand	2:1

Source: Timmons and Youngs (1991).

Special Types of Raceways

A culture unit configuration that merits attention for intensive culture use is the *silo* or *vertical raceway*, illustrated in Figure 9-18. A basic silo consists of a deep tank with the culture water entering near the bottom and exiting out the top. Water may be pumped downward through a pipe centrally located in the culture unit (Figure 9-18a), or it may enter through a side wall of the culture unit (Figure 9-18b). The water flows vertically upward through the unit and discharges into a trough constructed around the tank perimeter at the top. Other configurations are possible. Some silos are partially buried in the ground. Buss et al. (1970) reported that a silo unit 5 m high by 2.3 m in diameter, with a flow-through rate of 28.4 L/s was capable of supporting 2,800 kg of rainbow trout.

AQUACULTURE IN TANKS

The propagation of fish in tanks has received much attention in recent years. The pros and cons of tank culture are still being researched and are hotly debated. Tanks are commonly used for fry production or as temporary holding facilities for fingerling, brood fish, or food fish. Tanks are also used for aquarium fish production and as public display aquaria. However, except for small operations, tanks are not routinely used for commercial food fish production. Production in ponds is considered to be more economical and better suited for large-scale fish production.

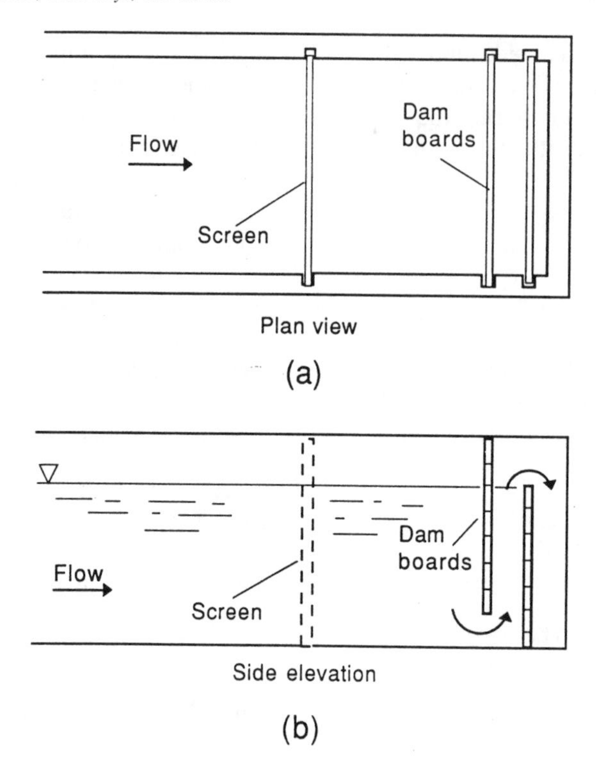

Figure 9-17. Dam boards for drawing fish wastes from bottom of raceway.

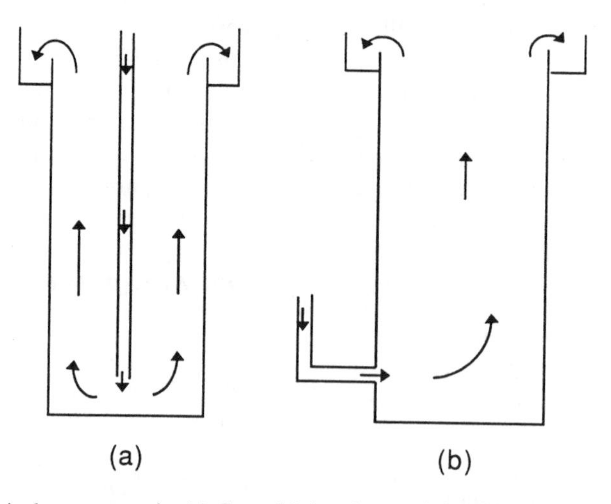

Figure 9-18. A silo or vertical raceway unit: (a) Central inlet pipe; and (b) side-wall inlet pipe.

This philosophy is slowly changing, however, as land costs rise, good quality water becomes scarce, pumping water becomes more expensive and advances in technology make tank culture more attractive. Many universities use tank facilities for aquaculture research (Figure 9-19) since ponds are expensive to build and maintain. In addition, experimental parameters are more easily controlled in closed tank environments.

Tanks can either be custom-fabricated on site or mass produced by a variety of manufacturers.

Figure 9-19. An aquaculture tank farm at Louisiana State University (photo by T. Lawson).

They are fabricated from many materials and are available in an infinite number of shapes and sizes. Decisions influencing these choices are the subject of the remainder of this chapter.

Construction Materials

Tanks are constructed from a variety of materials, including wood, concrete, plastic, fiberglass, metal, and glass. Aquaculture tanks should have the following characteristics: smooth interior surface to prevent abrasion, nontoxic surfaces, durability and portability, long life, ease of cleaning and sterilizing; noncorrosive, and affordability. The choice of material should reflect these characteristics.

Inexpensive tanks can easily be fabricated from wood. Besides being light in weight, wood is typically less expensive than most other construction materials and easy to work with. Marine plywood is often used to construct wooden tanks, but other grades can be used as well. Plywood having a minimum thickness of 19 mm (0.75 in) should be used since the material is rigid enough so that tanks won't flex excessively when filled. The tank sides should be braced to resist the static forces exerted by the water. Treated woods should not be used since many contain substances that are toxic to fish. All exposed surfaces should be painted to seal the wood against rot. Paints containing lead or other heavy metals that may leach into the water should never be used for aquaculture purposes. Tank interiors can be sealed with non-toxic materials such as epoxy or fiberglass resin paint. These materials take one to two days to cure, and, once cured, they form a hard, smooth surface. In lieu of paints or sealers, waterproof liners can be used in wooden tanks. Wooden tank fabrication is very site-specific, therefore, no design concepts will be presented here, although Wheaton (1977) provides a brief overview of small wooden tank construction.

Concrete is widely used for constructing large tanks or pools. It is easy to work with and can be formed into any shape. Properly reinforced with steel bars, concrete will last indefinitely for use with either fresh- or salt water. Because of its weight and expense, however, concrete is used for the construction of permanent facilities. The interior surfaces of concrete tanks should be smoothed to avoid abrasion of fish skin or scales. Also, rough surfaces are harder to clean and disinfect. Interior surfaces can be coated with a sealer, many of which are commercially available. Uncoated concrete surfaces are suitable for most aquaculture purposes, but only after a sufficient curing period. Harmful substances can leach from newly poured concrete. Thus, concrete tanks should be flushed with clean water for several days prior to use.

Gunite is a strong, durable concrete-like material that has an indefinite life. It is often used to construct small spas, swimming pools, and ornamental fish ponds. Gunite material is very compact and can be blown under high pressure over a support framework fabricated from steel reinforcing rods and small mesh poultry fencing. The material is usually applied in a 5–10 cm (2–4 in) thick layer. Gunite is more expensive than liners but has an indefinite useful life, whereas liners must be replaced every few years. The utility of gunite as a construction material for production tanks has yet to be proven.

The term plastic refers to a number of polymers including polypropylene, polyethylene, polybutylene, polyvinyl chloride (PVC), acrylics and vinyl. Each has its own set of good and bad features. A major advantage to plastic tanks is that they are lightweight and portable. Repairs to plastic tanks are also easier. Tanks of plastic manufacture are available in a variety of shapes and sizes, from small fish bowls to aquaria up to several hundred liters in volume. Most plastics are nontoxic, but some, like polyethylene, are initially toxic and should be conditioned with clean, running water for at least two weeks prior to use, particularly for marine culture (Huguenin and Colt 1989).

Fiberglass is the material of choice for most aquaculture tank construction since it is lightweight, strong, durable, modestly priced, inert to both fresh and salt water, and can withstand the effects of UV rays if used outdoors. Fiberglass tanks are normally gel-coated on the inner surfaces to provide a smooth surface that may be easily cleaned and disinfected. The tanks can be easily drilled for installation of drains and other plumbing fixtures. A variety of tank sizes are available from a few hundred liters up to several hundred cubic meters. Some models are available with legs, skirts, or stands so that they can be elevated above the floor or ground. Tanks smaller than about 1.8–2.4 m (6–8 ft) in diameter are usually available in a single molded unit, but larger tanks may come unassembled in two or more sections that must be bolted or glued together. Tanks are manufactured so that they can be nested when two or more tanks are transported, thus saving on shipping costs. Some manufacturers will assemble tanks on-site for an additional fee. Fiberglass tanks are available from numerous manufacturers world wide.

Aluminum and steel are two metals that are commonly used in the fabrication of small (several hundred liters and smaller) tanks popular in hatcheries for rearing small fry and fingerling. If used with caution, some aluminum alloys can be used for brackish or saltwater culture. Water with a pH well into the acidic range will cause aluminum to become soluble and leach into the water. Metals become more toxic in waters that are poorly buffered. Sectional steel tanks that bolt together are readily available in many countries. They are easily erected and dismantled, making them completely portable. Some are rubber-coated but most are intended for use with a liner. Small stainless steels are generally safe to use with either fresh- or salt water, particularly 316 stainless (Huguenin and Colt 1989). However, stainless steel tanks are too expensive for large-scale use.

Galvanized metal tanks should not be used for aquaculture purposes unless they are coated or lined. Zinc leaches from the galvanized coating and can cause heavy metal poisoning.

Salmonids are generally the most susceptible. The 96-hour LC_{50} concentrations for zinc range from 0.09 mg/L to 41 mg/L for various species (Piper et al. 1982). It is difficult to get most paint coatings to adhere to galvanized metal; therefore, it is best to avoid use of this material altogether or a liner should be used.

Glass culture units are found almost exclusively in the aquarium trade or for public display aquaria. Aquaria range in size from small fish bowls to over 400 liters in volume. Glass is not practical for use in the fabrication of large tanks because of its high cost, excessive weight and the ease with which it can be damaged. In the aquarium manufacturing industry acrylic is rapidly replacing glass for tank construction, but acrylics are too expensive for large tank construction.

Waterproof liners can be used with tanks fabricated from virtually any material. They are commonly used in place of expensive coatings or sealers. They eliminate the hazard of toxicity by heavy metals, paints, treated wood, or other substances. Liners can be custom-made to fit any tank size or shape. They must be carefully handled to avoid tearing and have a useful life of about 5–10 years, depending on use. The life of a liner is significantly reduced if water pH is too extremely acidic or alkaline. Plastic liners become brittle in acid waters and soft in alkaline waters. Many modern tank or pond liners are resistant to UV rays, and therefore, have a relatively long life when used outdoors.

Physical Features

Culture tanks can be custom manufactured into virtually any shape desired, however, some tank shapes and configurations are more advantageous than others, and the aquaculturist must be aware of this fact. Most culture tanks are round, but rectangular, square,and oval tanks are also used.

Shape. Round tanks are commonly used for nursery or grow-out purposes. Square and rectangular tanks make better use of available floor space since they can be placed side by side, but round tanks typically have better hydraulic characteristics (Wheaton 1977). Velocity, circulation and mixing are such that oxygen and food particles are more evenly distributed, and water quality tends to be more uniform. Thus, it is reasoned that round tanks are able to support greater densities of fish than other shapes. On the negative side, oxygen consumption per unit weight is greater in round tanks, probably due to the increased energy demand created by the faster circulatory currents (Piper et al. 1982). The additional oxygen consumption must be accounted for in the system design.

The water inlets and outlets in round tanks can be positioned in such a way as to create a vortex that sweeps the tank bottom and moves detritus and wastes toward a central drain (Figure 9-20). Tanks having a rapid turnover time are often referred to as *circular raceways* or

Figure 9-20. Water circulation patterns in a round tank.

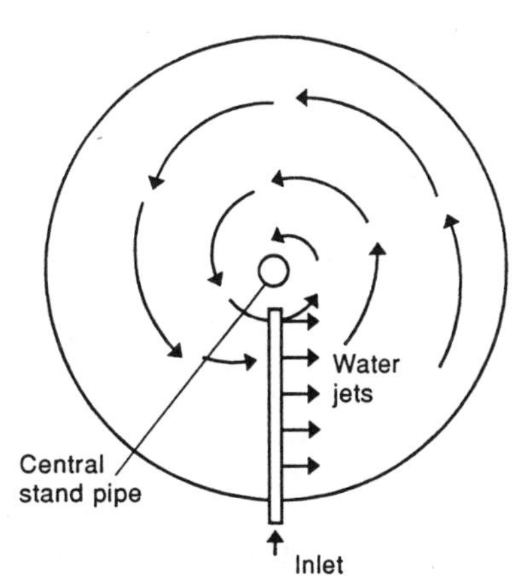

(a) Plan view of round tank

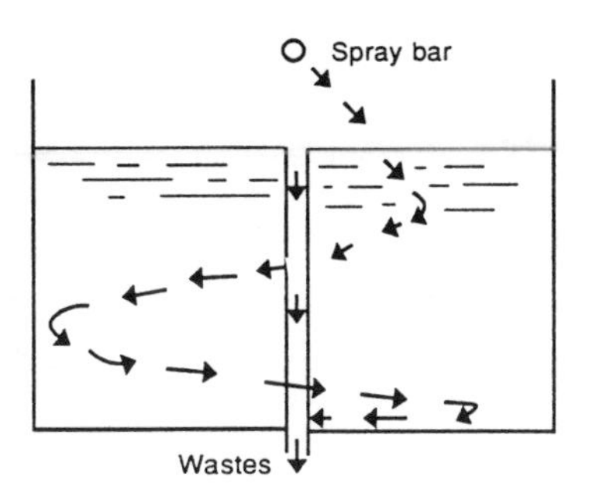

(b) Elevated view of round tank

circulating ponds. A compromise must often be reached between: (1) Water velocity and the flow pattern required for even feed distribution; (2) the self-cleaning function of the tank; and (3) the swimming energy required of the fish. Flow velocity should not be so great as to cause the fish to expend all of their energy swimming against the current, leaving none for growth. For this reason flow velocities in culture units containing fry or young fingerling are often so low that the tanks are not self-cleaning. In these cases wastes must periodically be vacuumed from the culture units or manually pushed toward the drain.

Round tanks are manufactured with either a flat or a sloping (conical) bottom. Tanks with sloping bottoms are more expensive, but they are more efficient at removing wastes. Flat-bottomed tanks will self-clean if adequate water velocities are maintained. Where bottom-dwelling fish like catfish, sturgeon, or eels are maintained, the activity of the animals will keep bottom sediments in suspension so that they may more easily be removed with the water flow.

Rectangular tanks and troughs are most often used for rearing fry and small fingerling in hatchery buildings. The bottom can slope toward one end (Figure 9-21a) or toward the center (Figure 9-21b). In tanks that drain at one end the water enters one end and exits at the other, much like a raceway. Flow patterns are often characterized by short-circuiting and the formation of dead areas. Wastes tend to collect in the corners (Figure 9-22). Fish may become stressed or death may result if metabolic wastes

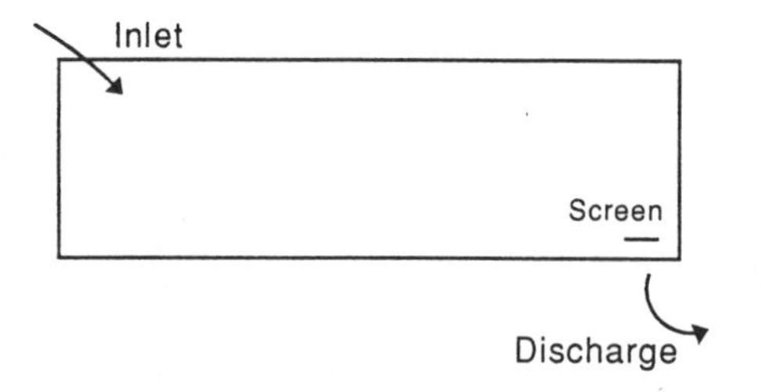

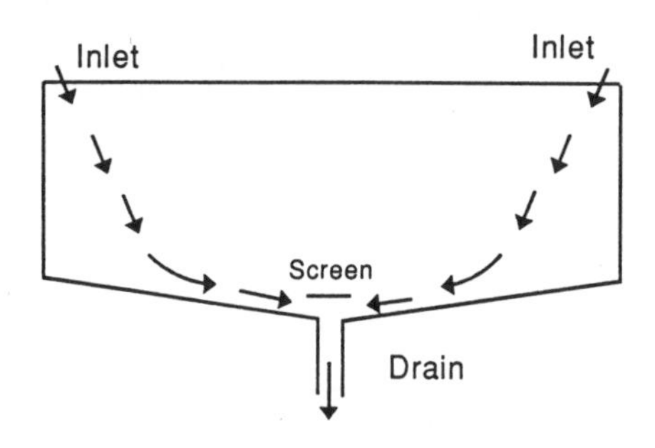

Figure 9-21. Rectangular tanks with sloping bottoms: (a) Toward drain at one end; and (b) toward center drain.

accumulate to dangerous levels. Fish collected from the wild and then placed into rectangular or square tanks tend to crowd into a corner where they may locally deplete the oxygen, or they may bash into the sides of the tank, causing serious injury in some instances. Fish raised in tanks from the egg stage rarely exhibit this behavior (Wheaton 1977).

Figure 9-22. Fish wastes collecting in corners of rectangular tank.

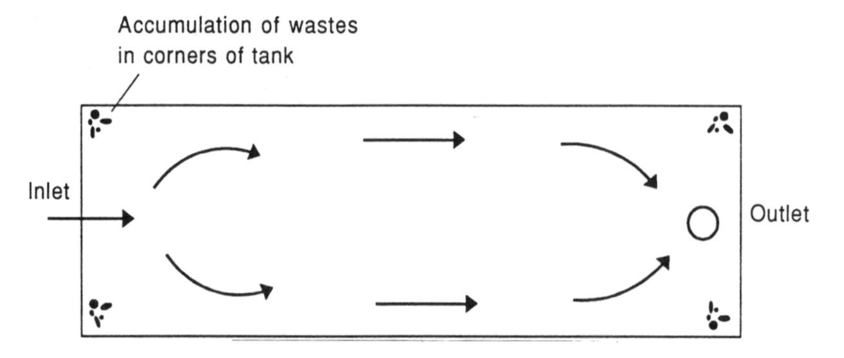

Flow patterns and short-circuiting are functions of the water inlet and outlet design. To minimize short-circuiting in rectangular tanks the water at the point of entry should flow as nearly uniform as possible across the width of the unit to achieve plug flow (Wheaton 1977; Klapsis and Burley 1984). This can be accomplished by applying the water through a perforated pipe, over a shallow trough or over a weir at the inlet. Oxygen must be evenly distributed otherwise fish may collect at the inlet end of the culture unit, not fully utilizing all available tank volume.

Wastes can be removed from rectangular culture units by increasing the water velocity to where the tank becomes self-cleaning. However, as previously discussed, this is not always practical. An alternate method of removing wastes from rectangular culture units was described by Wheaton (1977) and is illustrated in Figure 9-23. Tanks may be constructed with a waste collection trough built into the floor. A flat plate is positioned above the trough at the tank bottom as shown to prevent fish from escaping. As wastes accumulate in the trough, they are removed by periodically opening a valve at one end of the trough. A flat, slotted screen may be used in lieu of a plate, however, the self-cleaning action may not be as efficient.

Square culture units have disadvantages similar to rectangular tanks in that wastes tend to accumulate in the corners. The hydraulic characteristics of square tanks can be made similar to circular tanks by constructing them with rounded corners (Figure 9-24). A central stand pipe or flat screen with external stand pipe may be used for drainage as previously described for round culture units. The hydraulic characteristics of square tanks with rounded corners are discussed in greater detail in Burley and Klapsis (1985).

To take advantage of limited floor space, square or rectangular tanks can be stacked. The number of tanks in the stack depends on tank depth and ceiling height, keeping in mind that adequate bracing must be provided to support the tanks and water, and an adequate space must be provided between tanks for stocking, feeding, harvesting, and other activities.

A compromise between the round and rectangular tank is the oval tank shown in Figure 9-25. These tanks are constructed with rounded ends and a partition that divides the tank. Water flow is generated by spraying the inlet water through a nozzle or series of water jets in the direction of circulation. In lieu of water jets or nozzles, a small paddle wheel can be installed on one side of the partition to develop water circulation. The outlet standpipe is normally placed at the center of the partition. A special type of oval tank, the Foster-Lucas pond, was discussed in an earlier section.

Draining. Round culture tanks constructed with flat bottoms are sometimes preferred because they rest directly on a flat surface, such as a sand or gravel foundation, or on a concrete floor. Those tanks with sloping bottoms must be elevated. Elevated tanks can be drained completely by removal of a central standpipe, but

Figure 9-23. A waste collection trough in a rectangular tank.

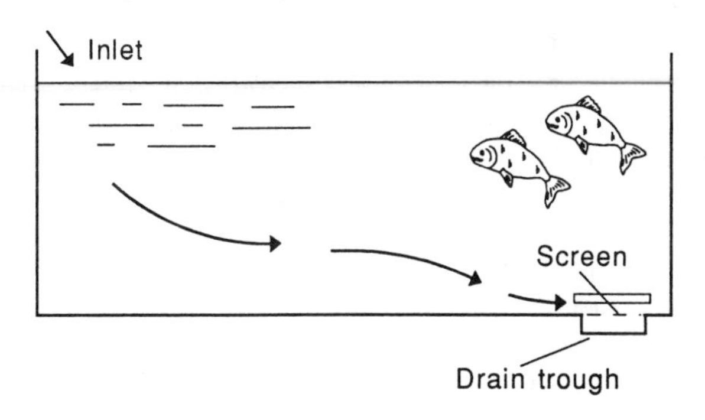

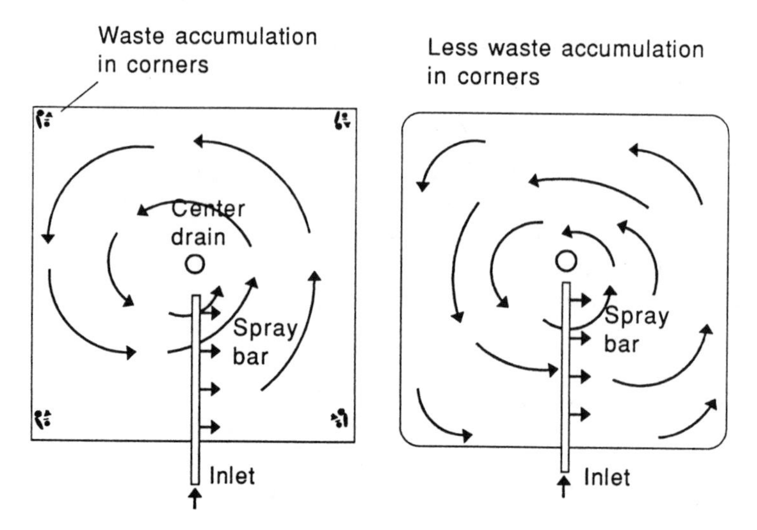

Figure 9-24. Flow patterns in rectangular tanks.

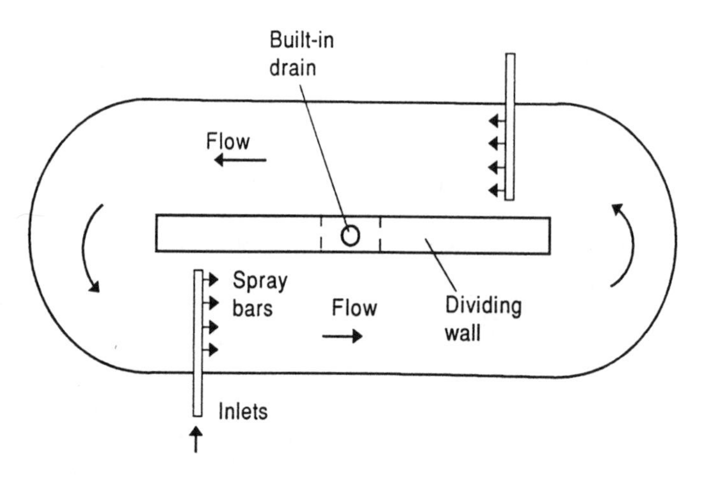

Figure 9-25. An oval fish tank.

those that rest directly on a flat surface must be pumped to drain completely.

Round tanks and pools should be drained with a central standpipe fitted with a second outer pipe or sleeve with openings near the tank bottom to facilitate waste removal (Figure 9-26). Short-circuiting will almost certainly occur, leaving wastes on the bottom of the tank unless some type of mixing action is used to keep wastes in suspension. Alternatively, a flat center screen and outside standpipe can be used. Horizontal slots in the screen offer more screen area and better cleaning action than round holes and are not as easily clogged (Piper et al. 1982). An emergency overflow system is advisable should the drain eventually become clogged. The outer standpipe should be constructed with a watertight collar that can be rotated so that any desired water level may be maintained in the culture unit. A more detailed discussion for a central screen with outer standpipe system is presented by Parker (1980).

Harvesting. Removal of fish from large circular tanks or pools is difficult without the

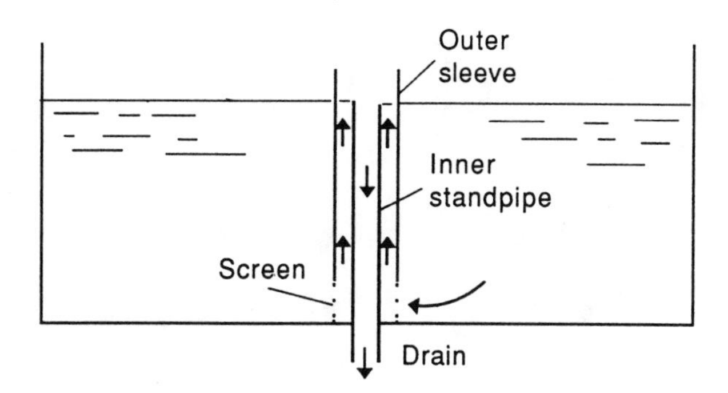

Figure 9-26. A central drain with outer sleeve for removing wastes from the bottom of a round tank.

proper equipment. Screens or grader bars that are rotated around the center standpipe facilitate the removal of fish by crowding them into a small area where they may be netted. Some tanks are constructed with inside wells for the accumulation of wastes and removal of fish (Piper et al. 1982). If it is necessary to harvest all of the fish from the culture unit, the screen mesh size should be selected to harvest only those fish above a desired size, leaving the smaller ones in the tank.

CHAPTER 10
RECIRCULATING AQUACULTURE SYSTEMS

The relatively new agribusiness of aquaculture is receiving considerable attention in both the public and private sectors as an enhancement to the diversification of agricultural and fishery economies in the United States. To date, most land-based production of species of major commercial importance, like catfish, hybrid striped bass, trout, tilapia, etc., has been in ponds, raceways, or estuarine impoundments. Traditional production practices of these and other species require large amounts of good quality water, a limited resource in many areas. In our modern world the condition of many water sources in some areas no longer favors these production practices. The decline of clean water has shifted focus to high-density aquaculture in water reuse and recirculating systems. While the technology for such systems is costly, high profit potential has captured the attention of new entrepreneurs and has made high density culture more attractive. Economic analyses of recirculating systems indicate that key operating parameters are: (1) System operation with minimal energy input; and (2) proper water purification (Kugelman and Van Gorder 1991). Aquaculture engineers are challenged to develop system designs that optimize these criteria.

EFFECTS OF FISH ON WATER

It is common knowledge that fish produce wastes that cause water quality degradation. Contaminants having a negative effect on water quality are ammonia, urea, CO_2, organic fecal material and other metabolic wastes. Organic material is further degraded to produce additional ammonia, nitrite and nitrate. In the close confines of a recirculating system the effects of these substances are to depress water pH, deplete dissolved oxygen, increase turbidity, and make the water more inhospitable to the fish. In addition, many minerals and essential trace elements are lost. The more intensive the culture practice, the greater the impact. The rate of waste production in an aquaculture system is dependent on the fish species and life stage, system biomass, and the type and amount of feed given to the fish. The rate and degree of water quality degradation can be managed with proper water treatment, which is the focus of this chapter.

BASIC RECIRCULATING SYSTEM COMPONENTS

In recirculating systems fish are confined at high densities. Normal culture density is 0.5–1

Table 10-1. Physical, chemical, and biological processes used to recondition water for recycle.

Physical	Chemical	Biological
Screening	Aeration	Nitrification
Settling	Pure oxygen injection	Denitrification
Sand filtration	Alkalinity and hardness control	
DE filtration	Carbon adsorption	
Centrifuging	pH control	
Temperature control	Reverse osmosis	
UV sterilization	Degassing	
Cartridge filtration	Foam fractionation	
Bag filtration	Ion exchange	
	Ozonation	

lb/gal, which is about 3.8–7.5 lb/ft^3 (61–122 kg/m^3), but densities in small-scale experimental systems have exceeded 34 lb/ft^3 (545 kg/m^3) (Huguenin and Colt 1989). The water is treated by several processes before it is recirculated to the culture units. In a typical recirculating system a small amount of water is exchanged daily with fresh water for nitrate control and to replace water lost through evaporation and filter backwashing. Treatment options are numerous, as shown in Table 10-1. Most of these processes have been tested individually and in combination in small-scale experimental recirculating systems at some point in the past, but many have not proven to be economically viable in commercial production systems. The principal treatment processes used in recirculating systems are screening, settling (sedimentation), granular media filtration, biological filtration, aeration, and disinfection.

A generic recirculating aquaculture system is illustrated in Figure 10-1. *Primary clarification* in this example refers to solids removal. Removal can be accomplished by one or several processes in combination including screening, sedimentation and granular media filtration. It is important to remove solids prior to *biological filtration*, which is the heart of any recirculating system. Biological filters are used for nitrogen control (i.e., ammonia and nitrite). *Secondary clarification* usually follows biological filtration to remove the biological floc that frequently sloughs from the filter media. It is important that this material not be allowed to remain in suspension. Secondary clarification includes sedimentation but may also include screening. Finally, aeration is added for basic system life support. The system is driven by a water pump (not shown).

Water in recirculating systems completes the entire circuit many times daily. Depending on culture intensity, filtration may be required from two to four times per hour. At a minimum, the water should receive complete treatment one to two times per hour. A daily partial water exchange is necessary to control nitrate, remove pollutants and replenish minerals and trace elements. Figure 10-1 demonstrates a 90% daily recycle rate with a 10% water exchange rate. This means that 90% of the water is reconditioned for reuse and 10% is replaced with fresh water. In this manner, the total system volume is replaced once every 10 days. Lightly loaded systems may fair well with a 3–5% daily water exchange. Systems employing 100% recycle are rare. Technical and management difficulties increase as one approaches a 100% recycle rate.

SOLIDS IN WASTEWATER

All contaminants in wastewater, except for dissolved gases, contribute to the solids load. Solids may consist of both organic and inorganic constituents. Solids removal is critical in recirculating aquaculture systems because they can

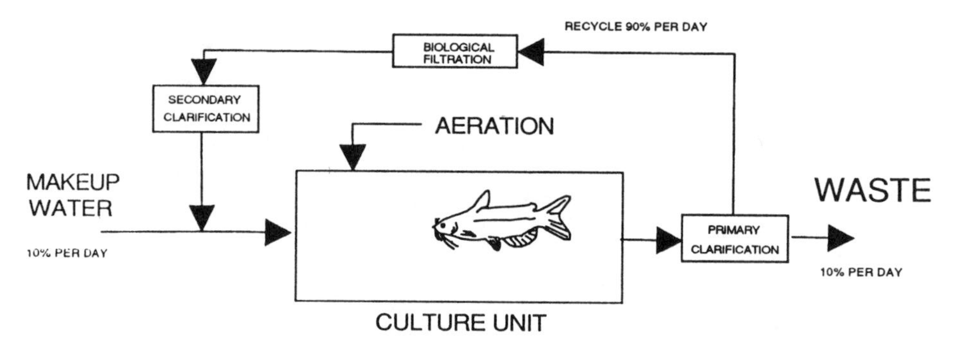

Figure 10-1. A recirculating aquaculture system utilizing 90% recycle per day.

physically block pipes, pumps, and filtration equipment. Just as important, as they decompose, organic solids consume oxygen, produce ammonia, and can exert a heavy oxidative load on biological filters. Liao and Mayo (1974) estimated that 70% of the NH_3-N in fish wastewater is associated with organic solids.

Fish wastes are quite different from domestic sewage and organic wastes from other industries. Fish wastes also differ between species. Channel catfish, for example, produce a waste that seemingly is more gummy and harder to handle than most fish wastes (Lomax 1976). Fecal material typically contains both digested and undigested material bound together by a mucus coating. The waste is produced in long strings by most species, but may vary. The nature of the mucus coat varies among species. Fish fecal material is normally more dense than water, but small fecal particles are often observed to float, particularly in systems with high water turbulence or oxygen-saturated waters where trapped gas bubbles cause the material to become buoyant.

Many of the water treatment and filtration methods commonly used in recirculating aquaculture systems are similar to those that have been used for years to treat domestic sewage. However, since fish waste is so radically different from sewage, the use of off-the-shelf water treatment technologies may be limited and will require modification. In order to make the right choices when selecting treatment processes it is essential that the designer be knowledgeable of solids characterization.

Solids Characterization

Solids in aquaculture wastewater can be classified similarly to organic wastes from other industries. They are classified using three criteria: size and state, chemical characteristics and size distribution (Tchobanoglous and Schroeder 1985). Solid and semisolid materials in wastewater are classified by size and state as shown in Figure 10-2. They are categorized as either settleable, suspended, dissolved, or colloidal. The difference between settleable and suspended solids is purely a matter of practicality. Settleable solids are classified as those having a particle diameter of 10^{-2} mm and larger. They will generally settle in an Imhoff cone in one hour. Suspended solids do not settle in one hour and, therefore are not expected to be removed by conventional gravity settling. Colloidal particles have diameters between 10^{-3} and 10^{-6} mm, which includes fine clay particles, some viruses, and unspecified bacteria. Dissolved particles have diameters less than 10^{-6} mm. They consist of both organic and inorganic ions and molecules that are present in solution. They may include metals like copper, iron, aluminum, or zinc, as well as ions of ammonia, nitrite, and nitrate. Both dissolved and colloidal particles are too small to settle by gravity and normally require more advanced methods for removal.

The term *colloidal solids* is rarely used today. Rather, solids are grouped as either *dissolved* or *suspended*. The dissolved portion includes both colloidal and small suspended particles (see Fig-

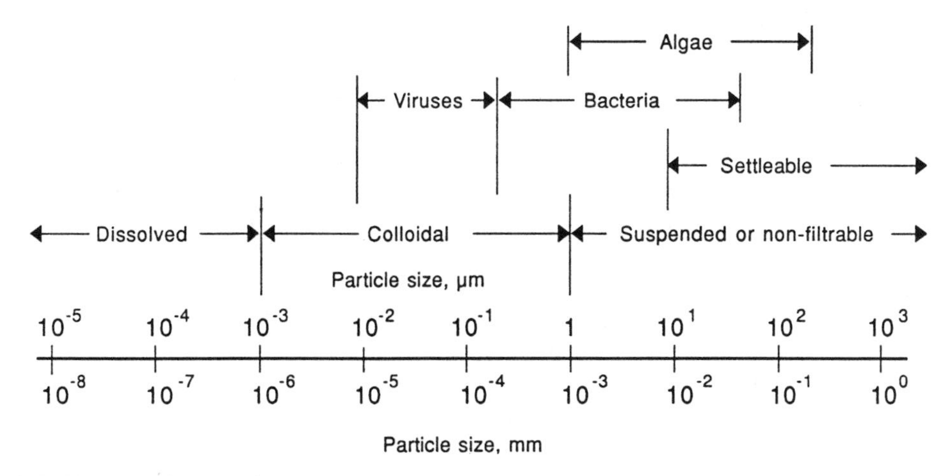

Figure 10-2. Size classification of particles in water.

ure 10-2). The suspended portion includes those solids that are settleable. The differentiation is made using a membrane filter having a pore size of about 1.2 µm. Particles passing through the filter are dissolved solids, and those trapped on the filter are suspended solids. Many scientists prefer to use the terms *filtrable* and *nonfiltrable* to describe solids that pass through the filter and those that are trapped on the filter, respectively. The sum of dissolved and suspended solids is referred to as *total solids*.

Chemically, solids are classified as *volatile* or *nonvolatile*. Volatile solids, by definition, are those organics that volatize at 550°C (1,022°F) (Tchobanoglous and Schroeder 1985). Thus, the test for volatile solids is often used to determine the organic fraction of the sample.

Size distribution is determined by examining the mass of solids retained on a 0.1-µm membrane filter. The solids collected will comprise a broad range of particle diameters. These diameters are grouped into size ranges and are expressed as a percentage of the total sample.

PHYSICAL TREATMENT METHODS

Solids can be removed from aquaculture wastewater using physical methods and processes borrowed from other industries. These processes include screening and gravity separation (sedimentation, centrifuging, and hydrocloning). They also include physical interception or adsorption between particulate beds (filters). Table 10-2 lists a number of physical processes and the particle sizes for which they are commonly used. In the wastewater literature, processes are identified as either primary, secondary, or tertiary (advanced) treatment. Primary treatment can consist of one or more of the above gravity separation processes. Secondary treatment refers to biological filtration. Tertiary, or advanced, treatment techniques, including ion exchange, reverse osmosis, foam fractionation, and carbon adsorption, are often used to remove fine dissolved and suspended organic solids. In some cases tertiary treatment may include disinfection.

Screens. Screening is the simplest and the oldest wastewater treatment process. It is often used as a pretreatment process before primary treatment. Screens are placed across the path of the waste stream and trap a broad range of macro and micro solids. Coarse screens are used to treat the raw effluent from the culture unit. Another application of coarse screening is to trap the biofloc that sloughs from biological filter media. Fine screens (microscreens) are used as a tertiary or advanced treatment process since coarse solids will rapidly clog the small openings in fine screens. Screens are constructed from a variety of materials including carbon

Table 10-2. Solids removal processes and the particle size ranges and flow rates over which they are applicable[1].

Particle size µm	Flow rate L/min (gal/min)			
	3.8 (1) or less	3.8–38 (1–10)	38–380 (10–100)	380–3800 (100–1,000)
1 or less	Cartridge filt. DE filters	Cartridge filt. DE filters	DE filters	DE filters
1–10	Cartridge filt. Centrifuges and cyclones	Cartridge filt. Centrifuges and cyclones Sand filters	Centrifuges and cyclones Sand filters	Sand filters
10–75	Filter bags Centrifuges and cyclones	Filter bags Centrifuges and cyclones Sand filters	Filter bags Centrifuges and cyclones Sand filters	Filter bags Sand filters
75–150	Filter bags Microscreens	Filter bags Microscreens Sand filters	Filter bags Sand filters Sand filters	Microscreens
150–1,000	Screen bags Microscreens	Screen bags Microscreens Sedimentation	Screen bags Sedimentation Sedimentation	Microscreens

Source: Huguenin and Colt (1989) with permission.

[1] Intended as a general guide.

steel, brass, stainless steel, nylon fabric, or other fabrics. Cost increases accordingly as mesh size decreases and whether cleaning is manual or automated.

Screens are either static (stationary) or rotary. The simplest form of static screen is shown in Figure 10-3. These screens are cleaned manually by removing the screen and hand-washing it or scraping it with a brush or other tool. The cleaning schedule should be frequent enough to prevent significant head loss. Improperly sized screens will seriously impair flow, making frequent cleaning necessary. A mechanically cleaned static screen is shown in Figure 10-4. This device consists of a single flat-angled screen made of nylon fabric or stainless steel mesh. Wastewater flows over the screen, solids are trapped on the screen surface, and the clean water passes through. As debris collects on the screen, the wastewater must flow further down the screen until it reaches a point where a sensor activates a clean water spray that washes the trapped solids into a sump for later disposal.

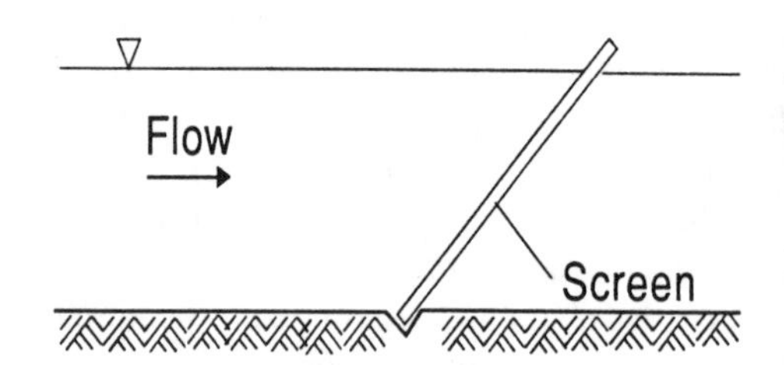

Figure 10-3. A static screen placed across a fluid's path.

Static screens used in wastewater treatment have screen openings from 0.01 to 0.06 in. (0.25 to 1.5 mm). They are typically loaded at about 4–16 gpm/in screen width (840–3400 m^3/day/m) and have a solids removal efficiency of 5–25% (Corbitt 1990).

Rotary screens are used to reduce labor costs

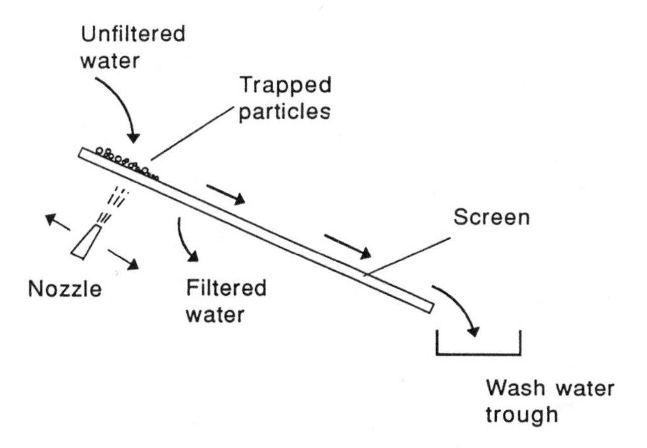

Figure 10-4. Side view of a static microscreen utilizing a water spray backwash setup.

associated with cleaning or where frequent screen clogging is a problem. A variety of rotary screen types are available, the choice being largely a matter of cost and personal preference. Chain-type rotary screens consist of side chains with attached screens, the total assembly being rotated by a series of motor-driven sprockets (Figure 10-5). This type of screen is used primarily for rough filtering of source water rather than as an integral component in a recirculating system. Material screened from the waste flow is mechanically scraped into a side trough for disposal. In general, these screen types are much more expensive than stationary screens, but labor requirements are less. Other types of rotary screens may be either axial or radial flow devices. In axial flow screens the wastewater flows through the screen in an axial direction. Axial flow screens are not widely used in aquaculture.

A variety of rotary screens are manufactured for aquaculture use. The most common is shown in Figures 10-6 and 10-7. Wastewater enters the rotating screen drum axially from one end, and the clean water exits radially through the mesh. The water level is maintained such that about 40% of the screen is submerged at a time. A backwash spray, located either inside or outside of the drum, periodically washes the collected debris into a fixed trough inside the drum where it is carried away. Coarse rotary screens typically have screen openings of about

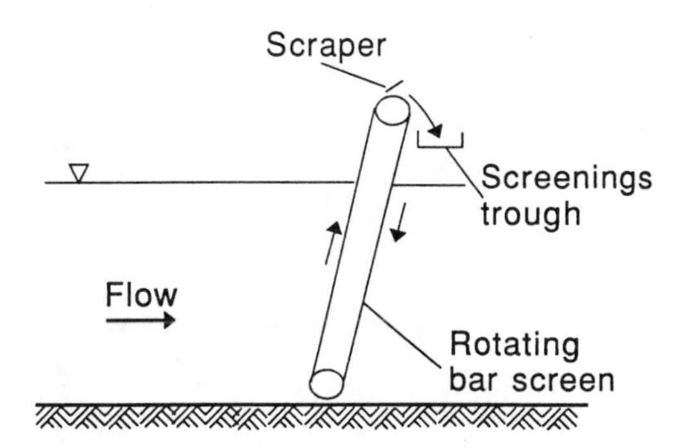

Figure 10-5. A rotary microscreen.

0.01–0.06 in. (0.25–1.5mm). Removal efficiencies are 5–25%. Rotary microscreens, used for fine solids removal, typically have screen openings of 15–60 μm and are 50–70% efficient (Corbitt 1990).

Gravitational Separation of Solids

SEDIMENTATION. If screens are not used, wastewater in recirculating systems is first treated by simple sedimentation (primary treatment). Sedimentation is the process by which suspended and settleable materials are separated from the wastewater by gravity settling. Common applications in recirculating system aquaculture include: (1) Settling of discrete particles, colloidal solids, and flocculent suspensions; and (2) settling of biological floc that sloughs from biological filter media. The principal design considerations are the basin's cross-sectional

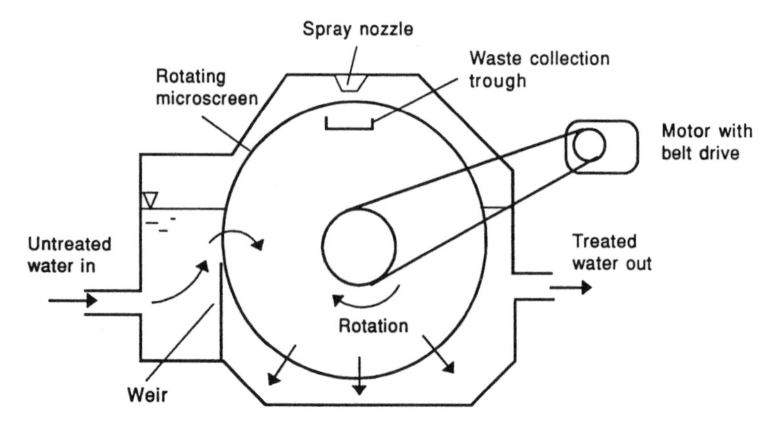

Figure 10-6. A radial flow microscreen.

Figure 10-7. A rotary microscreen (photo courtesy of Zeigler Brothers, Inc., Gardners, PA).

area, detention time, depth, and overflow rate. Efficiency of particle removal is determined by the characteristics of the wastewater, flow variations, and general maintenance procedures. Site conditions and method of sludge handling also influence sedimentation basin design and efficiency (Corbitt 1990).

Particle settling velocity is determined by considering the ideal sedimentation tank as illustrated in Figure 10-8. Four zones are identified: (1) Inlet zone; (2) settling zone; (3) sludge zone; and (4) outlet zone. The inlet zone is designed to reduce wastewater velocities and to distribute the flow evenly both horizontally and vertically to avoid turbulence and short-circuiting. In the settling zone it is essential that quiescent conditions be maintained for effective settling. Improperly designed inlet conditions can cause high velocities and turbulence which keep particles in suspension. Baffling is often necessary to reduce turbulence. The sludge zone should be deep enough to provide temporary storage of settled solids. Basin outlets are usually constructed with an overflow weir for flow control. Exit velocities should be kept low by minimizing weir hydraulic loadings. Weir loadings of 125–185 m^3/day/m of weir length (10,000–15,000 gpd/ft) are used by wastewater regulatory agencies. Higher rates may be required under certain circumstances.

In the ideal settling basin particles move horizontally with the flow and vertically under the influence of gravity. Equation 10-1 can be used to describe the settling rate of discrete particles (Tchobanoglous and Schroeder 1985):

$$V_s = \sqrt{\frac{4}{3}\frac{g(\rho_p - \rho_w)d_p}{C_D\rho_w}} \qquad (10\text{-}1)$$

where V_s = particle settling velocity, m/s; g = gravity term, m/s^2; ρ_p = density of particle, kg/m^3; ρ_w = density of water, kg/m^3; d_p = particle diameter, m; and C_D = drag coefficient, dimensionless. The drag coefficient can be calculated with Equation 10-2:

$$C_D = \frac{24}{RN} + \frac{3}{\sqrt{RN}} + 0.34 \qquad (10\text{-}2)$$

where RN = Reynold's number, dimensionless,

$$RN = \frac{V_s d_p \rho_w}{\mu} \qquad (10\text{-}3)$$

and μ = liquid viscosity, kg/m·s.

When RN < 0.3, the first term in Equation 10-2 predominates, and the discrete particle settling velocity becomes Stokes law (Tchobanoglous and Schroeder 1985):

$$V_s = \frac{g(\rho_p - \rho_w)d_p^2}{18\mu} \qquad (10\text{-}4)$$

For RN > 0.3, Equation 10-1 must be used to solve for settling velocity, and the problem must be solved by trial and error. The viscosity of water is about 0.001 N-s/m^2 (0.0000505 lb-s/ft^2) at 20°C (68°F), but varies considerably with temperature (see Table 6-2).

Discrete particle settling analysis in the ideal

Figure 10-8. Four zones within a typical sedimentation chamber: inlet zone, settling zone, sludge zone, and outlet zone.

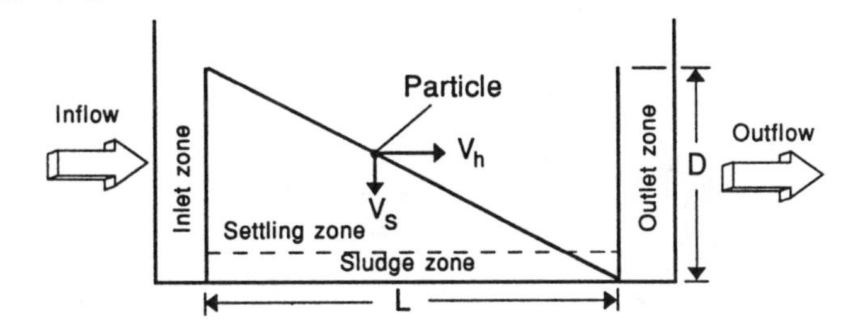

sedimentation basin is used to define the basin overflow rate, Q/WL. Assuming plug flow, particles entering the settling zone follow a straight line path that is the vector sum of the horizontal fluid velocity, V_h, and the particle settling velocity, V_s. Short-circuiting should be prevented since turbulent flow may result, causing resuspension of settled particles. Complete settling occurs when

$$V_{sc} > DV_h/L \qquad (10\text{-}5)$$

where V_{sc} = critical settling velocity, m/s; V_h = horizontal fluid velocity, m/s; D = basin depth, m; and L = basin length in horizontal flow direction, m. The horizontal fluid velocity V_h is equal to the fluid flow rate/basin cross-sectional area, $V_h = Q/DW$, and W = basin width, m. Combining this definition with Equation 10-5 results in

$$V_{sc} > Q/LW = Q/\textit{basin surface area} \qquad (10\text{-}6)$$

Particles having a velocity $V_s < V_{sc}$ will be removed in the proportion V_s/V_{sc}. The term to the right of the > sign in Equation 10-6 is the basin overflow rate and has the units m^3/day/m^2 (gpd/ft^2). Typical overflow rates used in wastewater treatment range from 24–49 m^3/day/m^2 (600–1,200 gpd/ft^2) combined with detention times of 1.5–2 hours (Corbitt 1990).

Equation 10-6 shows that the particles settling velocity is independent of basin depth. However, basin depth defines the volume associated with the required area for settling and is used to determine the detention time for wastewater flow. The depth of the basin must be sufficient to provide volume for storage of settled solids but also prevent the formation of high scouring velocities along the bottom and resuspension of the solids. Typical basin depths are 2.4–4.5 m (8–15 ft) (Corbitt 1990).

Both rectangular and circular sedimentation basins are common. Important features of each are shown in Figures 10-9 and 10-10. Basins should have mechanical sludge and surface skimmers as shown. Skimming troughs collect floating and settled substances. Typical dimensions for sedimentation basins used in wastewater treatment are given by Corbitt (1990). Rectangular basins range up to 91 m (300 ft) long and 6 m (20 ft) wide. Circular basins range from 3–60 m (10–200 ft) in diameter. The depth of each ranges from 3–4.5 m (10–15 ft).

In the real world, the ideal sedimentation basin doesn't exist. Sediment suspensions in aquaculture wastewater typically contain a mixture of particle sizes having a range of settling velocities. Particle size distribution and settling velocity information for aquaculture wastewater is lacking. Thus, it may be necessary to conduct laboratory tests with settling columns to predict settling basin performance. Once the particle settling velocity is known, basin dimensions

Figure 10-9. A rectangular sedimentation basin.

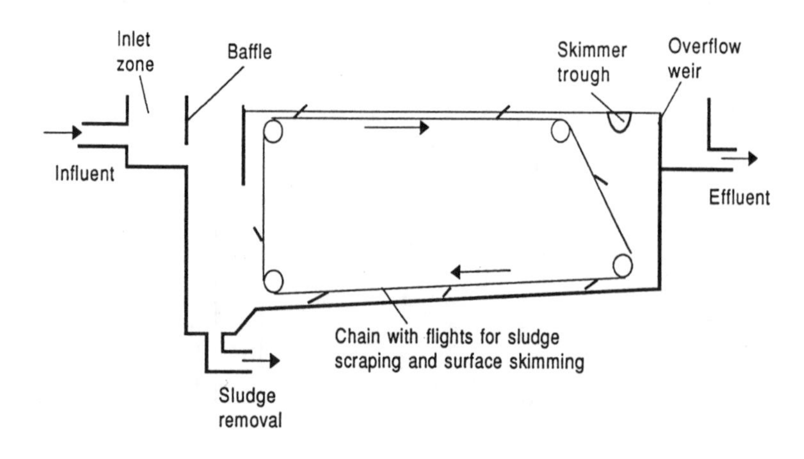

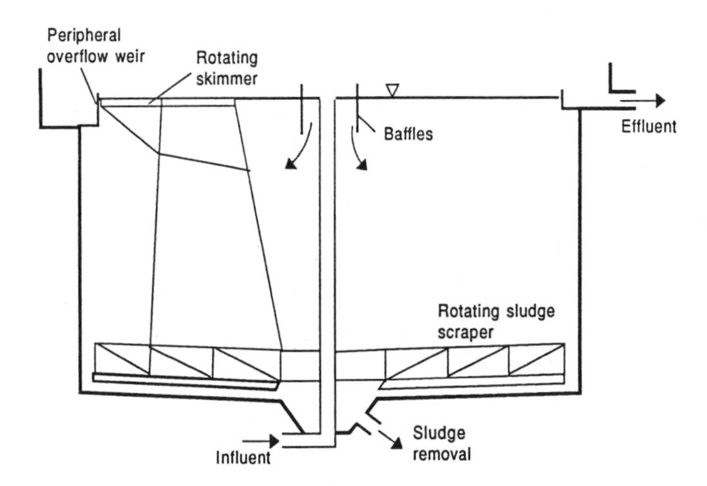

Figure 10-10. A circular sedimentation basin.

can be estimated with Equation 10-6 for a given discharge rate. It is often necessary to conduct full-scale tests to obtain sedimentation rates since particles do not always behave in the field like they do in laboratory conditions.

Settling of extremely fine suspended and dissolved solids can often be enhanced with the addition of polymers or other chemicals. The process is similar to that for chemical treatment of drinking water supplies. Chemicals are added for coagulation, that is, the formation of a floc of small particles. The flocculated mass, having a density slightly greater than water, then settles in a similar manner as discrete particles. Some of the more commonly used chemicals are alum (aluminum sulfate), ferric and ferrous chloride, ferric and ferrous sulfate, and sodium aluminate. Activated silica or bentonite clay particles are sometimes used as coagulant aids. Jar tests should be performed to determine the best coagulant and optimum dosage for a specific wastewater. Some of the coagulants are reactive within specific pH ranges, which may preclude their use for treating aquaculture wastewater. The jar test is described in Tchobanoglous and Schroeder (1985).

The chemical coagulants are delivered to the site in either a liquid or dry powder form. A metering system must be used to add the coagulant to the wastewater and a rapid-mix tank followed by a gentle mixing process is used to disperse the mixture. These processes must be included as an integral component of the sedimentation basin design. Finally, the treated wastewater is discharged to a settling basin. Figure 10-11 illustrates a circular sedimentation basin for chemical flocculation.

The quantity of sludge produced in the chemical flocculation process is much greater than that produced through plain sedimentation processes. Therefore, sludge handling and disposal must be an integral part of the overall waste treatment plan. Sludge may require dewatering to reduce its bulk and weight before transporting long distances for disposal. Aquaculture wastewater sludges are generally disposed of by land application.

PLATE AND TUBE SETTLERS. Lamellar plate (tilted plate) and tube settlers have been developed to enhance the settling characteristics of traditional sedimentation basins. They are shallow settling devices consisting of modules of either flat parallel plates or small plastic tubes of various geometric shapes. In wastewater treatment they are used in either primary, secondary, or tertiary sedimentation applications, but with limited success (Metcalf and Eddy 1991). Normal practice is to insert the modules of plates or tubes into either a rectangular or circular sedimentation basin. The wastewater flows upward through the modules and exits

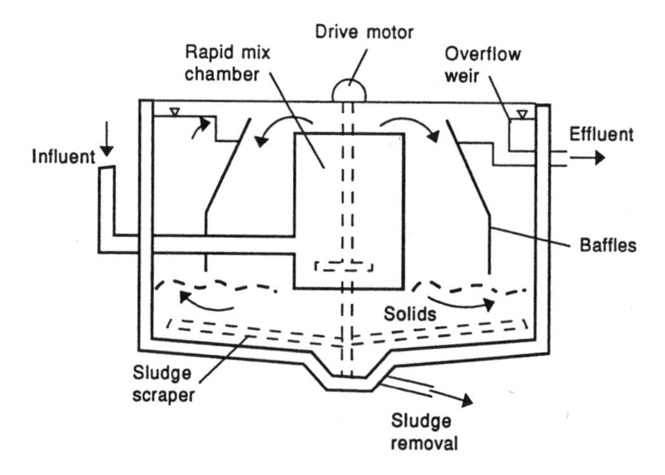

Figure 10-11. A circular basin for chemical mixing and sedimentation.

above them, as shown in Figure 10-12. These settlers have a very short settling distance, and turbulence is dampened within the small settling spaces. Typical tube sizing or spacing between plates is 25–50 mm (1–2 in). Solids that settle out gravity-flow downward through the spaces and fall to the bottom of the basin. To be self-cleaning the plates or tubes are set at an angle θ varying from 45–60° with the horizontal. If the plates or tubes are angled at less than 45°, sludge will accumulate in the spaces, and efficiency decreases if the angle is increased beyond 60°. A major problem is clogging caused by buildup of biological growths in the small spaces formed by the plates or tubes. The growths tend to impede water flow through the spaces, and often are accompanied by obnoxious odors. Biological growth accumulation is controlled by periodically flushing the spaces. The need for flushing poses problems where these devices are used. Plate or tube settlers can be used in specially designed new sedimentation basins or existing basins can be retrofitted to accommodate the settlers.

The critical settling velocity in plate or tube settlers is defined in a similar manner as the ideal sedimentation basin. The critical velocity becomes

$$V_{sc} = \frac{h}{\theta_H} = \frac{hu}{L} \tag{10-7}$$

$$V_{sc} = V_s \cos\theta \tag{10-8}$$

where V_{sc} = critical settling velocity for tube or plate settler, m/s; h = distance across tube space, m; θ_H = detention time, s; u = fluid velocity through tube, m/s; L = length of tube, m; V_s = normal settling velocity of particle, m/s; and θ = angle, degrees.

Lamellar plate separators have been used with success at the Bozeman National Fish Hatchery in Montana and the Quinault National Fish Hatchery in Washington (McLaughlin 1981). The Bozeman facility separator is designed to handle a flow of 315 L/s at an overflow rate of 0.34 L/s/m^2. The Quinault facility separator treats 1,387 L/s at an overflow rate of 0.34 L/s/m^2.

Biodek (The Munters Corp., Fort Meyers, Fl.) is specially designed material manufactured from corrugated plastic sheets that are attached such that slanted tubes are formed. The material is available in 0.3 × 0.3 × 1.9 m (1 × 1 × 6 ft) modules (Figure 10-13). The modules can be arranged to create any size tube settler. Similar materials are available from other manufacturers.

CENTRIFUGES AND HYDROCLONES. Centrifuges increase the gravitational force on particles with a spinning motion, causing increased settling rates. There are many types of centrifuges offering a range of g values (i.e., the num-

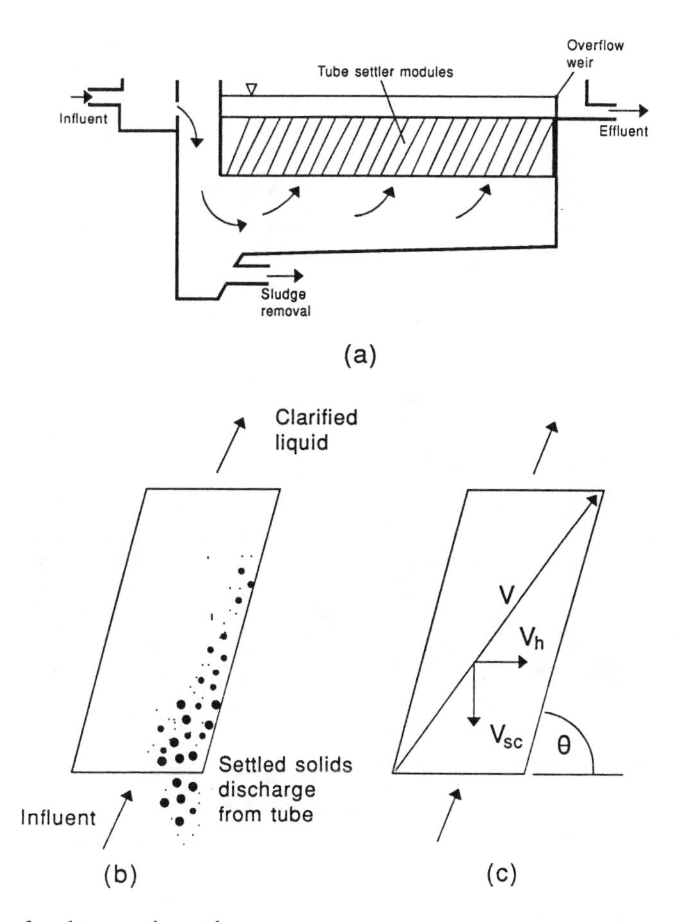

Figure 10-12. Diagram of a plate or tube settler.

ber of times the centrifugal force exceeds normal gravitational force). Higher rotational speeds are required for seawater systems since many organic particles have densities close to that of seawater (Huguenin and Colt 1989). Centrifuges are available in batch and continuous flow units, but only the continuous flow models have significant application in aquaculture. These devices are expensive and complicated and have had limited success in aquaculture applications. Design details are discussed in greater detail in Wheaton (1977).

Hydroclones, cyclones, or swirl separators are conical-shaped structures in which particles settle under the influence of rotational forces similar to centrifuges. Figure 10-14 illustrates the basic design and operating principles of a simple hydroclone. Wastewater enters tangentially near the top of the unit, creating a spiral. The rotational flow causes heavier particles to move outward toward the wall of the unit and downward. A small percentage of the flow and particulates exit through the bottom of the unit. This flow is called underflow and is denoted by F_u. A low pressure area created at the center of the spiral carries the clean water flow out the top of the unit. This flow is called overflow and is denoted by F_o. Like centrifuges, these units are expensive and have limited application for aquaculture wastewater treatment.

GRANULAR MEDIA FILTERS. Another type of filter used to remove solids prior to biological filtration is the granular media or ''sand'' filter. Many types are used in aquaculture applications. Their performance is dependent on filter

Figure 10-13. Fabricated tube settler module (photo by T. Lawson).

type, operating procedures, waste characteristics, and the characteristics of the media. The physical differences between sands can be critical, and each of their characteristics is important to filter performance. Other critical design factors are cycle time before clogging, ease of backwash, and pressure head required for operation (Huguenin and Colt 1989).

Sand filters are classified in a variety of ways. They may be classified according to the number of different sand sizes or types as single- or dual-media filters. Single-media filters

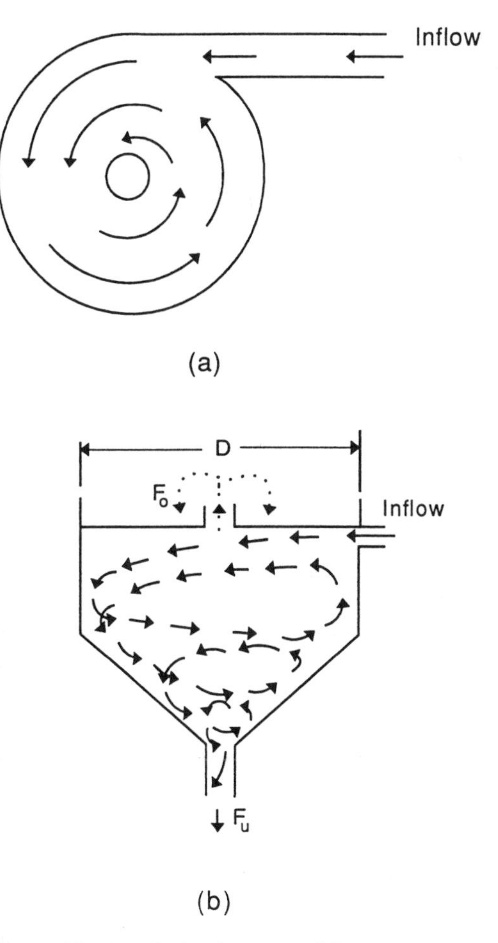

Figure 10-14. A hydroclone or swirl separator.

use a single sand size throughout the filter bed depth, whereas dual-media filters use more than one sand size or they may contain a layer of sand and a layer of anthracite. Dual-media filter beds are normally graded with media particle sizes arranged from coarse to fine in the direction of flow to increase filtering capacity before clogging occurs. Typical media characteristics for wastewater treatment are given in Table 10-3. The media are arranged in layers according to density so that they will settle in the correct order after backwashing, particularly if the bed is fluidized during backwashing. Typical backwash rates for single- and dual-media filters are also given in the table.

Sand filters are also classified as slow or rapid filters, or they can be pressurized or open (nonpressurized). Both slow and rapid filters may use a foundation of coarse gravel layer with a high specific weight (2.5 or greater) over a manifold with perforated laterals (Figure 10-15). Alternate underdrain systems may be used.

Slow sand filters are not normally found in off-the-shelf models and are either custom-built or the aquaculturist may have to build his own. They are usually open to the atmosphere and have low hydraulic loadings about 0.68 Lps/m^2 or less. Since they have low flow rate, they require more floor space. Head losses in slow sand filters are generally low, only a few centimeters, allowing the filter to be used in gravity flow situations. Clogging can be corrected by scraping off and replacing the first few centimeters of sand or the filter may be designed with a backwashing feature. Filters prone to clogging may be designed with an overflow drain or a cutoff switch. The smallest size particles removed in slow sand filters are about 30 μm (Spotte 1970; Wheaton 1977).

Rapid sand filters are usually closed units, and are either pressure- or vacuum-operated. A number of styles and sizes are commercially available for a number of applications. An advantage of pressurized filters is that they require much less floor space than slow sand filters. Pressure-type filters are used much more frequently than vacuum filters in aquaculture applications, therefore only pressure filters will be discussed in this text.

Figure 10-16 illustrates a common pressure sand filter. These typically handle high flow rates (1.4 L/s/m^2 or 20 gpm/ft^2). For larger flows, multiple units may be used in parallel. Disadvantages are more frequent backwashing requirements and high head losses. Head losses for pressurized sand filters are typically 9.1–27.4 m (30–90 ft) (Huguenin and Colt 1989). Swimming-pool-type sand filters (Figure 10-17) have been used with limited success for treating aquaculture wastewater. Although they are readily available, inexpensive and are easily retrofitted into virtually any existing recirculating system, they are subject to rapid clogging.

Table 10-3. Typical design data for granular-medium filters used for wastewater treatment[1].

	Single medium[2]		Dual medium	
Parameter	Range	Typical	Range	Typical
Sand				
Depth, mm	500–900	600	150–300	300
Effective size, mm	0.45–0.7	0.5	0.4–0.7	0.55
Uniformity coefficient	1.3–1.7	1.5	1.4–1.7	1.6
Anthracite				
Depth, mm	900–1,800	1500	300–600	500
Effective size, mm	0.8–1.8	1.4	0.8–1.8	1.2
Uniformity coefficient	1.4–1.8	1.6	1.4–1.8	1.6
Filtration rate (L/m²/min)	80–400	160	80–400	160
Backwash rate (L/m²/min)	360–1,000	500[3]	500–1,600	800

[1] Effective size is defined as the 10% size by mass, d_{10}. The uniformity coefficient is defined as the ratio of the 60% to the 10% size by mass ($uc = d_{60}/d_{10}$).

[2] Separate sand and anthracite single-medium filters.

[3] For single-medium sand filter only.

Source: Tchobanoglous, G. and E.D. Shroeder (1987) with permission.

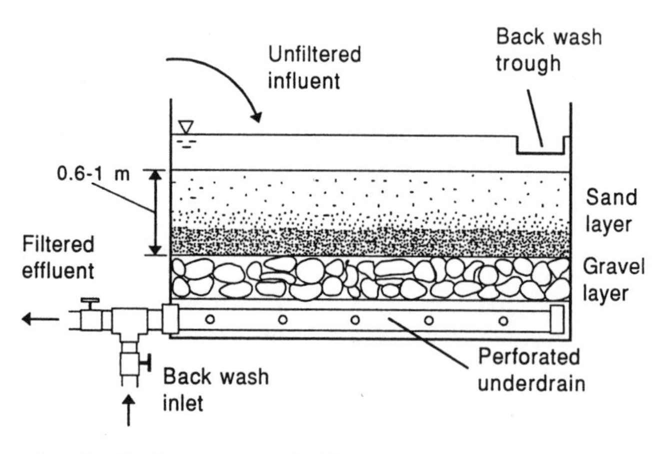

Figure 10-15. A vacuum (gravity flow) granular media filter.

However, they can be used for lightly loaded systems such as outdoor ornamental fish ponds.

The media size must be carefully selected for rapid sand filters to avoid failure. In situations where the wastewater contains extremely heavy solids loads the hydraulic loading may be reduced because the filter may rapidly clog. Because of their nature for removing fine particulates their main function is for secondary treatment following sedimentation, coarse screening, or some other coarse solids removal process prior to biological filtration. Single-media rapid filters are capable of removing particles as small as 20 μm.

Rapid sand filters are available with automatic backwash valve assemblies. Automatic

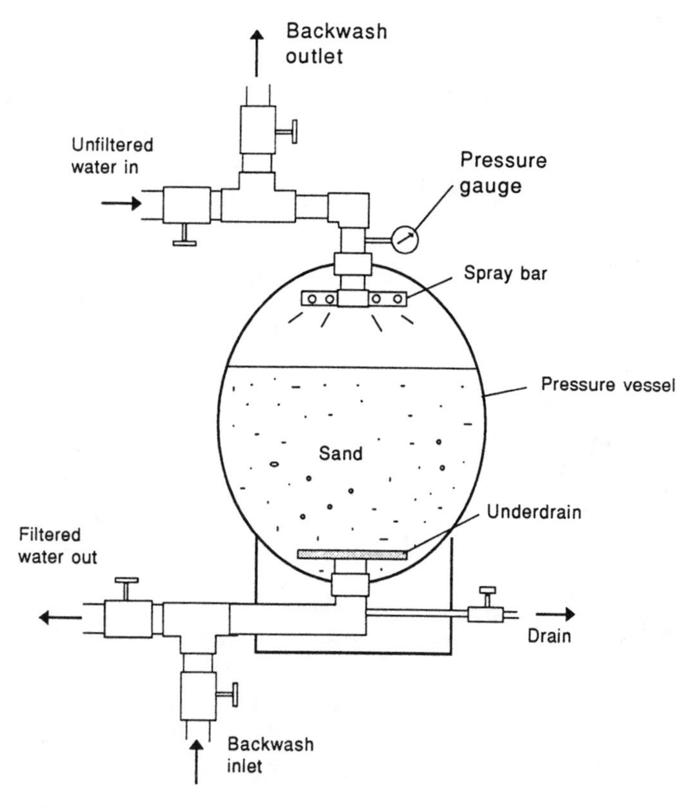

igure 10-16. A pressurized sand filter.

ackwashing adds to capital costs but reduces naintenance and labor costs in the long run. rratic service requirements due to flow variaons are reduced with automatic backwashing. he effectiveness of backwashing is critical. ackwash rates and frequency depend on charcteristics of the wastes and filter sand as well s filter design. Typical rates are 0.5–1 $m^3/m^2/$ in or 720–1,440 m^3/m^2/day (Muir 1982). ackwash design must match filter characteriscs since backwash flow rate must be sufficient) fluidize the filter media but not so great as to ush the media from the filter. The backwash ow must be evenly distributed over the filter ross section to avoid short-circuiting. Pressure cross the filter must be monitored to determine hen backwashing is required or backwashing ay routinely be done one or more times daily.

CARTRIDGE FILTERS. Cartridge filters consist f two parts: (1) A canister; and (2) a replaceable cartridge or filter element (Figure 10-18). They function much like the oil filter in automobiles. Cartridge filters are usually placed directly in line and should be located downstream from a pump since considerable back-pressure is developed. A single filter is used for low flows, but multiple filters in parallel can be used for large flows. Head losses are high and increase dramatically as the filter cartridge clogs. When this happens swift action should be taken to clean or replace the cartridge to avoid having the water pump burn out.

Cartridge filters are available from a number of manufacturers who specialize in filters of this type, and many cartridges are interchangeable between different manufacturer's models, providing considerable operational flexibility. Replacement cartridges are available in a wide variety of sizes, materials, and filtration performances, for particulate removal as small as 0.01 µm. Some canisters are fabricated from clear

Figure 10-17. Swimming pool sand filter (photo by T. Lawson).

plastic so that the cartridge can be viewed and the system operator can tell when it requires replacement. An in-line pressure gauge should be used to indicate when the cartridge requires cleaning or replacement. These filters have application in systems where a high degree of water clarity is desired, such as in public display aquaria or educational systems. In commercial food-fish systems, however, clarity is usually not a consideration. Due to high head losses and associated energy costs, maintenance, and frequent element replacement, cartridge filters are generally not recommended for most aquaculture applications.

DIATOMACEOUS EARTH FILTERS. Diatoma ceous earth (DE) is a granular material com posed of the graded skeletal remains of diatoms Filters composed of diatomaceous earth ar used as an alternative to cartridge filters in sys tems where a high degree of clarity is desired The material is easily clogged by fine sus pended matter, therefore, the wastewater mus be relatively free of suspended materials. Som of the finer grades of diatomaceous earth ar capable of removing particles as small as 0. µm (Spotte 1970).

The performance of a DE filter is determine by a number of factors including the character

Figure 10-18. A cartridge filter (photo by T. Lawson).

stics of the diatomaceous earth used since many different grades are available, filter operating conditions, and filter maintenance schedule. DE filters are composed of three basic components: the central core, filter sleeve, and the filter cake. The central core is a porous structure, usually fabricated from a rigid plastic material. In large DE filters several core elements are attached to a manifold to handle large flows. The filter sleeve, a thin, tightly woven clothlike material (usually polypropylene), fits over the central core. The sleeve is often in the shape of a long tube or a two-sided flat plate (leaf). The central core and filter sleeve together comprise the filter element.

Figure 10-19 illustrates how a DE filter functions. In Figure 10-19b, clean water and diatomaceous earth are pumped through the system, and the earth coats the filter elements. Unfiltered water is then mixed with more diatomaceous earth, normally of a coarser grade, and the mixture is pumped through the filter (Figure 10-19c). The new diatomaceous earth material and the pollutant particles are trapped on the coated filter elements, and the filtered water passes through the filter cores of the elements and out of the filter. The thickness of the filter cake continually increases. When the cake become so thick that water cannot pass through without great difficulty, the filter must be backwashed (Figure 10-19d). During the backwash cycle the flow of water through the filter is reversed, much like in a granular media filter, and the filter cake is flushed to waste. A new filter cake must be established after backwashing to complete the cleaning cycle. Very little effort is required to backwash since the filter cake is held to the substrate by operating pressure or a slight vacuum. Thus, DE filter backwashing requires little pressure.

DE equipment can be expensive both to purchase initially and to operate. Therefore, DE filter units are only recommended for large systems of at least several thousand gallons to make their purchase and use economical. Some DE filters are produced as *cartridge filters* where disposable cartridges already coated with DE are available. This feature reduces initial and operating costs for DE filter units and may make their use in recirculating systems more attractive.

CHEMICAL TREATMENT METHODS

Only three chemical filtration methods have been identified as feasible in large commercial recirculating aquaculture systems: activated carbon, ion exchange, and foam fractionation. Reverse osmosis is a process which in the aquarium trade for the production of high-priced ornamental fish. However, due to its extremely high cost, this process is not used in large-scale production of food fish. Reverse osmosis will receive no further discussion.

Activated Carbon

Activated carbon is prepared by first making a char from coal; pecan; coconut, or walnut shells; wood; or the bones of animals by heating the material in the absence of air at about 900°C (1,652°F). The charred material is then acti-

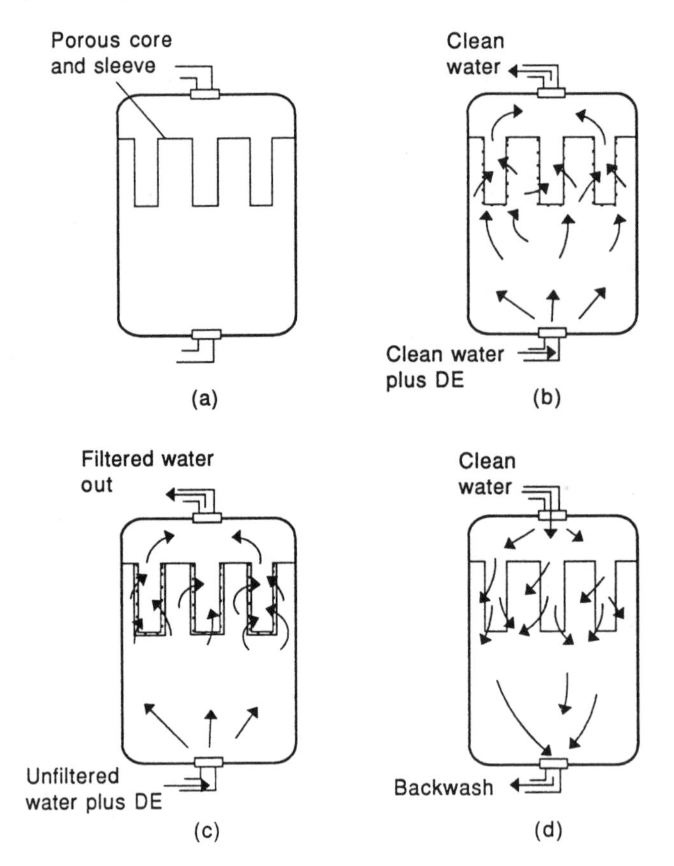

Figure 10-19. A diatomaceous earth (DE) filter.

vated by exposing it to an oxidizing gas at high temperature. The gas creates a highly porous structure in the char and thus creates a large internal surface area as shown in Figure 10-20. The adsorption surface area has been estimated at 1,000,000 m^2/kg (4.9×10^{-6} lb/ft^2) material (Wheaton 1977). Activated carbon is used to remove volatile organics, color, odor, and turbidity from wastewater. Carbon filters function more efficiently at low organic loadings. Influent suspended solids concentrations greater than 20 mg/L will form deposits on the carbon granules, causing clogging, short-circuiting, high head losses, and loss of adsorptive capacity (Metcalf and Eddy 1991). Once the adsorptive capacity is reached the filter then functions as a simple granular media filter. Carbon filters are often used in addition to biofilters as a polishing stage to remove persistent nonbiodegradable organic materials. However, they also remove cer tain essential trace elements and therapeuti compounds.

After activation, the carbon material can b prepared into different sizes with different ad sorption capacities. The two most commo sizes are powdered activated carbon (PAC) an granular activated carbon (GAC). PAC is finel ground, loose carbon having a diameter les than 200 mesh, and GAC has a diameter greate than 0.1 mm. PAC has not received much at tention in the aquaculture industry except in th aquarium trade. Because of its greater surfac area, adsorption is quicker, but it is also mor expensive. The GAC type is probably the on most used in the aquaculture industry. A fixed bed column is often used as a means of contact ing the wastewater with GAC (Figure 10-21) The water is applied at the top of the colum

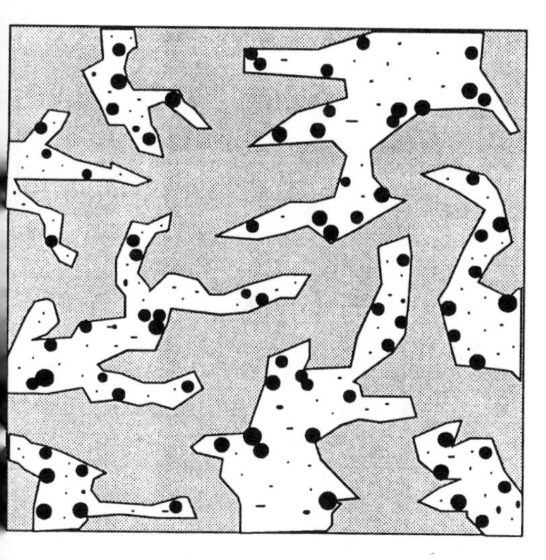

Figure 10-20. Pore spaces within a carbon granule.

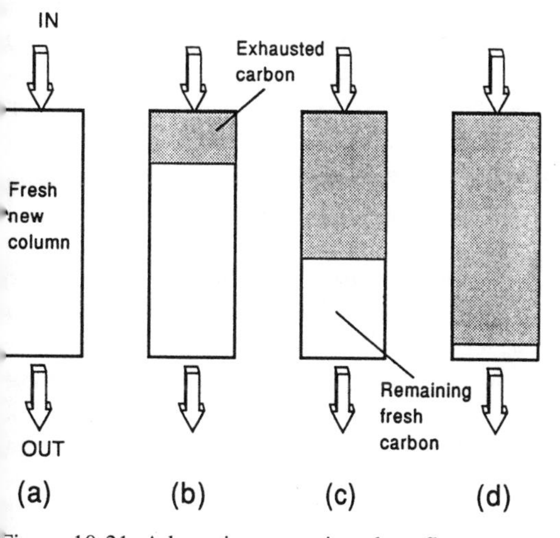

Figure 10-21. Adsorption zones in a downflow granular activated carbon (GAC) bed.

and flows downward through the adsorbent. Adsorbed material accumulates at the top of the bed until it reaches a maximum. The maximum amount of material that can be adsorbed by the GAC occurs when the adsorbed material on the carbon is in equilibrium with the concentration of the material in the liquid solution surrounding the carbon (Corbitt 1990). The adsorbent is thus loaded to capacity, and that portion of the bed is said to be exhausted. The exhausted zone moves downward through the column with time until the entire adsorbent bed is exhausted as illustrated by the shaded areas in Figures 10-21b through 10-21d. The concentration of contaminant in the effluent from the column remains near zero until the exhausted zone reaches the bottom. The contaminant concentration then abruptly increases to become equal to that of the influent when the entire GAC bed is exhausted.

The economics of the activated carbon process depend upon the efficient regeneration of the carbon after its adsorptive capacity has been exhausted. GAC is regenerated by heating at high temperatures to oxidize the organic material and remove it from the adsorptive process. About 5–10% of the carbon is destroyed in this process and must be replaced with fresh carbon. The adsorptive capacity of the carbon is reduced with each regeneration. The methodology for regenerating PAC is not well defined. For this reason, PAC is usually replaced with new carbon when its adsorptive capacity is exhausted (Metcalf and Eddy 1991).

Carbon filters are typically not found in food-fish production systems because of their expense. However, they may have a niche in other areas, such as in ornamental fish production, in public aquaria where water clarity and, hence, visibility is important, and for production of rare or high-valued species.

Ion Exchange

Ion exchange is a unit process in which certain ions are displaced from an insoluble exchange material (resin) by ions of a different species dissolved in the wastewater. Ion exchange resins are manufactured in the form of tiny porous beads about 1 mm in diameter. The ion exchanger can be operated either in a batch process or in a continuous mode. In the batch process the resin is mixed in a reactor with the water to be treated. Once the reaction is complete, the spent resin is removed by settling and is subsequently regenerated and reused. This process is normally too complicated to be incorporated into an aquaculture system. The more common approach is to use ion exchange resins in a continuous process. The exchange resins are put into a filter bed or a packed column, and the wastewater enters the top of the column and down through the resin. When the resin exchange capacity is exhausted, the column can be backflushed to remove trapped solids, and the resin can be regenerated. High concentrations of suspended solids in wastewater can clog ion exchange resin beds, causing reduced effectiveness and high head losses. Pretreatment may be necessary to remove solids before ion exchange.

The resins used exchange either anions (negatively charged ions) or cations (positively charged ions). They can be either strongly or weakly acidic or strongly or weakly basic. Cation exchange resins are usually colored gray, brown, or dark brown while anion resins are white or tan in color. The choice of a resin depends on the properties of the substance to be extracted form the wastewater. The resins must be regenerated or replaced when their exchange capacity is exhausted. The weaker cationic or anionic resins are preferred since they are easier to regenerate (Moe 1989).

Cationic exchange resins are regenerated with a strong acid such as sulfuric or hydrochloric acid. Sodium hydroxide is commonly used to regenerate anionic exchange resins. Those resins can be regenerated by soaking in laundry bleach followed by a rinse with dechlorinator. All of the regeneration materials mentioned can cause unfavorable water quality changes, particularly pH, if they are not thoroughly rinsed from the exchange resin before use.

The ammonium ion, NH_4^+, is typically the ion removed from the waste stream for nitrogen control. Ammonia removal is achieved with strongly acidic cation exchange resins in the sodium form. Strongly basic ammonia exchange resins in the chloride form are effective in the removal of nitrate and phosphate ions. Both of these resins can be regenerated with a sodium chloride solution. It is obvious that these resins are not effective in marine systems since the high chloride concentration in seawater (19,000 mg/L) will continually flush the trapped ions from the resin beads.

Both natural and synthetic ion exchange resin beads are available. Synthetic resins are more durable and are consequently used more often. Some natural resins (zeolites) have been successfully used for removing ammonia from aquaculture wastewater. One naturally occurring zeolite, clinoptilolite, is one of the best natural ion exchange resins and is effective for removal of the ammonium ion (Figure 10-22). Deposits are found throughout the United States. Its purity, strength, and adsorptive capacity varies from one deposit to another as does its density and granule size (Huguenin and Colt 1989). It has been reported that clinoptilolite will remove up to 10 times its weight in ammonia before its exchange capacity is exhausted. It is also inexpensive in comparison to many synthetic resins.

Clinoptilolite is regenerated by flushing with a sodium chloride solution, and, hence, is not effective in marine systems. Several researchers have reported on its performance for ammonia removal in freshwater aquaculture systems (Jorgenson et al. 1979; Berka et al. 1980). Mayo (1976) observed a 98% ammonium ion removal at a hydraulic loading rate of 2.6 m^3/m^3 resin day.

Although a handful of researchers have reported success with ion exchange on a pilot scale, it remains questionable if ion exchange is economically feasible in large commercial aquaculture production systems because of the expense and complexity involved in its use and regeneration.

Foam Fractionation

Foam fractionation is a process that removes dissolved organic carbon (DOC) and particulate organic carbon (POC) from fish culture water by adsorbing them onto the surfaces of air bubbles rising in a closed contact column. The bubbles create a foam at the top of the liquid column, and accumulated organic wastes are thus discarded along with the foam produced. The substances removed are called *surfactants* because they are surface active. Other names given to this process include protein skimming, air stripping, and froth flotation. The reported

igure 10-22. A naturally occurring zeolite material, clinoptilolite (photo by T. Lawson).

enefits of foam fractionation are: (1) Reduced logging of pipes, filters, pumps, etc.; (2) removal of proteins and high molecular weight ompounds: (3) increased water clarity through he removal of humic substances: (4) increased eration; and (5) pH stabilization through the emoval of organic acids. The process works, lthough the exact nature of all surfactants involved and their effects on aquatic organisms re not clearly defined.

A surfactant molecule is polar at one end and ipolar at the other. The polar end is attracted to vater molecules and therefore becomes *hydrohilic*. The dipolar end repels water molecules nd is *hydrophobic*. The surface-active moleules tend to congregate at the air-water interace with their hydrophobic ends in contact with ir, as demonstrated in Figure 10-23.

To be effective, the foam fractionation process requires a column in which the air and surfactant-laden water can interact. Air-water interfaces are provided by diffused aeration at the bottom of the column. Surfactants migrate to the bubble-water interfaces, and the rising bubbles acquire coatings of the surface-active material. The bubbles burst at the top of the column, and thick layers of foam develop. If the foam is stable enough, the surface-active material accumulates and can later be removed with the foam.

The performance of a foam fractionator depends on the chemistry of the bulk liquid (i.e., organic load and composition), surface tension, temperature, viscosity, pH, salinity, bubble size, air-to-liquid ratio, and contact time (Huguenin and Colt 1989). These factors are generally beyond the control of the aquaculturist with the exception of bubble size. Water temperature

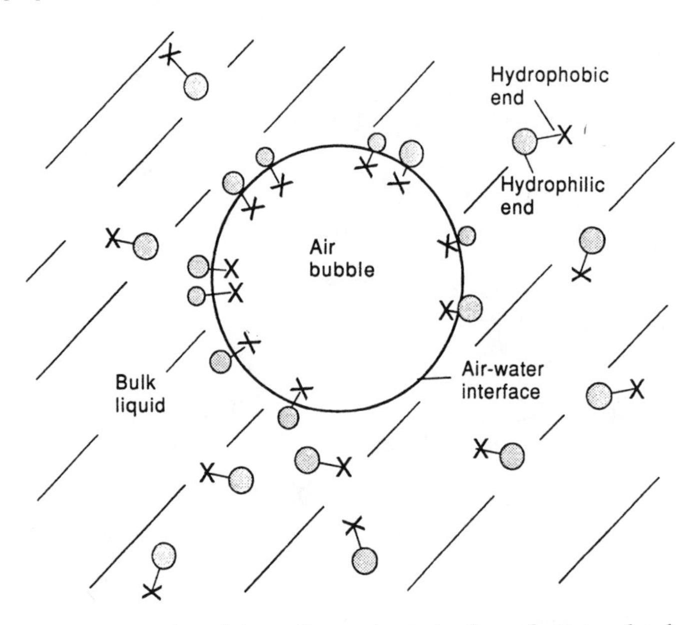

Figure 10-23. Illustration of the attraction of the surface-active ends of a surfactant molecule to an air bubble rising in a column of liquid.

and pH can be controlled to a great degree, but these parameters must be adjusted to meet the physiological needs of the fish and not foam production. Bubble size can be controlled depending on the type of air diffuser used. Smaller air bubbles have greater surface area for surfactant attachment, plus they rise through the bulk liquid at a slower rate as a result of fluid friction, thus increasing the air-liquid contact time. Higher air-to-liquid ratios also tend to improve the foaming process (Wheaton et al. 1976). In a study by Wace and Banfield (1966), the ideal air-liquid ratio was obtained at a gas flow rate of 1.8 $cm^3/s\ cm^2$ of column cross-sectional area. They also determined the ideal bubble diameter to be 0.8 mm (0.03 in.).

The design of effective foam fractionation devices for aquaculture purposes is more of an art than a science, mainly due to the chemical complexity of the process. Foam fractionation was originally developed as a chemical engineering process for separating multiple liquids from a common fluid and for separating solids from liquids (Lawson and Wheaton 1980). There has been considerable mathematical modeling of the removal process (Lemlich 1966; Dwivedy 1973; Chen et al. 1993), but actual column design data is lacking. Foam fractionators are commercially available, but aquaculture applications may require pilot-scale experimentation or trial-and-error methodology.

The important parameters to consider in the design of foam fractionators are: (1) the air-to-water ratio; (2) air bubble diameter; (3) column height and diameter; (4) foam height; and (5) air-water contact time. Small variations in just one of these parameters can make a significant difference in foam fractionation efficiency.

Two fractionator designs are commonly used for aquaculture applications: the *cocurrent* and the *countercurrent* designs. In the cocurrent design (Figure 10-24) compressed air is introduced through a diffuser (1). As the bubbles rise the untreated culture water flows into the column (2) much like an air lift. The bubbles contact the water to produce a foam, which collects at the top of the column (3), where it can be collected in a receptacle for later disposal. The treated culture water then exits the column at (4).

In the countercurrent design the air bubbles

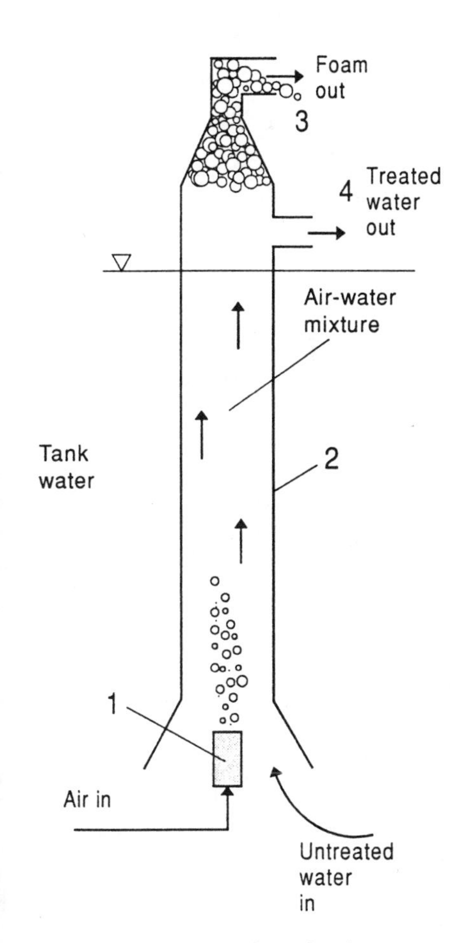

Figure 10-24. A cocurrent foam fractionator.

and water move past each other in opposite directions. Friction and drag forces on the bubbles are increased, reducing the rate at which the bubbles rise through the bulk liquid. This tends to increase the air-water contact time. In the cocurrent design, the air bubbles and water rise together, and the contact time is shortened. This observation suggests that the countercurrent design is more efficient, although no data directly exists to support this assumption.

Two countercurrent designs are illustrated in Figure 10-25. In the first, untreated water enters the column through openings (1) near the waterline (Figure 10-25a). The water moves downward through the column (2), through a connecting elbow (3) and is air-lifted back into the tank via a smaller-diameter tube (4). The velocity of the air-water mixture is greater in the air lift tube than in the foam column, thus causing water to flow through the system. Foam is removed from the fractionator at (5). The mechanism in Figure 10-25b is similar except that the contact column (1) is fitted with a larger-diameter outer tube (2) to draw the untreated water into the column from greater depths below the surface. Water in the tank is thus better mixed, and the water cannot be forced backed into the tank by injected air as sometimes happens at (1) in the design shown in Figure 10-25a. This design is reported to be slightly more efficient (Spotte 1992).

BIOLOGICAL TREATMENT METHODS

The term *biological filtration* in the broad sense refers to filtration techniques that utilize living organisms to remove a substance from a liquid solution. This definition includes systems utilizing algae and higher green plants to filter the water. These systems are commonly referred to as hydroponic systems. The term *biological filter*, or *biofilter*, as used here refers to the removal of ammonia and nitrite by bacteria. It is generally accepted that, after oxygen, ammonia often becomes the limiting factor for successful closed recirculating system operation.

Filter Kinetics

Nitrification is the oxidation of ammonia to nitrate with nitrite formed as an intermediate product. Un-ionized ammonia (NH_3) and nitrite (NO_2) are toxic to fish at very low levels while nitrate (NO_3) is considered relatively nontoxic in concentrations up to about 400 mg/L (see Chapter 2). The conversion of ammonia to nitrate is an aerobic process. If anaerobic conditions develop, denitrification occurs, and nitrate is converted back to ammonia. Also, denitrification may cause nitrogen loss from culture water. Consequently, denitrification processes have been suggested as possible alternatives for ammonia removal from aquaculture systems;

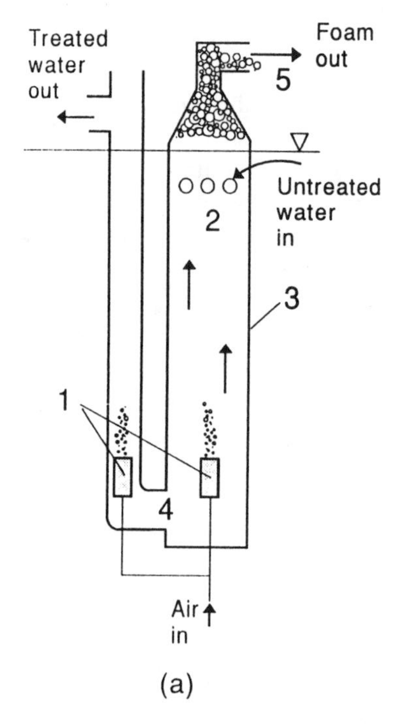

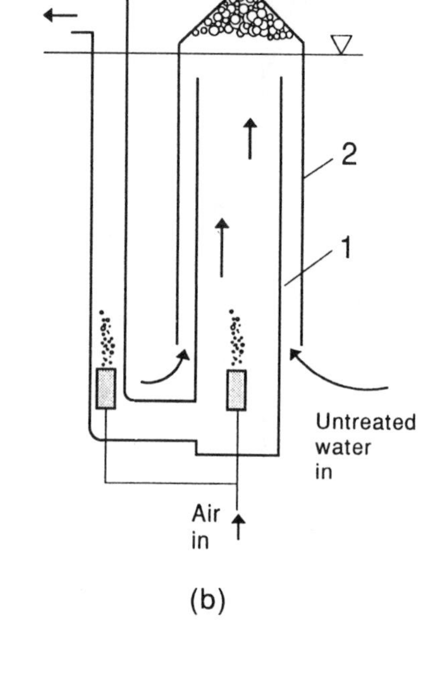

Figure 10-25. Countercurrent foam fractionators.

however, most aquaculture systems do not address denitrification (Hochheimer and Wheaton 1991). Thus far, economical denitrification processes for nitrogen removal from culture systems have not been developed.

Nitrifying bacteria are called *chemosynthetic autotrophs* or *chemolithotrophs*. Autotrophic bacteria derive their energy from inorganic compounds, as opposed to heterotrophs, which derive their energy from organic sources (WPCF 1983). In addition, autotrophs derive carbon from carbon dioxide, and their hydrogen requirements come from either ammonia, water molecules, or the atmosphere (Carpenter 1967).

Over 400 species of bacteria capable of oxidizing ammonia have been identified (Cutler and Crump 1933). The autotrophs, of which *Nitrosomonas sp*. are the major species, are best at ammonia oxidation. Painter (1970) identified at least five genera of bacteria capable of oxidizing ammonia to nitrite: *Nitrosomonas, Nitrosococcus, Nitrosospira, Nitrosocystis*, and *Nitrosogloea*. He felt, however, that the genus *Nitrosomonas* was the most valid, and two species that he isolated were *Nitrosomonas europea* and *N. monocella*. Painter also identified two genera of bacteria capable of oxidizing nitrite to nitrate: *Nitrobacter* and *Nitrocystis*. Two species that he isolated within the *Nitrobacter* genera were *N. agile* and *N. winogradskyi*. It is generally accepted that the conversion of ammonia to nitrite is done by *Nitrosomonas sp.* and the conversion of nitrite to nitrate is done by *Nitrobacter sp.*

Nitrification is a two-stage aerobic process. The first stage is the conversion of ammonia to nitrite by *Nitrosomonas* bacteria (USEPA 1975; WPCF 1983):

$$NH_4^+ + 1.5O_2 \rightarrow 2H^+ + H_2O + NO_2 \quad (10\text{-}9)$$

The process is a multistep one that produces hydroxylamine and other undetermined compounds as intermediates. *Nitrosomonas* can only use ammonia in the ionized form as an energy source. The process yields energy, which is used to assimilate carbon dioxide (Atlas and Bartha 1987). *Nitrosomonas* produces nitrite only when ammonia is used as an energy source. The process is inhibited in the presence of chelating agents (Lees 1952).

The oxidation of nitrite is a single-step process that uses oxygen from water to form nitrate and only molecular oxygen as an electron acceptor (Atlas and Bartha 1987). The chemical reaction is shown by the following equation (USEPA 1975; WPCF 1983):

$$NO_2 + 0.5O_2 \rightarrow NO_3 \quad (10\text{-}10)$$

The chemical reactions in Equations 10-9 and 10-10 release energy that is used by *Nitrosomonas* and *Nitrobacter* to produce new cell growth. The energy generation process is inefficient, and cell yield per unit of energy is low. Consequently, the growth of nitrifiers is very slow. The average generation time is 10–12 hours (Laskin and Lechevalier 1974).

Nitrifiers require considerable amounts of substrate to meet their energy demands. This is desirable for biofilter performance since little cell mass will be produced. Biofilters will then be slow to clog, and smaller volumes of sludge will be produced. Equations 10-11 and 10-12 describe stoichiometric requirements of nitrification and cellular growth (WPCF 1983; Gujer and Boller 1986):

$$NH_4^+ + 1.83O_2 + 1.98\,HCO_3 \rightarrow 0.021C_5H_7O_2N + 0.098NO_3 + 1.041H_2O + 1.88H_2CO_3 \quad (10\text{-}11)$$

$$NH_4^+ + 1.9O_2 + 2HCO_3 \rightarrow NO_3 + 1.9CO_2 + 2.9H_2O + 0.1CH_2O \quad (10\text{-}12)$$

In Equation 10-12 CH_2O represents cell biomass. These equations can be used to calculate the oxygen and alkalinity requirements and the cell mass production. Table 10-4 gives predicted values for these parameters, based on the oxidation of one gram of NH_4^+-N. For each gram of NH_4^+-N oxidized to NO_3-N, 4.57 g of oxygen and 7.14 g of alkalinity (as $CaCO_3$) are consumed. In addition, 8.59 g of carbonic acid, 0.17 g of cell mass, 4.43 g of nitrate, 3.73 g of water, and 5.97 g of carbon dioxide are produced. In the intermediate step of ammonia oxidation to nitrate, one mole of ammonia produces one mole of nitrite (1 g NH_4^+-N produces 3.29 g NO_2-N (Wheaton 1977; Wheaton et al. 1991; Hochheimer and Wheaton 1991). Nitrification is an acid-forming process. Therefore, water in closed systems must be buffered to prevent a decline in pH.

Table 10-4. Stoichiometric requirements when 1 gram NH_4-N is oxidized to NO_3-N.

	Equation 10-11		Equation 10-12	
Compound	g used	g produced	g used	g produced
O_2	4.18		4.33	
H^+		1.98		1.98
Alkalinity (as $CaCO_3$)	7.14		7.14	
$C_5H_7O_2N$ (cells)		0.17		
CH_2O (cells)				0.21
NO_3-N		4.34		4.43
H_2O		1.34		3.73
H_2CO_3		8.59		
CO_2				5.97

Source: Hochheimer and Wheaton (1991) with permission.

The rate of ammonia oxidation in biofilters has been described by the Monod equation (Eq. 10-13). This equation describes the relationship between the growth rate of bacteria and the substrate concentration (Benefield and Randall 1980).

$$\mu = \mu_m \times \frac{S}{K_s + S} \quad (10\text{-}13)$$

where μ = specific growth rate of nitrifying bacteria, L/day; μ_m = maximum specific growth rate of nitrifiers, L/day; S = residual growth-limiting substrate concentration, g/m^3; and K_s = saturation constant numerically equal to the substrate concentration at which $\mu = \mu_m/2$, g/m^3

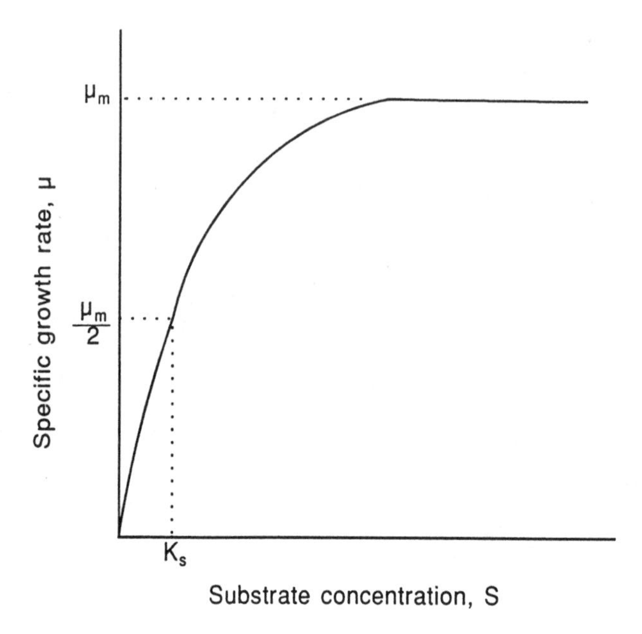

Figure 10-26. Monod curve.

of NH_4-N or NO_2-N. Monod (1949) observed that the growth rate of nitrifiers is a function of some nutrient concentration as well as the organism concentration. The relationship is shown in Figure 10-26.

If other nutrient requirements are not met, ammonium (NH_4^+) is the growth-limiting substrate for *Nitrosomonas* while nitrite (NO_2) is the growth-limiting substrate for *Nitrobacter*. The growth rate of *Nitrobacter* is greater than that of *Nitrosomonas*, but the limiting step in the conversion of ammonia to nitrate is ammonia oxidation (WPCF 1983). For this reason, the start-up of *Nitrobacter* in biofilters lags behind the establishment of *Nitrosomonas* by several days (usually about 10 days). Hochheimer (1990) summarized the literature values for the half-saturation constant K_s for *Nitrosomonas* and *Nitrobacter* (Table 10-5) and maximum specific growth rates for *Nitrosomonas* (Table 10-6).

Nitrifying bacteria are essential in closed recirculating aquaculture systems. Ammonia and other toxic metabolites must be removed from the system if the culture operation is to be successful. With the exception of ion exchange, no

Table 10-5. Half-saturation constants (K_s) for *Nitrosomonas* (NS) and *Nitrobacter* (NB) at various temperatures.

Species	Temperature (°C)	K_s (g NH_4-N/m^3)	Source
NS	5	0.13	Knowles et al. (1965)
NS	10	0.23	Knowles et al. (1965)
NS	15	0.40	Knowles et al. (1965)
NS	15	0.5	WPCF (1983)
NS	20	0.73	Knowles et al. (1965)
NS	20	1.00	Loveless and Painter (1968)
NS	25	1.30	Knowles et al. (1965)
NS	25	3.5	Ulken (1963)
NS	30	10.0	Hofman and Lees (1952)
NB	25	5.0	Ulken (1963)
NB	28	5.0	Gould and Lees (1960)
NB	30	6.0	Lees and Simpson (1957)
NB	32	8.4	Laudelout and van Tichelen (1960)

Source: Wheaton et al. (1991) with permission.

Table 10-6. Estimates of maximum specific growth rate μ_m for *Nitrosomonas*.

Temperature (°C)	μ_m (days^{-1})	Source
5	0.18	Gujer and Boller (1986)
10	0.29	Gujer and Boller (1986)
10	0.30	WPCF (1983)
15	0.47	Gujer and Boller (1986)
15	1.12	Williamson and McCarty (1976)
20	0.77	Gujer and Boller (1986)
20	0.65	WPCF (1983)
20	2.0	Kissel et al. (1984)
20	2.0	Williamson and McCarty (1976)
25	1.25	Gujer and Boller (1986)
30	1.20	WPCF (1983)

Source: Wheaton et al. (1991) with permission.

other chemical, physical, or biological processes have been shown to be as effective at removing ammonia (Hochheimer and Wheaton 1991). Ion exchange, however, has not been proven to be economically viable on a large scale.

Biofilter Operating Characteristics

Unwanted soluble substances can be removed from culture systems in the suspended media phase, such as activated sludge, or in the fixed media phase with filtration. In each process the substance or substances are reduced by colonies of specific bacteria (principally *Nitrosomonas* and *Nitrobacter sp.*), which have been encouraged to grow. Removal by activated sludge is common in the sewage treatment industry but is used infrequently in aquaculture since the low concentration of wastes in fish culture effluents makes them poorly suited for this process (Petit 1990). The potential of activated sludge was investigated by German researchers (Meske 1971; Scherb and Braun 1970, 1971).

In early recirculating system aquaculture research, submerged filters were the most common type. These were soon replaced, to a large degree, by trickling filters. Nitrification characteristics in trickling filters were described by Haug and MacCarty (1972). These same general principles may be applied to all types of biofilters.

Types of Biofilters

There are many types of biofilters, for example, submerged, trickling, rotating biodisks/biodrums, fluidized beds, and low-density media filters. When closed recirculating aquaculture research began in earnest about 20 years ago, the submerged biofilter was the most common type. The filter was relatively simple, and design characteristics could be copied from the wastewater or sewage treatment industries. In recent years, however, other types like rotating biological contactors and fluidized beds have been shown to be more efficient at ammonia removal. Biofilters that use a hard substrate for bacterial attachment are often referred to as *fixed film reactors*.

Submerged Filters. Submerged filters can be further categorized as downflow (top to bottom) or upflow (bottom to top). Figure 10-27a illustrates a simple downflow submerged filter. This was perhaps the earliest filter type used for aquaculture purposes, and is still used to some degree today, particularly by novice aquaculturists. Basically, the filter consists of a vessel filled with a media upon which nitrifying bacteria grow. Culture water is applied at the filter surface, usually through a spray bar, and gravity flows downward through the unit. The media remains completely submerged at all times. The media is supported by a perforated bottom through which the treated water drains and flows out of the unit.

Rock is the most common type of media used in submerged filters. Limestone rock is popular since it provides buffering against rapid pH changes. That is, until the media becomes coated with nitrifying bacteria, then the buffering capacity is diminished. Other calcareous materials offer buffering capacity, such as oyster shells (either crushed or whole), clam shells, and crushed coral rock. Numerous other materials are used for filter media including ceramic

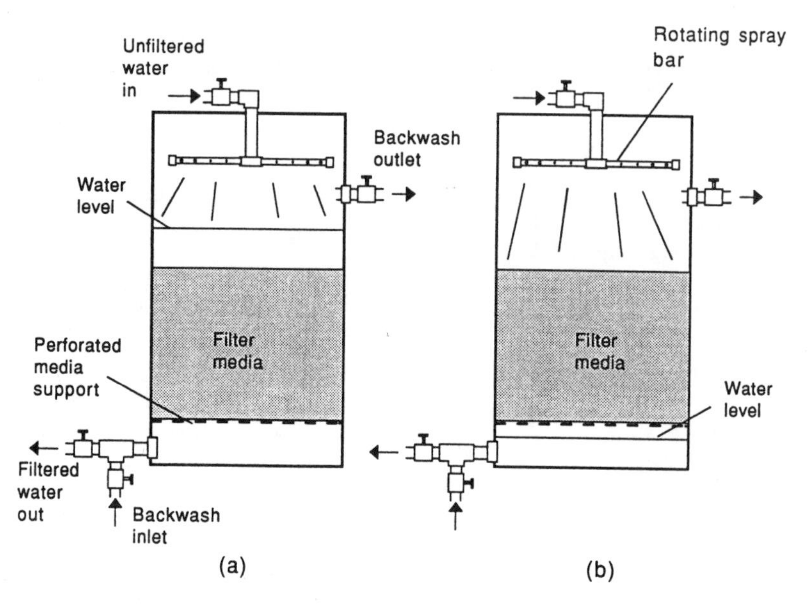

Figure 10-27. Granular media filters (a) Submerged; and (b) trickling.

and plastic modules of various shapes, glass or plastic beads, small pebbles, sand and wooden slats. Any material may be used for biofilter media so long as it is not toxic to the bacteria or the cultured animals.

Media size is important in submerged filters because of clogging. Downflow filters using media particle sizes smaller than 19–25 mm (0.75–1 in.) are generally not recommended because of clogging problems. The downflow of water through the filter forces solids deep into interstices between media particles where they are hard to remove by backflushing. As filters become clogged water flow through the filter is diminished, head loss increases, and short-circuiting will develop. Short-circuiting causes the formation of anaerobic cells within the filter where methane gas may form. If the culture water contains sulfur, hydrogen sulfide gas may form. Both of these gases are toxic to fish and nitrifying bacteria. The smaller the filter media, the more intense these problems become. Smaller media must be backflushed more often to avoid clogging problems. The net result is increased operating expenses.

Submerged filters can also be operated in the upflow mode. Upflow filters may be either pressurized (closed top) or unpressurized (open top). Energy requirements for pumping depend primarily on how high the water must be lifted. Upflow filters are not as prone to clogging as downflow filters since wastes must flow against the pull of gravity, and, therefore, energy costs for backflushing are less. Smaller media sizes, such as sand, may be used without problems associated with submerged filters.

The most restrictive limitation of submerged filters is that all oxygen must be delivered to the filter in the influent water since the bacteria are constantly submerged. Bacteria may become oxygen-starved if sufficient aeration or oxygenation is not provided prior to the water entering the filter.

Trickling Filters. Trickling filters basically function the same as submerged filters except that the media is not submerged. Culture water trickles downward through the media and keeps the bacteria moist but never completely submerged. Effluent collects beneath a perforated bottom that supports the media (Figure 10-27b). Since the void spaces between media particles are filled with air rather than water, the bacteria rarely become oxygen-starved. Air circulating throughout the media provides most oxygen needs of the bacteria, and aeration of biofilter influent is not as important as for submerged filters. Clogging is not as serious a problem in trickling filters.

Obviously, trickling filters can only function in the downflow mode. Water is normally sprayed over the surface of the filter with either fixed or rotating spray bars. Rocks were the media of choice in many early trickling filter designs, and are still used today in some wastewater and sewage treatment plants because of their low cost. Other media like plastic modules, Nor-pac (NSW Corp., Roanoke, Va.) and Bioblocks (Dynasty Marine Associates, Marathon, Fla.) are popular for media in trickling filters since they provide a large surface area and are relatively lightweight. These media are expensive in terms of volume but are cost effective when one calculates the cost on the basis filter surface area. Examples of these media are shown in Figure 10-28, and technical information is provided in Table 10-7. Many other types of filter media are available from various manufacturers. Sand is not recommended for use in trickling filters because the small void spaces between particles will rapidly become clogged.

Rotating Media Filters. Rotating media filters incorporate aspects of both fixed and suspended media operation; fixed media are rotated in vessels through which the culture water (suspended media) flows (Muir 1982). These filters are commonly referred to as revolving plate biofilters or rotating biological contactors (RBCs). They are further subclassified as either rotating biodisks or biodrums. A rotating biodisk consists of a series of flat or corrugated disks mounted on a horizontal shaft, as illustrated in Figure 10-29. The assembly rotates in a trough or other vessel through which the culture water flows. Approximately 40% of the disk surfaces are submerged at a time with the shaft and bearings above the water's surface. Biological attachment occurs on the disk surfaces, and as the disks rotate the biofilm is kept moist and aerated. Spacing between disks should be no less than about 13 mm (0.5 in.) to avoid bridging of the biofilm between disks, hindering the flow of water and air. Energy use is that required to rotate the disks. Basic biodisk design procedures have been presented by Steels (1974) and Lewis and Buynak (1976).

Disks are most commonly fabricated from flat or corrugated fiberglass or plastic sheet material. Small experimental RBCs are shown in Figure 10-30. The disk surfaces should be roughened to enhance attachment of biofilm, since it will more easily shear from smooth surfaces. RBCs are rotated anywhere from about 2 to 6 rpm, depending on size. Peripheral speed should be no greater than about 0.3 m/s (1 ft/s) to avoid excessive shearing of the biofilm. Solids that shear from disk surfaces normally settle in the vessel beneath the rotating disks if water flow is not so turbulent as to keep them in suspension. Provision should be made in the RBC design for removal of the floc that shears from the disks. Removed solids are then normally more easily settled in basins following biological treatment (Lewis and Buynak 1976).

A variation of the rotating biodisk is the rotating biodrum. Biodrums consist of cylindrical-shaped cages that are filled with media. The drums normally provide more surface area and greater aeration, but have greater energy requirements than biodisk systems (Muir 1982). The media selected is a matter of choice, but the drum should be kept as light-weight as possible to reduce energy costs. Many types of media have been used with varying success. The most popular are the plastic balls and rings because of their light weight and high surface area. The media-filled cage rotates about a horizontal shaft through its center (Figure 10-31). Water flows through the system in a manner similar to biodisk systems.

An experimental biodrum fabricated from Nor-pac (NSW Corp., Roanoke, Va.) is shown in Figures 10-32 and 10-33. Nor-pac is a perforated tubinglike material that is fabricated with a series on internal fins. The material has a high surface area and is available in a variety of diameters. Norpac is manufactured using an extrusion process and is, therefore, available in any desired length and in 25-, 38-, and 51-mm

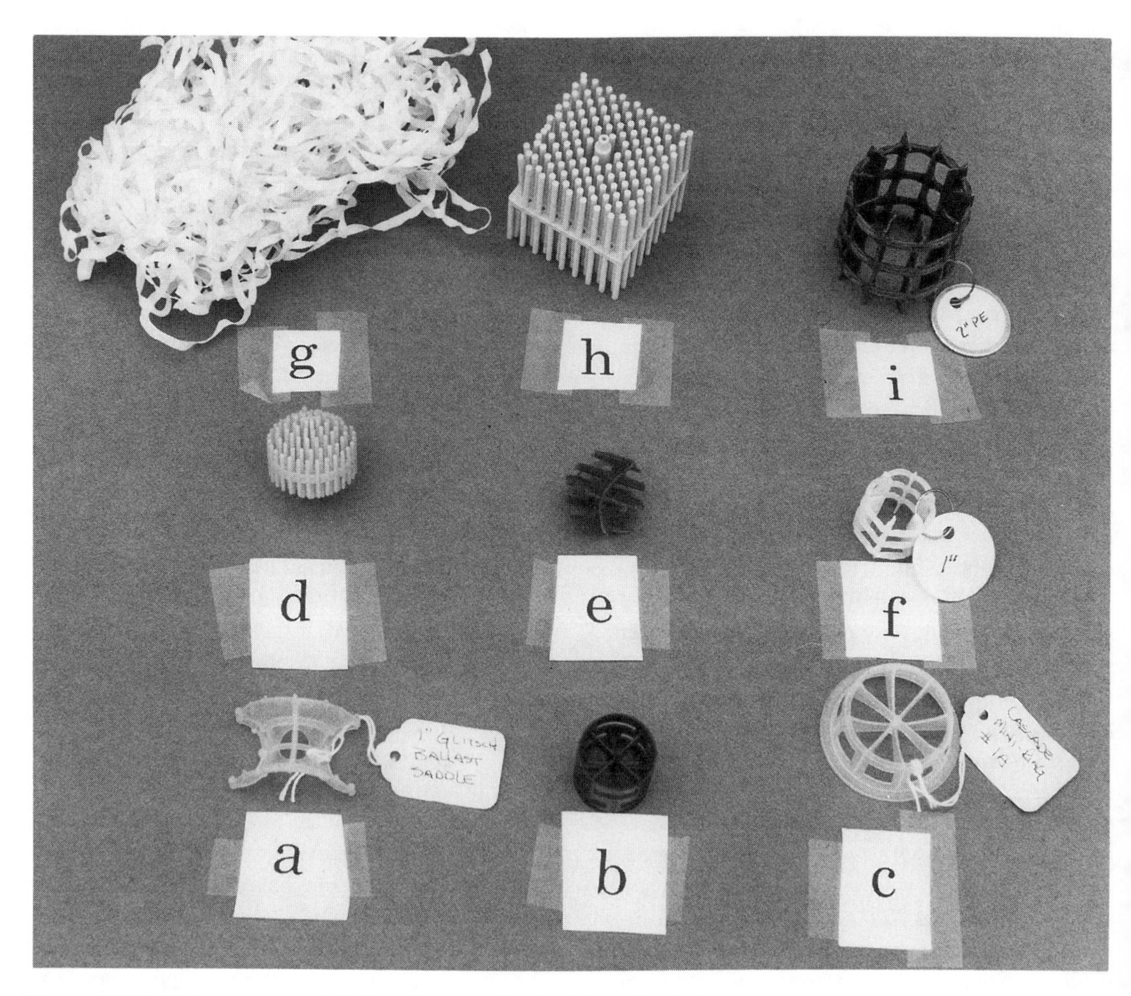

Figure 10-28. Various types of tricking filter media: (a) 1-in. Glitsch ballast saddle; (b) 1-in. Flexring; (c) 2- in. Cascade mini-ring; (d) Bio-ball; (e) Bl-Ox; (f) 1-in. Nor-pac random tower packing; (g) Bio-bale; (h) Bio-block; and (i) 2-in. Bio-Pac90.

(1-, 1.5-, and 2-in.) diameters. Table 10-8 gives specifications for the material. Biodrums fabricated from Nor-pac are currently being used in experimental aquaculture systems at Cornell, Virginia Tech, Illinois State, and Louisiana State universities.

A hybrid RBC unit fabricated from Biodek (The Munters Corp., Fort Meyers, Fla.) is shown in Figure 10-34. The filter is used in a 36.3 m^3 (9,600 gal) channel catfish culture system. Biodek offers a much greater surface area than flat or corrugated disks. The RBC in the figure is fabricated with 2 m^3 (71 ft^3) of Biodek material having a surface area of 137 m^2/m^3 (42 ft^2/ft^3). The total surface area of the filter in Figure 10-38 is 281 m^2 (3,024 ft^2). Many other types of material can be used to construct RBCs, so long as the material is nontoxic to nitrifying bacteria and fish.

Wastewater can be introduced into RBC filters either parallel or perpendicular to the axis of rotation (Figure 10-35). In systems where water enters from one end of the filter unit (parallel to rotating axis) as in Figure 10-35a, growth conditions in the anterior sections of the filter favor heterotrophic bacteria, and nitrifica-

Table 10-7. Examples of biological filter media shown in Figure 10-28.

Item	Description	Specific surface area	
		ft^2/ft^3	m^2/m^3
a	1-in. Glitsch saddle ballast	—	—
b	1-in. Flexiring	65	212
c	2-in. Cascade mini-ring	—	—
d	Biomax ball	165	539
e	Bl-Ox	71	232
f	1-in. Norpac tubing	75	245
g	Biobale	247	807
h	Bioblock	216	706
i	2-in. Bio-Pac90	31	101

tion will suffer in these sections. When water enters from the side of the unit (perpendicular to rotating axis) as in Figure 10-35b, heterotrophic and autotrophic bacteria will be more evenly distributed, and nitrification will be more uniform. Therefore, this water flow configuration is recommended. For large flows RBC units may be operated in series (Figure 10-35c) or parallel (Figure 10-35d).

Fluidized Bed Reactors. In fluidized bed reactors the media is contained in a vertical vessel with a cylindrical or square cross section. A layer of coarse gravel (19–25 mm or 0.75–1 in. diameter) is normally used to support the sand. A perforated plate is used to support the gravel layer and evenly distribute the incoming water to limit short-circuiting. The media is kept in suspension or fluidized by the upward flow of water (Figure 10-36). The degree of fluidization depends on the upward flux of water through the vessel. The most common media used is coarse sand, therefore, the reactors are sometimes called fluidized sand filters. These reactors can be operated either in the pressurized (closed-top) or unpressurized (open-top) modes.

Fluidized bed reactors are operated so that the sand bed expands, thus increasing the void spaces between the sand grains. In this manner, more water comes into contact with the sand grains, and the filter bed can be more easily oxygenated. The reactors are normally operated at 25–100% expansion. Figure 10-37 illustrates reactors operating at 0, 50, and 100% expansion. The greater the expansion, the greater the hydraulic requirements.

Fluidized bed reactors are recommended for ammonia removal only, since they do not have the capability to trap solids. Thus, they must be used in combination with a solids removal process, such as an upflow sand filter. Nitrifying bacteria growing on the fluidized media are continually sheared from the media and replenishing themselves. Therefore, fluidized reactors have a much higher capacity for ammonia and

Figure 10-29. A typical rotating biological contactor (RBC) unit.

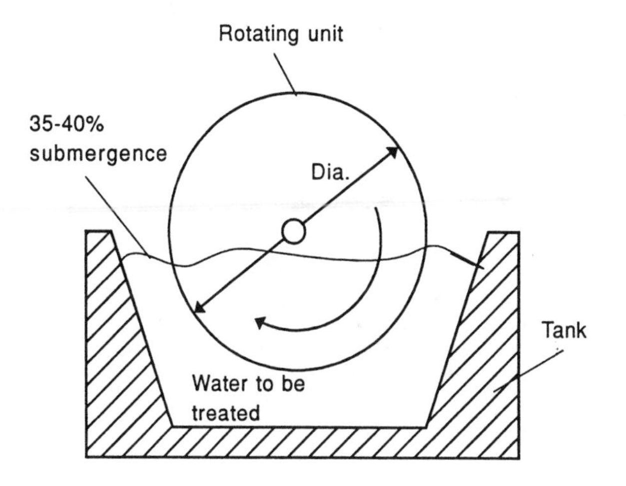

Figure 10-30. An experimental rotating biodisk unit fabricated from fiberglass sheets (photo by T. Lawson).

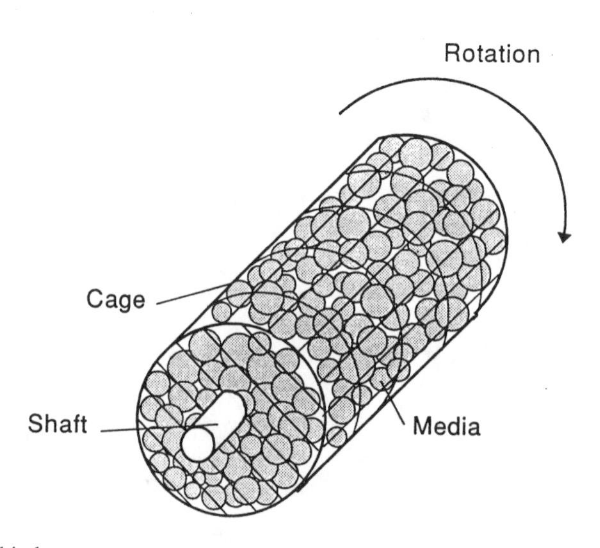

Figure 10-31. A rotating biodrum.

Figure 10-32. An experimental RBC at Cornell University (photo courtesy of NSW Corp., Roanoke, Va.)

nitrite removal than most other filter types, providing that the bacteria receive sufficient oxygen.

Malone and Burden (1988a, 1988b) optimized fluidized bed reactors for the soft-shell crawfish and blue crab industries. In studies at Louisiana State University, fluidized bed reactors using 1.19–2.38 mm (0.05–0.09 in.) diameter sand media were shown to have approximately 10 times the carrying capacity of limestone rock filters with 14.3–25.4 mm (0.55–1 in.) diameter media. The flux rates required for 0–100% filter bed expansion are quite high, however. Flux rates for three media sizes are shown in Table 10-9. Application of fluidized bed technology to aquaculture is questionable because of the high flux rates recommended (2,044–2,851 L/min/m^2 or 60–65 gpm/ft^2) (Wimberly 1990).

Low Density Media Filters. Low-density media filters, commonly referred to as *floating bead filters*, use polyethylene beads (3–5 mm diameter) as the filtering media in a pressurized, upflow mode, as illustrated in Figure 10-38. The beads are less dense than water, hence, they float above the influent injection point. A screen retains the beads at the top of the filter, preventing their loss in the effluent. The filtering zone is underlain by a cone-shaped settling zone. A motor-driven propeller is used to periodically agitate and clean the beads.

Figure 10-33. An experimental RBC fabricated from Nor-pac tubing (photo courtesy of NSW Corp., Roanoke, Va.).

Table 10-8. Specifications for Nor-pac biological filter media.

Tube diameter mm (in.)	Specific surface area m^2/m^3 (ft^2/ft^3)
25.4 (1.0)	137.8 (42)
38.1 (1.5)	196.8 (60)
50.8 (2.0)	246.0 (75)

Source: NSW Corp., Roanoke, Va.

Bead filters have a dual purpose: solids capture and biofiltration. Filtration of total suspended solids (TSS) is accomplished by trapping particles between the granular media matrix and then by settling the trapped particles. The filter traps suspended particles while at the same time provides a large specific surface area (1,145 m^2/m^3) for the growth of heterotrophic and nitrifying bacteria (Malone et al. 1993). Nitrification capacity is 25 mg of total ammonia nitrogen (TAN)/ft^2 per day (270 mg TAN/m^2/day). During a typical cycle (Figure 10-38a) bacteria attach to and grow on the floating beads and in the pore spaces between beads. The bacteria function to extract dissolved organics (BOD), ammonia, and nitrite from the water. As solids and the bacterial biomass accumulate, the filter progressively clogs, impeding the flow of water, oxygen, and nutrients through the bed. The propeller-driven backwash mechanism is then triggered by either a timer, pressure sensor, or computerized control unit (Fig 10-38b). Its function is to agitate the filter bed, separating

Figure 10-34. An RBC fabricated from Biodek (photo courtesy of G. Libey, Virginia Tech University, Blacksburg, Va.).

the trapped solids and biofloc from the beads. When the motor is switched off (Figure 10-38c) the beads float upward and re-form the filter matrix while the accumulated solids and biofloc sink to the bottom of the settling zone. The solids gradually compress, forming a concentrated sludge that is removed prior to the beginning of a new filtration cycle (Figure 10-38d). Water loss during backwashing is minimal.

It would follow that more organic solids can be removed from a culture system by increasing the number of backwash cycles. However, Malone et al. (1993) stressed that running several backwash cycles per day actually limits nitrification. To achieve optimal solids removal as well as nitrification they recommend a backwash interval of one or two days. The authors also recommended that propeller-washed bead filters should not be loaded in excess of 24 kg feed/m^3 day (1.5 lb feed/ft^3 day) unless the system is very closely monitored or unless additional nitrification support is provided, such as an RBC in series following the bead filter.

Malone et al. (1993) stated that one cubic meter of beads is capable of providing complete water treatment for the wastes generated from 12–16 kg of feed per day, assuming that a 35% protein feed, commonly used in warm-water fish culture, is provided. The filter effluent criteria is thus as follows: TAN < 0.5 mg N/L and BOD_5 < 10 mg O_2/L. Assuming a feeding rate of 3% body weight per day, the filter is capable of supporting 400–530 kg fish/m^3 media. Assuming a feeding regime of 2% body weight per day, the filter can support 600–800 kg fish/m^3. A commercial-scale propeller-washed floating bead filter (1 m^3 volume) is shown in Figure 10-39.

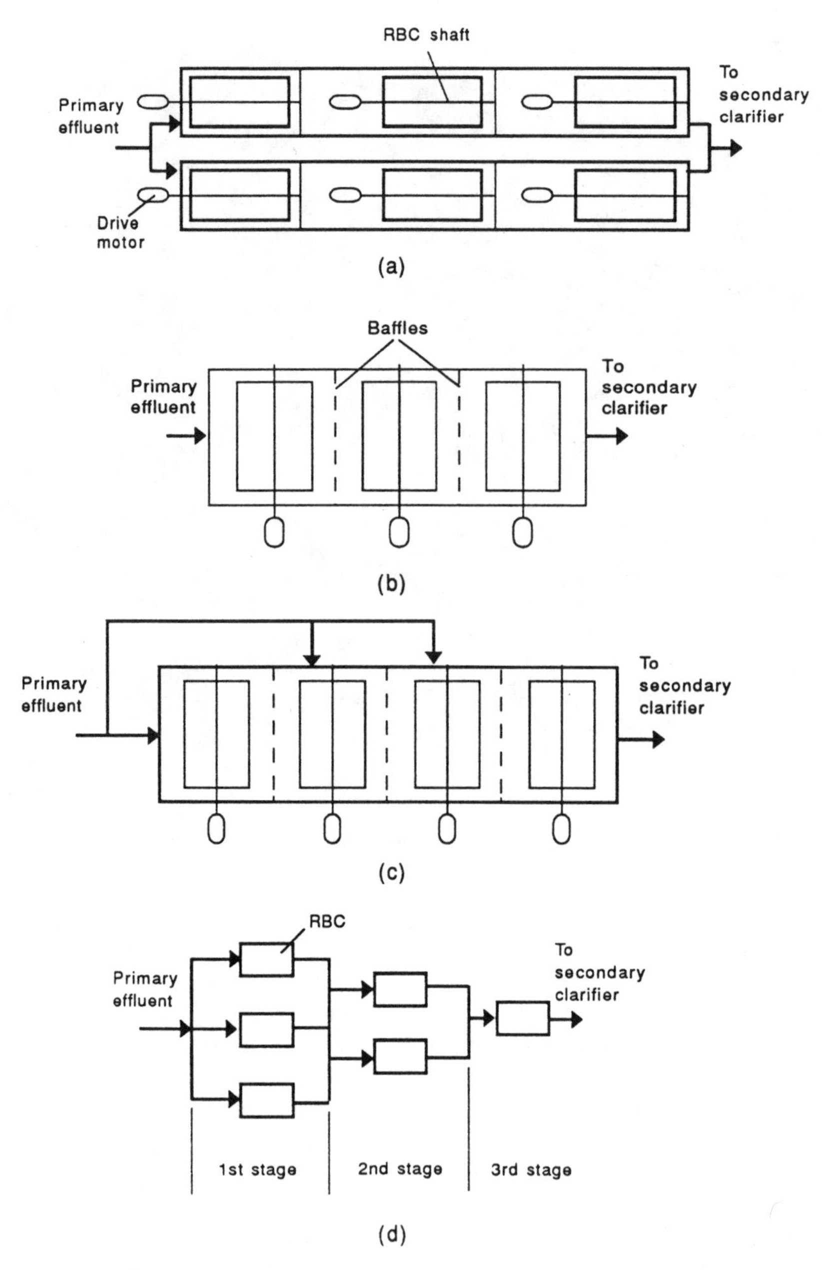

Figure 10-35. Typical RBC arrangements: (a) Flow parallel to shaft; (b) flow perpendicular to shaft; (c) step feed and; (d) tapered feed.

Variables Affecting Biofilter Performance

The effectiveness of biofilter performance in aquaculture is most often measured by the filter's ability to oxidize ammonia to nitrate. Filter efficiency is dependent upon many factors, which may be either chemical, physical, or biological in nature.

Chemical Factors

pH. Biofilters should operate in the pH range of 6–9. The range may be narrower for specific

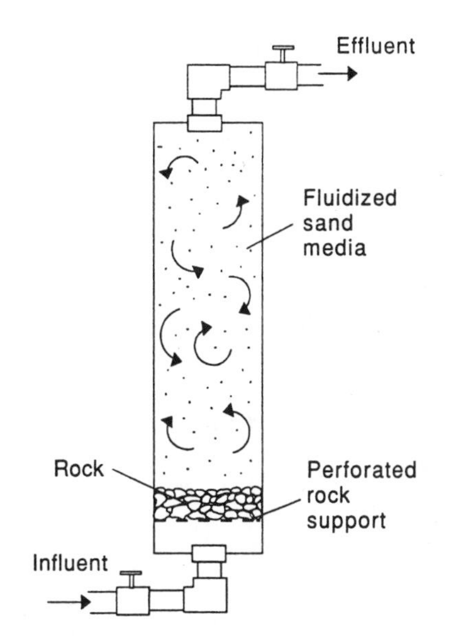

Figure 10-36. A pressurized fluidized bed reactor.

filters in which the bacteria are slow to adapt to rapid pH changes. Biofilters in which the bacteria are slowly adapted may operate within the range 5–10 (Wheaton et al. 1991). However, nitrification efficiency decreases the more the pH is allowed to drift toward either extremity. The minimum pH requirements for the development of *Nitrosomonas* and *Nitrobacter* is 6.5–7.0 (Petit 1990). Nitrification inhibition begins slightly below 7.0. The pH optima are normally on the alkaline side of neutrality (>7.0) but are not sharply defined. If culture water pH drifts too far into the acidic range, there is a risk of toxicity to free CO_2 for fish and possibly for nitrifying bacteria. Table 10-10 lists acceptable pH levels for both *Nitrosomonas* and *Nitrobacter* as reported by various authors.

Bacteria can adapt to gradual pH changes. The rate of adaptation is a function of temperature, degree of pH change, ammonia concentration, and other factors. Rapid pH changes of 0.5–1.0 units will significantly reduce biofilter ammonia and nitrite conversion efficiency. A rapid increase in ammonia will be observed in the system until bacteria adapt to the new conditions (Wheaton et al. 1991).

ALKALINITY. Alkalinity is a measure of the buffering capacity of aquatic systems. Nitrification is an acid-forming process, producing hydrogen ions that consume alkalinity and lower system pH (Bisogni and Timmons 1991). If culture water is poorly buffered, the system pH will continue to decline. Water with an alkalinity less than 20 mg/l as $CaCO_3$ is considered poorly buffered (Boyd 1982; Tucker and Robinson 1990).

A review of Equations 10-5 and 10-6 demonstrates that 7.14 g of alkalinity are removed for each gram of ammonia oxidized in biofil-

Figure 10-37. A fluidized bed reactor: (a) At rest; (b) at 50 expansion; and (c) at 100 expansion.

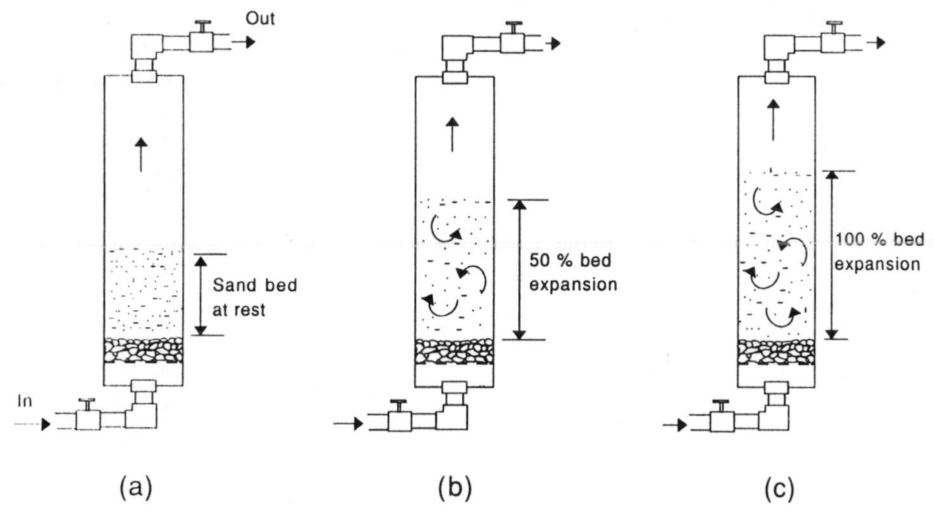

Table 10-9. The relationship between hydraulic flux rate and percent expansion for three filter media in clean, washed 38.1-cm (15-in.) deep filter beds.

	Hydraulic flux rate L/min/m^2 (gal/min/ft^2)		
Expansion, %	Medium river sand[2]	Coarse river sand[3]	Crushed dolomite[4]
0[1]	367 (9)	570 (14)	1,018 (25)
25	1,385 (34)	1,792 (44)	3,055 (75)
50	1,996 (49)	2,647 (65)	4,032 (99)
75	2,525 (62)	3,218 (79)	4,847 (119)
100	3,177 (78)	3,910 (96)	5,417 (133)

Source: Malone and Burden (1988a, 1988b).

[1] Maximum flow without bed expansion.

[2] 0.84–1.68 mm diameter.

[3] 1.19–2.38 mm diameter.

[4] 4.76–7.93 mm diameter.

ters. An alkalinity mass balance should be done on individual aquaculture systems and the alkalinity supplemented at the same rate as alkalinity destruction to maintain system pH. This is vitally important with a soft water source (water with low hardness), since it will have no buffering capacity. Efficiency of nitrification in biofilters will be lower if sufficient alkalinity is not available to the bacteria.

The minimum alkalinity for aquaculture should be 20–50 mg/l-$CaCO_3$ (Klontz and King 1975; Muir 1978). For *Penaeids* and *Macrobrachium* cultured at a pH less than 7, the level of calcium in system water should be at least 50% of the value found in the animal's natural environment to prevent adverse effects (Wickens 1976). Malone and Burden (1988a, 1988b) indicated that bicarbonate alkalinity may be critical for nitrifier growth. In soft crawfish and blue crab shedding systems they noted that nitrification was inhibited when system alkalinity fell below 100 mg/L-$CaCO_3$.

Table 10-11 lists potential alkalinity supplements and their characteristics. Compounds such as $MgCO_3$ should be avoided due to their slow rate of solubility at pH ranges normally found in aquaculture systems. Sodium bicarbonate (common baking soda—$NaHCO_3$) has a rapid rate of solubilization and is inexpensive in bulk quantities (Bisogni and Timmons 1991). Malone and Burden (1988a) recommended that daily additions of 1 kg/1,000 kg crawfish (1 lb/1,000 lb crawfish) of baking soda were necessary for pH control in poorly buffered soft crawfish shedding systems.

AMMONIA AND NITRITE. Excessively high ammonia and/or nitrite concentrations are toxic to nitrifying bacteria. *Nitrosomonas* and *Nitrobacter* are affected at differing levels of toxicity. The major inhibitory agents are un-ionized ammonia (NH_3) and nitrous acid (HNO_3) (Cole and Tchobanoglous 1976; Russo and Thurston 1977). NH_3 inhibits *Nitrosomonas* at 10–150 mg/l and *Nitrobacter* between 0.1 and 1.0 mg/l. Nitrous acid is inhibitory to both in the range 0.22–218 mg/l (Anthonisen et al. 1976).

Ammonia is more toxic at higher pH ranges since more of the total ammonia is in the unionized (NH_3) form. Nitrite becomes less toxic at high pH ranges since it shifts to the ionized form. At low pH values nitrite becomes more toxic and NH_3 less toxic. Nitrite oxidation is more sensitive to environmental stresses than

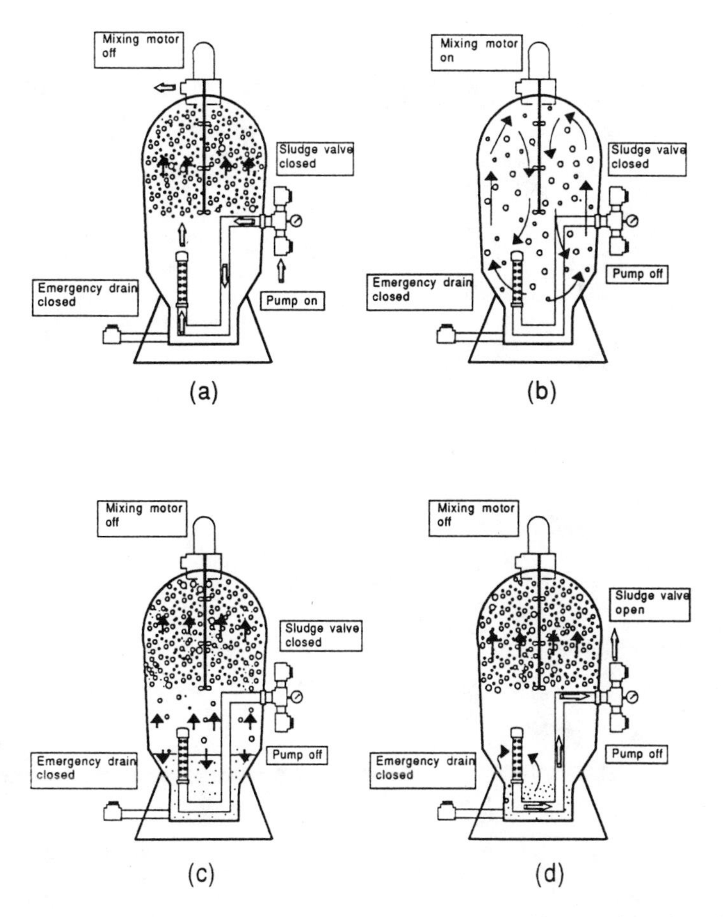

Figure 10-38. A floating bead reactor: (a) Filtering mode; (b) prop-washing mode; (c) settling mode; and (d) backwash (sludge removal) mode.

ammonia oxidation. Consequently, nitrite accumulation may result in systems with rapidly changing environmental conditions (Atlas and Bartha 1987). The toxicity of nitrite is reduced in waters containing high concentrations of monovalent ions such as chlorides or bicarbonates (Tomasso et al. 1979).

OXYGEN. The rate of nitrification in biofilters decreases as the available oxygen decreases. The limiting concentration of oxygen is dependent on temperature, salinity, concentration of impurities, system biomass, and lesser factors. Since the nitrogen cycle begins with organic compounds and proceeds through nitrite and nitrate formation, the part of a biofilter first receiving wastes from the culture unit(s) has a high percentage of heterotrophs. As the waste proceeds through the filter, the *Nitrosomonas* population rises to remove ammonia. Further into the filter the *Nitrobacter* population rises to convert nitrite to nitrate. Therefore, position in the filter dictates that the heterotrophs will receive oxygen first, followed by *Nitrosomonas* and *Nitrobacter*. Nitrifiers often become oxygen-starved in biofilters in which influent water is not rich in oxygen.

Most literature dealing with biofilter oxygen dynamics originated from research on wastewater treatment. It has thus been reported that, as DO concentration decreases to just above 1 mg/L, oxygen then becomes limiting rather than ammonia (WPCF 1983). The scientific evi-

Figure 10-39. A commercial-scale (1 m^3) propeller-washed bead filter (photo by T. Lawson).

dence is too limited to determine minimum oxygen concentrations in biofilters used in aquaculture applications, however, most researchers agree that biofilter effluent should contain a minimum of 2 mg/L of dissolved oxygen. In order to maintain that concentration in the effluent the filter influent should contain a minimum of 5–6 mg/L. It may be necessary to aerate water before entering biofilters, or liquid oxygen may be injected. If the flow rate through the

Table 10-10. Optimal pH for *Nitrosomonas* (NS) and *Nitrobacter* (NB) growth.

Species	Range tested	pH optima	Source
NS	-	8.0–9.0	Hofman and Lees (1952)
NS	7.0–9.0	8.0	Engle and Alexander (1958)
NS	-	6.0–9.0	Winogradsky and Winogradsky (1933)
NS	5.0–10.0	9.0	Kawai et al. (1965)
NS	-	7.2–7.8	Loveless and Painter 1968
NB	-	8.3–9.3	Meyerhof 1917
NB	7.0–8.6	7.8	Boon and Laudelout 1962
NB	-	6.3–9.4	Winogradsky and Winogradsky 1933
NB	5.0–10.0	9.0	Kawai et al. 1965

Source: Wheaton et al. (1991) with permission.

Table 10-11. Alkalinity supplements and their properties.

Formula	Common Name	Equivalent weight (g/eq.)	Solubility	Rate of solubilization
NaOH	sodium hydroxide	40	high	rapid
Na_2CO_3	sodium carbonate or sida ash	53	high	rapid
$NaHCO_3$	sodium bicarbonate or baking soda	83	high	rapid
$CaCO_3$	calcium carbonate or calcite	50	moderate	moderate
CaO	slaked lime	28	high	moderate
$Ca(OH)_2$	calcium hydroxide or hydrated lime	37	high	moderate
$CaMg(CO_3)_2$	dolomite	46	moderate	slow
$MgCO_3$	magnesium carbonate or magnesite	42	moderate	slow
$Mg(OH)_2$	magnesium hydroxide	29	moderate	slow

Source: Bisogni and Timmons (1991) with permission.

filter is too slow, or if influent DO is low, anaerobic cells may develop in biofilters with subsequent methane gas or H_2S formation. Both of these gases are detrimental to aquatic life.

SOLIDS. Suspended solids in recirculating systems may be organic (i.e., feces and food, etc.) or inorganic (i.e., sand and grit) in nature. Research on the effects of suspended solids in recirculating systems is scarce. However, Timmons et al. (1987) suspected that suspended solids were responsible for fish kills in recirculating systems containing brook trout. In addition to direct effects on fish health, solids can clog filters and other system components, and ammonia and/or BOD may be generated from the decay of organic particulates (Chen and Malone 1991). The effects of solids clogging in filters are increased head loss and short-circuiting or reduced flow rate through filters with subsequent potential development of anaerobic or dead zones in filters. Additionally, solids provide substrate where nitrifiers can grow. Particles in the range of 1.4–2.7 microns provide the best nitrification results (Kholdebarin and Oertli 1977b).

Another serious effect of solids has to do with disinfection processes. Solid particles may shield disease organisms from the effects of disinfecting agents like chlorine, ozone, or ultraviolet (UV) radiation.

SALINITY. Biofilters can function at almost any salinity range so long as nitrifying bacteria are slowly acclimated to salinity changes. Freshwater nitrifiers are strongly inhibited by seawater and vice versa. Constant salinity promotes the most rapid nitrification rate (Kawai et al. 1965). Abrupt changes in salinity of greater than 5 g/L will shock nitrifiers and decrease the rate of ammonia and nitrite removal until the bacteria slowly readjust (Bower and Turner 1981; Hochheimer 1990). Akai et al (1983), while studying the effects of rapid salinity changes on nitrification, found that, as salinity increased, the maximum specific growth rate of ammonia-oxidizing bacteria decreased from 0.4/hr in freshwater to 0.0028/hr in full-strength seawater.

GAS DIFFUSION RATE. Biofilter media in a properly functioning filter are coated with a bacterial slime layer (biofilm) of a given thickness. As waste loads and environmental conditions change, so does the biofilm thickness. O_2, NH_3, and NO_2 diffuse inward through the biofilm while NO_2, NO_3, and CO_2 diffuse outward. These gases must also diffuse through a thin film of water that surrounds the biofilm. Thus, under certain circumstances, the thicknesses and nature of these films control nitrification rate (Wheaton et al. 1991). Hochheimer (1990) developed equations describing diffusion coefficients for O_2 and NH_3 in both water and filter biofilm as a function of temperature.

Physical Factors

TEMPERATURE. Nitrifying bacteria can also adapt to a wide range of temperatures if acclimated slowly. As nitrifiers are slowly adapted to new temperatures, their optimum and lethal temperature values shift up and down accordingly (Jones and Morita 1985). Wortman and Wheaton (1991) studied the effects of the temperature range 7–35°C (45–95°F) on nitrification in a biodrum. They found a linear relationship between ammonia removal rate and temperature:

$$AMR = 140 + 8.5T \qquad R^2 = 0.89 \qquad (10\text{-}14)$$

and a linear relationship between nitrate production and temperature:

$$NRTP = 63 + 9.9T \qquad R^2 = 0.82 \qquad (10\text{-}15)$$

where AMR = Ammonia removal rate (mg NH_4-N/L filter volume/day); and $NRTP$ = nitrate production rate (mg NO_3-N/L filter volume/day). Temperature optima and inhibitory temperatures for nitrification as reported by various authors are shown in Table 10-12.

FILTER DEPTH. Depth for submerged and trickling filters refers to the physical dimension from top to bottom of the filter media. Depth for rotating biodisks and biodrums refers to the

Table 10-12. Optimal temperature for *Nitrosomonas* (NS) and *Nitrobacter* (NB) growth.

	Temperature (°C)			
Species	Range	Optimum	Inhibits	Source
NS	30–36	30–36		Buswell et al. (1954)
NS	10–40	30–35		Kawai et al. (1965)
NS			<5	Buswell et al. (1954)
NB	8–28	28		Nelson (1931)
NB	4–45	34–35		Deppe and Engle (1960)
NB		42		Laudelout and van Tichelen (1960)
NB	10–40	30–35		Kawai et al. (1965)
NB			<4	Deppe and Engle (1960)
NB			<45	Deppe and Engle (1960)

Source: Hochheimer and Wheaton (1991) with permission.

depth of submergence of the media. The optimum depth for submerged and trickling filters is a function of media size, ammonia and hydraulic loading rates, Reynolds number, and other factors. Filters require sufficient depth for *Nitrosomonas* and *Nitrobacter* to do their jobs. Additional depth is required for the reduction of organic substances by heterotrophs before nitrification begins. The smaller the filter media size the greater the surface area available for all of these conversions, and the filter media depth required is less. For example, if sand is the filter media, only a few centimeters of filter depth is required for heterotrophs and autotrophs to function. However, a filter using a larger media size, 7.3 cm (3 in.) for instance, may require a bed depth of several meters. Little design information exists on which engineers can base intelligent choices for filter bed depths. There is no fixed filter bed depth that is optimum for all filter media sizes.

CROSS-SECTIONAL AREA. Filter cross-sectional area refers to the area of filter bed material looking in the direction of water flow. This parameter is useful in designing submerged and trickling filters, fluidized beds, and upflow sand filters, but has little meaning when designing rotating biodisks or biodrums except to specify the diameter of disks or drums. Obviously, the smaller the cross-sectional area, the greater restriction to water flow through the filter bed. Therefore, this parameter should be kept as large as practical.

HYDRAULIC LOADING RATE. Hydraulic loading in submerged, trickling, upflow, or fluidized bed filters is a measure of the volume of wastewater flowing through the filter per unit of cross-sectional area of filter bed per unit of time, for example m^3/m^2/day (gal/ft^2/min). This is also referred to as the *flux rate* but is most often expressed as $m^3/min/m^2$ (gal/min/ft^2). Hydraulic loading in RBCs is expressed as volume per unit of media specific surface area per day (m^3/m^2/day) or as volume of flow per unit volume of filter material per day (m^3/m^3/day).

Maximum and minimum hydraulic loading rates apply to biofilters and vary depending on filter and media types. Minimum loading in a submerged filter is the lowest flow that will satisfy the filter oxygen demand, whereas minimum flow for trickling filters and RBCs is the lowest flow rate that will keep the media wet and supply all essential nutrients. If the media is allowed to dry the bacteria will die, and the filter will cease to function.

The maximum velocity for most filter types, except fluidized beds, is set by either the velocity at which the biofilm shears from the media or by excessive head loss through the filter. Biofilm will normally shear from the media periodically, but excessive velocity through the filter will cause shearing at too rapid a rate, with subsequent loss of nitrification. The biofloc shearing from filter media should be removed from the system by sedimentation or other processes and not allowed to enter culture units. For filters using manufactured plastic modules and the like, allowed velocities should follow manufacturer's guidelines.

In fluidized bed filters the maximum allowable flow rate is dependent upon the filter media particle size, particle density, fluid density, and the degree of expansion desired. As media particles become coated with bacteria, the density of the total mass decreases, and the floc becomes lighter than the working fluid. It may become necessary to reduce liquid flow rate to prevent the floc from blowing out the top of the filter.

Reynold's number (RN) is the ratio of inertial to viscous forces and is widely used to characterize filters. Biofilter modeling requires that RN be defined to describe flow in a filter, but this is difficult to do. RN contains a length term that is not well defined for biofilters and is an area ripe for research (Wheaton et al. 1991).

FILTER MEDIA. Any nontoxic, solid material may be used as a substrate or biofilter media. There are dozens of media types including sand, gravel, rock, oyster and clam shells, wooden pallets, and plastic and ceramic modules of various shapes. The selection of a media is largely dependent on cost, availability, weight, surface area desired, and system loading.

Sand offers more surface area for bacterial attachment, but the small spaces between particles clog rapidly. They require frequent backflushing to remove solids, resulting in high water and energy requirements. Rapid sand filters for solids removal with little nitrification are most often used prior to biological filtration or as a polishing filter following biological filters. The most successful use for sand filters has been demonstrated in fluidized bed filters used in soft crab and crawfish production systems (Malone and Burden 1988a, 1988b).

Rock offers less surface area but does not clog as readily as sand. It is generally accepted that rock with a 19–25 mm (0.75–1.0 in.) diameter is best since it offers good surface area but will not readily clog. Carbonate-based rocks, such as limestone or marble, provide pH buffering, that is, until they become coated with a bacterial slime layer, then the buffering capacity is diminished. Certain other materials such as whole or crushed oyster shells, clam shells, and crushed coral also provide pH buffering. However, like rock, the buffering capacity is diminished as the bacterial slime layer is established on exposed surfaces.

Biofilters containing sand, rock, oyster shells, and similar materials are typically bulky and heavy although they are probably the least costly of filter materials. Plastic and ceramic modules generally have higher void ratios and have an indefinite useful life but are expensive. In addition, plastic and ceramic media have no buffering capacity.

FILTER VOID RATIO. Void ratio refers to the ratio of the volume of void spaces between media particles to the volume of media plus voids. Empty filter volume above the level of media is not considered in the calculation. Media with high void ratios have larger open spaces between media particles that allow solids to pass. Clogging is more of a problem in filters with low void ratios.

SPECIFIC SURFACE AREA. The specific surface area is the total surface area available for bacterial attachment and is related to media size and void ratio. Small-diameter filter media particles have a higher specific surface area per unit volume and a lower void ratio than larger media. Thus, for a particular media type, the higher the specific surface area per unit volume, the higher will be the rate of ammonia conversion per unit volume. For example, a filter using sand with a mean particle diameter of 10-mm (0.4-in.) will have a larger specific surface area than one using 100-mm (4-in.) diameter media.

The larger media size requires that a much larger filter be constructed to obtain the same specific surface area, greatly increasing filter construction cost. However, since smaller media filters clog at a faster rate and require more frequent backflushing, a trade-off often results. The filter designer must be able to make the choice of which media particle size is best for his particular applications.

FILM THICKNESS. Film thickness refers to the thickness of the stagnant water film coating the bacterial colony growing on the filter media. It is dependent on water velocity through the filter plus water viscosity and temperature. Film thickness affects the rate of gas exchange into and out of the biofilm. Hochheimer (1990) studied film thickness between 1 and 100 μm.

LIGHT. Biofilters function best in total darkness (Horrigan et al. 1981). This is not always practical, however. Light intensities as low as 1% of the intensity of natural daylight inhibit nitrification. It is believed that light oxidizes cytochrome C in both *Nitrosomonas* and *Nitrobacter*. *Nitrobacter* are more sensitive to light, probably because they contain more cytochrome C (Olson 1981).

Biological Factors

CELL YIELD. Cell yield is a measure of the mass of cells produced per unit mass of ammonia oxidized to nitrate. There are 0.1 g of cells produced per gram of ammonia oxidized (Eq. 10-12). Nitrifiers grow very slowly, and filters adapt to changes slowly. Slow growth means that filter break-in periods are long (Wheaton et al. 1991).

BIOMASS DENSITY. This is a measure of the mass of bacterial cells per unit volume of biofilm. Biomass density is dependent on water velocity through the filter, the nutrient load carried by the water, and filter characteristics. In their research Hujer and Booler (1986) determined that cell density averaged 886 g/m^3.

Biofilter Start-up

Biological filters normally require several weeks to become completely functional (i.e., acclimated). As heterotrophs convert organic compounds to ammonia, the ammonia concentration in the system will continue to rise until a sufficient *Nitrosomonas* population becomes established and begins to oxidize ammonia (Figure 10-40). This normally requires about 10–14 days. As ammonia is converted to nitrite, *Nitrobacter* develop and begin converting nitrite to nitrate. The nitrate concentration will continue to rise in a culture system unless it is removed by green plants or by water exchanges. The total process of ammonia conversion to nitrate normally requires from 30–100 days (Kawai et al. 1965; Hirayama 1974; Mevel and Chamroux 1981).

Methods that can be used to accelerate filter acclimation are desirable, especially in commercial applications. Filter start-up usually consists of introducing live animals to the system in small quantities so that ammonia increases gradually. The animal load is increased in small amounts, allowing three to five days for the system to equilibrate between additions. The system is thus brought to full load gradually so that shock loadings do not cause system failure.

Filter start-up may also be accomplished with the addition of either water or wet filter media from an established biofilter. Either will contain live nitrifying bacteria. Also, several commercial inoculants are available that contain live bacteria. These shorten filter acclimation time when added to new culture systems. Bower and Turner (1981, 1984) compared filter start-up times using commercial inoculants and media from established biofilters to controls in seawater systems. The addition of 10% wet media from filters reduced start-up time by 81% (4 days compared to 21 days for controls) for ammonia removal and 89% (4 days compared to 37 days) for nitrite removal. The addition of either seawater, dry media from seawater filters, or wet media from freshwater systems produced significantly less reductions in filter start-up times than wet media from seawater filters. The addition of commercial inoculants had varied results, none of which were as rapid as the addition of wet media from seawater filters. Carmigiani and Bennett (1977) found that the addition of 3% wet filter media from an estab-

Figure 10-40. Nitrification curve showing the rise and fall of ammonia and nitrite and the rise of nitrate in closed aquaculture systems.

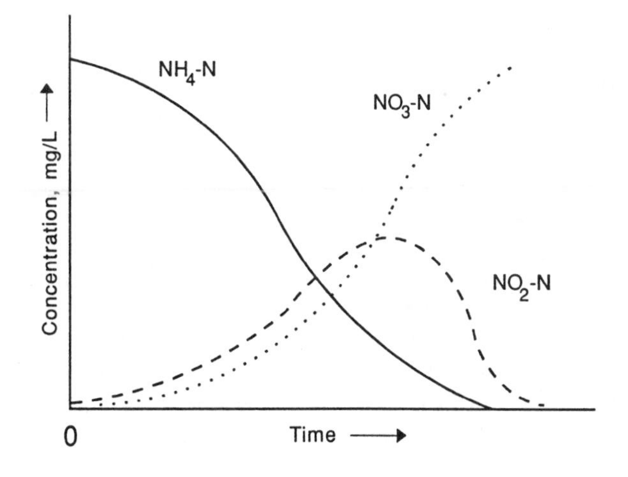

lished biofilter decreased start-up time by 48% compared to controls.

Manthe and Malone (1987) compared several methods to reduce filter start-up time in soft-shell blue crab production systems: addition of commercial inoculants, addition of live crabs, addition of an inorganic ammonia solution, and addition of inorganic ammonia and nitrite solutions combined. They also studied the effect of filter media size on acclimation time. Results indicated that the addition of commercial inoculants provided no advantage over the addition of live crabs and did not have a significant impact on shortening acclimation time. The addition of inorganic ammonia by itself properly conditioned the filter but did not shorten acclimation time. The addition of a solution consisting of 12 mg-N/L of ammonium chloride plus 12 mg-N/L of sodium nitrite reduced filter start-up time by 28% (10 days). Controls normally required 36 days for start-up.

The simultaneous addition of inorganic ammonia and nitrite solutions holds promise for reducing filter start-up times (Manthe and Malone 1987). In normal filter start-up with organic compounds, the autotrophs must compete for space on filter media surfaces with heterotrophs. By using an inorganic ammonia solution *Nitrosomonas* become established well before the addition of organic compounds and the subsequent development of heterotrophs. By adding nitrite simultaneously with ammonia *Nitrobacter* do not have to wait the normal 10–14 day period before they have enough nitrite for their development; they start up right away. The total result is reduced filter start-up time and lower concentrations of ammonia and nitrite.

Biofilter Design

The published literature provides few approaches to the analysis of aquaculture biofilter design. Most design procedures are based on pilot-scale studies or by trial and error. Wheaton (1977) presented a lengthy approach to biofilter design using procedures established by Speece (1973) for cold-water trout culture. This procedure is useful as a guide but is marginally applicable to biofilters operating at warmer temperatures, such as in catfish and tilapia culture. It is generally accepted that nitrification proceeds at a more rapid rate at higher temperatures. Wheaton et al. (1991) later presented a more direct approach for biofilter design that has application to a broader spectrum of fish species and water temperatures. Losordo (1991) demonstrated a mass balance approach to biofilter design. The filter design example that follows was developed using a combination of procedures and technical information contained in Wheaton et al. (1991), Losordo (1991), and others.

Mass Balance Approach. A materials or mass balance can be performed on a recirculating aquaculture system by: (1) Defining the system boundaries; (2) identifying stream flows which cross the boundaries; (3) identifying the material to be balanced; and (4) identifying the processes affecting the mass balance within the system boundaries. Once these steps have been taken, a mass balance equation can be written. The mass balance equation for non-steady-state conditions can be written as follows:

$$\begin{matrix}\textit{Rate of}\\ \textit{accumulation}\\ \textit{of material}\\ \textit{within system}\end{matrix} = \begin{matrix}\textit{Mass flow}\\ \textit{rate into}\\ \textit{system}\end{matrix} - \begin{matrix}\textit{Mass flow}\\ \textit{rate out}\\ \textit{of system}\end{matrix} + \begin{matrix}\textit{Net rate of}\\ \textit{conversion}\\ \textit{of mass within}\\ \textit{system}\end{matrix} \quad (10\text{-}16)$$

Or more precisely:

$$\textit{Accumulation} = \textit{Input} - \textit{Output} + \textit{Conversion} - \textit{Consumption} \quad (10\text{-}17)$$

Under steady state conditions no change is implied, hence no mass accumulation of material occurs within the system. The mass balance equation thus simplifies to:

$$0 = \textit{Input} - \textit{Output} + \textit{Conversion} - \textit{Consumption} \quad (10\text{-}18)$$

This design approach can be used as a basis to create mathematical models to simulate water quality in tank-based aquaculture production systems. The system boundary for a recirculating system should be drawn around the culture unit with water inputs and outputs crossing the oundary and the purification/filtration pro-esses lying within the boundary as illustrated ı Figure 10-41. If the system is to be operated ı a batch mode,the analysis should be for non-teady-state conditions. However, in a continu-us production unit it can be assumed that iomass doesn't change or changes by an insig-ificant amount. We thus assume *completely ixed tank* conditions (Losordo 1991).

Ammonia Mass Balance. The need to keep ıe un-ionized ammonia in the system in check sually dictates the recirculation rate of water in closed recirculating system. The mass balance or ammonia is generally based on total ammo-ia nitrogen, TAN = $(NH_3 + NH_4^+) - N$. If AN is known, the mole fraction of NH_3 then ecomes a function of pH and temperature. The ass balance equation for TAN under non-eady-state conditions is:

$$\frac{dC_{TAN}}{dt}V = Q \times C_{(TAN)i} - Q \times C_{(TAN)o} + PR_{TAN} - CR_{TAN} \qquad (10\text{-}19)$$

here C_{TAN} = concentration of TAN in system, g/L; V = system volume, L; Q = flow rate rough system, L/min; PR_{TAN} = production rate f TAN, mg/min; CR_{TAN} = consumption rate of AN, mg/min; t = time, min; and the subscripts and o denote concentrations flowing into and ut of the system, respectively.

The rate of production of TAN in the system is a function of fish metabolism rate and the biological degradation of unconsumed feed. PR_{TAN} can be estimated by the following equation:

$$PR_{TAN} = BM \times \%BW \times K \qquad (10\text{-}20)$$

where BM = fish biomass in the system, mg; $\%BW$ = feeding rate expressed as a percentage of fish body weight (decimal fraction); and K = constant (K = 30 g TAN/kg feed in SI units; K = 0.03 lb/lb feed in English units). The $\%BW$ value can be obtained from handbooks. Feeding rates take fish size and water temperature into consideration; therefore, these factors do not have to be accounted for in Equation 10-20.

CR_{TAN} in Equation 10-19 refers to the removal rate of the treatment process in the system. This can be expressed by:

$$CR_{TAN} = Q_f \times C_{TAN} \times E \qquad (10\text{-}21)$$

where Q_f = flow rate to the filter, L/min; and E = filter efficiency (decimal fraction). The filter efficiency is defined as the fraction of TAN removed in one pass through the biofilter. Predicting the efficiency of biofilters is very complex and is the focus of numerous past and present research efforts. It is generally accepted that biofilter efficiency is a function of many variables, including inflow ammonia concentration, hydraulic loading rate, water temperature and pH, filter surface area, type of biofilter, and

igure 10-41. System boundaries around the culture unit in a recirculating system.

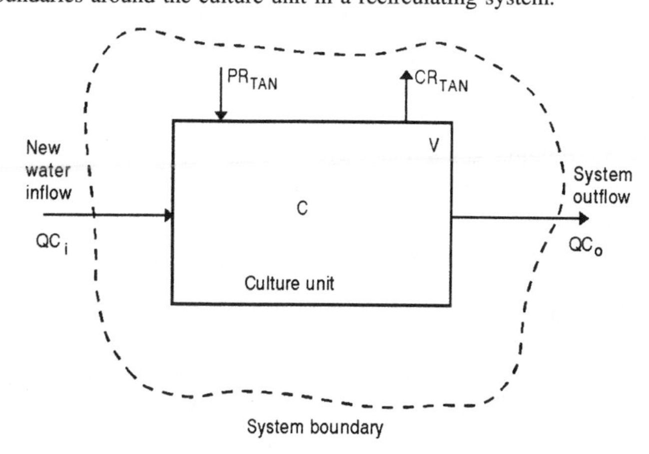

media type. It is reasonable to assume that biofilter performance is also influenced by parameters not yet identified. For design purposes biofilter efficiencies are normally conservatively estimated at about 25–50%.

Equations 10-19 and 10-21 can be combined to solve for the flow required to maintain steady-state conditions for a given set of system operating variables:

$$Q_f = \frac{Q \times C_{(TAN)i} - Q \times C_{(TAN)o} + PR_{TAN}}{C_{TAN} \times E} \quad (10\text{-}22)$$

Equation 10-22 can be simplified by assuming that the new water flowing into the system will have a very low TAN concentration (i.e., $C_{(TAN)i} = 0$). Hence, the term $Q \cdot C_{(TAN)i}$ can be eliminated from the equation. Also, by definition in a completely mixed system, $C_{(TAN)o} = C_{TAN}$. Thus, the equation to estimate flow to the biofilter simplifies to:

$$Q_f = \frac{PR_{TAN} - Q \times C_{TAN}}{C_{TAN} \times E} \quad (10\text{-}23)$$

In order to solve for flow to the filter (Q_f), the flow of new water through the system (Q) must be estimated. This can be done by conducting a mass balance on nitrate nitrogen since Q is used to limit the accumulation of nitrates in the system (Losordo 1991).

Nitrate Nitrogen Mass Balance. The flow rate of new water is a function of the NO_3 production rate within the culture unit and a given acceptable concentration of NO_3 during steady-state conditions. Nitrate is the end product of the biological degradation of ammonia in a biofilter. It is not known to have toxic effects on fish at relatively high concentrations, but it must be maintained at an acceptable level in the culture system regardless. Some closed recirculating systems may use a denitrification process for nitrate control, but most waste the nitrate from the system by making daily water changes. Thus, given a maximum acceptable level of nitrate in the system, a mass balance can be used to determine the flow of new water required. The mass balance of nitrate in a nonsteady-state system can be written as:

$$\frac{dC_{NO_3}}{dt} V = Q \times C_{(NO_3)_i} - Q \times C_{(NO_3)_o} + PR_{NO_3} - CR_{NO_3} \quad (10\text{-}24)$$

where C_{NO3} = nitrate concentration in the system, mg/L; PR_{NO3} = production rate of nitrate mg/min; and CR_{NO3} = consumption rate of nitrate, mg/min.

Given steady-state conditions, $Q_i = Q_o$, $C_{(NO3)i} = 0$, and there is no treatment process in the system to remove nitrate (i.e., $CR_{NO3} = 0$) the flow rate of new water to the system can be estimated by:

$$Q = \frac{PR_{NO_3}}{C_{(NO_3)_o}} \quad (10\text{-}25)$$

The rate of production of NO_3 can be estimated by stoichiometry from the rate of nitrification in the biofilter, recalling that 4.43 g of NO_3 are produced for each gram of ammonia nitrified. Thus,

$$PR_{NO_3} = CR_{TAN} \times 4.43 \quad (10\text{-}26)$$

Also, given steady-state conditions and assuming that the rate of loss of TAN due to the flow through the system is very small (i.e., $Q \cdot C_{(TAN)o} \approx 0$), then $CR_{TAN} \approx PR_{TAN}$. Thus,

$$PR_{NO_3} = PR_{TAN} \times 4.43 \quad (10\text{-}27)$$

and

$$Q = \frac{PR_{TAN} \times 4.43}{C_{(NO_3)_o}} \quad (10\text{-}28)$$

Equation 10-28 can then be used to calculate the required flow through the filter to maintain steady-state conditions (Q_f) in Equation 10-23.

Dissolved Oxygen Mass Balance. It is not sufficient to simply solve for the mass balance of ammonia in a recirculating system. Althoug ammonia accumulation is a primary limiting factor governing the carrying capacity of a system, another factor that is often limiting is dissolved oxygen (DO). The DO mass balance equation for nonsteady-state conditions can be written as:

$$\frac{dC_{DO}}{dt} V = Q \times C_{(DO)_i} - Q \times C_{(DO)_o} + PR_{DO} - CR_{DO} \qquad (10\text{-}29)$$

C_{DO} = the dissolved oxygen concentration in the system, mg/L; PR_{DO} = production rate of DO, mg/min; and CR_{DO} = consumption rate of DO, mg/min.

The production rate of DO depends solely on the aeration/oxygenation system designed as an integral part of the total recirculating system. The rate of oxygen addition is therefore a function of the many variables as discussed in Chapter 11. The consumption rate of DO in a recirculating system is the sum of three components: (1) The respiration rate of the fish; (2) the carbonaceous oxygen demand (BOD); and (3) the oxygen demand of the biofilter. The overall DO consumption rate can be estimated using Equation 10-30:

$$CR_{DO} = CR_{FISH} + CR_{BOD} + CR_N \qquad (10\text{-}30)$$

All of the components in the above equation have the units mg/min or mg/hr.

DO CONSUMED BY FISH. The oxygen consumed by the fish in the system is a function of fish size, water temperature, fish condition, species, total biomass, and other parameters. This component can be estimated with Equation 10-31:

$$CR_{FISH} = BM \times \%BW \times K \qquad (10\text{-}31)$$

A value for K of 200 g O_2/kg feed (0.2 lb O_2/lb feed) is recommended as a mean value for salmonids (Westers 1981; Colt 1987). A more conservative value of 214 g O_2/kg feed is recommended for channel catfish (Boyd and Watten 1989) although some researchers use an even higher value of 300 g O_2/kg feed for warm-water species like catfish and tilapia.

CARBONACEOUS OXYGEN DEMAND. The carbonaceous oxygen demand (CR_{BOD}) is the demand placed on the culture system by the biological degradation of fish wastes and unconsumed feed. Depending on system design, the CR_{BOD} can be exerted either in the culture unit or in the biofilter. Culture unit design should incorporate mechanisms for the rapid removal of solid wastes, otherwise the BOD load will be exerted within the culture unit. Once solid wastes leave the culture unit they can be removed by screening, settling, sand filtration, ozonation, or other means before entering the biofilter. This greatly reduces the oxygen demand on the biofilter, otherwise BOD will be expressed as an oxygen demand within the biofilter (Losordo 1991). Thus, rapid removal of solids is essential for reducing the aeration requirements of a recirculating system.

Wimberly (1990) demonstrated that the average unfiltered BOD_5 excretion rate for channel catfish fed 1% BW of a 35% protein ration daily was 2,160 mg O_2/kg fish/day. BOD_5 excretion rate for catfish was about four times that of several other species. He also showed that the biofilter oxygen demand was about 2.3 times the BOD_5 production rate of the fish. Thus, the CR_{BOD} for the biofilter can be estimated as:

$$CR_{BOD} = 2.3 \times K_{BOD} \times BM \qquad (10\text{-}32)$$

where K_{BOD} = 2,160 mg O_2/kg fish/day.

NITRIFICATION OXYGEN DEMAND. The DO consumption of the biofilter during nitrification can be determined from basic stoichiometry since we know that, for every gram of TAN oxidized to nitrate, 4.57 g of DO are consumed (Wheaton 1977). Thus, the oxygen loss due to nitrification in the biofilter is:

$$CR_N = 4.57\,(PR_{TAN}) - Q \times C_{(TAN)_o} \qquad (10\text{-}33)$$

where $PR_{TAN} - Q \cdot C_{(TAN)o}$ is the net TAN flowing to the biofilter.

Rearranging Equation 10-29 and solving for

PR_{DO}, we can develop an expression for estimating the rate of oxygen addition required by the total system. By assuming steady state and completely mixed conditions, and by setting $C_{(DO)o}$ = desired value, the required rate of DO input to the system by an aerator or oxygenator is estimated by:

$$PR_{DO} = Q \times C_{(DO)_o} + CR_{DO} - Q \times C_{(DO)_i} \quad (10\text{-}34)$$

where CR_{DO} is defined by Equation 12-30. The results of this calculation can assist with the identification of an aerator or oxygenator design.

Dissolved Oxygen Mass Balance in Biofilter. An oxygen mass balance should be performed for the biofilter to ensure that the filter will have adequate DO to function properly. If the biofilter is oxygen-starved, nitrification will proceed at a reduced rate with the resulting buildup of ammonia in the culture unit. This situation can be mitigated by increasing the water flow rate to the system to flush wastes, but operating costs will be greater.

For a biofilter DO mass balance we can place the system boundaries around the filter only, as shown in Figure 10-42. The DO mass balance under steady-state conditions can be written as:

$$0 = Q_{f*} \times C_{(DOf)_i} - CR_{BODf} - CR_N - Q_f \times C_{(DOf)_o} \quad (10\text{-}35$$

where Q_{f*} = flow rate to biofilter to maintai oxygen, L/min; $C_{(DOf)i}$ = DO concentratio flowing into biofilter, mg/L; CR_{BODf} = carbon aceous oxygen demand in biofilter, mg/L; an $C_{(DOf)o}$ = DO concentration flowing out of bio filter, mg/L.

The DO concentration in the biofilter efflu ent must be at least 2.0 mg/L to ensure tha nitrification is not limited due to oxygen star vation (Malone and Burden 1988a, 1988b Equation 10-35 can be rearranged to solve fo Q_f, the required flow through the biofilter fo given influent and effluent DO concentrations

$$Q_{f*} = \frac{CR_{BODf} + CR_N}{C_{(DOf)_i} - C_{(DOf)_o}} \quad (10\text{-}36$$

The resulting value obtained for Q_{f*} should b compared to Q_f obtained with Equation 10-2 and the higher of the two values should be use in the system design. Alternatively, the DO con centration to the biofilter, $C_{(DOf)i}$, could be in creased or CR_{BODf} could be reduced by prefil tration or settling of solids. The biofilter oxyge demand can also be reduced by limiting the bi ological degradation within the filter. This ca

Figure 10-42. System boundaries around the biological filter in a recirculating system.

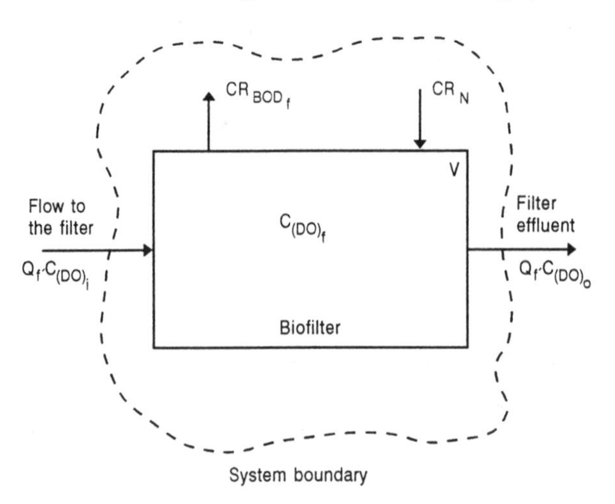

be done with frequent filter backwashing to remove trapped solids, as is done with submerged filters (Wimberly 1990).

The usefulness of the procedure outlined above for biofilter design can best be understood by use of an example. Catfish are used for the example due to the abundance of technical culture information for the species.

Example problem 10-1. A recirculating production system is to carry a catfish biomass of 1,000 kg. The fish have a mean weight of 567 g (1.25 lb) at harvest, and are fed 1% of their body weight daily. (Note: Biofilters should always be designed to handle the maximum load that the system will sustain, otherwise the filter will fail.) Other assumptions and pertinent information are listed in Table 10-13. Assume that settling and an ozonation process are built into the system for organic solids removal.

Solution: Five options will be explored providing five biofilter choices: upflow sand filter, fluidized bed sand filter, floating bead media filter, RBC (Biodek), and RBC (Nor-pac tubing).

Step 1. Estimate the ammonia production rate using Equation 10-20:

$$PR_{TAN} = 1{,}000 \text{ kg fish} \times 0.01 \times 30 \text{ g TAN/kg feed}$$

$$= 300 \text{ g TAN/day} = 12{,}500 \text{ mg TAN/hr}$$

Step 2. Estimate system flow-through rate using Equation 10-25, assuming that $C_{NO3} = C_{(NO3)o}$:

$$Q = \frac{4.43 \text{ g } NO_3\text{/g TAN} \times 12.5 \text{ g TAN/hr} \times 1{,}000}{300 \text{ mg/L}}$$

$$= 185 \text{ L/hr (0.8 gpm)}$$

Step 3. Calculate the recycle flow rate through the biofilter using Equation 10-23 and assuming $C_{(TAN)o} = C_{TAN}$:

$$Q_f = \frac{12{,}500 \text{ mg TAN/hr} - (185 \text{ L/hr} \times 2 \text{ mg/L})}{2 \text{ mg/L} \times 0.30}$$

$$= 17{,}329 \text{ L/hr (76 gpm)}$$

Step 4. Determine the oxygen utilization rate for the system using Equation 10-30:

$$CR_{DO} = CR_{FISH} + CR_{BODf} + CR_N$$

a. $CR_{FISH} = 1{,}000 \text{ kg} \times 0.01 \times 300 \text{ g } O_2\text{/kg}$ feed (from Equation 10-31)
$= 3{,}000 \text{ g } O_2\text{/day} = 125{,}000 \text{ mg } O_2\text{/hr}$

b. $CR_{BODf} = 2.3 \times 2{,}160 \text{ mg } O_2\text{/kg fish/day} \times 1{,}000 \text{ kg fish} \times (1\text{-}E_f)$
$= 1.49 \times 10^6 \text{ mg } O_2\text{/day} = 62{,}100 \text{ mg } O_2\text{/hr}$

c. $CR_N = 4.57 \times \{12{,}500 \text{ mg TAN/hr—}(185 \text{ L/hr} \times 2 \text{ mg/L})\}$
$= 55{,}434 \text{ mg } O_2\text{/hr}$

Table 10-13. Data for example problem 10-1.

Parameter	Notation	Value	Units
Fish biomass	*BM*	1,000	kg
Feeding rate	*% BM*	1.0	decimal fraction
Influent DO	$C_{(DO)i}$	7.4	mg/L
Influent TAN	$C_{(TAN)i}$	0	mg/L
Influent NO_3	$C_{(NO3)i}$	0	mg/L
Water temperature	T	25	°C
Water pH	-	7.8	pH units
System TAN concentration	C_{TAN}	2.0	mg/L
System NO_3 concentration	C_{NO3}	300	mg/L
Biological filter efficiency	*E*	30	%
Filter BOD removal efficiency	E_{BOD}	70	%

d. Thus, the total oxygen consumed in the biofilter is

$$CR_{DO} = 125{,}000 + 62{,}100 + 55{,}434$$
$$= 242{,}534 \text{ mg } O_2\text{/hr}$$

Step 5. Calculate the flow to the biofilter required to maintain an effluent DO concentration of 2 mg/L using Equation 10-36. Assume that a minimum C_{DO} = 7.0 mg/L will be maintained in the culture unit (i.e., $C_{DO} = C_{(DO)i}$):

$$Q_{f*} = \frac{62{,}100 \text{ mg } O_2\text{/hr} + 55{,}434 \text{ mg } O_2\text{/hr}}{7.0 \text{ mg/L} - 2.0 \text{ mg/L}}$$

$$= 23{,}507 \text{ L/hr (165 gpm)}$$

Step 6. Calculate the oxygen production rate necessary to maintain 7.0 mg DO in the system using Equation 10-34:

$$PR_{DO} = (185 \text{ L/hr} \times 7.0 \text{ mg/L}) + 242{,}534 \text{ mg } O_2\text{/hr—}(185 \text{ L/hr} \times 7.4 \text{ mg/L})$$

$$= 242{,}460 \text{ mg } O_2\text{/hr} = 0.24 \text{ kg } O_2\text{/hr}$$

The results of this mass balance indicate that the system oxygen balance is not greatly affected by water inflow and outflow. An aeration/oxygenation system must be capable of producing 0.24 kg (0.53 lb) of oxygen per hour to meet system demands.

Biofilter Sizing

OPTION 1: UPFLOW SAND FILTER. With a new water (flow-through) rate of 185 L/hr, the system will waste ($Q \times C_{(TAN)o}$) 370 mg TAN per hour. The amount that the biofilter must remove (TAN_f) can be calculated as follows:

$$TAN_f = PR_{TAN} - Q \cdot C_{(TAN)o}$$
$$= 12{,}500 \text{ mg TAN/hr} - 37\text{- mg TAN/hr}$$
$$= 12{,}130 \text{ mg TAN/hr} = 291 \text{ g TAN/day}$$

According to Malone and Burden (1988a, 1988b), the carrying capacity of an upflow sand filter using 1.2–2.4 mm diameter coarse sand is 541 g TAN/m^3 media surface/day. The volume of sand media required to meet this requirement is

$$\text{Sand media volume} = \frac{291 \text{ g TAN/day}}{541 \text{ g TAN/m}^3\text{/day}}$$

$$\text{Sand volume} = 0.54 \text{ m}^3$$

For the upflow sand filter the maximum recommended hydraulic flux rate (volume/cross-sectional area) is 570 L/min/m^2 (149 gpm/ft^2). In our problem the design recycle flow rate through the filter (Q_f) was 23,507 L/min. Thus, the total sand filter cross-sectional area required is

$$\text{Filter cross-sectional area} = \frac{392\text{L/min}}{570 \text{ L/min/m}^2}$$

$$= 0.69 \text{ m}^2$$

Therefore, the system requires approximately 0.69 m^2 of cross-sectional filter area. A 102-cm (40-in.) diameter upflow sand filter having a cross-sectional area of 0.81 m^2 and a media depth of 38 cm (15 in.) will provide a media volume of 0.31 m^3. Our design will require two of these filters, each operating at 462 L/min (121 gpm) to maintain the given design conditions. According to Malone and Burden (1988a 1988b), a hydraulic flux rate of 2,650 L/min/m^2 (65 gpm/ft^2) is needed to expand the sand media (50% expansion) to flush trapped solids from the filter. Thus, a flow of 2,147 L/min (545 gpm) is needed to each filter for flushing the filters.

OPTION 2: FLUIDIZED SAND FILTER. The carrying capacity of a fluidized sand filter is the same as an upflow sand filter, provided that the same size sand grain is used in each (Malone and Burden 1988). Advantages of fluidized bed filters are that they are better at oxidizing NO_2 and they are less subject to clogging. On the negative side, fluidized beds cannot trap solids requiring an additional component for solids removal, and fluidized bed filters require a much higher hydraulic flux rate. Fluidized beds using the same size sand grains as in the previous option require a hydraulic flux rate of 2,650 L/min/m^2 (65 gpm/ft^2) to maintain a 50% bed expansion. Higher flux rates are needed for greater expansion. Thus, fluidized bed filters having the same diameter and media depth as the previous example will require a pumping rate to each filter of 2,147 L/min (565 gpm). The application of fluidized bed filters to aquaculture systems is questioned because of their high hydraulic flux rate requirements.

OPTION 3: FLOATING BEAD FILTER. Using the results obtained by Malone et al. (1993) the floating bead filter has a nitrification capacity of 270 mg TAN/m^2/day (25 mg TAN/ft^2/day). Thus, the volume of beads required for our example is calculated as follows:

$$\text{Volume beads} = \frac{291 \text{ g TAN/day} \times \text{m}^3/1{,}145 \text{ m}^2}{0.270 \text{ g TAN/m}^2\text{/day}}$$

$$\text{Volume beads} = 0.94 \text{ m}^3$$

Thus, a filter having a bead volume of one m^3 will be sufficient to handle the wastes in this problem.

OPTION 4: RBC (BIODEK). Miller and Libey (1985) determined that the removal rate for a four-stage RBC in a recirculating system constructed from Biodek was approximately 1.37 g TAN/m^2 media surface/day. The fish density in their system was 227 kg fish /m^3 water (TAN ≈ 1.7 mg/L). The required surface area for a RBC constructed from this material can be estimated as:

$$\text{RBC surface area} = \frac{291 \text{ g TAN/day}}{1.37 \text{ g TAN/m}^2\text{/day}}$$

$$\text{RBC surface area} = 212.4 \text{ m}^2$$

The Biodek used in Miller and Libey's study has a surface area of 371 m^2/m^3. Therefore, for our problem the volume of material required is:

$$\text{RBC volume} = \frac{212.4\text{m}^2}{371 \text{ m}^2/\text{m}^3} = 0.57 \text{ m}^3$$

OPTION 5: RBC (NOR-PAC). Miller and Libey (1985) determined that their biofilter removed about 1.37 g TAN/m^2/day. Others obtained different figures using different types of biofilters. Speece (1973), using submerged rock filters, obtained a removal rate of approximately 1.0 g TAN/m^2/day in trout culture systems at 20°C. Filters removed 0.56 g TAN/m^2/day in systems for culturing freshwater prawns (*Machrobrachium rosenbergii*) (Rogers and Klemetson 1981). In a later study Rogers and Klemetson (1985) achieved a removal rate of 2.83 g TAN/m^2/day with an RBC used in prawn culture systems. If we use a conservative ammonia removal rate of 1.0 g TAN/m^2/day (Petit 1990), we will have an overdesigned biofilter, but one in which we will have reasonable confidence against failure. Thus, the RBC specific surface area (SSA) requirement for this example is:

$$SSA = \frac{\text{Ammonia production rate}}{\text{Nitrification rate}}$$

$$= \frac{219 \text{ g TAN/day}}{1.0 \text{ g TAN/m}^2\text{/day}}$$

$$= 219 \text{ m}^2$$

Nor-pac tubing having a diameter of 2.54 cm (1.0 in) provides a surface area of 246 m^2/m^3 (75 ft^2/ft^3). Therefore, our problem requires an RBC volume of:

$$\text{RBC volume} = \frac{219 \text{ m}^2}{246 \text{ m}^2/\text{m}^3} = 0.89 \text{ m}^3$$

Thus, a two-stage RBC constructed from 2.54 cm Nor-pac tubing could be used, each RBC drum having an approximate volume of 0.45 m^3. We could use two RBC drums, each having a diameter of 1 m and a length of 0.5 m.

Discussion. Once a biofilter is selected and designed, an aeration and/or oxygenation system must be designed that will supply the oxygen needs of the system and biofilter. There are many types of aeration or oxygenation systems from which one may select. The pros and cons of various types of aeration and oxygenation systems are the topics of Chapter 11.

TEMPERATURE CONTROL

Maintaining the proper water temperature in a recirculating aquaculture system is extremely important. The temperature desired is dependent on species and life cycle. Temperature control is required in systems used to overwinter fingerling or brood-stock, to induce spawning, for egg hatching and fry development, or for fingerling grow-out. There is an optimum temperature for most species at which growth is most rapid. However, not only must the cultured species be considered, but the temperature requirements of the microorganisms in the biological filter must also be considered. It is often a tricky business to provide the ideal temperature for both biological systems. Energy requirements for heating/cooling water contribute significantly to overall system operating costs. An advantage is gained in recirculating systems since the water is recycled within the system and not dumped, wasting dollars that went into heating or cooling.

Various methods have been used in the past for water temperature control in recirculating systems. Probably the most common method is to operate recirculating systems inside of enclosed structures and to control the air temperature inside the structure. Air is usually cooled with room air or central air conditioning units. Heating is done with either electric or gas-fired space heaters. Controlling room air temperature is very costly and inefficient. Also, the response time of changes in water temperature to environmental air temperature changes is often very slow, an undesirable situation where rapid water temperature changes are desired for breeding and spawning purposes.

A better approach is to heat the water mass directly. A variety of heat exchangers is available, but they are not always economical to operate. For heating purposes only an array of submersion water heaters are marketed from small 25 W aquarium models to commercial units of several kW. These may be placed directly into the culture unit or they may be operated in a sidestream configuration (Figure 10-43). Heating elements placed directly into culture units have the disadvantage that they may interfere with cleaning or harvesting operations. However, this is usually offset by the cost savings by using this approach over room air heating. The sidestream configuration requires an additional pump in the system. Additionally, when submersion heaters are used, a separate system must be employed whenever cooling is desired.

Lawson et al. (1989) used a ground-coupled water-source heat pump system for controlling water temperature in a 11,300 L recirculating system for spawning red drum (*Sciaenops ocellatus*). Similar systems have for several years been used for residential and commercial heat-

Figure 10-43. Heat and cooling culture units: (a) Heat exchanger in culture unit; and (b) sidestream heat exchanger.

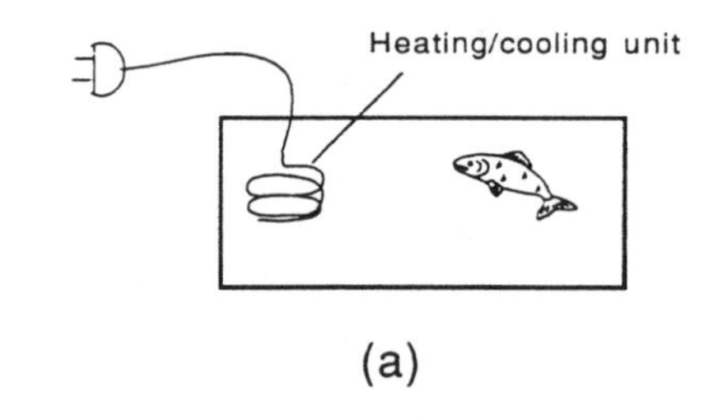

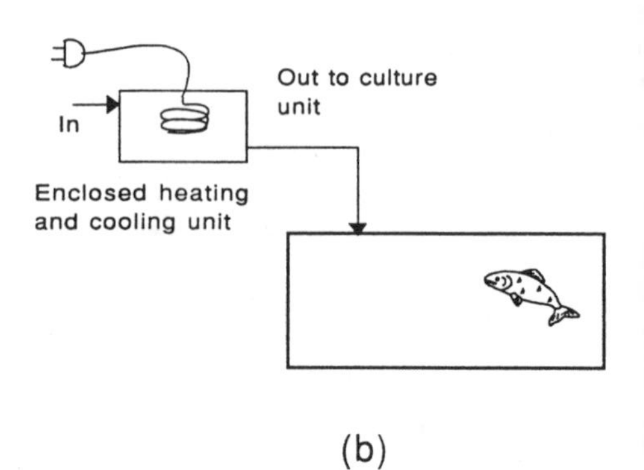

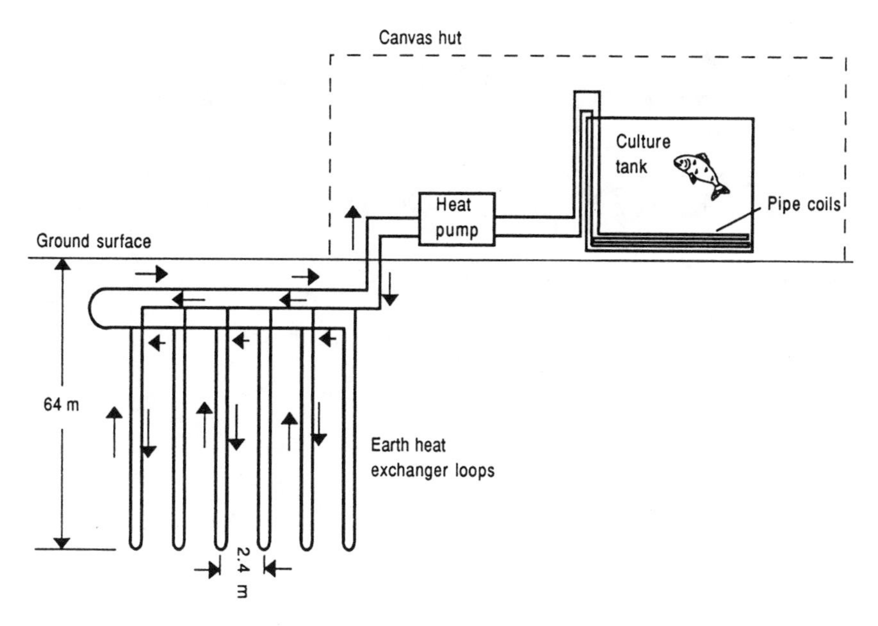

Figure 10-44. A parallel-reverse ground loop heat exchanger system for heating/cooling a red drum spawning tank.

ing and cooling in the United States and elsewhere. The experimental system used a heat pump unit linking closed pipe loops in the earth and spawning the tank (Figure 10-44). The tank loop consisted of four 11.8-m long coils of 2.5-cm inner diameter polyethylene pipe. The ground loop was fabricated by inserting 2.5-cm inner diameter polyethylene pipe into six wells. The tank and ground loops were closed units so that no water was exchanged between the two. Rather, heat was exchanged between the two loops via the heat pump unit. In South Louisiana the earth remains a constant 21°C (70°F) at a depth of 15.6 m (50 ft) and greater below the ground surface. The wells were drilled to a depth of 65.6 m (250 ft) to take advantage of the constant temperature by using the earth as a heat source/sink for the water circulated through the pipe loop. A canvas hut housed both the spawning tank and heat pump unit. Throughout the 190-day study the water temperature in the spawning tank was maintained at + 0.8°C of the target temperature. The fish were successfully spawned using a water temperature regime that simulated the seasonal cycles which the fish would have experienced in the wild.

Heat pumps have been used in salmon smolt hatcheries in Norway since 1979 (Johansen 1991). Heat pump systems show great promise for water heating and cooling in aquaculture systems. Although capital costs are relatively high, these units pay for themselves in a three to five year period in terms of energy saved for heating and cooling.

CHAPTER 11

OXYGEN AND AERATION

Oxygen is essential for the production of all species of fish and shellfish. Aeration is the most effective means of increasing the dissolved oxygen (DO) concentration of culture water. *Aeration* is the addition of oxygen or air to water. It is not a new process, but interest has increased tremendously over the past decade. Aeration can increase the carrying capacity of an aquaculture system when DO is the limiting factor. Unfortunately, many fish farmers, researchers, extension specialists, equipment manufacturers and others have a poor understanding of aeration principles. Consequently, they have unrealistic expectations of what benefits aeration can provide.

This chapter attempts to assemble basic oxygen and aeration facts and fundamentals. This chapter also discusses aerator tests, types of aerators, and application practices. In addition, relatively new concepts involving the application of pure oxygen are discussed.

DISSOLVED OXYGEN THEORY

In order to gain an appreciation for aeration, it is first necessary to acquire an understanding of dissolved oxygen dynamics in aquaculture systems.

Solubility

The atmosphere is composed of 20.946% oxygen gas (O_2), 78.084% nitrogen (N_2), 0.934% argon (Ar), 0.032% carbon dioxide (CO_2), and a trace of other gases (Colt 1984). According to Dalton's Law, total atmospheric pressure, commonly referred to as barometric pressure (*BP*), is the sum of the partial pressures of the gases in the atmosphere. Hence,

$$BP = P(O_2) + P(N_2) + P(Ar) + P(CO_2) \quad (11\text{-}1)$$

The partial pressure of each gas in a mixture of gases is directly proportional to the mole fraction of the gas. For example, the partial pressure of oxygen in dry air is

$$P(O_2) = BP \times \chi(O_2) = 760 \times 0.20946 \quad (11\text{-}2)$$

$$P(O_2) = 159.2 \text{ mm Hg}$$

The partial pressures of the other gases in the atmosphere can be calculated in a similar manner.

The atmospheric solubility of a gas depends on both the solubility of the pure gas and the mole fraction of the gas (Colt and Orwicz 1991). The solubilities of the major atmospheric

Table 11-1. Solubility of major atmospheric gases as a function of pressure and gas composition at 15°C.

Pressure (atm)	Gas composition	Gas solubility (mg/L)			
		Nitrogen	Oxygen	Argon	Carbon dioxide
1	air	16.36	10.08	0.62	0.69[1]
1	pure gas	20.95	48.14	65.94	1,992.53
2	air	33.00	20.32	1.24	1.38[1]
2	pure gas	42.26	97.02	133.01	3,979.36

[1] Based on a mole fraction of 0.000350 for air.

Source: J. Colt and C. Orwicz. Aeration in intensive culture. IN: D.E. Brune and J.R. Tomasso, eds. 1991. *Aquaculture and Water Quality*, World Aquaculture Society, Baton Rouge, LA.

gases in freshwater are shown in Table 11-1 as a function of atmospheric pressure and gas composition.

The atmosphere also contains water vapor in addition to the gases mentioned. It is generally assumed that the atmosphere at air-water interfaces is saturated with water vapor. Thus, the partial pressure of a gas in a mixture of gases, including water vapor, is

$$P_i = \chi_i \frac{(BP - VP)}{760} \qquad (11\text{-}3)$$

where P_i = partial pressure of the *ith* gas in the mixture, mm Hg; χ_i = mole fraction of the *ith* gas; and VP = vapor pressure of water, mm Hg. Table 11-2 lists the vapor pressure of pure water at different temperatures.

When air is in contact with water, oxygen enters the water from the air until the *tension* (pressure) of oxygen in the water equals the partial pressure of oxygen in the air. We often find it more convenient to express the solubility of oxygen in mg/L rather than as oxygen tension (Boyd 1982). The concentration of oxygen is normally given as the concentration at equilibrium for water in contact with dry air at standard pressure (760 mm Hg) and at specific temperatures and salinities. Oxygen concentration is expressed as C_s. Table 2-4 gives C_s values for various temperatures and salinities.

Even though the atmosphere contains nearly 21% oxygen, oxygen gas is only slightly soluble in water. Increases in temperature and salinity cause a corresponding decrease in oxygen concentration (refer to Figure 2-6). Oxygen solubility is also influenced by atmospheric pressure. Values of C_s at different temperatures and elevations are presented in Table 11-3. Oxygen saturation decreases as elevation above mean sea level (MSL) increases, due to corresponding decreases in pressure. Values of C_s can thus be corrected for changes in BP with Equation 11-4:

Table 11-2. Vapor pressure of pure water at various temperatures.

Temperature (°C)	Vapor pressure (mm Hg)	Temperature (°C)	Vapor pressure (mm Hg)
0	4.579	18	15.477
1	4.926	19	16.477
2	5.294	20	17.535
3	5.685	21	18.650
4	6.101	22	19.827
5	6.543	23	21.068
6	7.013	24	22.377
7	7.513	25	23.756
8	8.045	26	25.209
9	8.609	27	26.739
10	9.209	28	28.349
11	9.844	29	30.043
12	10.518	30	31.824
13	11.231	31	33.695
14	11.987	32	35.663
15	12.788	33	37.729
16	13.634	34	39.898
17	14.530	35	42.175

Source: Boyd (1990) with permission.

Table 11-3. Solubility of oxygen in water exposed to water-saturated air at various temperatures and elevations above mean sea level (MSL).

Temperature (°C)	Oxygen solubility, mg/L						
	Elevation, ft						
	0	500	1,000	1,500	2,000	3,000	4,000
0	14.60	14.34	14.09	13.84	13.59	13.12	12.65
1	14.20	13.95	13.70	13.46	13.22	12.75	12.30
2	13.81	13.57	13.33	13.09	12.86	12.41	11.97
3	13.45	13.21	12.97	12.74	12.52	12.07	11.65
4	13.09	12.86	12.63	12.41	12.19	11.76	11.34
5	12.76	12.53	12.31	12.09	11.87	11.46	11.05
10	11.28	11.08	10.88	10.68	10.49	10.12	9.76
15	10.07	9.89	9.71	9.54	9.37	9.04	8.71
20	9.08	8.91	8.75	8.59	8.44	8.14	7.84
25	8.24	8.09	7.95	7.80	7.66	7.38	7.12
30	7.54	7.40	7.26	7.13	7.00	6.74	6.50
35	6.93	6.81	6.68	6.55	6.43	6.19	5.96
40	6.41	6.29	6.17	6.05	5.94	5.71	5.49

Source: Weiss (1970).

$$C_s' = C_s \frac{BP - VP}{760 - VP} \qquad (11\text{-}4)$$

where C_s' = oxygen saturation under given conditions, mg/L; and C_s = oxygen saturation at 760 mm HG, mg/L (from Table 2-4). For all practical purposes, the vapor pressure can be ignored (Boyd and Watten 1989), and the equation becomes

$$C_S' = C_s \frac{BP}{760} \qquad (11\text{-}5)$$

Equations 11-4 and 11-5 can be used for surface aerators. For diffused air systems, the pressure on air bubbles is greater than atmospheric, and C_s is commonly taken as the DO saturation concentration corresponding to the average $P(O_2)$ in the bubbles at mid-depth (APHA 1980).

Elevation is the major factor affecting barometric pressure. However, BP is also affected by the passage of storm fronts. The most accurate way to measure BP is with a *barometer*. However, lacking a barometer, BP can be calculated with the following relationship (Colt 1984):

$$\log_{10} BP = 2.880814 - \frac{E}{19{,}748.2} \qquad (11\text{-}6)$$

where BP = barometric pressure, mm Hg; and E = elevation above MSL, m.

An alternative relationship was presented in Soderberg (1982) that can be used to solve for the solubility correction factor, $BP/760$, in Equation 11-5:

$$\frac{BP}{760} = \frac{760}{760 + E/32.8} \qquad (11\text{-}7)$$

where E = elevation, ft. (This equation is left in its original English form since it is an empirical formula.)

Example problem 11-1. Calculate the barometric pressure at an elevation of 609.8 m (2,000 ft) above MSL. What is the oxygen solubility in water at 15°C at this elevation?

Solution:

a. *BP* is first calculated from Eq. 11-6:

$$\log_{10} BP = 2.880814 - \frac{609.8}{19{,}748.2}$$

$$\log_{10} BP = 2.849935$$

$$BP = \text{antilog } 2.849935$$

$$= 707.84 \text{ mm Hg}$$

b. Oxygen solubility is then calculated from Eq. 11-5:

$$C'_s = (C_s)_{15} \times \frac{\text{BP}}{760}$$

$$C'_s = (10.07) \times \frac{707.84}{760}$$

$$C'_s = 9.38 \text{ mg/l}$$

which agrees closely with the value for C_s in Table 11-3 at 15°C and 2,000-ft elevation.

In the United States, barometers normally read in inches of mercury. Recall that 760 mm Hg equals 29.92 in. Hg. Barometric pressure is reported by government weather service agencies for MSL at many stations across the United States. For station pressures at locations other than MSL, correction factors such as those in Table 11-4 can be used for conversion. This table was reported in English units in the original source and, therefore, was left in this form. However, the reader can convert from English to SI units recalling that 1 ft equals 0.3048 m and 1 in. Hg equals 25.4 mm Hg.

Example problem 11-2. The local weather service reports a *BP* of 755 mm Hg at a location where the air temperature is 21°C. What is the station pressure at an elevation of 450 m?

Solution: The elevation is 450/0.3048 = 1,476 ft. The air temperature in °F equals 9/5(21) + 32 = 69.8°F, and the pressure equals 755/25.4 = 29.7 in Hg. From Table 11-4, the approximate pressure correction factor is 0.010609 in Hg for each 10-ft change in elevation. The total pressure correction factor is thus

$$0.010609 \times \frac{1{,}476}{10} = 1.57 \text{ in Hg} = 39.77 \text{ mm Hg}.$$

Then, the station pressure is 755 − 39.77 = 715.23 mm Hg.

Table 11-4. Change in pressure (inches Hg) corresponding to a change in elevation of 10 ft.

Temp. (° F)	Pressure (inches Hg) 31.00	30.00	29.00	28.00	27.00	26.00	25.00	24.00	23.00
0	0.012632	0.012225	0.011817	0.011410	0.011002	0.010595	0.010187	0.009780	0.009372
10	0.012363	0.011964	0.011565	0.011166	0.010768	0.010369	0.009970	0.009571	0.009172
20	0.012105	0.011715	0.011324	0.010934	0.010543	0.010153	0.009762	0.009372	0.008981
30	0.011858	0.011475	0.011093	0.010710	0.010328	0.009945	0.009563	0.009180	0.008798
40	0.011621	0.011246	0.010871	0.010496	0.010121	0.009747	0.009372	0.008997	0.008622
50	0.011393	0.011025	0.010658	0.010290	0.009923	0.009555	0.009188	0.008820	0.008453
60	0.011173	0.010813	0.010452	0.010092	0.009732	0.009371	0.009011	0.008650	0.008290
70	0.010963	0.010609	0.010255	0.009902	0.009548	0.009194	0.008841	0.008487	0.008134
80	0.010759	0.010412	0.010065	0.009718	0.009371	0.009024	0.008677	0.008330	0.007982
90	0.010564	0.010223	0.009882	0.009541	0.009201	0.008860	0.008519	0.008178	0.007838
100	0.010375	0.010040	0.009706	0.009371	0.009036	0.008702	0.008367	0.008032	0.007697
110	0.010193	0.009864	0.009535	0.009207	0.008878	0.008549	0.008220	0.007891	0.007563

Source: Boyd (1990) with permission.

At some point below the water surface of a water body the pressure is greater than the *BP* acting at the surface because of the weight of the water above the point in question. Hence, the total pressure *TP* (Colt 1984) at the point in Figure 11-1 is

$$TP = BP + HP \tag{11-8}$$

where TP = total pressure at a given depth (mm Hg, in. H_2O, psi); BP = barometric pressure (mm Hg, in. H_2O, psi); and HP = hydrostatic pressure (mm Hg, in. H_2O, psi). Recall from Chapter 8 that the hydrostatic pressure *HP* is a function of the water depth as related by Equation 11-9:

$$HP = \gamma Z \tag{11-9}$$

where γ = specific weight of water, kg/m^3 (lb/ft^3); Z = water depth, m (ft); and

$$\gamma = \rho\, g \tag{11-10}$$

The units for *HP* must then be converted to mm Hg to be consistent with Equation 11-7.

Hydrostatic pressure is often used to increase the pressure at which gas transfer occurs in many aeration systems. At 20°C, *HP* is equal to 73.42 mm Hg for each meter of submergence. The solubility of a gas is approximately doubled at a depth of 10 m (Colt and Orwicz 1991). The saturation concentration of a gas at a depth *Z* is equal to

$$(C_s)_Z = C_s\frac{(TP - VP)}{760 - VP} \tag{11-11}$$

or, neglecting vapor pressure,

$$(C_s)_Z = C_s\frac{(BP + \rho g Z)}{760} \tag{11-12}$$

Figure 11-1. Pressure at a point below the surface of a water body.

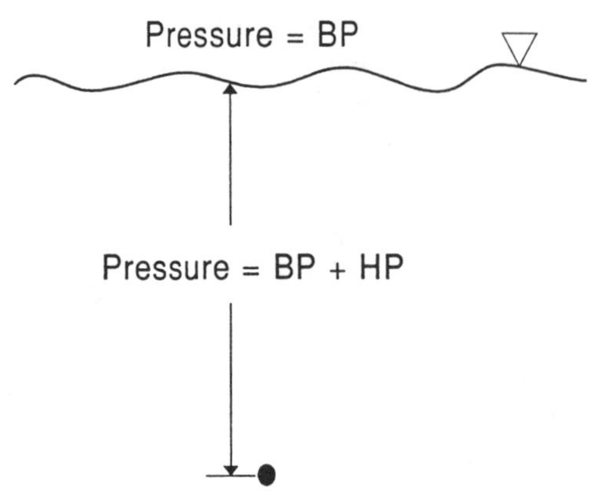

If water contains the amount of dissolved oxygen that it should theoretically hold at a given temperature, salinity, and pressure, then it is said to be saturated with oxygen. However, because of the numerous factors that can affect the amount of oxygen that it can hold, water can also be either *undersaturated* or *supersaturated* with oxygen. The percent saturation of water with oxygen can be calculated as follows:

$$\%\ \text{saturation} = \frac{C_m}{C_s} \times 100 \tag{11-13}$$

where C_m = measured oxygen concentration, mg/L.

For example, suppose brackish water (10 g/L salinity) at 10°C contains 5.50 mg/L of dissolved oxygen. From Table 2-4, the theoretical saturated *DO* value is 10.58 mg/L. Then, the percent saturation of the sample is

$$\%\ \text{saturation} = \frac{5.50}{10.58} \times 100 = 52\ \%$$

Total Gas Pressure

Gas supersaturation can cause problems in aquaculture systems, the most notorious of which is *gas bubble trauma* (*gas bubble disease*), which is discussed in Chapter 2. The difference between the total gas pressure (*TGP*) and the barometric pressure (*BP*) at any location is called ΔP. Thus,

$$TGP = BP - \Delta P \tag{11-14}$$

The value of ΔP can be measured with an instrument called a *saturometer*. Total gas pressure may be expressed as a percent of local barometric pressure by the following relationship:

$$\% \, TGP = \frac{BP + \Delta P}{BP} \times 100 \qquad (11\text{-}15)$$

The percent *TGP* is a measure of the percent saturation of water with atmospheric gases. A ΔP value of 76 mm HG represents a *TGP* of 110% when *BP* is 760 mm HG:

$$\% \, TGP = \frac{760 + 76}{760} \times 100 = 110 \,\%$$

Neglecting argon (since argon is considered biologically inert), the ΔP for surface water is the difference between *TGP* and *BP*:

$$\Delta P = P(O_2) + P(N_2) + P(CO_2) + VP - BP \qquad (11\text{-}16)$$

If ΔP is known, one could, for example, calculate the partial pressure of nitrogen in water using measured concentrations of *DO*, CO_2, *BP*, and tabular values of *VP* (from Table 11-2). Dissolved gas concentrations can be converted to partial pressures using conversion factors given in Table 11-5.

Some workers subtract the vapor pressure from the numerator in Equation 11-15 to get:

$$\% \, TGP = \frac{BP + \Delta P - VP}{BP} \times 100 \qquad (11\text{-}17)$$

The differences between Equations 11-15 and 11-17 become critical at low ΔP values and in warm water (Colt 1987).

The ΔP is measured with respect to the surface water, and the ΔP that a fish experiences is equal to the difference between *TGP* and local pressure (*BP* + *HP*). This is called the uncompensated ΔP and is equal to

$$\Delta P_{uncomp} = \Delta P - \rho g Z \qquad (11\text{-}18)$$

Table 11-5. Multipliers for converting gas concentrations in mg/L to partial pressure in mm Hg as a function of temperature.

Temperature, °C	O_2	N_2	CO_2
5	12.363	28.937	0.2695
10	13.938	32.312	0.3218
15	15.535	35.676	0.3793
20	17.130	38.974	0.9417
25	18.699	42.156	0.5084
30	20.218	45.175	0.5789

Source: Boyd (1990) with permission.

or,

$$\% \, TGP_{uncomp} = \frac{BP + \Delta P}{BP + \rho g Z} \times 100 \qquad (11\text{-}19)$$

The value of ρg is expressed in mm Hg per meter of water depth and is dependent on temperature and salinity (Colt 1987). Recalling that, at 20°C the value of ρg is equal to 74.3 mm Hg per meter of depth, then the values of ΔP_{uncomp} and $\% \, TGP_{uncomp}$ are decreased by 74.3 mm Hg and 10.7%, respectively, for each meter of submergence. Thus, the fish's position in the water column significantly impacts its tolerance of gas supersaturation. Submergence of only a few meters can greatly reduce the effect of ΔP on fish. When ΔP_{uncomp} within the body of a fish is > 0, fine bubbles may form in the blood and tissues. The degree of response at a given ΔP depends on species, life stage, water quality, and water depth (pressure). Comparative data for chronic exposure to gas supersaturation is lacking, but Colt and Orwicz (1991), report the following species to be sensitive: lake trout (*Salvelinus namaycush*), striped bass (*Morone saxatilis*), Atlantic salmon (*Salmo salar*) and mullet (*Mugil cephalus*). Eggs and sac fry are less sensitive than feeding fry. Lacking species and life-stage information, a hatchery criteria of 10 mm Hg is suggested for ΔP. This criteria will protect the most sensitive life stages of certain species, older fish, and fish in deep production systems. A higher criteria may be used in pure oxygen systems with nitrogen-to-oxygen ratios less than the atmospheric value of 3.77 (based on partial pressure).

FACTORS CAUSING GAS SUPERSATURATION

Gas supersaturation can be produced by a number of physical and biological processes. Colt (1987) identified eight mechanisms by which gas supersaturation can occur. Supersaturation may be caused by any one or a combination of these mechanisms.

Water Heating. Gas solubility decreases with increasing temperature. For example, the solubility of nitrogen is 18.14 and 16.36 mg/L at 10 and 15°C, respectively (Colt 1986). Therefore, if saturated water at 10°C is heated without letting the excess gas escape, the nitrogen gas then becomes saturated. The same can be said for other gases in water in equilibrium with the atmosphere.

Ice Formation. As water cools, gas solubility increases, and negative ΔP values will be produced unless gas is transferred into the water. As ice is formed, the dissolved gases are expelled and concentrated in the remaining water. As ice forms on the surface of a body of water, gas transfer to the atmosphere is prevented. Mathias and Barcis (1985) demonstrated the formation of lethal dissolved gas levels under the ice in shallow Canadian lakes. The magnitude of the resulting ΔP is dependent on the ice to total lake volume ratio, temperature at the time of ice formation, final temperature, and DO concentration.

Mixing Waters Of Different Temperatures. The variation of gas solubility is not linear (see Chapter 2). When waters of different temperatures are mixed, gas supersaturation can result, even if the waters are initially at equilibrium. However, relatively high water temperature differences are required to produce significant ΔP values.

Air Entrainment. Gas supersaturation can potentially be produced any time that air and water are in contact at pressures greater than atmospheric. Typical examples are water flowing over a waterfall or discharging from an open pipe and plunging into a receiving pool. The resulting ΔP of the receiving water depends on the depth of bubble submergence, the volume of air entrained and the degree of mixing. Values of ΔP typically range from 18–44 mm Hg per meter of submergence (Colt and Westers 1982).

The same mechanism will produce gas supersaturation when air is drawn into a pressurized water system through leaks on the suction side of the pump or through inadequately submerged intake structures. Submerged aeration devices can also cause gas supersaturation under certain conditions. Seawater has a higher surface tension than freshwater, causing a more rapid rate of dissolution of air and oxygen than in freshwater. Thus, small air leaks in marine systems can result in higher ΔP values than in similar freshwater systems (Bouck and King 1983).

Photosynthesis. Algae and vascular aquatic plants produce oxygen during the day and consume oxygen at night or during periods of low solar intensity. The net production is dependent on algae density, solar radiation, and water turbidity (Romaire and Boyd 1979). Water mixing also plays an important role. Gas supersaturation may be produced in the upper water layers through photosynthetic production, resulting in higher ΔP values. Oxygen is normally transferred to the atmosphere during the day, but during periods of intense solar radiation and little mixing by the wind, high levels of DO may accumulate. Various water blending devices are commercially available that can be used to break up thermal and oxygen stratification and distribute oxygen more evenly through the water column.

Pressure Changes. At any given *TGP*, a decrease in *BP* will increase the ΔP. The passage of storm fronts can cause changes in ΔP ranging from +5 to −17 mm Hg. More significant decreases occur when aquatic animals are transported by air freight. Modern jet aircraft fly at approximately 10,000 m (*BP* = 199 mm Hg) but pressurize the cabin and cargo storage areas to about 3,000 mm Hg (*BP* = 526 mm Hg)(Colt 1986). The variation of BP with elevation and the resulting ΔP are shown in Table 11-6.

Table 11-6. Variation of barometric pressure with elevation and resulting ΔP.

Elevation m (ft)	Barometric pressure mm Hg (psi)	ΔP mm Hg (psi)
−500 (−1,640)	806 (15.58)	−46 (−0.89)
0	760 (14.69)	0
500 (1,640)	716 (13.84)	44 (0.85)
1,000 (3,280)	674 (13.03)	86 (1.66)
1,500 (4,920)	634 (12.26)	126 (2.44)
2,000 (6,560)	593 (11.46)	167 (3.23)
2,500 (8,200)	560 (10.83)	200 (3.87)
3,000 (9,840)	526 (10.17)	234 (4.52)
3,500 (11,480)	493 (9.53)	267 (4.58)
4,000 (13,120)	462 (8.93)	298 (5.76)
5,000 (16,400)	405 (7.83)	355 (6.86)
6,000 (19,680)	354 (6.84)	406 (7.85)
7,000 (22,960)	308 (5.95)	452 (8.74)
8,000 (26,240)	267 (5.16)	493 (9.53)
9,000 (29,520)	231 (4.47)	529 (10.23)
10,000 (32,800)	199 (3.85)	561 (10.84)

Source: Colt and Watten (1988).

Physiological Changes. Gas bubbles occasionally form in the swim bladders or in the eyes of fish. The sensitivity of a particular species to gas supersaturation depends on the partial pressures in these organs (Colt 1987). Detailed physiological data is lacking; however, Wittenberg and Wittenberg (1974) reported that the maximum $P(O_2)$ in the retina ranges from 400 to 1,300 mm Hg. The animals would have to submerge to depths ranging from 3–10 m to compensate for these partial pressures.

Bacterial Activity. Bacterial activity can increase or decrease the ΔP. H_2S and CO_2 production will have little effect, O_2 and CH_4 have intermediate effects and H_2 can have tremendous impact on the ΔP. Bacteria can seriously alter gas levels as water passes through soil and in the bottom sediments of water bodies.

EFFECTS ON FISH

The lower lethal DO concentration depends on fish species, life stage, exposure time, environmental conditions, size, and physiological condition. Most warm-water species can survive long periods at DO concentrations of 2–3 mg/L (Boyd 1982; Stickney 1979), but cold-water species require a minimum of 4–5 mg/L (Liao 1971; McLarney 1984). A minimum criteria of 6.0 mg/L is recommended for all juvenile fish and crustaceans (Huguenin and Colt 1989). Practically all species can survive for short periods at DO concentrations less than optimum, but feeding and growth is poor, and they are more susceptible to infectious diseases.

The maximum sustained oxygen uptake rates of fish are 10–100 times less than that of other animals of comparable size (Tucker and Robinson 1990). The metabolic rate (rate of oxygen consumption) of fish is limited by low DO availability and also by such parameters as size, species, activity, condition, daily ration, and water temperature (Boyd and Watten 1989; Smart 1981).

Oxygen requirements for crustaceans are roughly the same as for finfish, However, bivalves may be able to tolerate lower DO concentrations for extended periods (Huguenin and Colt 1989). Vitually all species can survive lower DO concentrations for short periods at lower water temperatures. Species DO requirements are discussed in greater detail in Chapter 2.

Oxygen consumption rate increases for increasing water temperature. In general, a 10°C rise in temperature approximately doubles the respiration rate of aquatic animals (Boyd 1982). The effects of water temperature and fish size on oxygen consumption rates for channel catfish are shown in Table 11-7. The relationship between water temperature and oxygen consumption is illustrated in Figure 11-2.

Equations are available by which one can estimate the oxygen consumption of certain species of fish. These equations should be used only *as a guide*. Approximate estimates of fish respiration can be obtained with these equations.

The following relationship was developed for channel catfish (Boyd et al. 1978):

Table 11-7. Oxygen consumption rates (lb oxygen/lb of fish/hour) for channel catfish of different weights.

Temperature	Average fish weight (lb)					
°F (°C)	0.05	0.1	0.25	0.5	1.0	1.5
35 (1.67)	50	48	42	35	25	22
50 (10.0)	92	88	77	63	47	41
65 (18.3)	167	160	140	114	86	75
80 (26.7)	326	311	254	224	168	147
95 (35.0)	589	365	493	405	306	267

Source: Tucker and Robinson (1990) with permission.

$$\log O_2 \text{ consump} = -0.999 - 0.000957W + 0.0000006W^2 + 0.0327T + 0.0000087T^2 + 0.0000003WT \quad (11\text{-}20)$$

where O_2 consump = oxygen consumed by fish, mg O_2/g fish/hr; W = average weight of fish g; and T = water temperature, °C. The data used for preparing this equation ranged from 2 to 1,000 g for W, and T ranged from 24 to 30°C. The correlation coefficient for the equation was 0.99.

Equation 11-21 is a general equation that can be used for estimating the DO consumption of warm-water fish at 20–30°C (Schroeder 1975; Romaire et al. 1978):

$$Y = 0.001\ W^{0.82} \quad (11\text{-}21)$$

where Y = oxygen consumption per fish, mg O_2/hr; and W = fish weight, g.

An equation for estimating the respiration of rainbow trout (*Salmo gairdneri*) was developed by Muller-Feuga et al. (1978) as follows:

$$O_2 \text{ consump (mg } O_2\text{/kg fish/hr)} = A \times W^B \times 10^{CT} \quad (11\text{-}22)$$

where W = fish weight, g; T = water temperature, °C; A = 75 (for T = 4–10°C) or 249 (for T = 12–22°C); B = −0.196 (for T = 4–10°C) or −0.142 (for T = 12–22°C); and C = 0.055 (for T = 4–10°C) or 0.024 (for T = 12–22°C).

Fivelstad and Smith (1991) developed the following equation for Atlantic salmon (*Salmo salar* L.) cultured in single-pass land-based seawater systems:

$$M = 10^{-0.841}\ P^{-0.261}\ T^{1.378} \quad (11\text{-}23)$$

where M = oxygen consumption rate, mg/kg fish/min; P = mean wet weight, kg; and T = water temperature, °C.

The feed/oxygen consumption relationship is often more convenient to use than an equation. Experience has demonstrated that rainbow trout

Figure 11-2. Effect of water temperature on oxygen consumption of fish (reproduced from Boyd (1990)).

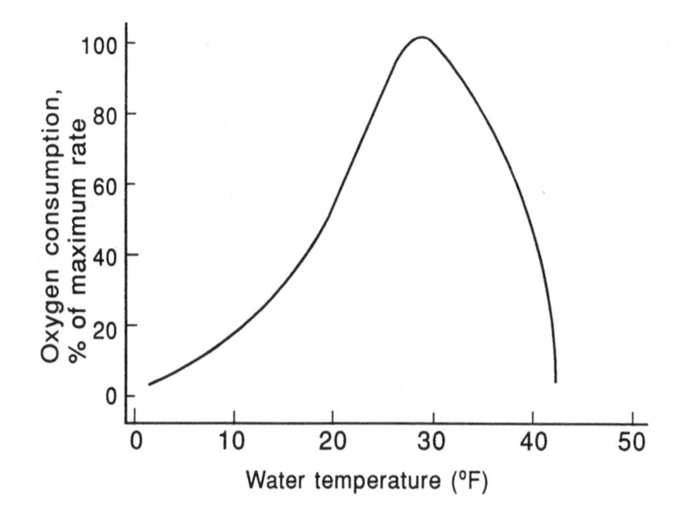

produced in raceways consume about 0.2 kg O_2 for each kg of feed (Colt and Huguenin 1989). This value is used to establish water flow requirements to maintain DO levels in flow-through systems. Fivelstad (1988) found that oxygen is usually the first limiting factor for salmonid culture in single-pass land-based seawater systems.

DISSOLVED OXYGEN DYNAMICS IN PONDS

The dynamics of dissolved oxygen in fish ponds is very complex. The decision of when and where to check DO in ponds varies among fish farmers, but experience dictates that DO be checked several times daily. Routine oxygen analysis can help to determine changing oxygen patterns and predict potential problems before they develop. The frequency of monitoring DO will depend on experience and being able to recognize certain signals and responses when oxygen problems are developing. New producers may want to check DO more frequently until they gain experience.

Dissolved oxygen measurements should be taken at two sites on opposite ends of ponds larger than 0.4 ha (1 ac) since concentrations can vary considerably within the pond. Most experienced farmers take routine measurements about 0.3 m (1 ft) below the surface. DO measurements should be made at least three times daily during the warmer months: at dawn, at dusk, and about four hours after dusk (Tucker and Robinson 1990). Additional measurements are then taken throughout the night as needed. Daily records of the DO changes in each pond should be maintained so that trends can be observed. DO should also be checked several times during the day during hot weather and during cloudy, overcast days since oxygen depletions can rapidly occur if photosynthesis is poor. It is especially important to check DO after treating ponds with herbicides or medications since these substances can lower photosynthetic activity. Oxygen concentrations should be checked whenever it is suspected that the fish are stressed as concentrations can lower unexpectedly.

Oxygen problems are less frequent during the late fall, winter, and early spring months when water temperatures are cooler. Fish standing crops are normally lower, and feeding rates are reduced. Many producers do not measure DO during the winter. However, since DO dynamics is a function of temperature, depletions can occur during periods of unseasonably warm weather during winter months. Tucker and Robinson (1990) recommend measuring DO concentrations at least once daily in ponds, even during the cooler months.

Boyd and Watten (1989) presented an equation by which the DO concentration may be computed in channel catfish ponds. This is shown by Equation 11-24.

$$DO_2 = DO_1 \pm DO_{df} + DO_{ps} - DO_{pr} - DO_{fr} - DO_{br} \quad (11\text{-}24)$$

where DO_2 = dissolved oxygen at time 2; DO_1 = DO at time 1; DO_{df} = DO gained or lost by diffusion; DO_{ps} = DO gained by photosynthesis; and DO_{pr}, DO_{fr}, and DO_{br} = the oxygen lost by planktonic respiration, fish respiration, and benthic respiration, respectively.

Predicting Dissolved Oxygen Concentrations

Equation 11-24 can be used to calculate DO concentration when all of the equation components are known. In reality, however, it is difficult to calculate fish respiration in ponds accurately. Every pond has different standing crops of fish, different densities of algae, and the sediment respiration is different. Therefore, respiration rates vary from pond to pond. Since DO decline at nighttime is the fish farmer's most important water quality management problem (Tucker and Robinson 1990), the key to successful DO management is the ability to identify those ponds where supplemental oxygen may be required before problems get critical.

Boyd et al. (1978) developed a computer

model based on an equation describing the components that contribute to the nighttime loss of DO. The equation has the form:

$$DO_t = DO_{dusk} \pm DO_{df} - DO_{pr} - DO_{fr} - DO_{br} \quad (11\text{-}25)$$

where DO_t = DO concentration after t hours of darkness; and DO_{dusk} = DO concentration at dusk. The gain or loss of oxygen by diffusion can be estimated from the DO concentration at dusk and Table 11-8.

Respiration by the planktonic community can be calculated from Equation 11-26 as follows (Boyd et al. 1978):

$$\text{Oxygen consumption (mg/l)} = 1.006 - 0.00148(COD) - 0.0000125(COD)^2 + 0.0766T - 0.00144T^2 + 0.000253CT \quad (11\text{-}26)$$

where (COD) = chemical oxygen demand, mg/L; and T = water temperature, °C. Data used in preparing the equation ranged from 20–160 mg/l for (COD) and from 20–30°C for T. The correlation coefficient was 0.92.

Benthic respiration can be taken as a constant value of 61 mg $O_2/m^2/hr$ (Mezainis 1977). This respiration rate represents a DO reduction of 0.72 mg/l over 12 hours of darkness for a 1-m deep pond, provided the entire pond bottom is aerobic.

Equation 11-26 provides highly accurate estimates of DO concentrations at dawn, but is probably too complex for general use on commercial fish farms (Boyd 1990). Chemical oxygen demand can be estimated from Secchi disk visibility, and there is usually a good correlation between average fish weight and total weight of fish in ponds. A series of tables were prepared from these data that can be used by fish farmers to predict dawn DO concentration from water temperature, DO concentration at dusk, total weight of fish in a pond, and Secchi disk visibility. An example of these tables is shown in Table 11-9. Other tables are published in Boyd (1990).

A more simple approach is to plot the DO concentration at dusk and another two or three

Table 11-8. Gains (+) and losses (−) of dissolved oxygen (DO) because of diffusion during the night (approximately 12 hours of darkness) in ponds with different DO saturation values at dusk. The values represent the average change in the water column for a pond of 1-m depth.

DO concentration at dusk (% of air saturation)	Gain or loss of DO during the night (mg/L)	DO concentration at dusk (% of air saturation)	Gain or loss of DO during the night (mg/L)
50	+1.69	160	−1.64
60	+1.49	170	−1.82
70	+1.18	180	−1.98
80	+1.00	190	−2.11
90	+0.77	200	−2.37
100	+0.44	210	−2.42
110	+0.16	220	−2.54
120	0.18	230	−2.67
130	−0.55	240	−2.76
140	−0.94	250	−2.91
150	−1.48		

Source: Boyd (1990) with permission.

Table 11-9. Minimum Secchi disk visibility (cm) to maintain a dissolved oxygen concentration above 2 mg/L at dawn in a 1-m deep pond containing 1,000 kg of channel catfish per ha.

Temperature (°C)	DO concentration at dusk (mg/L)									
	2	3	4	5	6	7	8	9	10	11
20	37									
21	58	26								
22	79	42	21							
23	90	58	32	16						
24	90	69	53	37	21					
25	90	79	53	37	21					
26	90	85	63	48	32	16				
27	90	90	69	53	37	26				
28	90	95	74	58	45	32	21			
29	90	95	79	63	53	40	29	18		
30	90	95	85	69	58	45	34	26	16	
31	90	95	87	74	63	50	40	32	21	
32	90	95	90	79	66	55	45	37	29	18

Source: Romaire and Boyd (1978).

hours later on a graph, as shown in Figure 11-3. A straight line drawn through the two points can be used to estimate the DO concentration at dawn (Boyd et al. 1978). This procedure is reliable and is popular with fish farmers because of its simplicity. The success of Equation 11-25 in predicting DO decline results from the large contribution of planktonic respiration and the accuracy for calculating planktonic respiration. The abundance of plankton, especially phytoplankton, is regulated primarily by feeding rate, the more feed, the more plankton produced (Boyd 1982).

Daily Dissolved Oxygen Changes

Dissolved oxygen concentrations in fish ponds usually follow diurnal cycles (Boyd 1982, 1990). During daylight hours DO concentration rises as photosynthetic production exceeds respiration losses. The DO concentration usually reaches a peak in early afternoon (Figure 11-4). The reverse is true at night and during periods of cloudy, overcast weather.

Effects of Phytoplankton

Photosynthesis is the major source of oxygen in fish ponds. Phytoplankton produce oxygen by photosynthesis during the day but are the major consumers of oxygen at night. The magnitude of daily changes in DO is influenced primarily by light intensity and phytoplankton density. Phytoplankton density continually waxes and wanes during a 24-hour day. Figure 11-4 illustrates how changes in phytoplankton density produce different oxygen levels. The density of the bloom restricts light penetration, and less oxygen is produced in deeper waters. Figure 11-5 demonstrates the effect of bloom density with respect to DO and water depth.

The phytoplankton population is composed of many different species of algae. Under given conditions, one particular species may predominate, usually a species of blue-green algae. If this species dies for some reason, DO in the pond may suddenly plunge, causing high mortalities. Certain pharmaceuticals, used to control aquatic weeds or fish diseases, may suddenly kill algae, causing oxygen depletions.

Wind Effects. When water is supersaturated with oxygen, oxygen diffuses from the water into the atmosphere. The reverse is true when water is undersaturated with oxygen. Wind greatly influences the rate at which oxygen diffuses to and from the atmosphere. Winds blow-

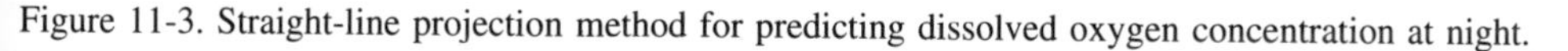
Figure 11-3. Straight-line projection method for predicting dissolved oxygen concentration at night.

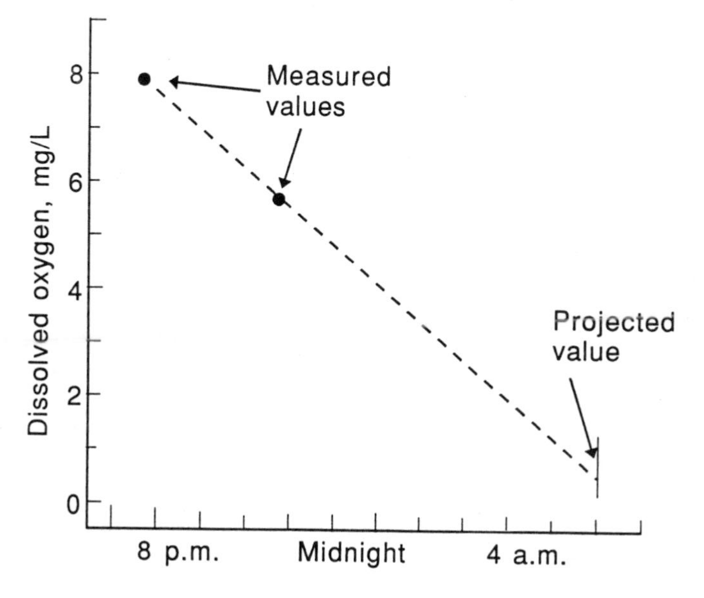

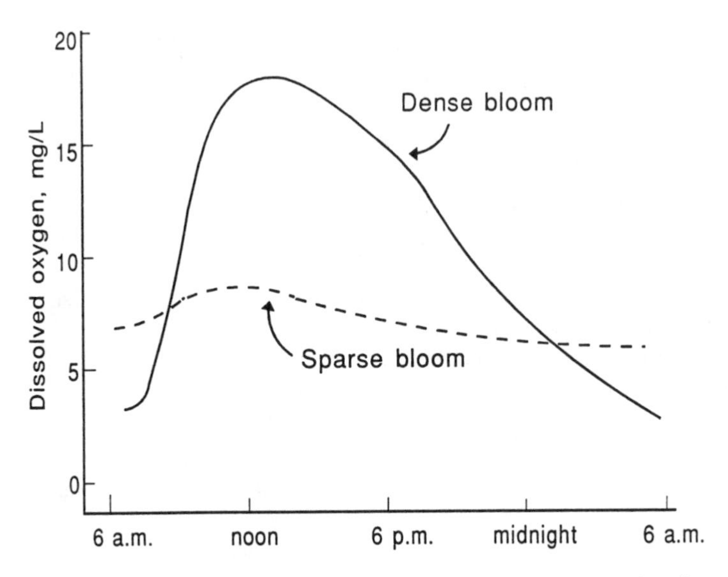

Figure 11-4. Changes in dissolved oxygen concentrations over a 24- hour period in pond surface waters with varying phytoplankton densities (reproduced from Tucker and Robinson (1990) with permission).

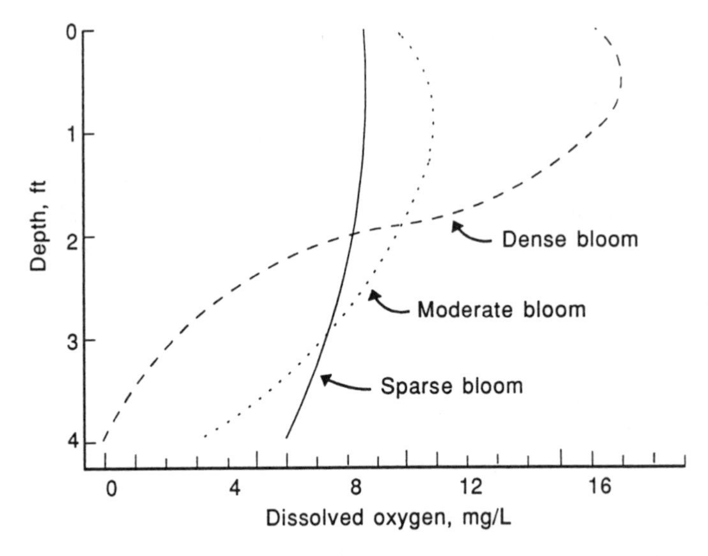

Figure 11-5. Changes in dissolved oxygen concentrations with depth in calm ponds having varying phytoplankton densities (reproduced from Tucker and Robinson (1990) with permission).

ing over a 24-hour period dampen the magnitude of oxygen fluctuations (Figure 11-6). Wind-induced turbulence also breaks up thermal and DO stratification in ponds. Consistently breezy conditions can keep DO levels high in shallow ponds, but sudden, strong winds on hot, calm days can cause turnovers, bringing low DO bottom waters to the surface, causing stress and/or mortalities. The consequences of turnovers are more severe in deep ponds (Tucker and Robinson 1990).

Temperature Effects. Dissolved oxygen concentrations in ponds change little during

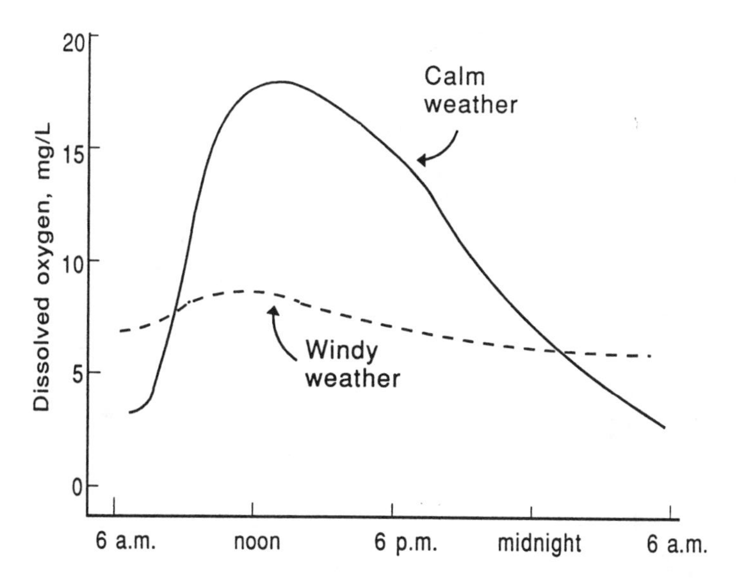

Figure 11-6. Effect of windy weather on dissolved oxygen concentrations in ponds (reproduced from Tucker and Robinson (1990) with permission).

cold weather since the rates of photosynthesis and respiration are decreased. Problems with low DO are rare when water temperature is below 13°C (55°F) and are common when water temperature gets above 23°C (75°F) (Tucker and Robinson 1990).

Effects of Cloudy Weather. Clouds reduce the amount of sunlight reaching ponds; therefore, photosynthesis is reduced. However, respiration rates remain unchanged, unless cloudy weather is accompanied by significant temperature changes. Extended periods of cloud cover during warm weather can result in oxygen depletions, as illustrated in Figure 11-7.

Feeding Rates. The density of phytoplankton blooms is directly correlated with feeding rates. A general relationship between maximum daily feeding rate and the amount of aeration required to keep fish alive is illustrated in Figure 11-8. Aeration is required when maximum sustained feeding rates exceed about 39 kg/ha/day (35 lb/ac/day) for catfish ponds (Tucker and Robinson 1990). These data are also indicative of feeding rates for other species cultured in ponds. When feeding rates remain below 39 kg/ha/da DO concentrations rarely fall to critical levels.

To increase fish yields and profits, many producers stock fish at high densities and feed at higher rates. The amount of aeration required thus increases along with energy and feed costs. A feeding rate of 56 kg/ha/day (50 lb/ac/day) is generally considered the maximum rate above which serious DO problems occur (Boyd and Watten 1989; Tucker and Robinson 1990). At some point the amount of oxygen consumed by the fish, plankton, and bottom sediments becomes so great that aeration is required for long periods nearly every night during warm weather. At very high feeding rates, besides problems with low DO, other factors limit fish production. The theoretical amount of fish production then becomes offset by slower growth rates, greater mortality, and higher costs for aeration equipment and energy consumption.

DISSOLVED OXYGEN DYNAMICS IN RACEWAYS

The rate of oxygen consumption by fish in single-pass raceway systems is, along with ammo-

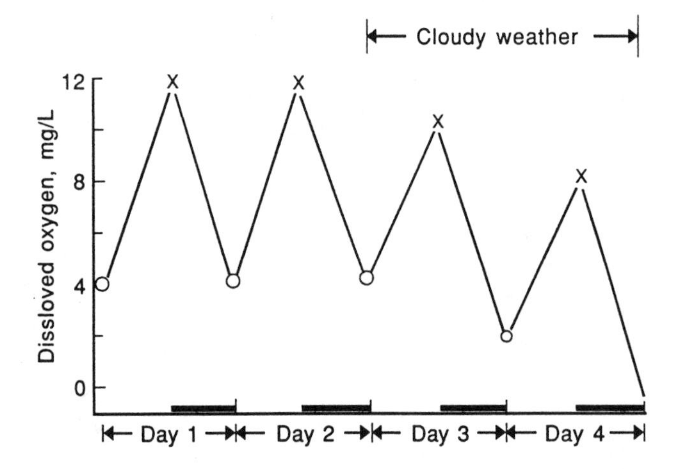

Figure 11-7. Effect of cloudy weather on concentrations of dissolved oxygen at early morning (O) and late afternoon (X) over a four-day period. Dark bars on the X-axis indicate periods of nighttime (reproduced from Tucker and Robinson (1990) with permission).

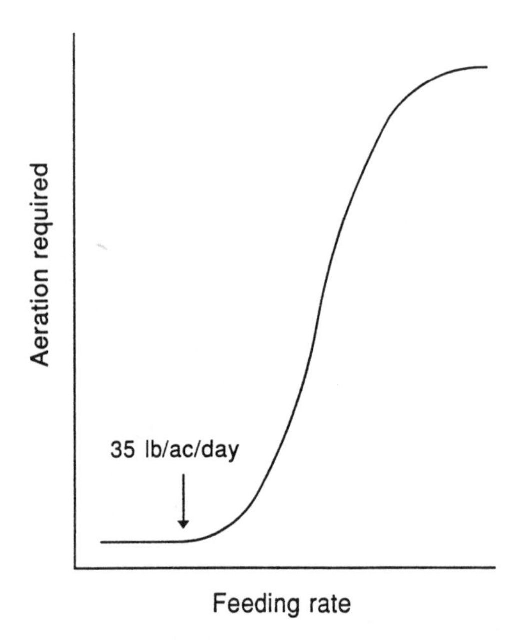

Figure 11-8. Relationship between maximum daily feeding rates (lb/ac/day) and amount of aeration (hours/night) required to keep fish alive in ponds (reproduced from Tucker and Robinson (1990) with permission).

nia criteria, one of the most important parameters to consider when designing these systems (Fivelstad and Smith 1991). Unlike in fish ponds, the dynamics of dissolved oxygen in raceways can be predicted with a high degree of accuracy (Boyd and Watten 1989). The retention time of water in single-pass systems is very short, therefore, the only influence on DO comes from the fish, as illustrated in Figure 11-9. Phytoplankton and sediment respiration are negligible. Also, oxygen diffusion is usually a very small amount. Dissolved oxygen supply is restricted to what is in the incoming water and what may be added by supplemental aeration within or between raceway units. Oxygen not consumed by the fish is lost in the effluent from individual units.

Fish respiration is correlated to feed by the proportionality constant K (g oxygen consumed per kg feed). The value of K equals 213 for channel catfish (Ray 1981) and 200–220 for salmonids (Willoughby 1968; Westers 1981). It should be kept in mind that the value of K is only an average. Values of K may increase by as much as 300% of these values just after feeding, during raceway cleaning, while fish are being handled, or following any stress-inducing situation (Boyd and Watten 1989). Given a K-value, influent DO, daily feed allotment, and the oxygen transfer rate of supplemental aeration equipment, the raceway effluent DO can be calculated:

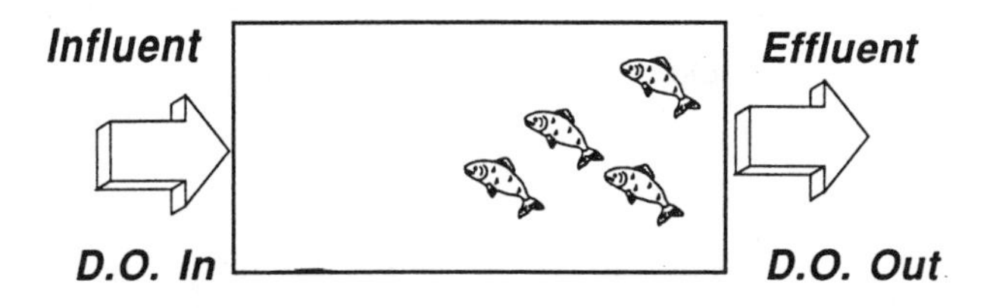

Figure 11-9. Mass balance of dissolved oxygen in a raceway unit.

$$DO_{out} = \frac{[(DO_{in} \times Q) + (OTR \times 24{,}000) - (F \times K)]}{Q} \tag{11-27}$$

where DO_{out} = effluent DO, mg/L; DO_{in} = influent DO, mg/L; Q = influent flow rate, m^3/day; F = total daily feed allotment, kg/day; K = grams of oxygen required per kg of feed; and OTR = oxygen added by supplemental aeration, kg O_2/hr. Oxygen transfer rate will vary with ambient conditions, as discussed in sections following.

The total daily feed allotment F is calculated as follows:

$$F = B\,R \tag{11-28}$$

where B = biomass of fish in rearing unit, kg; and R = ration in percent of body weight. Ration is related to feed conversion, fish length, and growth rate as follows (Boyd and Watten 1989):

$$R = \frac{300\,(FC \times \Delta L)}{L} \tag{11-29}$$

where FC = feed conversion (fish weight gain per amount of feed); ΔL = daily increase in fish length, cm/day; and L = fish length, cm. Feed conversion and fish growth data can be obtained from handbooks for certain species, for example Piper et al. (1982).

To avoid low oxygen problems in single-pass fish rearing units it is typically desirable to maintain the minimum effluent DO concentration at 6 mg/L. Knowing the influent and effluent DO concentrations, feed allotment F can be used to calculate flow requirement through a raceway culture unit to maintain the given oxygen criteria (Huguenin and Colt 1989):

$$Q_{w(oxy)} = \frac{K_1 \times F}{DO_{in} - DO_{out}} \tag{11-30}$$

or,

$$Q_{w(oxy)} = \frac{K_1 \times F}{DO_{available}} \tag{11-31}$$

where $Q_{w(oxy)}$ = water flow required for oxygen control, L/min.

Example problem 11-3. Determine the flow rate required to maintain an effluent DO concentration of 6 mg/L from a single-pass rearing unit containing 12,000 fish having a mean weight of 0.3 kg. The fish are fed 2.5% of their body weight daily, and the influent DO concentration is 12 mg/L.

Solution: The daily ration is found from Equation 11-28 as follows:

$$F = 12{,}000 \times 0.3 \times 0.025 = 90 \text{ kg/da}$$

The available DO is $DO_{in} - DO_{out} = 12 - 6 =$ 6 mg/L. Therefore, the required flow rate to required to maintain a DO concentration of 6 mg/L in the effluent is calculated from Equation 11-31 using $K = 200$:

$Q = (200)(90) \div 6 = 3{,}000$ L/min or 3 m^3/min.

Example problem 11-4. Note that, if DO_{in} were reduced, a higher flow rate would be required to maintain the given DO criteria. For example, if DO_{in} = 10.5 mg/L,

$$Q = (200)(90) \div 4.5 = 4{,}000 \text{ L/min or } 4 \text{ m}^3\text{/min.}$$

PRINCIPLES OF AERATION

There is no net transfer of oxygen between air and water if the water is at equilibrium with atmospheric oxygen. When water is undersaturated with oxygen, oxygen will transfer from air to water, and the reverse is true when water is supersaturated with oxygen. The driving force causing oxygen transfer is the difference in oxygen tension in the air and water. At equilibrium, the oxygen tension in air and water are the same, and there is no oxygen transfer. The oxygen deficit (OD) and the oxygen surplus (OS) can be expressed as (Boyd 1990):

$$OD = DO_s - DO_m \qquad (11\text{-}32)$$

and,

$$OS = DO_m - DO_s \qquad (11\text{-}33)$$

where DO_s = theoretical oxygen saturation concentration under given conditions; and DO_m = measured oxygen saturation concentration.

For static water, the net oxygen transfer depends upon OD or OS, the area of contact of the air-water interface, the temperature, and the time of contact (Haney 1954). Oxygen enters or leaves water at the interface between the air and water. Therefore, for the thin film of water in contact with the air, the greater the OD or OS, the faster oxygen will diffuse through this interface. Turbulence increases the rate of transfer by increasing the contact area of the air and water.

Theory of Gas Transfer

The dissolution of a gas in water involves four major steps, and each has the potential of being rate limiting (Wheaton 1977; Boyd and Watten 1989). The transfer process is illustrated in Figure 11-10. In step 1, oxygen moves from the bulk gaseous phase into the gas-liquid interface. In steps 2 and 3, the oxygen must diffuse through laminar gas and laminar liquid films, respectively. In step 4, the oxygen then enters the bulk liquid phase. Under normal conditions, gas transfer resistance occurs primarily in steps 2 and 3 (Boyd and Watten 1989). The transfer of highly soluble gases, such as ammonia, is restricted in the gas film while the transfer of less soluble gases, such as oxygen and nitrogen are restricted within the liquid film. In the latter case, the transfer rate is proportional to the differential between existing and saturated concentrations of a gas in solution. this relationship can be expressed by the Lewis-Whitman gas transfer model (Lewis and Whitman 1924):

$$\frac{dc}{dt} = \frac{D}{\Delta}\frac{A}{V}(C_s - C_m) \qquad (11\text{-}34)$$

where dc/dt = gas transfer rate, mg/L/hr; D = diffusion coefficient; Δ = the liquid film thickness; A = area of gas-liquid interface; V = volume of liquid into which the gas is diffusing; C_s = saturation concentration of the gas; and C_m = measured gas concentration at time t. Note that the driving force for Equation 11-34 is differential between theoretical saturated concentration and actual measured concentration of the gas. When C_s—C_m is positive, gas will transfer from the atmosphere into the bulk liquid, and when C_s—C_m is negative, gas will transfer from the liquid to the atmosphere. It is difficult to measure Δ and A, therefore, the ratios A/V and D/Δ are usually combined into a composite term called the *overall gas transfer coefficient* (K_La) such that

$$\frac{dc}{dt} = K_La\,(C_s - C_m) \qquad (11\text{-}35)$$

The overall gas transfer coefficient represents conditions in a specific gas-liquid contact system (Boyd 1990). The rate of gas transfer can be

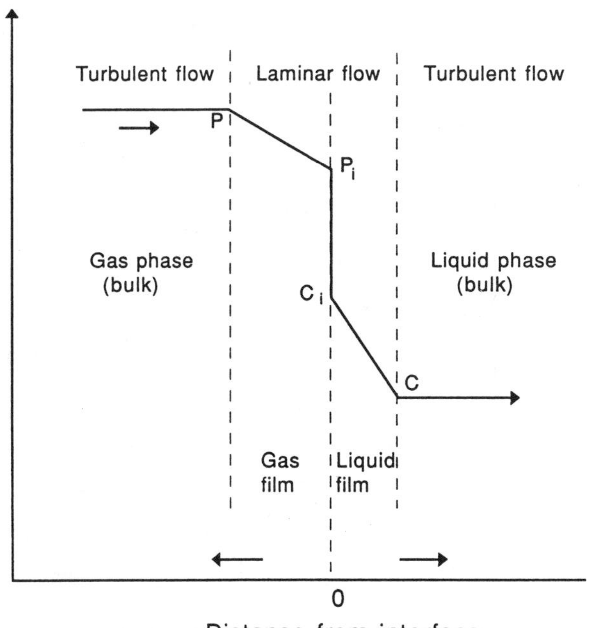

Figure 11-10. The four-step gas transfer process.

increased by reducing the thickness of the interface film (Δ), by increasing the surface area of contact (A) or by increasing the value of ($C_s - C_m$).

Turbulence and/or mixing is required for rapid transfer of oxygen to or from the atmosphere. All aerators (except for some pure oxygen aeration systems) increase the rate at which oxygen is transferred to water by creating turbulence and/or mixing. Turbulence is difficult to estimate, but it is possible to accurately the value of K_La empirically between two points. Integration of Equation 11-35 yields

$$K_La = \frac{ln(OD)_1 - ln(OD)_2}{t_2 - t_1} \qquad (11\text{-}36)$$

where K_La = overall oxygen transfer coefficient, hr^{-1}; OD_1 and OD_2 = oxygen deficit at times 1 and 2, respectively, mg/L; and t = time. K_La is calculated by plotting time versus the natural logarithm of the oxygen deficit. The slope of the straight line generated is K_La (Figure 11-11).

Temperature affects viscosity, which in turn affects surface tension and diffusion rate. Thus, for oxygen transfer, values of K_La can be corrected for temperature effects (Stenstrom and Gilbert 1981) by use of a *theta factor* (Θ). The temperature variation of K_La is:

$$(K_La)_T = (K_La)_{20}\ \Theta^{(T-20)} \qquad (11\text{-}37)$$

where $(K_La)_T$ = oxygen transfer coefficient at temperature T; $(K_La)_{20}$ = oxygen transfer coefficient under standard conditions; and T = temperature, °C. Theta typically varies from 1.016 to 1.047. In freshwater, a value of 1.024 is recommended for Θ. Slightly higher values for Θ are recommended for seawater, especially for temperatures above 20°C (Colt and Orwicz 1991).

For two specific gases, their K_La values and their molecular diameters are inversely proportional (Tsivoglou et al. 1965):

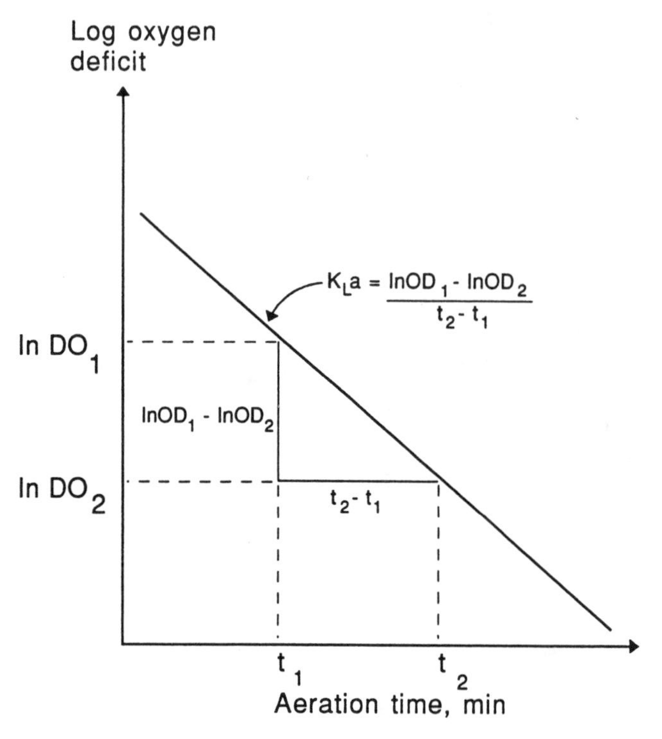

Figure 11-11. Graphical method for determining the overall oxygen transfer coefficient K_La.

$$\frac{(K_La)_1}{(K_La)_2} = \frac{d_2}{d_1} \quad (11\text{-}38)$$

where d_1 and d_2 = the diameters of the respective gas molecules. Thus, once oxygen K_La values are experimentally determined, these can be used to predict K_La values for other gases.

Aerator System Requirements

The addition of oxygen by aeration is one of the most common practices in aquaculture. Oxygen addition is usually required for two conditions: (1) To add oxygen to water already low in DO; and/or (2) to increase a system's carrying capacity. Aerator design is based on (1) The amount of oxygen needed; and (2) the minimum DO concentration (Huguenin and Colt 1989). The average and maximum daily oxygen demand for fed fish is equal to:

$$AOD = 0.20\ F \quad (11\text{-}39)$$

$$MOD = 0.29\ F \quad (11\text{-}40)$$

where AOD = average oxygen demand, kg/day or lb/day; MOD = maximum oxygen demand, kg/day or lb/day; and F = daily feed allotment, kg/day or lb/day from Equation 11-28). Aeration and water reuse can increase the carrying capacity in systems where ammonia is not limiting. However, due to high pH, ammonia may be more limiting than oxygen in some systems. Under such conditions, aeration and water reuse will result in increased disease and mortality problems.

Aerator Performance Tests

There are many types of mechanical aerating devices from which to choose. These devices are either electrically- or PTO-driven and are available in a broad range of sizes. Because of the confusing array of data in the literature describing aerator performance, commercially manufactured aerating devices are usually evaluated under a rigid set of standard conditions to determine their ability to transfer oxygen (APHA 1980; ASCE 1983; Brown and Baillod 1982). These tests are referred to as *standard tests* or *clean water tests*. Test results are normally included in the manufacturer's promotional literature, and it is the best way to compare two or more manufacturer's aerators under identical conditions. Standard tests were developed to eliminate many of the variables that affect tests under field conditions (in actual fish ponds). An accurate measurement of K_La is difficult in fish ponds since biological processes and complex circulation patterns (due to pond geometry) are continuously affecting test results. However, field-testing is valuable for use as a guide to give one an idea of how an aerator will perform in a fish pond.

Colt and Orwicz (1991) identified two types of testing procedures used to rate aerators: unsteady-state and steady-state procedures.

Unsteady-State Test Procedure. Standard tests for aerators are conducted in basins of clean tap water at standard temperature and pressure (20°C and 760 mm Hg). Some basins are as large as 3,000–6,000 m^3 and can be used to test aerators as large as 50–100 kW (67–134 hp) (Colt and Orwicz 1991). Basins are typically of concrete or cinder block construction. The water is first deoxygenated a sodium sulfite (Na_2SO_3) solution using cobalt chloride ($CoCl_2 \cdot 6H_2O$) as a catalyst (APHA 1980). The change in DO concentration is measured as the water is re-oxygenated with the aerator being evaluated. This procedure is termed *unsteady-state testing* since the amount of oxygen transferred and the DO concentration change during the test.

The theoretical requirement for deoxygenation is 7.88 mg/L of sulfite (as Na_2SO_3) per 1.0 mg/L of O_2 removed. The cobalt chloride should be used at a concentration of 0.10–0.50 mg/L (as Co). A slightly higher concentration may be required if the water temperature is below 10°C. The concentration of cobalt should never be allowed to fall below 0.10 mg/L. To ensure an adequate concentration, it is suggested that a cobalt addition equivalent to 0.10 mg/L be added to the test water for each test run. However, cobalt should be used cautiously since too much can affect the DO reading (ASCE 1983).

Chemical slurries are first made by mixing the respective chemicals with small amounts of tap water. The aerator is turned on and the chemical slurries are mixed until the basin DO drops to zero. The cobalt chloride catalyst is added to the basin water first, and the aerator should run for a minimum of 30 minutes to ensure complete mixing. The sodium sulfite solution is then splashed into the basin and mixed with the aerator. Sufficient sulfite should be added to depress the basin DO below 0.5 mg/L. Dissolved oxygen readings are then taken simultaneously at timed intervals (or at approximately every 1.0 mg/L increase in DO) while the DO increases to at least 90% saturation. The saturation concentration is normally reached when the run lasts for a period equal to the ratio of $6/K_La$ or the saturation concentration can be considered to be the concentration that remains unchanged for 15 minutes after the test time has reached $t = 5/K_La$. At least 10–15 measurements should be taken at equal time intervals. As a minimum, at least one run should be extended for each test condition (temperature, aerator operation, geometry) to obtain a DO concentration at saturation for the basin water (ASCE 1983).

The oxygen deficit (OD) is computed for each interval that DO was measured during reaeration from Equation 11-32:

$$OD = DO_s - DO_m$$

The natural logarithms of OD (Y) are plotted versus the aeration time (X), and the best fit line is computed with regression analysis. The oxygen transfer coefficient is computed using the points representing 10% and 70% oxygen saturation (Boyd and Watten 1989) as follows:

$$(K_La)_T = \frac{\ln OD_{10} - \ln OD_{70}}{(t_{70} - t_{10})/60} \qquad (11\text{-}41)$$

where $(K_La)_T$ = overall oxygen transfer coefficient (hr^{-1}) at temperature T, °C; OD_{10} and

OD_{70} = oxygen deficit at 10 and 70% saturation, respectively, mg/L; t_{10} and t_{70} = time DO reaches 10 and 70% saturation, respectively, minutes. The $(K_La)_T$ must then be adjusted to 20°C using Equation 11-37.

SAMPLING STATIONS. A minimum of four sampling points is recommended in the test basin. Sampling points are recommended at several depths (near the surface and bottom, and at middepth) within the basin. The sampling points should be located at least 0.6 m (2 ft) from the walls and floor of the tank. A greater number of sampling points may be needed for complex aeration systems (ASCE 1983).

VARIABILITY OF K_LA VALUES. Basin geometry and type of aeration equipment can significantly affect the value of K_La. A typical test basin is rectangular in shape with a depth of 1.0–1.5 m (3.2–4.8 ft). The power applied to unit volume ratio is often used to compare aeration systems. APHA (1980) recommends that this ratio be in the range of 0.01–0.04 kW/m^3 (0.05–0.20 hp/1,000 gal.). However, Boyd and Watten (1989) stated that reliable results can be obtained with power-to-volume ratios up to 0.1 kW/m^3. Accurate measurements of aerator power and water volume are necessary to obtain reliable results.

A K_La-value should be obtained for each sampling point for accuracy, and the mean K_La-value should be used to calculate the oxygen transfer rate. For each test run the K_La-values should not vary by more than ± 10% from the mean for each test run. For acceptable testing, at least 67% of the individual K_La values should be within ± 10% (ASCE 1983). Greater variations are considered to be indicators of poor mixing and/or analytical errors. Tests may vary by ± 15% in large test tanks (greater than 378.5 m^3 or 100,000 gal.). If the individual K_La plots are parallel, any deviation is due to the travel time between sampling points, unless all DO readings are taken simultaneously. However, nonparallel slopes indicate poor mixing.

STANDARD OXYGEN TRANSFER RATE. The overall oxygen transfer coefficient is used to estimate the *standard oxygen transfer rate* for an aerator:

$$SOTR = (K_La)_{20}\,(C_s)_{20}\,(V)\,(10^{-3}) \quad (11\text{-}42)$$

where $SOTR$ = standard oxygen transfer rate, kg O_2/hr; $(C_s)_{20}$ = DO concentration at saturation and 20°C, g/m^3 = mg/L; V = volume of water in test basin, m^3) and 10^{-3} converts g to kg. By definition, the SOTR is the amount of oxygen that an aerator will transfer to clean water at 0 mg/L DO and 20°C. The SOTR was referred to as N_o in the older literature (Colt and Tchobanoglous 1981). The value used for $(C_s)_{20}$ should first be corrected for barometric pressure with Equation 11-5. Standard oxygen transfer rate (SOTR) can be estimated in pounds of oxygen transferred per hour by multiplying Equation 11-42 by the factor 2.205 (Note: 2.205 lb = 1 kg).

STANDARD AERATION EFFICIENCY. The standard aeration efficiency, by definition, is an expression of the amount of oxygen transferred by the aerator per unit of energy consumption under standard conditions (0 mg/L, 20°C, and clean water).

Power input has customarily been reported in terms of power applied to the aerator shaft. However, if an aerator is driven by a belt, chain, etc. from an electric motor, the process of converting electrical power to mechanical power is not 100% efficient because of (1) The inefficiency of the electric motor; (2) the loss of energy between the motor and the aerator shaft; and (3) friction losses in the bearings. Therefore, all power is not transferred to the water for aeration purposes. A more realistic approach is to express standard aerator efficiency (SAE) in terms of power used (Boyd and Watten 1989). For electrically-powered aerators, SAE can be expressed as kilograms of oxygen transferred per kilowatt per hour (pounds of oxygen transferred per horsepower per hour) as shown by Equation 11-43:

$$SAE = \frac{SOTR}{\text{Power input}} \quad (11\text{-}43)$$

where SAE = standard aerator efficiency, kg O_2/kW-hr or lb O_2/hp-hr. The power input is expressed in kW or hp. In aerator tests a wattmeter can be used to measure wire power input to the electric motor. Where aerators are powered by internal combustion engines, the fuel consumed by the engine can be measured during the aerator test and the SAE expressed in terms of oxygen transferred per unit of fuel used per hour (kg O_2/L/hr or lb O_2/gal/hr).

The energy content of gasoline and diesel fuel is 2.26 kw-hr/L and 3.28 kw-hr/L, respectively (Jones and Alfred 1980). Therefore, SAE for internal combustion engine-powered aerators can be expressed in terms of kg O_2/kW-hr for direct comparison to electrically powered aerators.

STANDARD OXYGEN TRANSFER EFFICIENCY. The standard oxygen transfer efficiency is equal to the oxygen transferred divided by the mass flow rate of oxygen supplied to the aerator (Colt and Orwicz 1991):

$$SOTE = \frac{SOTR}{m} \tag{11-44}$$

where $SOTE$ = standard oxygen transfer efficiency, decimal fraction; and m = mass transfer of oxygen, kg/hr.

EXAMPLE OF AN AERATOR TEST. The following example shows how calculations of SOTR and SAE can be made from actual aerator test data. A standard test conducted on a 1.5-kW (2-hp) aerator yielded the data shown in Table 11-10. The test tank contained 37,000 gal. (140 m^3) of clean tap water. The test was run sufficiently long to determine that $(C_s)_{25}$ of the basin water was 6.8 mg/L. The data was plotted as shown in Figure 11-12. From the figure, using 20% and 80% saturation,

$$C_1 = DO \text{ at 20\% saturation} = 1.36 \text{ mg/L}$$

$$C_2 = DO \text{ at 80\% saturation} = 5.44 \text{ mg/L}$$

Thus, from Equation 11-41

Table 11-10. Standard test results for 2-hp aerator.

Time (min)	DO (mg/L)	ln $(C_s - C_t)$	Time (min)	DO (mg/L)	ln $(C_s - C_t)$
0	0.00	1.92	48	5.31	0.40
4	0.25	1.88	52	5.65	0.14
8	0.76	1.80	56	5.88	–
12	1.05	1.75	60	6.09	–
16	1.50	1.67	64	6.29	–
20	2.15	1.54	68	6.42	–
24	2.98	1.34	72	6.58	–
28	3.31	1.25	76	6.72	–
32	3.83	1.09	80	6.79	–
36	4.48	0.84	84	6.80	–
40	4.68	0.75	88	6.80	–
44	5.20	0.47	92	6.80	–

$$(K_La)_T = \frac{\ln(6.8 - 1.36) - \ln(6.8 - 5.44)}{53 - 11.2)/60}$$

$$= \frac{\ln 5.44 - \ln 1.36}{0.697}$$

$$= 1.98/\text{hr}$$

and from Equation 11-37,

$$(K_La)_{20} = 1.98/(1.024)^{25-20} = 1.76 \text{ /hr}$$

The SOTR is found from Equation 11-42:

$$SOTR = (1.76)(9.08)(140)(0.001)$$
$$= 2.24 \text{ kg } O_2/\text{hr } (4.94 \text{ lb } O_2/\text{hr})$$

Finally, the SAE is found from Equation 11-43:

$$SAE = 2.24/1.5 = 1.5 \text{ kg } O_2/\text{kW-hr}$$
$$(2.47 \text{ lb } O_2/\text{hp-hr})$$

Steady-State Test Procedure. Both the influent and effluent oxygen concentrations can be directly measured for a number of gravity aerators, for example, packed column aerators. The SOTR, SAE, and SOTE are easily computed.

$$SOTR = K\,Q\,(C_{out} - C_{in}) \tag{11-45}$$

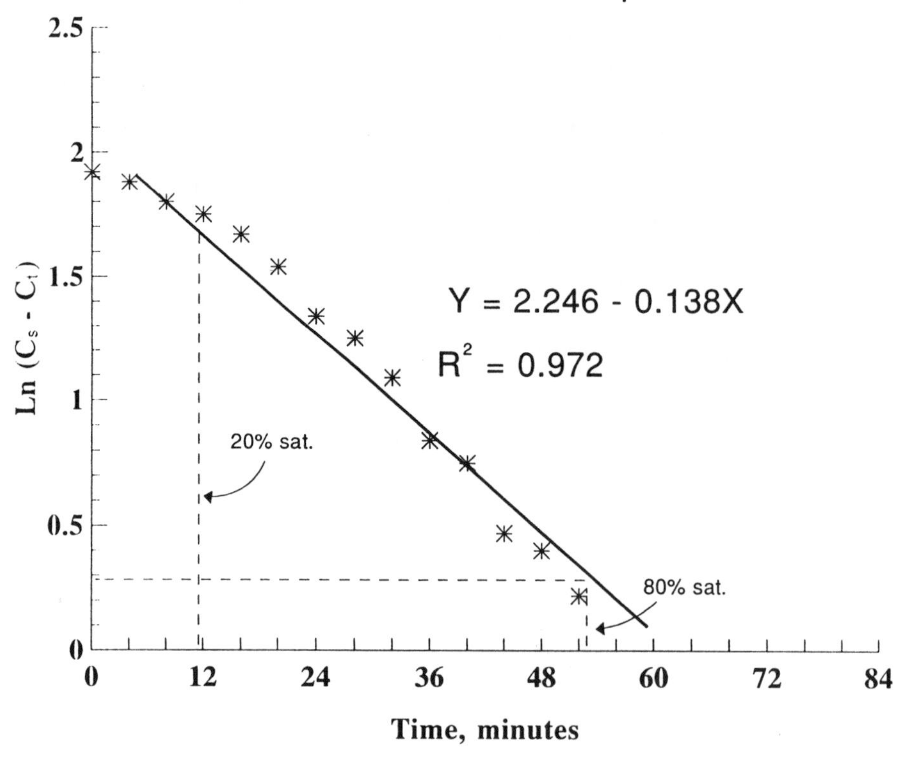

Figure 11-12. Plot of data for example problem.

where Q = water flow rate, m^3/s; C_{out} and C_{in} = DO concentration exiting and entering the aerator, respectively, mg/L; and K = conversion factor = 3,600. The SAE and SOTE are computed the same as for unsteady-state test conditions.

Rating Aerators Under Field Conditions

Clean water test results are valuable for comparing aerators under standard conditions, but they cannot be directly applied to oxygen transfer under field (pond) conditions because of a number of factors: pond water may be at some temperature other than 20°C; the DO concentration in pond water is greater than zero; and many factors either put oxygen in or remove oxygen from pond water during testing. The rate of oxygenation for tap water is usually greater than for pond water because of impurities in pond water, and clean water holds more oxygen at saturation than pond water. Ponds that contain fish, phytoplankton or rooted aquatic plants and bottom sediments are often referred to as *respiring systems*, and they must be treated quite differently than test basins. The rating of aerators in ponds requires the computation of a field oxygen transfer rate $(OTR)_f$, field aeration efficiency $(AE)_f$, and field oxygen transfer efficiency $(OTE)_f$.

For testing aerators in the field, alpha (α) and beta (β) factors must be determined. Alpha is the ratio of K_La for field conditions to K_La for clean water conditions:

$$\alpha = \frac{(K_La)_{20} \text{ pond water}}{(K_La)_{20} \text{ clean water}} \tag{11-46}$$

The value of α depends primarily on the concentration of surfactants in the water, which reduce the surface tension of the water and decrease bubble size in diffused aeration (Colt and Orwicz 1991). Alpha is usually determined in the laboratory by aerating containers containing equal amounts of pond and clean water for the same length of time at the same temperature. Plots of K_La are drawn for both the clean water and pond water samples, as previously described.

In production catfish ponds, Shelton and Boyd (1983) reported that α ranges from 0.66 to .07, with 0.94 considered an average in 43 channel catfish ponds in Alabama. Alpha values as low as 0.36 have been reported following feeding in recycle systems (Colt and Orwicz 991). The lower values appear to be related to the leaching of soluble compounds from feed. Alpha values are lower for a given water in packed column aerators. Salinity also has an effect on α. The transfer of both oxygen and air increases in seawater due to the decrease in bubble size (Bouck and King 1983).

Beta is a ratio of the oxygen saturation concentration (C_s') for field conditions to C_s for clean water:

$$\beta = C_s'/C_s \qquad (11\text{-}47)$$

The beta factor is influenced by dissolved solids, dissolved organics, and suspended solids Colt and Orwicz 1991). The magnitude of β in wastewaters typically ranges from 0.95 to 1.00 Stenstrom and Gilbert 1981). Shelton and Boyd (1983) reported that β factors ranged from .92 to 1.00 in channel catfish ponds, and the average was 0.98.

Once α and β have been determined, the field oxygen transfer rate can be calculated by:

$$(OTR)_f = \frac{\alpha (SOTR)_{20}\, \Theta^{T-20}\, (\beta C_s - C_m)}{9.092} \qquad (11\text{-}48)$$

where $(OTR)_f$ = field oxygen transfer rate, kg/hr; C_s = clean water DO saturation at test temperature and pressure; C_m = measured DO concentration in pond for test conditions, mg/L; and .092 is the value for C_s at 20°C and 760 mm Hg.

Table 11-11. Factors for rapid solution of the aerator performance equation[1].

Dissolved oxygen concentration (mg/L)	Water temperature, °C 10	15	20	25	30	35
0	0.71	0.71	0.72	0.74	0.76	0.79
1	0.64	0.64	0.64	0.65	0.66	0.67
2	0.57	0.56	0.55	0.55	0.55	0.55
3	0.51	0.49	0.47	0.45	0.44	0.43
4	0.44	0.41	0.38	0.36	0.33	0.30
5	0.37	0.34	0.30	0.26	0.22	0.18
6	0.31	0.26	0.21	0.17	0.12	0.06
7	0.24	0.19	0.13	0.07	0.01	0.0
8	0.17	0.11	0.04	0.0		
9	0.11	0.04	0.0			
10	0.04	0.0				
11	0.0					

[1] Multiply the oxygen transfer rate obtained for standard conditions (manufacturer's oxygen transfer rating) by the appropriate factor to get the oxygen transfer rate for pond conditions.

Source: Shelton and Boyd (1983).

Shelton and Boyd (1983) developed a table of factors (Table 11-11) that can be used for rapid solution of Equation 11-48 in channel catfish ponds. Knowing the water temperature and the oxygen concentration in a given pond, Equation 11-48 reduces to:

$$(OTR)_f = f \times (SOTR)_{20} \qquad (11\text{-}49)$$

where f is the factor from Table 11-11 and $(SOTR)_{20}$ is the standard oxygen transfer rate obtained from the aerator manufacturer. This method should also be applicable for obtaining estimates of $(OTR)_f$ in ponds for culturing species other than channel catfish.

Example problem 11-5. The XYZ Company claims to have a surface aerator with a $(SOTR)_{20}$ of 19 kg O_2/hr. What is the field oxygen transfer rate and the field aeration efficiency if the pond DO is 4.0 mg/L, the water temperature is 25°C and the power consumption is 8.60 kW?

Solution:

a. The factor from Table 11-11 at 25°C and 4.0 mg/L DO is 0.36. From Equation 11-49, the field oxygen transfer rate is

$$(OTR)_f = 0.36 \times 19 = 6.84 \text{ kg } O_2/\text{hr}$$

b. The field aeration efficiency is therefore

$$(AE)_f = 6.84 \div 8.6 = 0.8 \text{ kg } O_2/\text{kW-hr}$$

TYPES OF AERATORS

There are many different types of aerating devices that are used in aquaculture. These fall into the broad categories of mechanical aerators, gravity aerators, and air diffusion systems.

Mechanical Aerators

Mechanical aeration is achieved by imparting mechanical energy to water in order to break it up into droplets. Oxygen transfer is thus enhanced by increasing the air-water interface area. Mechanical aerators are either vertical- or horizontal-shaft aerators. In vertical-shaft aerators, oxygen transfer is achieved by exposing water droplets to the atmosphere, by turbulence of the water, and by air entrainment (Tchobanoglous and Schroeder 1985). Horizontal-shaft aerators transfer oxygen by surface turbulence, air entrainment, and by horizontal pumping. Mechanical aerators can be powered by either electric motors or internal combustion engines.

Many different hybrids of mechanical aerators exist. For example, pump sprayers use the mechanical energy of a pump to force water under pressure into a discharge pipe or nozzle. Turbine aerators use a combination of a turbine or impeller and an air source. The many different types of mechanical aerators are individually discussed.

Vertical Pumps. Vertical pump aerators consist of a submersible electric motor with an impeller attached to its shaft. The motor is positioned so that the shaft points vertically upward. The whole assembly is attached to a float and the impeller jets water into the air for aeration. A vertical pump aerator is shown in Figure 11-13. These aerators are manufactured in sizes ranging from 1 kW to over 100 kW, but those used for aquaculture are seldom larger than 3 kW (Boyd 1990).

Pump Sprayers. A pump sprayer consists of a high-pressure pump that discharges water through a series of orifices or slots in a pipe manifold (Figure 11-14). Many different orifice designs are used. Sizes typically range from about 7.5 to 15 kW. Pump sprayer aerators can have an attached power source or they can be PTO-driven. The speed of the device can be changed according to the speed of the engine.

Propeller - Diffuser Aerators. Propeller diffuser aerators, sometimes referred to as propeller-aspirator pumps (Boyd and Martinson 1984), consist of a rotating hollow shaft attached to an electric motor (Figure 11-15). A diffuser and an impeller are located at one end of the shaft and are submerged. The unit is supported at the water surface with a float assembly so that a hole in the opposite end of the shaft near the motor remains above the water surface. The motor usually operates at 3,450 rpm, and the impeller accelerates the water sufficiently to cause a pressure drop within the hollow tube that forces atmospheric air into the tube. The air passes through the diffuser and enters the water as fine bubbles that are mixed with the impeller. These aerators provide water circulation in addition to aeration.

Propeller-diffusers are manufactured in a variety of motor sizes from 0.37 to over 11 kW. In practice the float supports the motor and shaft at an angle with the water surface, and the angle can be adjusted to operate in either deep- or shallow-water conditions. Boyd and Martinson (1984) found that 30° was the best angle for maximum oxygen exchange.

Paddle Wheel Aerators. Paddle wheel aerators consist of a rotating hub to which a series of paddles are attached. The hub plus paddle assembly is called a *drum*, and as the drum ro-

Figure 11-13. A vertical pump aerator (photo by T. Lawson).

Figure 11-14. A pump sprayer aerator (photo by T. Lawson).

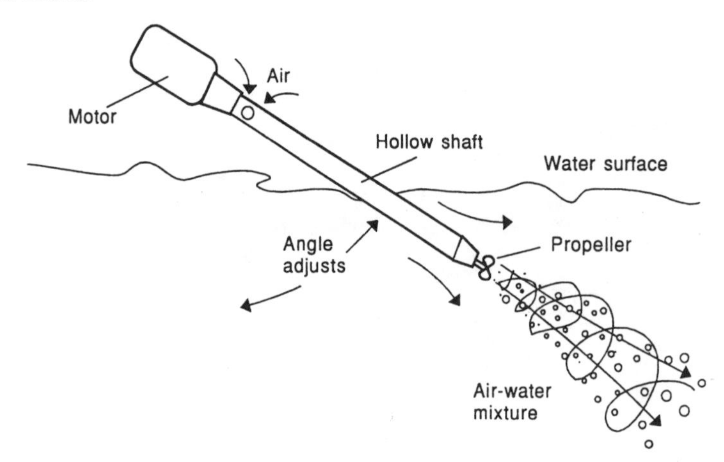

Figure 11-15. Illustration of a propeller-diffuser aerator.

tates, water is splashed into the air for aeration. The drum assembly and drive motor are supported at the water surface with floats, and the drum assembly on most aerators can be raised or lowered to adjust paddle depth. In addition to aeration, paddle wheels cause horizontal movement of the water. The paddles can be set deeper into the water to increase water circulation while sacrificing oxygen exchange. Decreasing paddle depth has the opposite effect. Drum rotational speed can be changed on many models. Each specific aerator model has a given paddle depth and rotational speed at which aeration is optimum. Most paddle wheel aerators operate with a paddle depth of 10–15 cm and a drum rotational speed of 80–90 rpm. Power requirements are on the order of 1 kW for each 50 cm of drum length (Boyd and Watten 1989).

Paddle width is not important for oxygen transfer (Boyd and Watten 1989). Fabrication of paddle wheels with wider paddles (10–15 cm) is easier than those with narrow paddles because fewer paddles have to be made and attached to the hub.

The most common paddle wheel design is the unit where the paddles are attached to the hub in a spiral configuration (Figure 11-16). This configuration dampens vibration forces since paddles are continuously in contact with the water. Aerators fabricated with the paddles attached to the drum in straight lines (Figure 11-17) frequently experience early bearing failure and other mechanical problems associated with the intermittent changes in torque as each row of paddles strikes the water.

Floats for paddle wheels consist of either aluminum tanks (Figure 11-18a) or foam blocks (Figure 11-18b). The foam blocks are usually coated with a protective paint that safeguards the block from deterioration, cracking, and chipping.

Paddle wheels powered by electric motors are manufactured in sizes ranging from 0.37 kW to 7.5 kW. Newer units are powered by diesel or gasoline engines.

Most paddle wheel aerators are constructed from steel with steel paddles and aluminum or styrofoam floats. However, some smaller paddle wheels , such as the popular Taiwanese model pictured in Figure 11-19, are fabricated with plastic paddle blades. Another model of U.S. manufacture is constructed almost wholly from polyurethane plastic (Figure 11-20).

PTO-Driven Aerators. Aerators powered by the PTO of farm tractors are manufactured for use in ponds where electricity is not available. They are also useful in emergency situations because of their portability. Therefore, the SOTR values for PTO-driven aerators are more important than the SAE values. A PTO-driven paddle wheel aerator is shown in Figure 11-21. Other types of PTO-driven aerators are discussed in Boyd and Watten (1989).

Figure 11-16. Aerator paddles arranged in spiral configuration around rotor (photo by T. Lawson).

Figure 11-17. Aerator paddles arranged in straight line on rotor (photo by T. Lawson).

Figure 11-18. Floats for paddle-wheel aerators: (a) Aluminum tanks; and (b) styrofoam blocks (photos by T. Lawson).

Figure 11-19. Taiwanese model paddle-wheel aerator (photo by T. Lawson).

Gravity Aerators

Gravity aerators are often referred to as *waterfall aerators* or *cascades*. Gravity aerators utilize the energy released when water loses altitude to transfer oxygen. Figure 11-22 illustrates the exchange of gases between the gas-liquid interface in a natural waterfall. Gravity fall is the most simple way to aerate in flowing-water aquaculture systems if a sufficient gradient exists. Man-made gravity aerators consist of weirs, splashboards, lattices, and screens. Transfer efficiencies of 1.2–2.3 kg O_2/kW-hr are possible under standard conditions (Chesness and Stephens 1971). They are used individually or in combination, or other devices, such as rotors or brushes, are used in combination with the above structures to enhance aeration.

Inclines. Figure 11-23 illustrates examples of gravity aerators including a simple weir (A), a weir and splashboard (B), a corrugated sheet (C), a corrugated sheet with holes (D), a lattice (E) and a cascade (F). These aerators are usually placed on an incline as shown. An almost infinite number of design combinations of these structures are available. These structures are used for aerating flowing water in raceways, but

Figure 11-20. Paddle-wheel aerator fabricated from polyurethane (photo by T. Lawson).

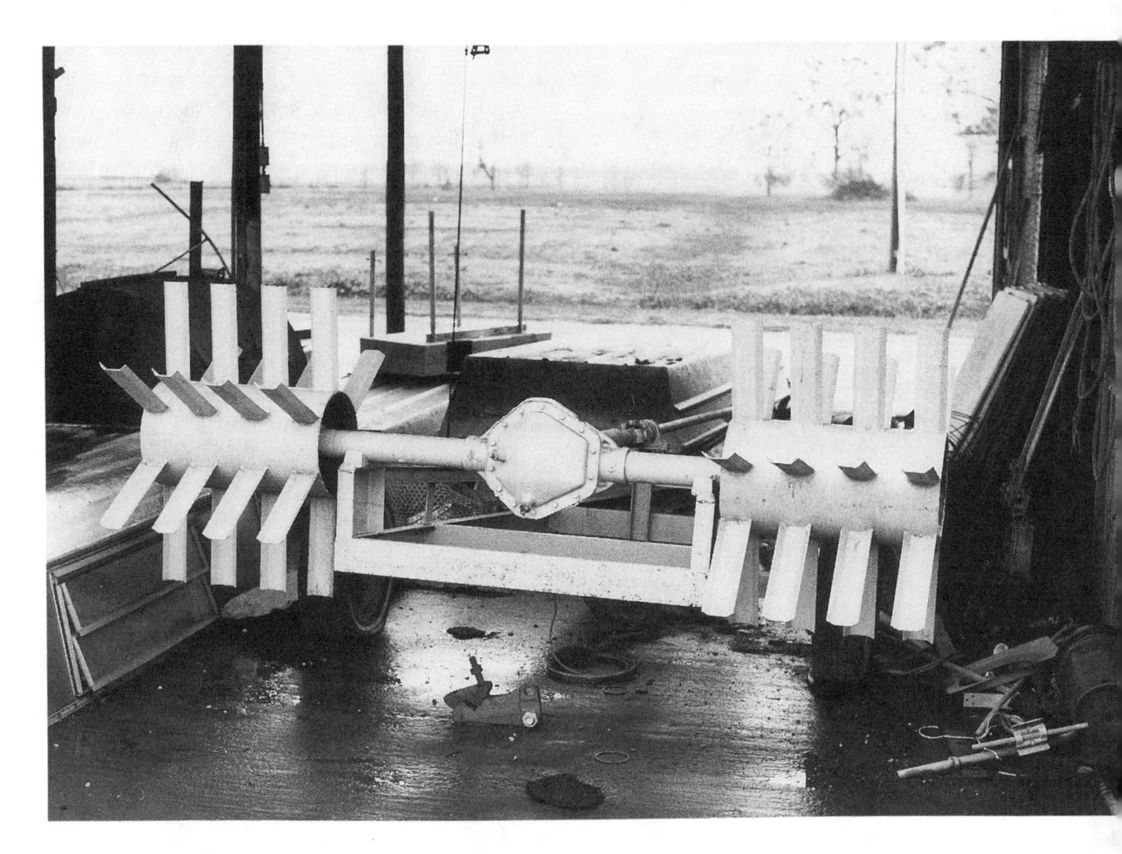

Figure 11-21. A PTO-driven paddle wheel (photo by T. Lawson).

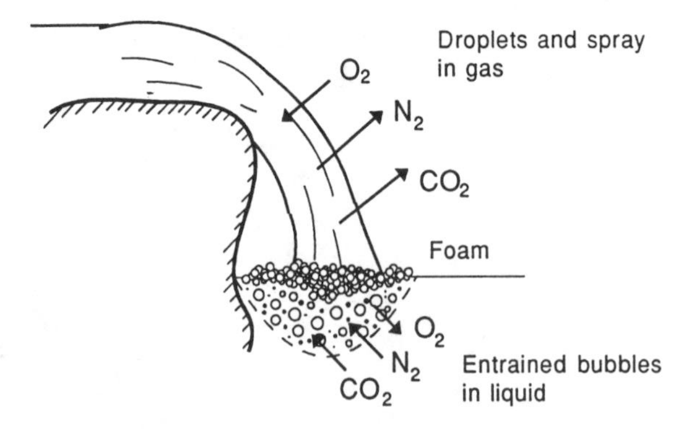

Figure 11-22. The gaseous exchanges occurring in a natural waterfall.

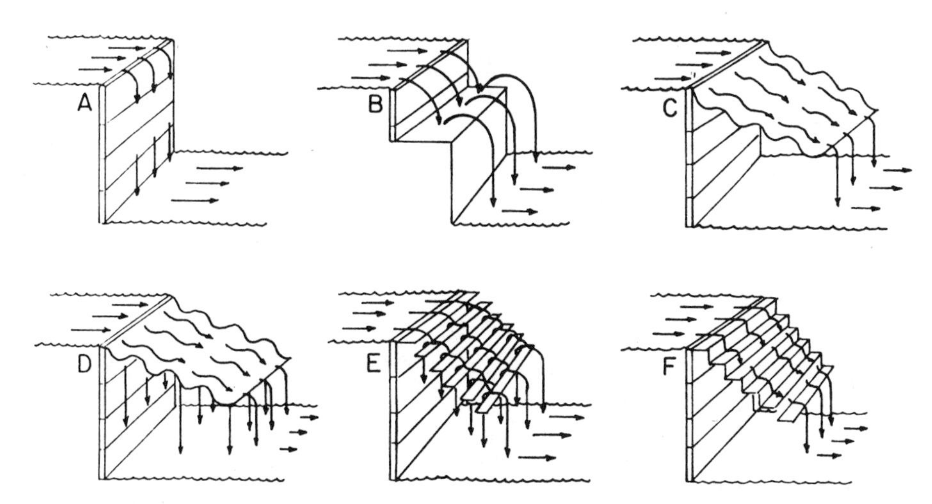

Figure 11-23. Various types of cascade aerators (reproduced from Soderberg (1982) with permission from the American Fisheries Society).

they may occasionally be seen aerating pond water. These aerators enhance oxygen transfer by spreading the water out so that it flows in a thin layer. Corrugations, perforations and screens are used to further enhance oxygen transfer by mixing the absorbed oxygen in the water and by exposing new water surfaces to the atmosphere.

Perforated Trays and Screens. As shown in Table 11-12, the lattice and perforated sheet gravity aerators are more efficient than simple weirs. A series of screens or perforated trays can be stacked one above the other to enhance oxygen transfer as the water falls from screen to screen. A typical spacing between screens or trays is 10–25 cm (4–10 in.) and 4–10 trays are common (Wheaton 1977). An 1-m tall aerator can usually aerate water to 70–80% saturation. For a given tray spacing, the amount of oxygenation increases with the number of trays, but usually at a decreasing rate (Boyd and Watten 1989). Likewise, oxygenation increases with tray spacing, but at a decreasing rate. These types of aerators are occasionally used to aerate water entering ponds, and can also be used in fish tanks, outdoor garden pools, and fountains.

A variation of the perforated tray aerator is the riser pipe with perforated trays or screens, as depicted in Figure 11-24, originally described by Chesness and Stephens (1971). Horizontal perforated aprons, screens, etc. are welded to a central riser pipe, and water discharging from the pipe is aerated as it cascades downward through the screens. Dissolved oxygen concentration is increased to 52% of saturation with a 0.6-m height of fall.

Figure 11-25 shows a gravity aerator constructed at Louisiana State University for aerating water discharging to a small crawfish pond. The tower assembly consists of a 7.6-cm (3-in.) diameter PVC riser pipe and six 63.5-cm (24-in.) diameter screens constructed from 1.27-cm (0.5-in.) mesh expanded metal. Screen spacing is 15.2-cm (6-in.). The riser pipe and screens are enclosed within an expanded metal cylinder. The total height of lift is 1.2-m (4- ft). Well water dissolved oxygen was increased from 0 mg/L to 85% of saturation (Lawson 1989). Larger versions of this type aerator are commonly used for aerating water in larger crawfish ponds.

Spray Aerators. The spray aerator is a variation of gravity aerator where the water is sprayed into the air, much like a fountain, and allowed to fall back to earth. Energy must be used in the form of a pump to force the water into the air.

Table 11-12. Selected data on measured efficiencies of some gravity aerators over various distances of water fall.

Device and distance of water drop (cm)	Efficiency (%)
Simple weir	
22.9	6.2
30.5	9.3
61.0	12.4
Inclined corrugated sheet	
30.5	25.3
61.0	43.0
Inclined corrugated sheet with holes	
30.5	30.1
61.0	50.1
Splashboard	
22.9	14.1
30.5	24.1
61.0	38.1
Lattice	
30.5	34.0
61.0	56.2
Cascade	
25.0	23.0
50.0	33.4
75.0	41.2
100.0	52.4

Source: Soderberg (1982) with permission from the American Fisheries Society.

Aeration Efficiency. To evaluate and compare gravity aeration devices, the DO concentration of the water above and below the aerator can be measured directly. For these types of aerators the SOTR can be computed as follows (Colt and Orwicz 1991):

$$SOTR = 3.6\ Q_w\ (Cb - Ca) \qquad (11\text{-}50)$$

where Q_w = water flow rate, m^3/s; C_a and C_b = DO concentration above and below the aerator, respectively, mg/L. Recalling that the definition of SOTR requires that initial DO concentration be zero, the SOTR at one atmosphere (760 mm Hg) pressure cannot exceed

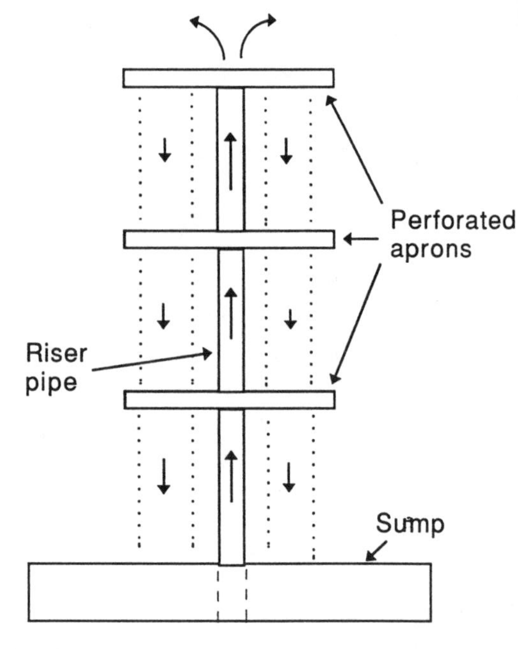

Figure 11-24. A type of perforated tray aerator.

$$SOTR = 3.6\ Q_w\ (9.092 - 0)$$

or,

$$SOTR = 32.7\ Q_w \qquad (11\text{-}51)$$

Thus, C_b is limited to a maximum of 9.092 mg/L (at 20°C).

Field oxygen transfer rate, field aeration efficiency, and field oxygen transfer efficiency are then defined by the following equations:

$$(OTR)_f = 3.6\ Q_w\ (Cb - Ca) \qquad (11\text{-}52)$$

where C_a and C_b = the actual DO concentrations above and below the aerator, respectively,

$$(AE)_f = \frac{(OTR)_f}{\text{Power input in kW}} \qquad (11\text{-}53)$$

and

$$(OTE)_f = \frac{(OTR)_f}{m} \qquad (11\text{-}54)$$

Figure 11-25. A small gravity aerator used for aerating crawfish ponds (photo by T. Lawson).

Field Aerator Effectiveness. For some types of atmospheric gravity aerators m cannot be measured and a term called the *aerator effectiveness* is used as a rating parameter. The equation describing aeration effectiveness for gravity aerators was given by Chesness and Stephens (1971) as:

$$E = \frac{C_b - C_a}{C_s - C_a} \times 100 \qquad (11\text{-}55)$$

where E = aeration effectiveness, %; and C_a, C_b, and C_s = DO concentration above the aerator, below the aerator, and at saturation, respectively, mg/L. A practical application of Equation 11-55 requires rearrangement so that the expected DO concentration below an aerator can be calculated knowing the aerator effectiveness:

$$C_b = E\frac{(C_s - C_a)}{100} + C_a \qquad (11\text{-}56)$$

Example problem 11-6. A simple weir is used to re-aerate water flowing from raceway 1 into raceway 2. The vertical drop between raceways is 30.5 cm (12 in.). If we know that the fish loading in raceway 1 is such that the effluent DO is 5 mg/L, what will be the DO concentration entering raceway 2 if the water temperature is 10°C?

Solution: At 10°C, C_s is 11.288 mg/L (from Table 2-4) and $C_a = 5$ mg/L. Table 11-12 shows the efficiency for a weir with a vertical drop of 30.5 cm to be 9.3%. Thus, from Equation 11-56 the influent DO concentration in raceway 2 is:

$$C_b = \frac{9.3\,(11.288 - 5)}{100} + 5$$

$$C_b = 5.58 \text{ mg/L}$$

Note that, if a lattice structure had been used instead of the weir, the efficiency would be 34%, and C_b would be increased:

$$C_b = \frac{34\,(11.288 - 5)}{100} + 5$$

$$C_b = 7.14 \text{ mg/L}$$

Mechanical Efficiency. Mechanical efficiency can be estimated using Equation 11-57 (Chesness and Stephens 1971):

$$ME = \frac{3600\,Q_w\,(C_b - C_a)}{\text{Power supplied by pump}} \qquad (11\text{-}57)$$

where ME = mechanical efficiency of aerator, kg O_2/kW-hr; Q_w = water flow rate, L/s; C_a = DO concentration above the aerator, mg/L; and C_b = DO concentration below the aerator, mg/L. The power supplied by the pump can be expressed as:

$$P_p = Q_w\,\gamma\,h/e \qquad (11\text{-}58)$$

where P_p = power supplied by pump, kw; h = the height of fall, m; γ = specific weight of water, N/m^3; and e = pump efficiency expressed as a decimal fraction. Putting Equations 11-57 and 11-58 together we get:

$$ME = \frac{3600\,Q_w\,(C_b - C_a)\,e}{Q_w\,\gamma\,h} \qquad (11\text{-}59)$$

Chesness and Stephens (1971) determined e and ME for several gravity aerators. The results of their efforts are shown in Table 11-13. They used flow rates ranging from 0.8 to 3.2 L/s and fall heights of 0.3 and 0.6 m. Results showed that aeration efficiency increases with increasing head loss (increasing height of fall), but mechanical efficiency decreases as head loss increases. Mechanical efficiencies for gravity aerators ranged 1.2–2.3 kg O_2/kW-hr.

Packed Column Aerators. The packed column is a variation of a gravity aerator where water is pumped to the top of a column and flows downward through packing media, which is designed to break up the water and maximize the air-water interface area. A typical packed column is shown in Figure 11-26. The SAE for such aerators ranges about 1.5–2.0 kg O_2/kW-hr (Hackney and Colt 1982). The depth of packing media required for a given DO change can be calculated from Equation 11-60:

$$\ln\frac{(C_s - C_a)}{(C_s - C_b)} - KZ - 0.4 = 0 \qquad (11\text{-}60)$$

where K = mass transfer coefficient for packing media, m^{-1}; and Z = depth of packing media, m. The mass transfer coefficient K is related to type of packing media, size, and operating temperature. Table 11-14 provides K-values (at 20°C) for various types of packing media. Values of K can be calculated for other tempera-

Table 11-13. Average aeration effectiveness (*E*) and mechanical efficiency (*ME*) for seven gravity aerator types.

	0.3 m head loss		0.6 m head loss	
Aerator type	*E* (%)	*ME* (kg O_2/kW hr)	*E* (%)	*ME* (kg O_2/kW hr)
Weir with splashboard	24.1	1.59	38.0	1.37
Weir with paddle wheel	24.2	1.68	39.0	1.23
Weir with rotating brush	23.9	1.62	34.9	1.22
Corrugated inclined plane, no holes	25.3	2.32	43.0	1.43
Corrugated inclined plane, with holes	30.1	1.89	50.1	1.67
Lattice	34.0	2.27	56.2	1.91
Riser with perforated apron	31.7	2.05	52.0	1.77

Source: Chesness and Stephens (1971).

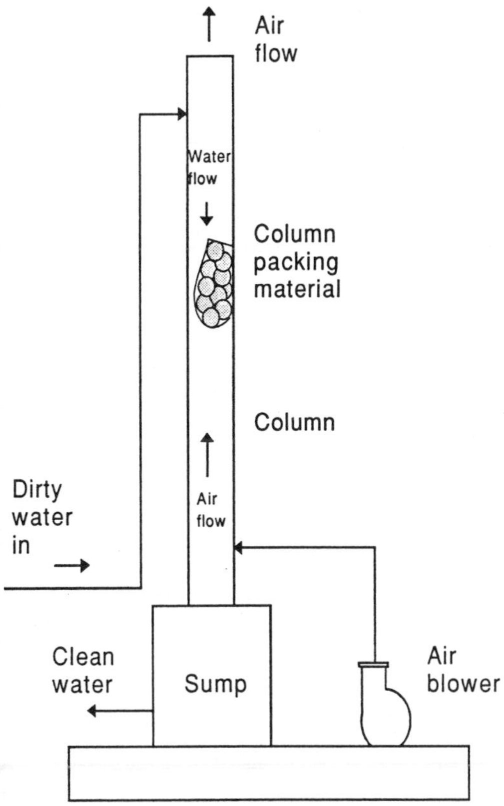

Figure 11-26. Illustration of a packed column with air blower.

tures using Equation 11-37. Table 11-14 also contains recommended hydraulic loading rates for the various packing media.

Water Discharging from Pipes. Efficiencies of various devices for oxygenating water discharging from pipes are shown in Table 11-15. The various devices are not described here but are described in Moore and Boyd (1984). Note that a simple gate valve has a very good oxygen transfer efficiency.

Diffused-Air Aeration Systems

Diffused-air aeration systems use air compressors or blowers to supply air and diffusers, porous pipe, or other devices to release air bubbles into the water. Figure 11-27 illustrates diffused-air systems. Pure oxygen can be used in place of atmospheric air in diffused-air systems. Pure oxygen will be discussed in a later section.

In diffused-air systems, air is introduced at the bottom of a pond or tank or at a given point in the water column, and oxygen is transferred as the bubbles ascend through the water column. The amount of oxygen transferred depends on the number, size and relative velocity of the ascending bubbles; the dissolved oxygen deficit; and the water depth at which the bubbles are released. Air bubble size varies from extremely fine to coarse, depending on the diffusion device used. Some common diffusers are porous diffusers (air stones), porous diffuser pipe, nonporous diffusers, perforated pipe, airlift pumps, and U-tube systems. In practice, more air can be released through coarse bubble diffusers, but oxygen transfer is highest for the

Table 11-14. Recommended design parameters for packed column aerators.

Media size (cm)	K at 20°C (m^{-1})	Recommended loading ($m^3/m^2/hr$)	Maximum loading ($m^3/m^2/hr$)	Minimum column diameter (m)	Relative cost of media
2.54	2.50	100	150	0.2	4.0
3.81	1.71	220[1]	300	0.3	2.4
5.03	1.58	220[1]	> 340	0.4	2.0
8.89	1.05	220[1]	> 340	0.7	1.0

[1]Highest loading rate tested.

Source: Boyd and Watten (1989).

Table 11-15. Efficiencies of various devices for aerating water from pipes.

Device	Discharge (L/min)	Pressure at discharge (m of water)	Efficiency (%)	Oxygen transfer (kg/hr)
Half-open gate valve	298	10.4	75.9	109.5
3.8-cm tee aspirator	294	10.6	71.7	101.9
3.8-cm ell aspirator	254	12.7	82.5	101.4
5.1-cm tee aspirator	304	9.9	66.4	97.5
Screen and rocks	304	9.9	63.3	93.1
Slotted cap	286	10.9	64.8	89.6
Screen	304	9.9	58.9	86.7
Alfalfa valve	287	11.0	60.9	84.5
Splashboard with holes	304	9.9	52.9	79.2
Screen cover	304	9.9	52.4	76.6
Splashboard	304	9.9	51.1	75.2
Screen extension	304	9.9	51.1	61.3
2.5-cm tee aspirator	154	17.0	82.1	61.3
Straight pipe	304	9.9	25.1	36.8

Source: Moore and Boyd (1984).

smallest bubble diameter. In addition, the greater the water depth at the point of bubble release, the higher the oxygen transfer because the air bubble-to-water contact time is greater.

Most diffused-air systems release large volumes of air at low pressure. The minimum operating pressure increases with increasing water depth above the diffuser, since enough pressure must be provided to force air from the diffuser against the total pressure (atmospheric plus hydrostatic) at the discharge point (Boyd 1990).

Diffused aeration is not efficient in shallow ponds commonly used for fish production (depth = 1.0–1.5 m) because the contact time of the air bubbles with the water is not great enough for sufficient oxygen transfer. With diffuser depths of about 0.9–1.2 m (3–4 ft), average SAE values are only about 0.5 kg O_2/kW-hr (1.6 O_2/hp-hr) (Boyd and Ahmad 1987). Average operating costs for diffused-air systems were $0.095/kg O_2 compared to $0.053/kg for paddle wheels, $0.059/kg for propeller-aspirator pumps, $0.079/kg for vertical pumps, and $0.079/kg for pump sprayers. The diffuser pores also clog very easily with use in fish ponds and culture tanks, requiring frequent maintenance for cleaning. In addition, diffuser networks in ponds interfere with seining oper-

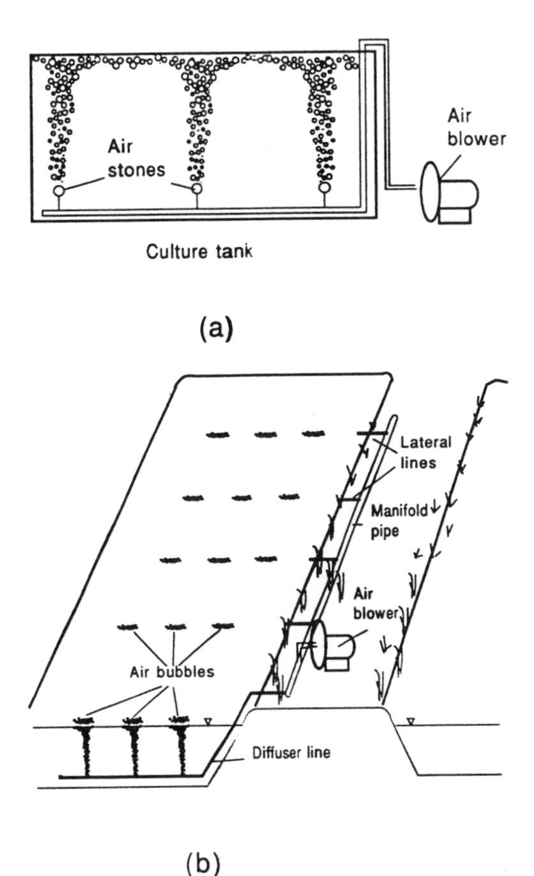

Figure 11-27. Diffused air systems: (a) In a culture unit; and (b) in a fish pond.

ations. Nonporous diffusers, such as water jets, orifices, venturis, and rotating discs, were developed to overcome some of the clogging problems experienced with fine bubble diffusers. Nonporous diffusers typically are less efficient than porous diffusers.

Comparative Performance of Aerators

Standard aerator efficiencies (SAE) for several types of aerators commonly used in aquaculture systems are listed in Table 11-16. The SAE for all aerators shown ranges from 0.6 to 3.0 kg O_2/kW-hr. Paddle wheels, in general, are more efficient at transferring oxygen and circulating water than other types of aerators, based on performance in clean water tanks (Boyd and Ahmad 1987; Rappaport et al. 1976). Many types of paddle-wheel aerators were tested at Auburn University (Boyd 1990), and SOTR values ranged from 17.4 to 23.3 kg O_2/hr. Of course, some paddle wheels were not as efficient as individual aerators of other types.

For aerators of 1 kW and larger in size, paddle wheels are equal to or less in cost than other types of aerators. However, small paddle wheels are more expensive than other electric aerators of similar size because the gear motors used in small paddle wheels are very expensive (Boyd 1990). For this reason vertical pumps, propeller aspirators, and diffused-air aerators are better suited for small ponds than paddle wheels even though the latter are more efficient to operate.

AERATION PRACTICES

Basically, four strategies are used to manage dissolved oxygen in aquaculture ponds: (1) No supplemental aeration; (2) emergency aeration; (3) supplemental aeration; and (4) continuous aeration. In years past, culture intensity was not nearly what it is today. Cultural practices have changed over the last decade, and cultural intensity has increased to the point where many farming operations require frequent if not continuous aeration to maintain adequate DO levels.

No Supplemental Aeration

If low enough stocking rates are used, supplemental aeration is not required. For channel catfish production, stocking at rates of 7,410 fish/ha (3,000 fish/ac) or less and low feeding rates (less than 39 kg/ha/day or 35 lb/ac/day) were common 20 years ago. This type of low-input aquaculture may be desirable in some circumstances, but few commercial enterprises currently use this approach to fish farming (Tucker and Robinson 1990).

Emergency Aeration

Aerators can be used in emergency situations to prevent occasional DO depletions that can kill or severely stress fish. Sick fish and high

Table 11-16. Relative efficiencies of various aeration devices.

Aeration device	Standard aerator efficiency (kg O_2/kW hr)	Reference
Gravity aerators		
Cascade weir (45°)	1.5–1.8	Colt and Orwicz (1991)
Corrugated incline plane	1.0–1.9	Colt and Orwicz (1991)
Horizontal screen	1.2–2.6	Colt and Orwicz (1991)
Lattice	1.8–2.6	Colt and Orwicz (1991)
Packed column		
Zero head	1.2–2.4	Colt and Orwicz (1991)
0.5–1.0 m head	10–80[a]	Colt and Orwicz (1991)
Surface aerators		
Low speed surface	1.2–2.4	Colt and Orwicz (1991)
Low speed surface with draft tube	1.2–2.4	Colt and Orwicz (1991)
High speed surface	1.2–2.4	Colt and Orwicz (1991)
Vertical pump	0.7–1.8	Colt and Orwicz (1991)
Pump sprayer	0.9–1.9	Colt and Orwicz (1991)
Paddle wheel	1.1–3.0	Colt and Orwicz (1991)
Subsurface aerators		
Aeration cone	2.5	Colt and Orwicz (1991)
Air-lift pump	2.0–2.1	Colt and Orwicz (1991)
Air diffuser	0.6–3.9	Boyd (1990)
Air diffuser		
Fine bubbles	1.2–2.0	Colt and Orwicz (1991)
Medium bubbles	1.0–1.6	Colt and Orwicz (1991)
Coarse bubbles	0.6–1.2	Colt and Orwicz (1991)
Propeller aspirator	1.7–1.9	Colt and Orwicz (1991)
Propeller aspirator	1.3–1.8	Boyd (1990)
Nozzle	1.3–2.6	Colt and Orwicz (1991)
Static tube	1.8–2.4	Colt and Orwicz (1991)
U—tube		
Zero head	0.7–2.3	Colt and Orwicz (1991)
0.5–1.0-m head	10–40[b]	Colt and Orwicz (1991)
Venturi	2.0–3.3	Colt and Orwicz (1991)

[a] Does not include power to the pump.

[b] Estimated.

concentrations of ammonia, nitrites, or carbon dioxide also require the need for additional aeration. In these situations, additional aeration is most commonly supplied by portable aeration units powered by tractor PTO shafts. Emergency aeration is probably the most common DO management strategy practiced today in fish ponds. It is also the only practice that has proven profitable for pond culture of channel catfish (Tucker and Robinson 1990).

In emergency aeration, aerators are turned on only when the DO has dropped substantially to threaten fish survival or when it is anticipated that DO concentrations will fall below 2–3 mg/L. When feeding rates exceed 50 kg/ha/day, emergency aeration may be needed quite frequently (Boyd 1990). In a typical commercial catfish pond in the South, about 45–90 kg (100–200 lb) of oxygen is consumed per hour at night by the fish, phytoplankton, and bottom sediments (Tucker and Robinson 1990). Most aerators used in commercial fish ponds transfer about 13–23 kg (30–50 lb) of oxygen per hour, under optimal conditions. It is obvious that one aerator cannot possibly meet the oxygen demand for a whole large pond. One aerator, however, can meet the needs of the fish. Thus, the usual strategy is to use a single aerator to create a *zone of oxygenated water* immediately around the aerator where the fish can congregate. In the immediate area around the aerator enough oxygen is added to the water to sustain life, but the overall DO concentration in the pond is not increased significantly.

Emergency aeration is actually detrimental to fish culture in that, although it saves their lives, the fish are continually exposed to suboptimal oxygen concentrations because aeration is used only when DO concentrations are very low. Repeated exposure to this practice results in added stress, which ultimately results in poorer feed conversion, slower growth, and increased susceptibility to disease infections.

Tractor-powered aeration units were the first effective emergency aerators used by fish farmers. However, in recent years the trend has been to replace the PTO-powered aerators with permanently installed electric aeration units in each pond where electric power is available. Large electrically powered aerators are not as expensive to purchase as large tractors and PTO-driven aerators, and they are less expensive to operate for each unit of oxygen transferred. Farmers can still maintain portable units that can be moved between ponds to supplement permanent aeration as needed. One portable aerator for every three or four production ponds is usually adequate for most large fish farms (Jensen and Bankston 1988).

Nightly and Continuous Aeration

Since DO concentrations in fish ponds typically drop at night, the purpose of nightly aeration is to begin aeration at some fixed time every night well before DO concentrations fall to stressful levels and continue aerating until photosynthesis begins to add oxygen to the pond the next day. For some ponds, this may require aeration for only three or four hours, but other ponds may need aeration all night long.

Many research studies were conducted to compare the results of nightly and/or continuously aerated ponds to control ponds receiving no aeration or only emergency aeration. Most of the studies were conducted with channel catfish, but the results should be applicable to other species. In one study by Hollerman and Boyd (1980), aeration was provided in channel catfish ponds with feeding rates of 90 kg/ha/day for six hours each night. The nightly aeration prevented pond DO concentrations from dropping below 3 mg/L. Net fish production averaged 5,100 kg/ha in the aerated ponds. Feed conversion ratio was 1.7, and survival averaged 92%.

In another study, channel catfish ponds were stocked at 10,000 fish/ha and fed to a maximum daily rate of 53 kg/ha (Lai-Fa and Boyd 1988). Three ponds were aerated for six hours nightly, and three ponds were unaerated controls, but emergency aeration was used as needed. Dissolved oxygen usually remained above 4 mg/L in nightly aerated ponds, and DO concentrations below 2 mg/L were common in the control ponds. Fish weight at harvest averaged 4,813 kg/ha in aerated versus 3,659 kg/ha in unaerated ponds. Feed conversion ratio was 1.32 in aerated ponds and 1.75 in unaerated ponds.

By contrast, in another study by Steeby and Tucker (1988), nightly aeration did not improve either feed conversion or total production when compared to ponds receiving emergency aeration when DO concentrations fell below 2 mg/L. The ponds were stocked at 12,350 fish/ha (5,000 fish/ac) and were fed up to 112 kg/ha/day (100 lb/ac/day).

The practice of continuous aeration requires that the aerators be operated for 24 hours per day during all or part of the growing season. A diffused-air aeration system was used in a recent study at Auburn University (Boyd 1990). Triplicate channel catfish ponds were aerated at the equivalent of 0, 0.9, 1.8, and 3.6 kW/ha (0, 0.5, 1.0, and 2.0 hp/ac, respectively). Aeration was applied continuously from May to October. No emergency aeration or water exchange was used in the aerated ponds. Aerated ponds were stocked at 15,000 fish/ha, and unaerated ponds were stocked at 10,000 fish/ha. Maximum daily feeding rates of 60 kg/ha were used in unaerated ponds versus 100–130 kg/ha in aerated ponds. Low DO concentration was common in unaerated ponds, and emergency aeration was used regularly. Dissolved oxygen never dropped below 2.0 mg/L in the aerated ponds. Fish production averaged 4,877 kg/ha in the aerated ponds versus 3,310 kg/ha in unaerated ponds. Feed conversion ratio and survival averaged 1.91 and 85.7%, respectively, in the aerated ponds versus 4.1% and 88.0%, respectively, in the unaerated ponds. Survival was slightly higher in the unaerated ponds, but feed conversion was much better in the aerated ponds. Thus, one would have to figure the cost of the feed used versus the total weight of fish produced in the two scenarios to determine if aeration had an economic advantage in this study.

In another experiment, Plemmons and Avault (1980) were able to produce 12,800 kg/ha (11,400 lb/ac) of channel catfish in small (0.4-ha), experimental ponds receiving continuous aeration with a combination of floating, electric aerators, emergency aeration with PTO-driven paddle wheels, and water exchange. It is unrealistic to expect production of over 10,000 kg/ha (8,900 lb/ac) of channel catfish annually in large ponds because most farms do not have a sufficient water supply to permit high rates of water exchange (Boyd 1990).

A major disadvantage of both nightly and continuous aeration is the large amount of oxygen that must be supplied over and above the requirements of the fish being produced. Fish are responsible for only a small fraction of the DO consumed in ponds (Tucker and Robinson 1990). Several times the oxygen requirement of the fish must be supplied to meet the demands of phytoplankton and sediments. Thus, most of the oxygen requirements are supplied to meet the needs of a crop (phytoplankton) that has no economic value.

Aerator Size Requirements

Current commercial aquaculture practice is to supply each fish pond with an aerator having an SOTR of at least 1.8 kg O_2/kW-hr (3 lb O_2/hp-hr). This requires about 1.8 kW/ha of electric power (1 hp/ac). However, many producers aerate with electric units at 1.0–1.8 kg/ha (0.5–1.0 hp/ac). In addition, one mobile, tractor PTO-driven aerator is recommended for every three to four ponds. The mobile aerators should have an SOTR of at least 2.2 kg/hr/ha (2 lb/hr/ac). When oxygen demands are high, or when conditions cause the fish to become stressed, additional aeration will be required.

Emergency aerators must provide sufficient oxygen to exceed the oxygen requirements of the fish. For nightly or continuous aeration, the aerator should ideally meet the oxygen demands of the fish, phytoplankton, and pond sediment. This normally requires three to four times the aeration capabilities needed for emergency aeration (Tucker and Robinson 1990).

Aerator Placement

Finfish become conditioned to the high DO concentrations around an aerator and thus will tend to congregate around the aerator when low DO conditions develop in other areas of the pond. Some farmers feed their fish around the aerator to take advantage of the higher DO concentrations and to encourage the fish to stay in this area. Permanently installed aerators should be located where they are easily maintained and replaced, should replacement be necessary. They should be located near an all-weather road for access during operation and maintenance. They should also be as close as possible to electric utility hookups to minimize the length of power lines required.

Aerators used in an emergency should be placed in the same area as the regular aerator so that the fish do not have to travel through oxygen deficient water to reach the aerated zone. Aerators that discharge water over a broad area are more desirable since they provide a larger aerated zone in which the fish can gather. Paddle wheel aerators are perhaps the most efficient at providing horizontal water movement in addition to oxygenation, and, thus, create larger zones of aeration. They are especially beneficial in crawfish ponds since the animals are bottom dwellers and cannot travel great distances to the aerators as can finfish.

In long, rectangular ponds the recommended location for a paddle wheel aerator is at the middle of one of the long sides pointing toward the middle of the pond (Figure 11-28a). If placed along one of the short sides, the aerator would mix only one end of the pond (Figure 11-28b). The worst-case scenario is to place a paddle wheel in one corner so that it directs water diagonally across the pond (Figure 11-28c) (Boyd 1990). For large rectangular or square ponds requiring more than one paddle wheel, they should be positioned so that they all direct water in a clockwise or counterclockwise direction and do not work against one another (Figure 11-29). In this manner, a circular pattern is developed in the pond. This can occasionally have detrimental effects, however, by the deposition of sediments in the center of the pond (Boyd 1990).

Crawfish ponds can be constructed in a recirculating mode using a variety of configurations involving a network of earthen *baffle levees* within the pond's main perimeter levees (Lawson et al. 1989). Figure 11-30 shows two such configurations where either one or two paddle wheel aerators are used to direct water flow through the pond. Approximately 0.55–0.92 kW/ha (0.3–0.5 hp/ac) is sufficient to produce circulation in most crawfish ponds.

Aeration Scheduling

When should ponds be aerated? It is apparent from previous discussions that a diligent program of dissolved oxygen monitoring should be practiced in commercial aquaculture production

Figure 11-28. Circulation patterns produced by paddle wheel aerators placed in different locations within a pond.

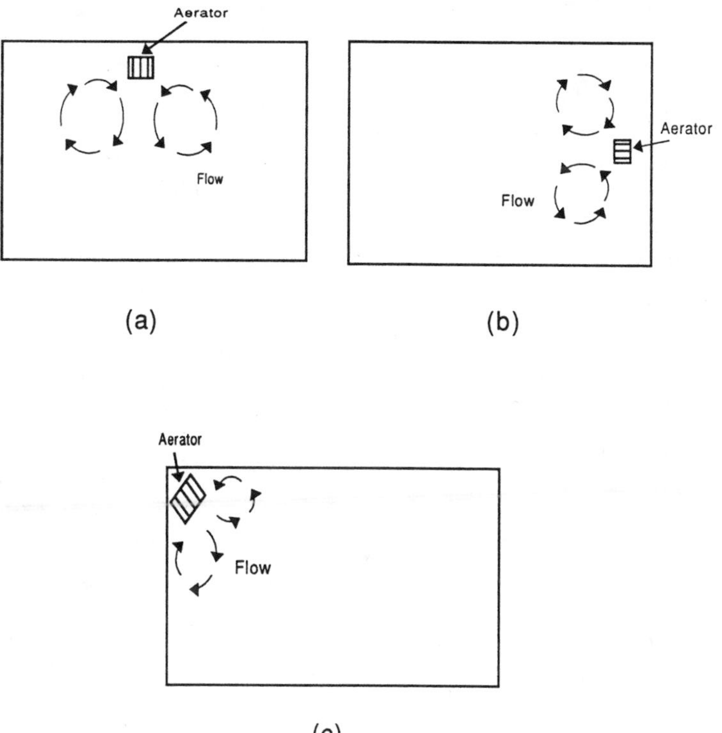

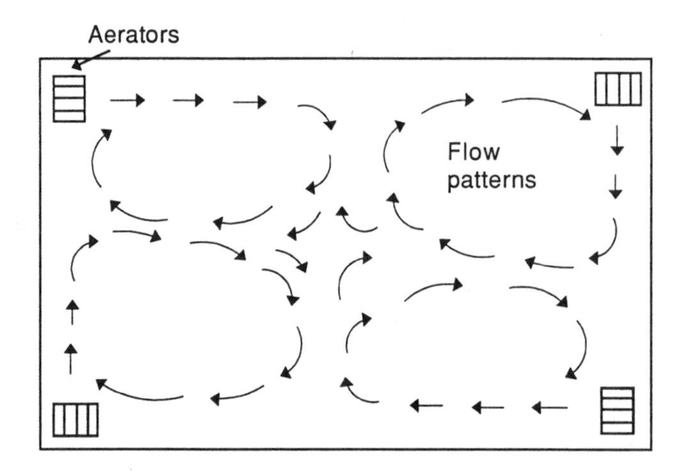

Figure 11-29. A four-paddle-wheel aerator configuration in a pond showing corresponding circulation patterns.

or crops worth many thousands of dollars can be lost in a matter of only a few hours.

On those farms where emergency aeration is practiced, trained personnel monitor the DO concentration in ponds at frequent intervals throughout the night. Aeration is initiated when DO concentrations fall to some predetermined value, usually between 2 and 4 mg/L. Monitoring is continued throughout the night, and additional portable aerators are used when necessary. Portable aerators should first be placed in those ponds where the DO concentration is falling the quickest. In many large, modern facilities, pond DO concentrations are monitored by computer, and aerators are automatically turned on when the DO drops to a preset value.

On those farms where nightly or continuous aeration is practiced, aeration is usually required only during the warmest months, when oxygen demands are highest. Nightly aeration is initiated at dusk or shortly thereafter, and continued until photosynthesis begins reoxygenating the water the following day.

The use of nightly aeration does not eliminate the need for regular DO monitoring. Dissolved oxygen levels should routinely be checked at least three to four times daily to ensure the proper operation of aeration equipment. If fish are diseased or if some event occurs to cause additional oxygen demand, additional aeration may be required.

WATER EXCHANGE AND CIRCULATION

Before mechanical aeration devices became so widely used, pond water was often aerated using pumps to exchange low DO pond water with water from another pond, or pumps were used to lift water so that gravity aerators could be employed. Aeration was effected by the turbulence created by the water stream striking the pond surface or by breakup during gravity aeration. In extreme cases some producers have used outboard motors to churn pond waters, however, this is only effective in small ponds.

Water exchange is mentioned in several places in this book since it serves several functions: it can be used to flush nutrients and algae from ponds to prevent eutrophication; it removes toxic metabolic wastes, such as ammonia and nitrites, from the pond; it can be used to dilute saline ponds during times of low rainfall; and it can be used as a substitute for aeration.

When water is exchanged in fish ponds, the water should be removed from the bottom to remove stale, poorly oxygenated water. New water should be added at the surface at the opposite end of the pond from which it was drained. The most beneficial means of water exchange is by first draining out a portion of the volume and replacing it with an equal volume. In this manner wastes such as ammonia can be

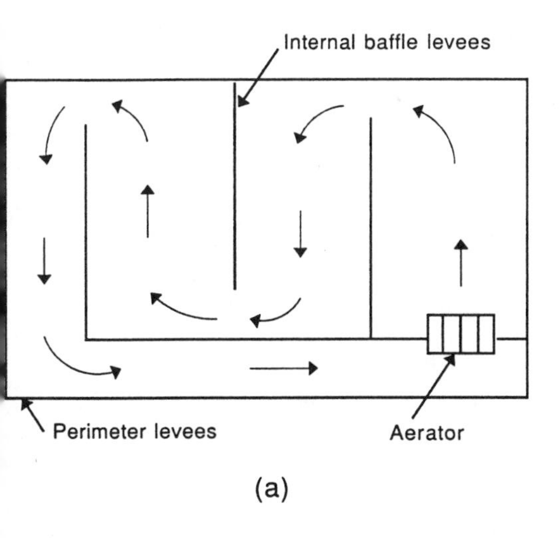

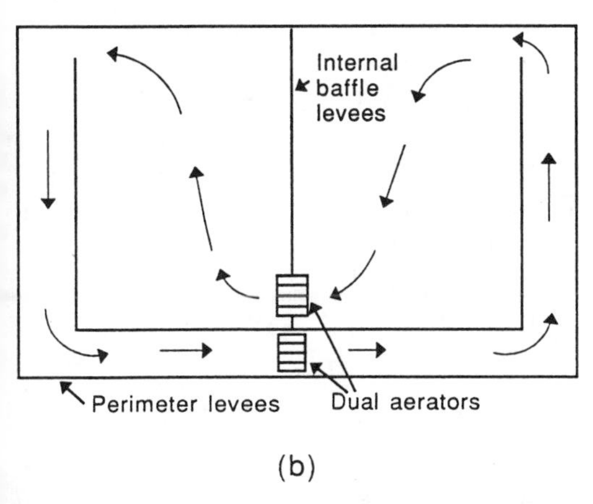

Figure 11-30. Paddle-wheel aerators used with baffle levee in crawfish ponds: (a) A single-aerator configuration; and (b) a dual-aerator configuration.

diluted. However, the most common method used by farmers is to pump fresh water into a pond and allow an equal amount to drain from the pond at the same time. Water should never be exchanged when the makeup water is of a lower quality than the original pond water.

In brackish water ponds, farmers typically use a daily exchange rate of 8–10% of the pond volume (Boyd 1990). Shrimp farming operations require a minimum daily exchange rate of about 10%. Past crawfish pond management practices recommended daily water exchanges of about 20–33% (Lawson et al. 1989). However, wasteful and inefficient water exchange practices formerly used in crawfish pond management have been discarded in favor of water recirculation, as discussed in a following section.

The water exchange rate in ponds can be estimated as follows, assuming that the pond is full of water (Boyd 1990):

$$ER = \frac{[(PR \times T) + P] - (S + E)}{V} \times 100 \qquad (11\text{-}61)$$

where ER = exchange rate, percent of pond volume per day; PR = pumping rate, m^3/hr; T = time of pumping per day, hr; P = precipitation, m^3/day; S = seepage, m^3/day; E = evaporation, m^3/day; and V = pond volume, m^3.

Example problem 11-7. A pond has a water surface area of 10 ha and a depth of 0.8 m. Water is pumped at the rate of 25 m^3/min for six hours per day. Evaporation is 0.6 cm/day and seepage is 0.25 cm/day. Rainfall is 0.5 cm/day. Assume that rainfall is zero. What percentage of the pond volume is exchanged?

Solution:

$$V = 10 \text{ ha} \times 10{,}000 \text{ m}^2/\text{ha} \times 0.8 \text{ m} = 80{,}000 \text{ m}^3$$

$$PR = 25 \text{ m}^3/\text{min} \times 60 \text{ min/hr} = 1{,}500 \text{ m}^3/\text{hr}$$

$$E = 100{,}000 \text{ m}^2 \times 0.006 \text{ m/da} = 600 \text{ m}^3/\text{da}$$

$$P = 100{,}000 \text{ m}^2 \times 0.005 \text{ m} = 500 \text{ m}^3/\text{da}$$

$S = 100{,}000$ m^2 (assume that bottom area is the same as the water surface area) $\times$ 0.0025 m/day = 250 m^3/day

$$ER = \frac{[(1{,}500 \times 6) + 500] - (600 + 250)}{80{,}000} \times 100 = 10.8\ \%\ \text{daily}$$

Exchange rate in saline ponds must be treated slightly different. Evaporation can drastically alter the salinity of a pond because, as water evaporates, the salts are left behind, and the salinity of the water remaining in the pond becomes concentrated. Freshwater addition is essential in some ponds to prevent salt buildup. Seepage does not affect salinity since the salts leave with the seeping water. However, precipitation will dilute saline water already in the pond.

Aquaculturists generally agree that *water circulation* in ponds is beneficial. Water circulation moves oxygenated water out away from the aerator, and circulation can eliminate thermal and DO stratification (Szyper and Lin 1990) and make the entire pond volume habitable. It also eliminates the danger of seasonal overturns and overturns following a rainfall event.

The main function of a water circulator is to mix surface water with subsurface water to eliminate stratification. Most water circulation devices create surface turbulence, which provides a limited amount of aeration, which is an added benefit but not the primary purpose. Some water circulators are referred to as *water blenders*. During daylight hours pond surface waters are often saturated or supersaturated with oxygen while subsurface waters are often low in DO. By mixing the water from both levels, a uniform oxygen gradient is established, and more of the water column is usable by the fish. Water circulation is also responsible for mixing and distributing nutrients that stimulate phytoplankton growth and ultimately increase DO production.

Air-lift pumps can be used to provide circulation in small fish ponds, tanks, and raceways. These are discussed in Chapter 7. Figure 7-19 illustrates an air-lift pump designed and tested by Parker and Suttle (1987). The air-lift tube was fabricated from PVC pipe, and the elbow was PVC. Air from an air blower was released into the pipe as shown. Air spargers or diffusers can be used to obtain smaller air bubbles and increase oxygen transfer, if desired. Parker and Suttle (1987) demonstrated that two 10-cm (4-in.) diameter air-lift tubes with 1.2-m (4-ft) long vertical risers can pump the entire volume of a 0.2-ha (0.5-ac) pond in three days using an air flow rate of 0.14 m^3/min (37 gpm). The air was injected at a point 0.76 m below the water surface. If greater mixing is desired, larger or an excessive number of air lifts would be required.

Paddle wheel aerators are excellent for circulating as well as aerating water (Figure 11-31). For those units having adjustable paddle depths, greater water circulation can be obtained by setting the paddles at their maximum recommended operating depth. Greater water circulation is obtained at the expense of losing some of the device's oxygenation capacity. Setting the paddles at a greater depth than that recommended by the manufacturer will cause excessive power consumption and possible motor damage.

In studies conducted in crawfish ponds, Lawson et al. (1994) reported that a 2.2-kW (3-hp) paddle wheel circulated 4.8 m^3/min (1,260 gpm) in 1.8-ha (4.5-ac) research ponds. Two 2.2-kW (3-hp) aerators in the same ponds roughly doubled the water circulation rate to 9.5 m^3/min (2,510 gpm). In an 8.9-ha (22 ac) crawfish pond a 3.7-kW (5-hp) paddle wheel created a circulation rate estimated at 5.1 m^3/min (1,330 gpm). In 12-ha (30-ac) crawfish ponds, 7.5-kW (10-hp) paddle wheels generated circulation rates of 8.6–9.8 m^3/min (2,280–2,600 gpm).

Another type of water circulator is shown in Figure 11-32. The device consists of a fan blade enclosed within a cylindrical tube. The blade is belt-driven by an electric motor from above. The whole assembly is supported by plastic floats. The circulators are available in three sizes: 0.75, 1.5, and 2.2 kW (1, 2, and 3 hp) respectively. Circulation rates generated by the devices are estimated at 6.8, 17.4, and 38 m^3/min (1,800, 4,600, and 10,000 gpm), respectively.

The mixing rate of circulation devices in ponds can be estimated by conducting tests using either a dye or salt solution. In the dye test, a given quantity of dye is poured into the pond just downstream of the circulator, and the dye concentration is measured at timed intervals at several locations and depths in the pond. When the dye concentration at all stations is the same, the pond is considered completely mixed. Dye concentrations can be measured with a fluorometer. The procedure is basically the same when salt is used except specific conductance is measured at timed intervals until specific conductance values are essentially equal. The mixing

Figure 11-31. A paddle-wheel aerator in action (photo by T. Lawson).

Figure 11-32. A 1.5-kW (2-hp) water circulator (photo by T. Lawson).

rate can then be estimated using the following equation (Boyd 1990):

$$MR = \frac{A \times D}{P \times T} \quad (11\text{-}62)$$

here MR = mixing rate, m^3/kW-hr; A = water surface area, m^2; D = pond depth, m; P = power consumption of the aerator, kW; and T = time for complete mixing, hr.

PURE OXYGEN ABSORPTION SYSTEMS

The discussion thus far has primarily centered around aeration methods for pond aquaculture. Many of the mechanical and gravity aeration devices also have application to flow-through (raceway) and closed recirculating system aquaculture. In recent years pure oxygen has been used for aquaculture production. Pure oxygen has been applied successfully in raceway systems both in the commercial and research sectors. Thus far, pure oxygen does not appear to be economically feasible for use in aquaculture ponds, because of its high cost. Additionally, oxygenation practices in recirculating systems are very complex, and additional research is needed in this area. Therefore, the discussion of pure oxygen that follows is intended primarily for application in flow-through systems.

Dissolved oxygen is often the limiting factor restricting production in intensive raceway systems. The DO constraint has been circumvented somewhat with the use of gravity and mechanical aeration devices, but DO concentration cannot exceed the saturation value for the given operating conditions with these devices. Three important benefits are realized when pure oxygen is practiced in intensive culture. First, pure oxygen contactors have the unique ability to increase the DO concentration in culture waters to 100% saturation or greater. Second as pure oxygen is added to the water, dissolved nitrogen (DN) is forced out, thus reducing DN concentrations to or below the saturation value for the purposes of controlling gas bubble trauma. Third, pure oxygen can greatly increase the production capacity of a given culture system.

Earlier discussions revealed that the rate of oxygen transfer into water using either air contact or pure oxygen contact equipment is regulated by the area of the gas-liquid interface, the thickness of the laminar film present in the gas-liquid interface, and by the dissolved oxygen deficit (C_s-C_m). The latter refers to the difference between the oxygen concentration at saturation and the ambient DO concentration and is the major driving force for oxygen transfer. In pure oxygen contactors the DO concentration can be increased to above the saturation value. Hence, the magnitude of the driving force is larger, and oxygen transfer is more rapid. For example, the saturation concentration of oxygen at 15°C and zero salinity is 10.08 mg/L (from Table 2-4). Exposing the same water to pure oxygen results in a saturation concentration of 48.14 mg/L.

Sources of Pure Oxygen

Pure oxygen is available from three sources: high-pressure (compressed) oxygen gas, liquid oxygen, and on-site generation. In many aquaculture facilities, at least two of these sources are used; one is used as a backup for the other. The most economic source will depend on power costs, transportation charges, and the mode of operation. High-pressure oxygen and liquid oxygen are readily available and either may be the most economical to use in highly industrialized areas since they are commonly used in the welding and glass-blowing industries, among others. In remote areas it may be more economical to generate oxygen on-site rather than pay exorbitant transportation costs. Both compressed oxygen gas and liquid oxygen are usually priced in terms of 100 ft^3 in the United States and m^3 in Europe (Colt and Watten 1988). Useful data for oxygen gas and liquid oxygen are reported in Table 11-17.

Compressed Oxygen Gas. Compressed oxygen gas (98–99 % pure) is available in metal cylinders containing 3–7 m^3 (100–250 ft^3) of gas at 17.6 MPa (2,550 psi) pressure. Oxygen in

Table 11-17. Physical properties of gaseous and liquid oxygen.

Parameter	English units	SI units
	Gas phase	
Density	0.08309 lb/ft^3	1.331 kg/m^3
Specific volume	12.04 ft^3/lb	0.7513 m^3/kg
Weight/unit volume	8.31 lb/100 ft^3	1.331 kg/m^3
	Liquid phase	
Density	9.52 lb/gal	1.141 kg/L
Specific volume	0.105 gal/lb	0.877 L/kg

Gas—liquid relationships: 1 gal liquid oxygen = 115 ft^3 gaseous oxygen; 1 L liquid oxygen = 0.857m^3 gaseous oxygen.

this form is very expensive. The 1988 price reported by Colt and Watten (1988) was \$8–10 per 100 ft^3 in the United States. Due to its expense, oxygen in this form is normally used as a backup for liquid oxygen or on-site oxygen generation. Used as a backup, a number of cylinders can be connected together using standard manifold assemblies.

Liquid Oxygen. Liquid oxygen (LOX) is about 98–99% pure. It is produced on a large scale by distilling liquefied air. Each m^3 of LOX provides 860 m^3 of gas with a specific volume of 0.7513 m^3/kg (Watten 1991). Liquid oxygen is delivered to the site in bulk storage trucks and pumped in Dewar's type storage vessels (Figure 11-33), which are designed to maintain a temperature of −182.9°C (−297°F). Tank pressure ranges from about 1,000–4,000 kPa (150–200 psi). Prior to use, the LOX is vaporized by directing it through heat exchanger coils. A typical LOX system consists of a storage tank, evaporator, filters, and a pressure regulator. A typical scheme showing the basic components of a LOX system is shown in Figure 11-34. A flowmeter or rotameter is used to measure the gas flow rate. Storage tanks range from 100 L (30 gal.) to over 40 m^3 (10,000 gal.) in volume. Some of the smaller tanks are portable and can be used for fish transport. Large cryogenic storage tanks can be purchased but most often are leased. The 1988 price for LOX was \$0.25–3.000/100 ft^3 in the United States (Colt and Watten 1988).

Oxygen Generation. Pure oxygen can be generated on-site using a process called *pressure swing adsorption* (PSA). PSA systems produce 0.5–10 m^3 (15–400 ft^3) of oxygen gas per hour. The systems require a source of dry, filtered air at 600–1,000 kPa (90–150 psi) pressure to produce pure oxygen that is 85–95% pure. PSA units operate on a demand basis and produce oxygen only when needed. They have been proven reliable and require little maintenance, but are not as reliable as LOX or compressed oxygen gas systems.

Figure 11-35 illustrates a typical PSA system. Compressed air flows through a pressure vessel packed with a granular material that functions as a molecular sieve. The sieve material is usually *clinoptilolite*, one of several types of naturally occurring zeolites. The packing material selectively absorbs nitrogen from the air, providing an oxygen-enriched gas. When the nitrogen-absorption capacity of the filter bed is reached, the column is depressurized, nitrogen is removed, and the bed is subsequently reused. Usually the system is designed with two or more filter beds in parallel, so that they may be alternately used, and oxygen is continuously generated. The enriched oxygen flow represents about 5% of the required air flow (Watten 1991).

PSA systems are expensive. An air compressor is required in addition to the molecular sieve beds. Also, it is a good idea to have a standby generator, in the event of power failures. Installation may require the construction of a separate building to house the compressor so that heat and noise are eliminated from the main working area.

Types of Pure Oxygen Contactors

Many different types of pure oxygen contactors (POCs) are used in aquaculture facilities. Contactor selection should be based on performance indicators previously discussed as well as on site-specific resources and limitations. For example, the presence of a hydraulic gradient

Figure 11-33. Bulk liquid oxygen storage tank (photo courtesy of Ziegler Brothers, Inc., Gardners, Pa.).

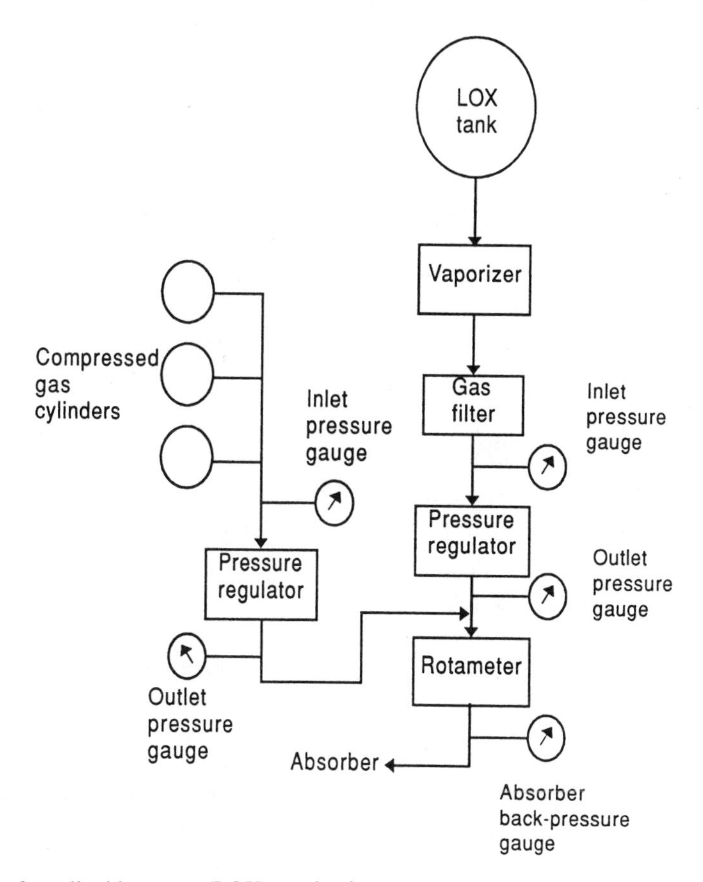

Figure 11-34. Layout for a liquid oxygen (LOX) production system.

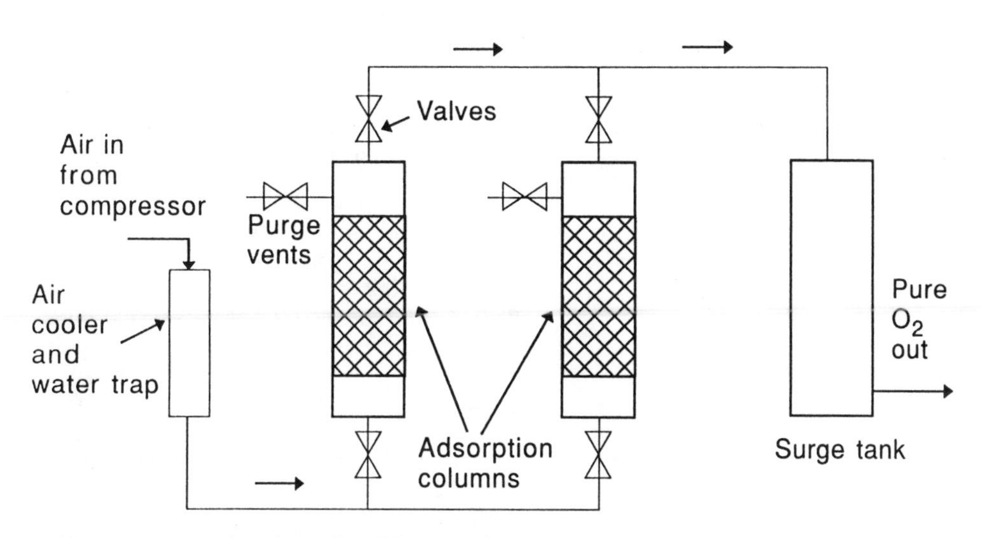

Figure 11-35. A pressure swing absorption (PSA) system.

may influence the decision to use one type of contactor over another, as will be shown later. As another example, the presence of particulate matter in the water being treated usually precludes the use of packed columns due to excessive fouling of the packing material. Compared to air, pure oxygen is relatively expensive. The payback comes in the forms of (1) Better growth and feed conversion; and (2) increased production due to the higher DO concentration. Therefore, only systems with high absorption efficiencies are generally economical to use. Oxygen contactors have the ability to increase DO concentrations in treated waters to high supersaturated values. The operating characteristics of several types of POCs are shown in Table 11-18.

U-tubes. The *U-tube contactor* consists of either a u-shaped vertical tube sunk into the substrate (Figure 11-36a) or two concentric pipes (Figure 11-36b), a gas diffuser, and in some cases, an off-gas collection and recycling system used to capture and recycle unabsorbed oxygen gas (Figure 11-36b). Oxygen is sparged at the upper end of the down-leg part of the contactor, and water velocity is maintained above the buoyant velocity of the entrained gas bubbles. As the gas-liquid mixture moves downward through the contact loop, a temporary increase in hydrostatic pressure forces the oxygen into solution at a faster rate. Water velocity in the down-leg part of the tube is maintained slightly above the buoyant velocity of the entrained oxygen gas bubbles. Oxygen absorption is influenced by the depth of the conduit, inlet gas flow rate, water velocity, diffuser depth, inlet DO concentrations, and the rate of off-gas recycle. Simulation studies have shown the operating costs to be minimal with U-tube depths of 25–60 m and water velocities of 1.8–3.0 m/s. Off-gas recycling increases the oxygen absorption efficiency (Colt and Watten 1988; Watten 1991). Principle advantages of the U-tube are (1) The ability to operate at a low hydraulic head; and (2) particulate matter is not a problem. If 2–3 m of head is available, the system can operate without pumping (Figure 11-36a).

Aeration Cones. The *aeration cone* (*downflow bubble contactor*) consists of an inverted cone that is partially submerged in the receiving water (Figure 11-37a and b). Water and oxygen gas enter the top of the cone and proceed with reduced velocity toward the bottom, or discharge end, of the cone. As the gas bubbles are carried downward toward the discharge, the water velocity is reduced until it just equals the upward buoyant velocity of the bubbles. Thus,

Table 11-18. Typical operating characteristics of pure oxygen systems.[a]

System	Absorption efficiency (%)	Transfer efficiency (kg oxygen/kW-hr)	Effluent concentration (mg/L)
U-tube	30–50	1.0–1.5	20–40
U-tube with off-gas recycling	60–90	2.0–3.0	20–40
Packed column	40–50	0.5–1.0	10–15
Spray tower	40–55	0.5–1.0	10–15
Pressurized packed column	95–100	1.0	30–90
Aeration cone	80–90	4.0	5–25
Oxygen injection: high-pressure venturi	15–70	< 0.5	50–80
Oxygen injection: bubble generator	20–40[b]	0.2–0.5	30–50
Diffused aeration	3–7	-	10–20

[a] Power requirements for oxygen generation not included.

[b] Overall absorption efficiency may approach 90–100 % when absorption in the rearing unit is included.

Source: Colt and Watten (1988).

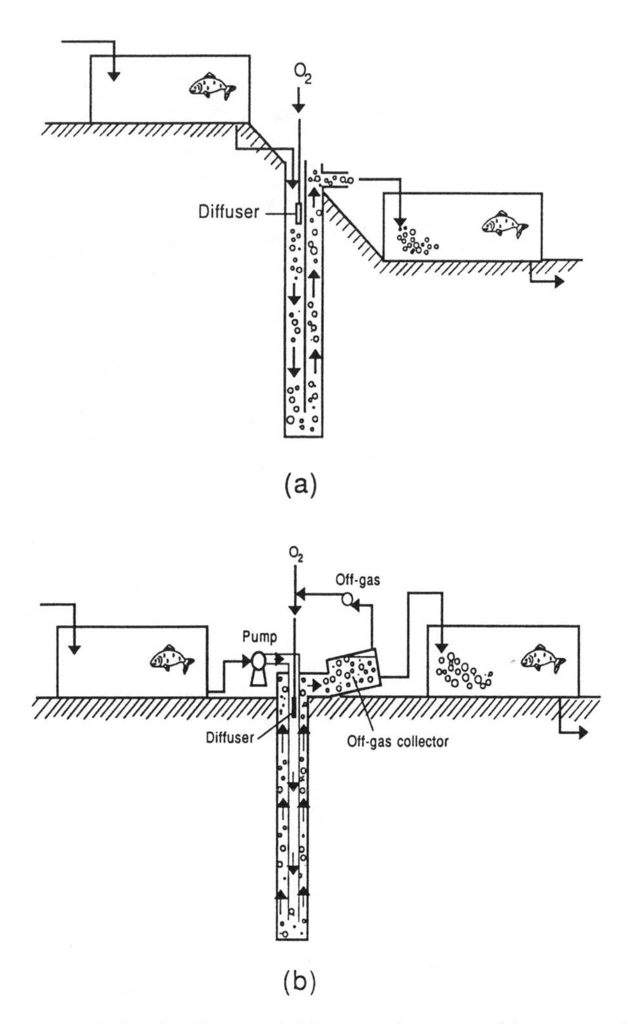

Figure 11-36. U-tube systems: (a) Gravity flow; and (b) pumped system with gas recycle.

the bubbles are held inside the cone for an indefinite time period. The inlet velocity is designed to be about 1.8 m/s, and the exit velocity should be kept below 0.15 m/s in order to keep the bubbles in suspension (Watten 1991). A portion of the gas should be vented for nitrogen stripping. Figure 11-37a illustrates an open-bottom cone that can be mounted in the rearing unit. Pressurized, or closed, cones are also used (Figure 11-37b).

The performance of aerator cones is determined by gas and water flow rates, influent DO concentration, the off-gas vent rate, cone geometry and operating pressure. Operating pressure is in turn determined by cone depth. Use of deeply submerged cones will increase oxygen absorption and decrease nitrogen stripping (Watten 1991). These units are commercially produced and are used widely in Europe.

Packed Columns. *Packed columns* (packed towers) consist of a vertical column filled with media having a high specific surface area. Water is evenly distributed at the top of the media over a perforated plate or through a spray bar, and trickles by gravity downward through the media. Oxygen is injected near the bottom of the media strata at an established volumetric gas-liquid ratio of 0.3–5%. The column may be either open or closed at the top. Plastic packing

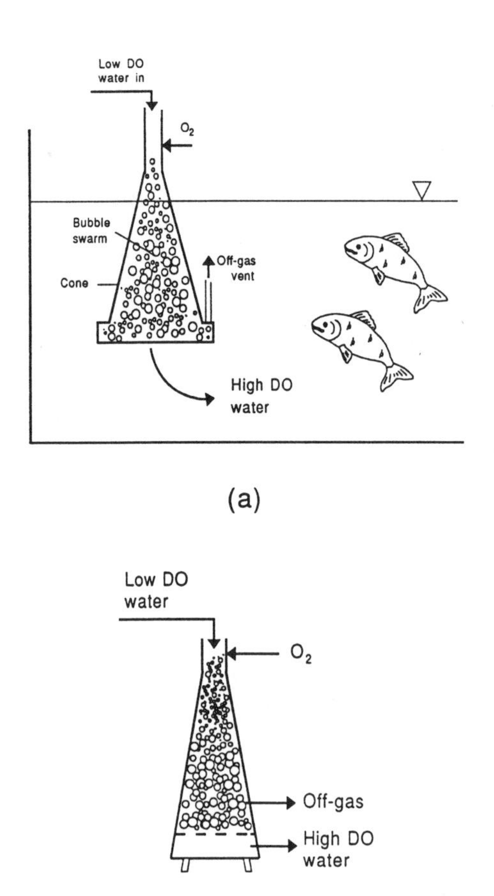

Figure 11-37. Aerator cones (downflow bubble contactors): (a) Mounted in rearing unit; and (b) a closed unit.

media is having a particle size of 25–50 mm and a specific surface area of 80–360 m^2/m^3 is typically used (Watten 1991). An application of a packed column in a raceway system is illustrated in Figure 11-38.

Pure oxygen packed towers are efficient oxygenators and nitrogen strippers (Colt and Bouck 1984). Their performance is influenced by liquid distribution method, media characteristics, media bed depth, gas-liquid loading, inlet DO concentration, and operating pressure. A disadvantage is that the media usually becomes clogged with particulates over time.

Spray Towers. *Spray towers* (spray columns) are similar to packed columns except that they do not contain media. A spray nozzle is positioned at the top of the tower Fig (11-39a), or the spray can be created by a perforated water distribution plate. The water spray generated provides the surface area for oxygen transfer. Little oxygen is transferred in the water flowing down the sides of the tower; therefore, a large-diameter column is desirable (Watten 1991). The column can be either pressurized or open at the bottom (Figure 11-39b). Pressurized spray columns can be designed for 100% oxygen absorption efficiency when nitrogen stripping is not required (Colt and Watten 1988).

Spray towers can be designed to produce a vacuum of 20–50 mm Hg (*vacuum degassers*), and, therefore, are excellent as nitrogen strippers (Colt and Watten 1988). They also have the advantage that they can handle water containing particulate matter.

Enclosed Surface Agitators. This contactor employs a rotating impeller powered by a submersible motor to provide agitation for increasing the area of the gas-liquid interface within a sealed chamber (Figure 11-40a). Figure 11-40b illustrates a variation where a surface aerator is enclosed by a hood that is open at its bottom. The impeller can be placed directly in the liquid stream, eliminating the need for a hydraulic gradient (Watten 1991). In addition, this contactor can handle liquids containing particulate matter. Gas transfer is influenced by inlet DO concentration, inlet gas-to-liquid ratios, chamber geometry, and the design, speed, and operating depth of the impeller.

Multi-stage Low Head Oxygenators. The *low head oxygenator* (LHO) was developed to reduce the head required by packed and spray column contactors to about 0.3–1.0 m, which is about the head that exists between the water supply and normal water level in raceway systems (Watten 1991). This contactor consists of an inlet trough and a series of contact chambers (Figure 11-41). Low DO water enters the trough and drops through a perforated plate into the chambers. Oxygen gas enters at one side of

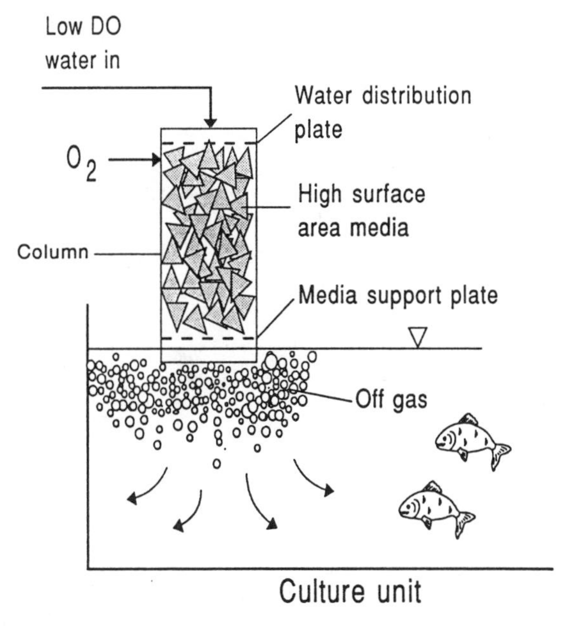

Figure 11-38. A packed column mounted in a culture unit.

the unit and is carried across the chamber by the pressure variations within the unit. The repeated contact of oxygen with the water provides a high mean gas deficit for both oxygen and nitrogen which, in turn, accelerates gas transfer. The head requirement reduction results in lower energy costs and minimizes the potential for system failure by eliminating the need for pumping. Performance is influenced by head (height of fall), water and gas flow rates, influent DO concentration, the number of chambers, and whether or not packing media is used within the chambers. An LHO system constructed within a raceway is shown in Figure 11-42.

Oxygen Injection. In oxygen injection systems, high operating pressures are used to increase the oxygen transfer rate. A typical system incorporates a sidestream of water, which is pumped at high velocity through an aspirator device, such as an orifice or venturi nozzle, into which oxygen is injected (Figure 11-43). The aspirator establishes a fine bubble suspension in the water being treated. The water-bubble mixture then traverses a closed conduit loop, which provides turbulence and contact time necessary for efficient absorption. Expansion nozzles are used to inject the mixture into the receiving water. A dissolved oxygen monitoring system can be used to control the flow of oxygen gas.

Typically, pressures of 190–860 kPa (28–125 psi) are required to achieve satisfactory absorption (Watten 1991). The closed contact loop should be sized to provide a contact time of 6–12 s with a fluid velocity of 3.5–4.5 m/s (11.5–14.8 ft/s). These contactors have worked well in tanks as well as in flow-through systems.

A second method of oxygen injection, described by Colt and Watten (1988), is shown in Figure 11-44. This system is designed to produce very fine oxygen bubbles in a mixing chamber. The water-bubble mixture then enters a pressurized water line or is directly injected into the rearing unit. The majority of oxygen transfer takes place in the rearing unit, but little oxygen is lost from the water because of the fine bubble size and turbulence.

Performance of Pure Oxygen Contactors

The operating characteristics of POCs vary widely. Several of the more important parame-

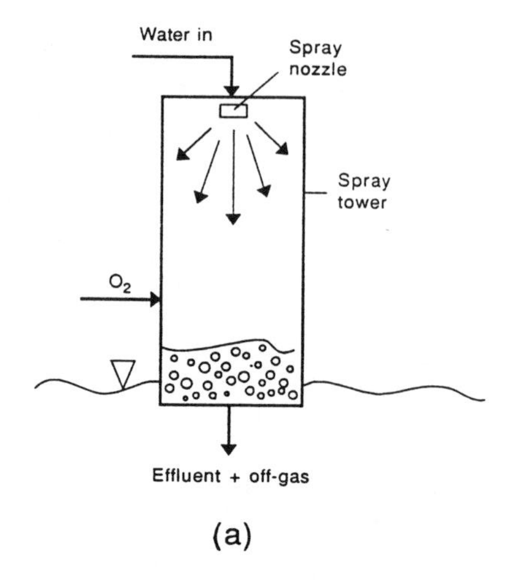

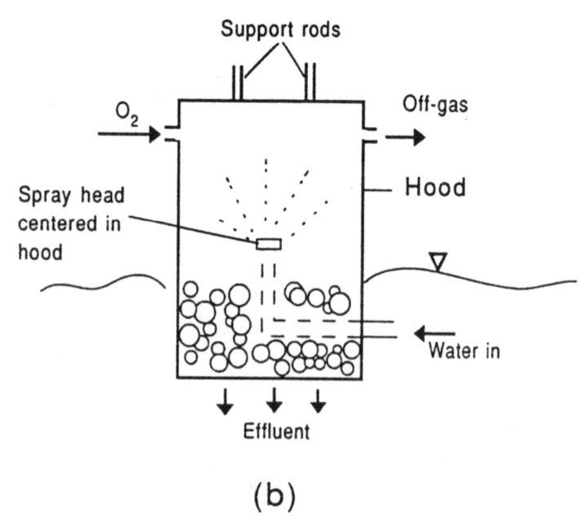

Figure 11-39. Spray towers: (a) Pressurized; and (b) open bottom.

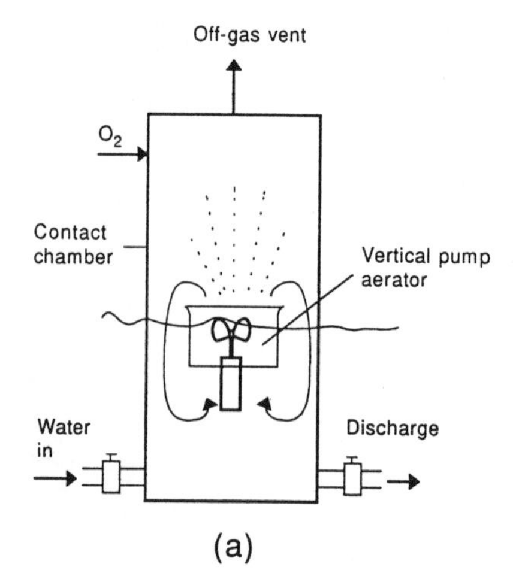

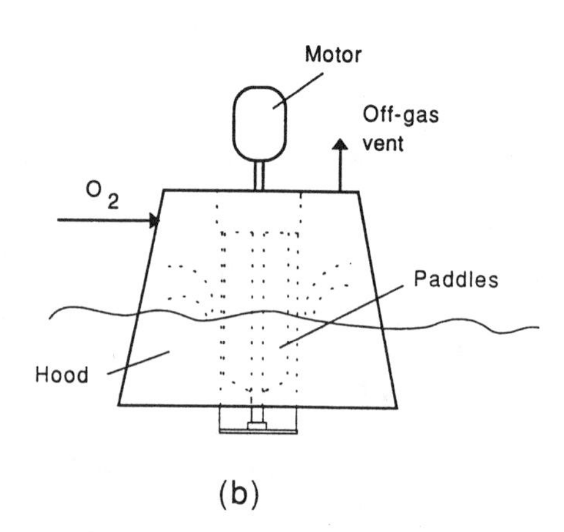

Figure 11-40. Enclosed surface agitators.

ters used to compare their performances are discussed in the sections that follow.

Oxygen Transfer Rate. Oxygen transfer rate (OTR) is calculated based on water flow rate and the change in dissolved oxygen across the system at equilibrium. In equation form, OTR is as follows (Boyd and Watten 1989):

$$OTR = Q_w\,(DO_{out} - DO_{in}) \times 10^{-3} \quad (11\text{-}63)$$

where OTR = oxygen transfer rate, kg/hr; DO_{in} and DO_{out} = system influent and effluent DO, respectively, mg/L; and Q_w = system water flow rate, m^3/hr.

Absorption Efficiency. The oxygen absorption efficiency (AE) represents the ratio of mass oxygen absorbed to total mass oxygen applied and can be defined as follows (Colt and Watten 1988):

$$AE = M_{abs}/M_{tot} \times 100 \quad (11\text{-}64)$$

where AE = absorption efficiency, %; M_{abs} = mass of oxygen transferred into the water, kg/s; and M_{tot} = total mass of oxygen used, kg/s.

Equation 11-64 can be written in another form as follows (Boyd and Watten 1989):

$$AE = \frac{Q_w\,(DO_{\text{increase}})}{1{,}000\,Q_o\,P_o} \times 100 \quad (11\text{-}65)$$

where Q_o = volumetric flow rate of oxygen, m^3/hr; and P_o = mass density of oxygen, kg/m^3.

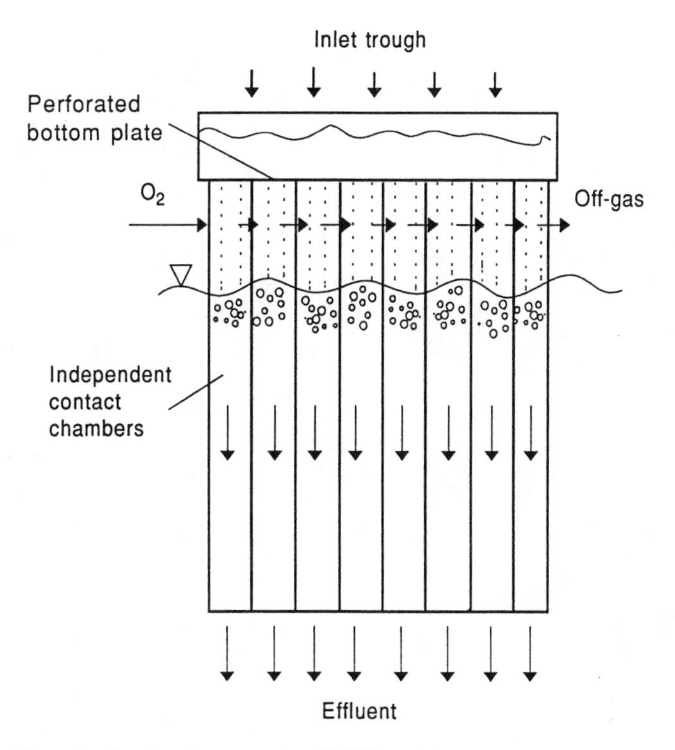

Figure 11-41. Cross section of a low head oxygenator (LHO) system.

Absorption efficiencies of pure oxygen systems range from about 40% to 100%. The product of Q_o and P_o is the mass flow of oxygen entering the contactor, which is usually measured with a rotameter. Corrections for temperature and pressure can usually be obtained from the rotameter manufacturer.

Example problem 11-8. Calculate the oxygen absorption efficiency when DO_{in} = 4 mg/L, DO_{out} = 20 mg/L, water flow rate = 2,200 L/min, and oxygen is added at 50 L/min. (Hint: One liter of O_2 weighs 1.43 g at STP.)
Solution:

$$AE = \frac{2.2 \times (20 - 4)}{1{,}000 \times 0.050 \times 1.43} \times 100 = 49.2\%$$

Transfer Efficiency. Similar to air-contact aeration equipment, the transfer efficiency (TE) of pure oxygen contactors represents the mass of oxygen absorbed per unit of energy input. The total energy input is the sum of the energy used to pump or agitate the water plus the energy to compress the gas. Transfer efficiency is calculated as follows (Boyd and Watten 1989):

$$TE = \frac{Q_w\,(DO_{out} - DO_{in})\,10^{-3}}{P_{pump} + P_{comp}} \qquad (11\text{-}66)$$

where TE = transfer efficiency, kg/kW-hr; and P_{pump} and P_{comp} = the power used in pumping and compressing, respectively, kW. The power used in compressing gas where oxygen is generated on-site using PSA equipment may be substantial. Gases also may be compressed in the recycle of off-gases, and this component must be included in the denominator of Equation 11-66.

The transfer efficiency for pure oxygen systems using LOX ranges from 0.2 to 4.0 kg O_2/kW-hr (Colt and Watten 1988).

Figure 11-42. An LHO system mounted in a raceway (photo courtesy of Ziegler Brothers, Inc., Gardners, Pa).

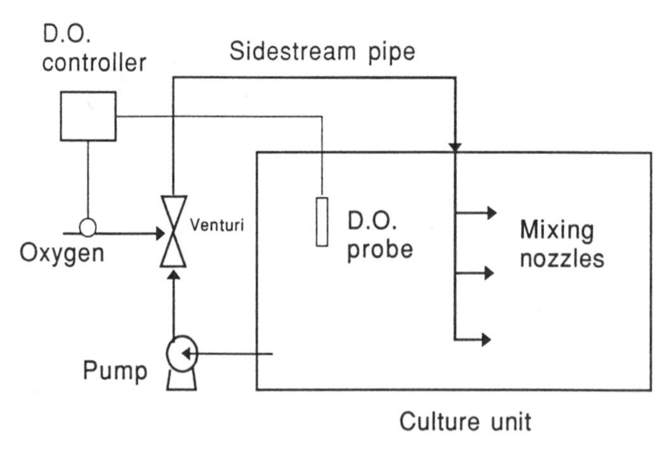

Figure 11-43. A sidestream pure oxygen injection system.

Transfer Cost. The total cost of oxygen transfer will include equipment transportation, energy use, and oxygen costs and can be calculated as follows (Boyd and Watten 1989):

$$TC = \frac{AC + EC + OC}{OTR \times 8760} \quad (11\text{-}67)$$

where TC = total cost of oxygen transfer/kg; AC

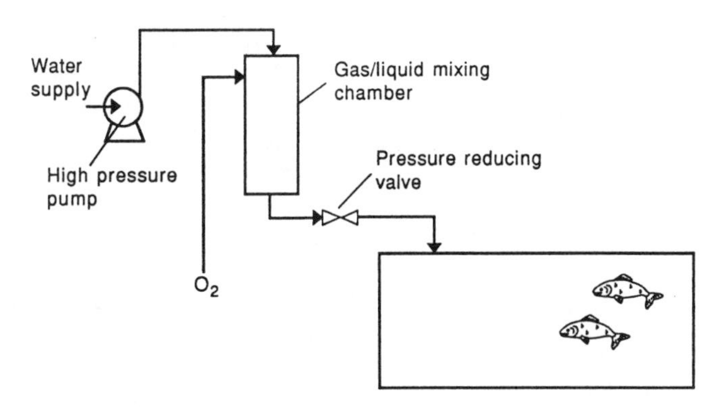

Figure 11-44. A high-pressure oxygen bubble generator.

= annualized equipment cost/year; EC = annual energy cost/year; and OC = annual oxygen cost.

Utilization Efficiency. The utilization efficiency (UE) is the percentage of the total oxygen passing through the system that is used by the fish (Colt and Watten 1988). For a single-pass system without aeration, the UE is calculated as follows:

$$UE\,(\%) = \frac{DO_{in} - DO_{out}}{DO_{in}} \times 100 \qquad (11\text{-}68)$$

In flow-through systems using pure oxygen, UE is equal to:

$$UE\,(\%) = \frac{Q_w \cdot DO_{in} + AE \cdot M_{tot} - Q_w \cdot DO_{out}}{Q_w \cdot DO_{in} + AE \cdot M_{tot}} \times 100 \qquad (11\text{-}69)$$

The utilization efficiency of nonaerated systems ranges from 20% to 40%, depending primarily on temperature and elevation. Pure oxygen systems have utilization efficiencies in the range of 50% to 70%, depending primarily on system layout.

Application of Pure Oxygen in Flow-Through Systems

Several methods for supplementing oxygen in flow-through systems have been described so far, including gravity and surface aeration and pure oxygen injection. The method most suitable for a specific facility is determined by the quantity of oxygen supplementation required, site specifics, and the equipment operating characteristics. Indications of the latter include OTR, AE, TE, and TC (Equations 11-63 through 11-67). Other factors that should be considered are reliability of equipment, nitrogen stripping needs and capability, safety, noise production, and performance flexibility.

It was previously mentioned that system production can be increased through the addition of pure oxygen. A mass balance approach can be used to establish the potential for increased production. The procedure is based on the assumption that production is limited by either DO or ammonia criteria. In most cases the first limiting factor is DO. The relationship expressing the required water flow for oxygen control was introduced earlier as Equation 11-30:

$$Q_{w(oxy)} = \frac{K_1 \cdot F}{DO_{in} - DO_{out}}$$

For salmonids, a minimum DO_{out} of 6.0 mg/L is recommended. Therefore, Equation 11-30 is reduced to:

$$Q_{w(oxy)} = \frac{K_1 \cdot F}{DO_{in} - 6} \qquad (11\text{-}70)$$

For catfish and other warm-water fish, a slightly lower value of DO_{in} such as 5.0 mg/L can be used safely. At this point, detailed information must be known about the amount of DO entering the system.

The flow rate is sometimes governed by un-ionized ammonia, and the mass balance relationship for ammonia control is as follows (Huguenin and Colt 1989):

$$Q_{w(amm)} = \frac{1319 \cdot \alpha \cdot K_2 \cdot F}{(NH_3 - N)_{out}} \qquad (11\text{-}71)$$

where $Q_{w(amm)}$ = flow required to maintain a given un-ionized ammonia criteria, L/min; α = mole fraction of un-ionized ammonia (mole fraction = NH_3-N/TAN); K_2 = ammonia production rate per unit of feed ≅ 30 g TAN/kg feed; F = total ration kg/day; and $(NH_3\text{-}N)_{out}$ = effluent un-ionized ammonia criteria, μg/L. Equation 11-71 assumes that the concentration of ammonia in the influent water is zero. A truly safe, maximum acceptable un-ionized ammonia concentration is unknown because of the number of parameters involved in fish culture. Westers (1981) has proposed a maximum un-ionized ammonia concentration of 10.0 μg/L for salmonids. For catfish and crawfish, a maximum un-ionized ammonia criteria of 50 μg/L is considered acceptable.

Reuse Ratio. At this point, another term called the *reuse ratio* (RR) requires introduction. The reuse ratio is the number of times that water can be reused before the un-ionized ammonia criteria is exceeded. The RR is defined (Colt and Watten 1988) as follows:

$$RR = \frac{\text{flow required for DO control}}{\text{flow required for ammonia control}} = \frac{Q_{w(oxy)}}{Q_{w(amm)}} \qquad (11\text{-}72)$$

If RR is less than 1.0, the maximum allowable ammonia criteria has already been reached, and supplemental oxygen is of little value. For RR values greater than 1.0, supplemental oxygen can increase the carrying capacity of the system. The reuse ratio is typically greater than 1.0 under conditions typical for salmonid culture, but, in warm water or marine culture, ammonia may be the most limiting parameter. The RR decreases as pH increases due to the increase in mole fraction of un-ionized ammonia. At a fixed pH, the RR first decreases as temperature increases, then increases as the temperature increases above 25°C. For marine applications, the reuse ratio varies from 1 to 2 for typical pH and temperature values (Huguenin and Colt 1989). The RR versus pH and temperature for a marine (salinity = 35 g/kg) system is shown in Figure 11-45.

Oxygen Demand and Supplemental Requirements. To size pure oxygen systems one must have estimates of the total oxygen demand of the fish, the available oxygen supplied by the flow of water through the system, and the amount of supplemental oxygen required. The total oxygen demand varies according to fish size and number and water temperature; therefore, it is necessary to estimate these parameters on a weekly or monthly basis throughout the production cycle. Calculations should be done

Figure 11-45. Reuse ratio as a function of pH and temperature. Based on K_{OXY} = 200 g/kg; K_{AMM} = 30 g/kg; $(NH_3 - N)_{OUT}$ = 10 mg/L; DO_{in} = 6 mg/L; and salinity of 35 g/kg (reproduced from Huguenin and Colt (1989).

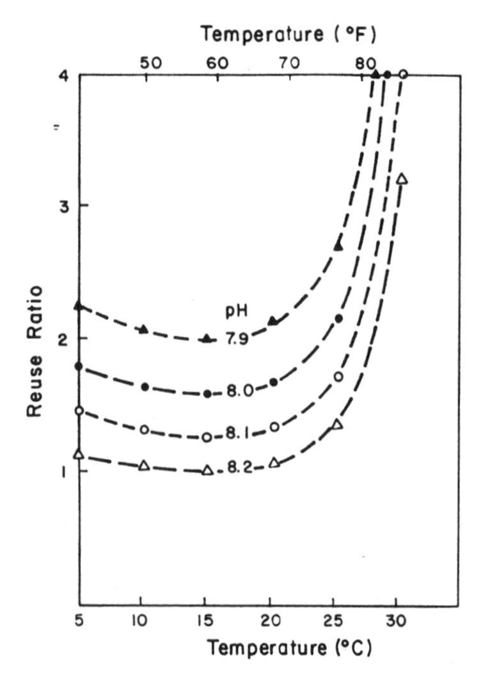

based on average and extreme maximum water temperatures.

The method used to define the average and maximum daily oxygen requirements and the supplemental oxygen requirement was described by Colt and Watten (1988).

AVERAGE DAILY TOTAL OXYGEN DEMAND. Fish use up a certain amount of oxygen in consuming and processing food. The oxygen requirement is based on fish size, feeding rate, ration size, composition, and digestibility. The oxygen-feed ratio (OFR) is calculated with Equation 11-73

$$OFR = M_e/Q_{ox} \quad (11\text{-}73)$$

where OFR = oxygen-feed ratio, oxygen required per unit mass of dry feed ingested, kg O_2/kg feed; M_e = metabolizable energy, the energy available from food less exogenous excretion (fecal plus nonfecal), kcal/g feed; and Q_{ox} = oxycalorific equivalent, the number of calories released by the substrate per unit mass of oxygen consumed, kcal/g O_2. Brett and Groves (1979) recommend a Q_{ox} equal to 3.24 kcal/g O_2. Assuming fish feed contains 2.6–3.6 kcal/g feed of metabolizable energy, the OFR for feeding at a maintenance level would be 0.80–1.11 kg O_2/kg dry feed. For feed having an M_e of 2.6 kcal/g and having moisture levels of 10% and 40%, the OFRs would be 0.72 and 0.48 kg O_2/kg wet feed, respectively.

Once the OFR is known, the average daily oxygen demand can be calculated from Equation 11-74:

$$\text{average daily } O_2 \text{ demand (kg/day)} = OFR \cdot R \quad (11\text{-}74)$$

where R = total ration, kg/day.

In actuality, feeding levels in most production systems would be greater than maintenance levels and would probably be closer to the maximum feeding levels. Westers (1981) reported that the OFR for salmonid culture is about 0.20 kg O_2/kg wet feed. The OFR for warm-water catfish cultured under laboratory conditions is 0.20–0.40 kg O_2/kg wet feed, but the OFR reported for commercial high-density culture of warm-water fish in general was 1.00 kg O_2 /kg wet feed. The higher values of OFR for warm-water fish may be due to higher levels of M_e and/or lower moisture levels in the feed, reduced oxygen transfer, higher bacterial oxygen demand from oxidation of organics and ammonia, different activity levels, and different feeding behavior (Colt and Watten 1988).

MAXIMUM DAILY TOTAL OXYGEN DEMAND. Westers (1981) suggested a peaking factor to compensate for the maximum oxygen consumption which occurs four to six hours after feeding (Figure 11-46). Thus, Equation 11-74 should be multiplied by 1.44 to account for the additional oxygen consumption:

$$\text{maximum daily } O_2 \text{ demand (kg/day)} = 1.44\ OFR \cdot R \quad (11\text{-}75)$$

OXYGEN SUPPLIED BY WATER FLOW. The oxygen supplied by water flow is:

Figure 11-46. Variation of oxygen demand during a normal production day.

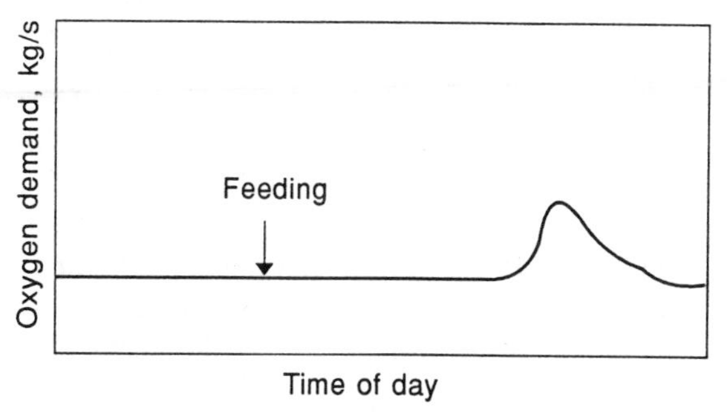

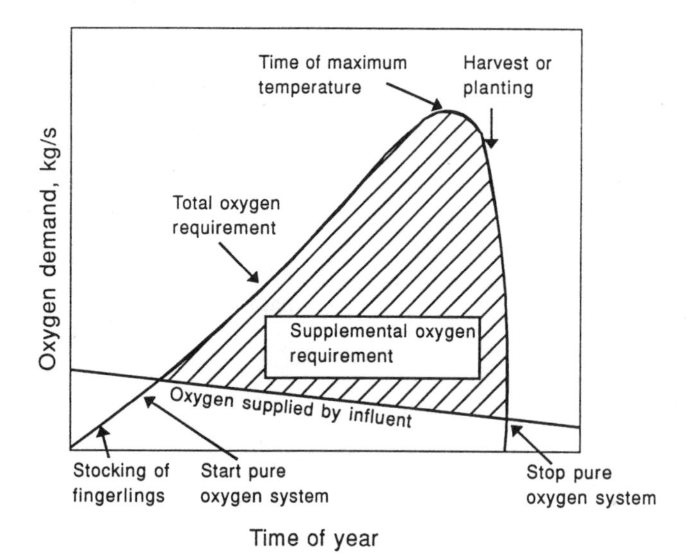

Figure 11-47. Variation of oxygen demand during production cycle.

$$O_2 \text{ supplied by flow (kg/day)} = 86.4\, Q_w\, (DO_{in} - DO_{out})\, (10)^{-6} \quad (11\text{-}76)$$

Detailed DO data for the facility should be known to get a reliable estimate of the oxygen supplied by the flow. Where reliable data is lacking, a DO_{in} value of $0.90C_s$ can be assumed for undeveloped streams, rivers, or other highly oxygenated water sources. If detailed DO data is unknown in situations where wells or springs are used for the water source, a DO_{in} value of zero should be assumed (Colt and Watten 1988).

SUPPLEMENTAL OXYGEN DEMAND. The amount of supplemental oxygen required is the difference between Equations 11-75 and 11-76. The supplemental oxygen requirement will increase throughout the production cycle as the fish biomass and feeding rates increase. Thus, for design purposes, the supplemental oxygen requirement should be based on weekly or monthly biomass and feeding levels as shown in Figure 11-47. Depending on the water temperature and harvest schedule, the maximum supplemental oxygen demand may occur near the end of the production cycle (Colt and Watten 1988).

Increased Production Examples. Fish production capacity, based on oxygen criteria, can be obtained by rearranging Equation 11-31 as follows:

$$\frac{DO_{avail}}{K_1 \cdot R} = \frac{B}{Q_{w(oxy)}} \quad (11\text{-}77)$$

Recall that R, the percent body weight (% BW), is the average fish weight divided by 100. Thus, if % BW is inserted into the equation as a whole number rather than a fraction, and if K_1 is divided by 100, the *loading formula* becomes:

$$\frac{\text{kg fish}}{\text{L/min}} = \frac{DO_{avail}}{2.0 \times \%\ \text{BW}} \quad (11\text{-}78)$$

Because Equation 11-78 includes % BW, water temperature and fish size are automatically taken into consideration (Westers 1987). The following examples demonstrate the effect of increased oxygen on production.

Example Problem 11-9. Rainbow trout are to be reared in a flow-through culture unit in 10°C water. The incoming water is 90% saturated. The feeding level is 2% BW. What fish produc-

tion (kg/L/min) can be achieved with this setup?
Solution:
a. At 10°C the DO_{in} is 0.90 × 11.288 mg/L (from Table 11-3). For salmonids, a minimum DO_{out} of 6.0 mg/L should be used. Thus, according to Equation 11-78, maximum production is 1.04 kg fish/Lpm flow.

$$\text{kg/L/min} = \frac{10.16 - 6.0}{2 \times 2} = \frac{4.16}{4} = 1.04$$

b. If the oxygen level is increased to saturation (11.288 mg/L) by the addition of pure oxygen, maximum production is now increased to 1.32 kg fish/L/min, a 28% increase.

$$\text{kg/L/min} = \frac{11.288 - 6.0}{2 \times 2} = \frac{5.29}{4} = 1.32$$

c. Suppose the oxygen level is increased to 150% saturation (16.92 mg/L) by the introduction of pure oxygen. This results in a production potential of 2.73 kg fish/Lpm, a 169% increase over the original production figure.

$$\text{kg/L/min} = \frac{16.92 - 6.0}{2 \times 2} = \frac{10.92}{4} = 2.73$$

Configuration Of Pure Oxygen Systems

Pure oxygen systems can be operated in a number of configurations. Some of the more common configurations are illustrated in Figures 11-48 through 11-50. The most common is the full-flow configuration. In this configuration all of the water to be processed passes through the oxygen absorber, either in a single-pass system with no reuse (Figure 11-48a) or with serial reuse (Figure 11-48b). Figure 11-49 illustrates a full-flow, series/parallel configuration. Full-flow configurations are used when existing facilities are retrofitted with either packed column or jet aerators (Colt and Westers 1982). This configuration is beneficial in situations where there is sufficient head to operate the oxygenator without additional power input. The minimum carrying capacity is limited by DO_{avail} (Colt and Watten 1988). The single-pass system is the most desirable from a disease transmission point of view.

A second configuration is the sidestream where a portion of the flow passes through the oxygen absorber (Figure 11-50). The sidestream configuration is more flexible than the former in that the oxygenated water can be mixed at several points to reduce gas supersaturation and bubble formation problems (Colt and Watten

Figure 11-48. Single-pass, pure oxygen systems: (a) Full-flow configuration; and (b) full-flow configuration with reuse.

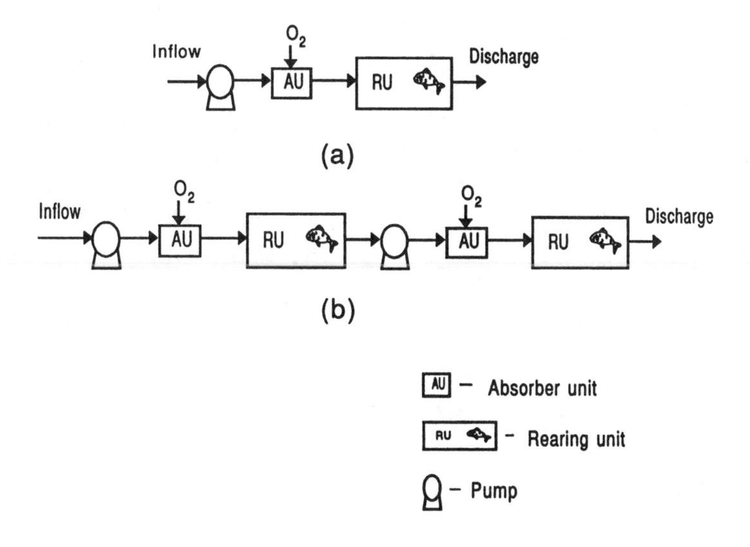

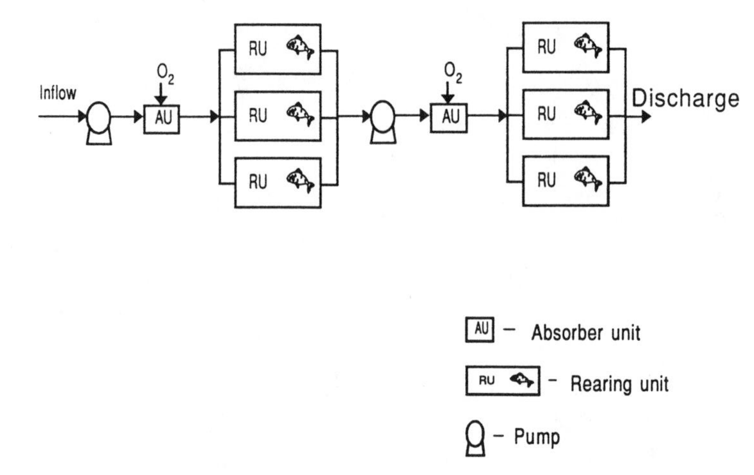

Figure 11-49. A full-flow, series/parallel pure oxygen system.

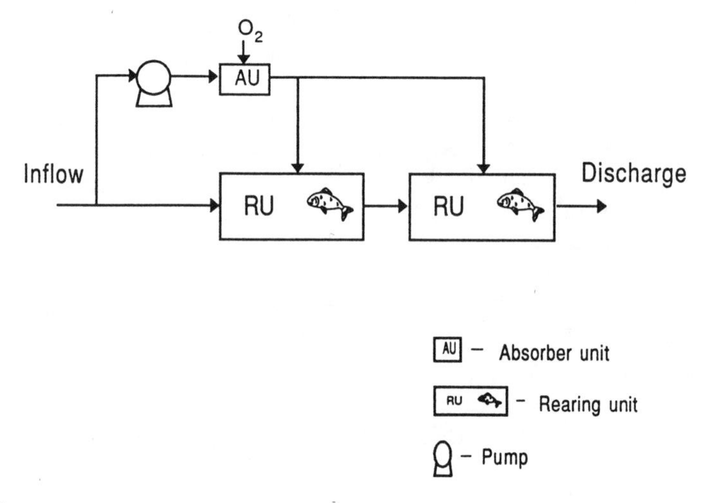

Figure 11-50. A sidestream pure oxygen system.

1988). The oxygen capacity of an individual rearing unit can be increased with the use of multiple oxygenator units in this configuration. A pressurized packed column can be used to produce supersaturated conditions. Oxygen loss to bubble formation when the water is redistributed into the main stream can sometimes be a problem. The cost of system distribution piping may be considerable, depending on the size of the system.

CHAPTER 12

STERILIZATION AND DISINFECTION

Aquaculture production systems must provide good water quality and disease control in order to be productive and profitable. Disease can occur at all levels of production, but problems increase accordingly as the size of the facility increases and culture intensifies. In today's world the difficulties encountered are magnified as good-quality surface water and groundwater supplies decrease and effluent regulations become more demanding. These factors encourage the use of more intensive culture methods to conserve water. However, under more intensive conditions the risk of disease transmission is greatly increased. Disinfection on some scale is usually necessary to minimize transmission of disease organisms and parasites.

Disease and parasitic organisms of concern to fish culturists include bacteria, viruses, fungi, protozoans, and worms. These organisms can enter culture systems through untreated incoming water, by the introduction of new fish into the system or through recycled culture water that has received insufficient disinfection. Normally harmless organisms can become infectious when environmental changes cause a sudden increase in their numbers. The environment and health of the host determine whether the organisms remain latent or become infectious. Heavy infestations often occur on seemingly healthy fish despite carefully maintained and controlled environmental factors (Spotte 1970).

Temporary declines in dissolved oxygen; rapid pH; and/or temperature changes; and increases in nitrogenous compounds, organics, or carbon dioxide can individually or in combination produce stresses that weaken the animals and target them for infections. Fish cultured in closed recirculating systems are usually hit hardest because of crowded conditions and because the water is reused. Most epizootic outbreaks of disease occurring in recirculating systems are caused by bacteria and protozoans. They can usually be held in check with stringent water quality control including disinfection.

It may be more desirable and economical to control only harmful organisms. The technique of selectively controlling organisms that enter aquaculture systems is called *disinfection*. Inline disinfection is commonly used in recirculating culture systems as a separate water treatment. *Sterilization* is the destruction of all organisms, both harmful and beneficial. It is used more for treating nets, boots, seines, and other gear that may easily transmit infectious organisms. In addition, it is a good idea to sterilize culture units before putting fish in or after a disease incident.

Disinfection and/or sterilization can be accomplished using one or a combination of readily accepted methods. It is not the intent of this author to discuss the characteristics or treatment of individual infectious organisms, but to address measures that may be used to prevent or treat infections by a broad range of organisms. For a detailed discussion of fish disease organisms and treatments the reader is referred to texts on aquatic pharmacology.

DISINFECTION OPTIONS

Basically, four types of disinfection processes are normally used at some level of aquaculture production: heat, chlorination, ultraviolet irradiation, and ozone. For various reasons, mostly economical, other disinfection processes have not been proven to be feasible for aquaculture purposes. The effectiveness of a given disinfecting agent or method depends on concentration used, contact time, temperature, turbidity, particulate concentration, and characteristics of the target organism.

Heat

Heat in the form of steam or elevated culture water temperature are two options available for use in aquaculture systems. Small steam generators are commonly used to sterilize boots, nets, seines, tanks, sumps, and other equipment that contact fish and are common avenues of infection. Water can also be disinfected by elevating its temperature. The percent kill of the microorganisms depends on the final temperature, holding time, and the target species. Since water has a high specific heat considerable energy is required to heat it to the proper disinfecting temperature. It then must be cooled to temperatures comfortable to the fish. Thus, the cost of heating and cooling water can often preclude its use for disinfection purposes.

Chlorine

Of the numerous chemical disinfectants available, chlorine is perhaps the cheapest and most readily available. Chlorine is used in the form of either chlorine gas (Cl_2), calcium hypochlorite ($Ca(OCl)_2$), or sodium hypochlorite ($Na(OCl)$). In all cases the active disinfecting agent is either the hypochlorite ion (OCl^-) or hypochlorous acid ($HOCl$). These species exist in a pH-dependent equilibrium as shown in Equation 12-1 (Piedrahita 1991):

$$HOCl \leftrightarrow OCl^- + H^+ \quad (pK_a = 7.54 \ @ \ 25°C) \qquad (12\text{-}1)$$

Calcium hypochlorite is a granular solid, is more stable and occupies less space than the liquid sodium hypochlorite (common household bleach). However, sodium hypochlorite is more readily available and is cheaper and easier to use (Huguenin and Colt 1989). Chlorine gas, $HOCl$ and OCl^- are strong oxidizing agents. In addition to microorganism destruction they react with a variety of organic and inorganic materials. A major compound in aquaculture systems that readily combines with chlorine is ammonia. $NH_3 - HOCl$ reactions may produce mono-, di-, and/or trichloramines as shown in Equations 12-2 to 12-4:

$$NH_3 + HOCl \rightarrow NH_2Cl + H_2O \qquad (12\text{-}2)$$

$$NH_3 + 2HOCl \rightarrow NHCl_2 + 2H_2O \qquad (12\text{-}3)$$

$$NH_3 + 3HOCl \rightarrow NCl_2 + 3H_2O \qquad (12\text{-}4)$$

The equilibrium constants for these reactions are dependent on pH, water temperature, and relative concentrations of NH_3 and $HOCl$. Chloramines can also be formed from reactions of $HOCl$ with organic amines.

The $HOCl$ and OCl^- compounds are called *free residual chlorine* (FRC), and the chloramines are referred to as *combined residual chlorine* (CRC). *Total residual chlorine* (TRC) refers to the sum of the free and combined forms. The proportions of free and combined chlorine present following a chlorine dosage depend mainly on the ammonia concentration of the receiving water that is available to combine with the chlorine (Wickens 1981).

Traditional chlorination nomenclature used

for freshwater is inappropriate for seawater since the chemistry of seawater chlorination is very complex and not well understood (Block et al. 1977). Chlorine hydrolyzes to hypochlorite, which, in turn, reacts with bromide in seawater to form hypobromite (Wong and Davidson 1977). Some of the hypobromite produced reacts with organic material to produce brominated organic compounds (Bean et al. 1978; Blogoslawski 1991). The rate at which these reactions take place is somewhat influenced by sunlight (Macalady et al. 1977; Wong 1980).

Chlorine concentrations recommended for disinfection are presented in Table 12-1. These concentrations are based on 65% available chlorine for calcium hypochlorite and 3.7% for chlorine bleach. Since the disinfecting power of chlorine is strongly dependent on pH, the time-concentration recommendations in Table 12-1 are based on the addition of an acid solution to produce a pH of 6.0. If it is not practical to lower system pH to 6.0, a contact time of 24 hours is recommended (Huguenin and Colt 1989).

The toxicity of the free and combined forms of chlorine may be different for different microorganisms and higher life forms. Sufficient contact time must be allowed for chlorine to be effective in killing microorganisms. It first dissociates into the water before it becomes toxic. It then diffuses through cell walls and kills by inactivating certain enzymes. Chlorine toxicity to microorganisms varies widely and is dependent on water conditions, temperature, and the species of target organism.

Table 12-1. Recommended chlorine dosages for disinfection.

		Chlorine form	
Description	Contact time (min)	Bleach (mg/L)	Calcium hypochlorite (mg/L)
Nets, buckets boots, etc.	5	0.70	40
Transport equipment	30	2.64	150
Rearing containers	60	3.51	285

Source: Huguenin and Colt (1989) with permission.

The use of chlorine-disinfected water in aquaculture systems carries a certain element of risk since fish and invertebrates are highly sensitive to low concentrations of both the free and combined forms (Heath 1977; Dupree 1981). Some authors suggest that there is little difference between free and combined chlorine compounds, although free chlorine appears to react faster. Holland et al. (1960) reported that di- and trichloramines are much more toxic than either free chlorine or monochloramine. Chloramines are not as easily removed from solution as the free forms and are often left as a residual with longer-lasting disinfecting qualities.

The toxicity of either free or combined chlorine can begin as low as 0.1–0.3 mg/L (Dupree 1981). Giles and Danell (1983) reported behavioral changes in rainbow trout exposed to 0.005 mg/L free chlorine. Sublethal TRC concentrations of 0.07 mg/L caused a 50% reduction in growth in marine plaice and Dover sole (Alderson 1974). Chloramines (0.022 mg/L TRC) reduced growth in juvenile coho salmon (Larson et al. 1977). Van Olst et al. (1976) reported a 96-hour LD_{50} of 0.22 mg/L TRC for juvenile *Homerus americanus*. The growth of *Mytilus edulis* was markedly depressed by exposure to 0.2 mg/L TRC over a 10-week period (White 1966).

Heath (1977) studied the effects of free chlorine and monochloramine on five species of fingerling-size freshwater fish. He noted that the toxicity of each form varied with water temperature and exposure time. Table 12-2 shows data extrapolated from Heath's study. Toxicities were evaluated for several exposure times, but only the 96-hour LC_{50} results are shown in the table since this is the most common exposure time reported in the scientific literature. The results show that rainbow trout, coho salmon, and channel catfish were more sensitive to free and combined chlorine than either common carp or golden shiner. Carp were the most resistant. Free chlorine was three to 14 times more toxic than monochloramine, depending on species.

The reported safe level of TRC in aqua-

Table 12-2. Toxicity (LC_{50}) of intermittent free and combined chlorination to fingerling-size freshwater fish.

			Chlorine form	
Species	Exposure time (min)	Temp. (°C)	Free (mg/L)	Combined (mg/L)
Rainbow trout *(Salmo gairdneri)*	96	5	0.082	-
		12	0.062	-
		17	0.095	-
	120	5	0.074	-
		12	0.052	0.750
		17	0.089	-
Coho salmon *(Onchorynchus kisutch)*	72	6	0.181	-
		12	0.119	-
	96	6	-	0.677
		12	-	0.640
	144	6	-	0.537
		12	0.093	0.553
Channel catfish *(Ictalurus punctatus)*	96	5	0.082	0.275
		24	0.064	0.260
	120	5	0.050	0.234
		24	0.032	0.246
Carp *(Cyprinus carpio)*	96	6	0.538	1.72
		24	0.219	-
	120	6	0.400	1.60
		24	0.219	-
Golden shiner *(Notemigonus crysoleucas)*	96	5	0.269	0.724
		24	0.193	0.930
	120	5	0.205	0.763
		24	0.182	0.921

Source: Heath (1977).

culture systems is 0.5–3 μg/L (Piedrahita 1991). Free chlorine levels of 0.001–0.002 mg/L are considered safe for intermittent or continuous exposure to most freshwater fish (Brungs 1973). The European Inland Fisheries Advisory Council in 1973 proposed a maximum HOCl concentration of 0.004 mg/L, which is equivalent to 0.016 and 0.005 mg/L TRC at pH 8.0 and 7.0, respectively (EIFAC 1973).

Chlorine compounds must be removed from water before it can be used for aquaculture purposes. Many city water supplies are chlorinated, and in many instances, city water is used as a water source for aquaculture. In some states chlorinated effluents from aquaculture systems must be treated and the chlorine removed before discharging to receiving streams in order to protect aquatic life. Chlorine removal can be difficult, but several dechlorination options are available: reducing agents, activated carbon, UV irradiation, aeration, and ozonation. Each of the methods individually has one or more disadvantages in terms of cost, chlorine concentration or variable chlorine residuals.

Reducing agents like sodium thiosulfate ($Na_2S_2O_3$), sodium sulfite (Na_2SO_3), and ferrous salts reduce free chlorine to inactive compounds, but they have little effect on chloramines (Dupree 1981). Complete dechlorination with sodium sulfite requires a 10.7:1 ratio by weight of sodium sulfite to chlorine (Giles and Danell 1983). Reducing agents should be used cautiously since they have toxic properties of their own. Piedrahita (1991) recommended that they not be used in strengths greater than 1–5 mg/L.

Activated carbon effectively removes free chlorine and is the only reported effective method for chloramine removal (Dupree 1981). It is safe for both fish and bacteria in biofilters, and can be used in a variety of ways in either batch, flow-through, or recirculating systems. However, activated carbon must frequently be replaced or cleaned and reactivated. This may be practical on a small scale but would be uneconomical on a large commercial scale.

Aeration for 24–48 hours is often recommended for dechlorination. However, it is doubtful if chloramine residuals are completely removed by this method. Aeration is particularly ineffective for dechlorination of water containing a high concentration of organic substances (Bedell 1971). Until better information is available, aeration by itself is not recommended for chlorine removal.

A limited amount of chlorine can be photochemically reduced by exposure to UV light (Giles and Danell 1983; Seegert and Brooks 1978) or removed by a combination of ozone and activated carbon (Seegert and Brooks 1978). Both UV and ozone effectiveness are inhibited by the presence of sodium thiosulfate. Ozone, however, must be used with caution. If not properly applied and managed, ozone residuals can be toxic to fish and invertebrates in culture units and to bacteria in biofilters. Also, exposure of ozone in air is risky to humans. Ozone toxicity is discussed in a later section.

Ultraviolet Irradiation

Ultraviolet irradiation (hereafter referred to as UV or UV light) was first used around 1910 for disinfecting the water supply of Marseilles, France. It was later used in Japan and Spain to sterilize seawater used to depurate bivalves (Blogoslawski 1991). Shellfish depuration with UV light was introduced into the United States in the early 1960s as an alternative to chlorination. By the 1980s about 20 depuration plants were certified in the United States, but this figure has since reduced to less than 10 (Herrington 1991). Briefly, in depuration systems, seawater is exposed to UV light either in a single-pass flow-through mode before exposure to the shellfish or water is recirculated after UV treatment. Today, UV light has become an indispensable tool for controlling fish pathogens in hatcheries (using both flow-through and recirculating technology), public aquaria, and in heated effluent aquaculture. UV light is effective in the control of certain viruses, bacteria, fungi, and many other microorganisms. The control of diseases in hatcheries has been shown to produce larger and more robust fish (Flatow 1981).

The effectiveness of UV light depends upon a number of factors (Moe 1989):

- UV bulb wattage, age, and cleanliness
- Temperature of the bulb and system
- Distance between the bulb and target organism
- Species and individual characteristics of the target organism.
- Duration and intensity of exposure
- Clarity of the culture water

UV Light Generation. UV light is derived from both natural and artificial sources. The sun is the most important natural source. Artificial UV light can be produced with a wide variety of arcs and incandescent lamps. The overwhelming percentage of UV light generated for disinfection in aquaculture systems is produced by low-pressure mercury vapor bulbs. These bulbs produce UV light as a result of a flow of electrons between the electrodes through a tube containing ionized mercury vapor. Mercury vapor bulbs produce UV radiation in such high doses that in fractions of a second greater UV irradiation is produced than is produced by the sun in several hours.

Mercury vapor bulbs create short-wave radiation at a wavelength of 254 nm (2,537 Angstrom units), which is the most effective wavelength (Spotte 1979; Rosenthal 1981). The biologically most active wavelengths are those between 1,900 and 3,000 Å, as illustrated in Figure 12-1. Wavelengths above 3,000 Å are significantly less effective, and wavelengths of about 1,849 Å produces excessive amounts of ozone. At a wavelength of 2,600 Å the maximum microbicidal action occurs for practically all microorganisms since DNA absorbs UV light most strongly at this wavelength (Deering 1962). UV light interacts with the purine and pyrimidine components of the DNA and has the greatest injurious and lethal effects on microorganisms (Shechmeister 1977; Carlson et al. 1985). No fish pathogens are known to survive UV light treatment when exposed to the proper intensity and wavelength (Flatow 1981).

Microorganism inactivation is proportional to UV light intensity per unit area times the exposure time in seconds. This is called the UV *dose* and is reported as microwatt seconds per square centimeter ($\mu Ws/cm^2$). Microwatts are determined by the output of the bulb, exposure time is determined by water flow rate, and contact area by the internal area of the UV bulb or other contact area (Moe 1989).

Reactor Types. Ultraviolet light reactor manufacture has improved over the years. Units are available in sizes small enough to be hung on the outside of an aquarium (costing under $100) to large units costing many several thousands of dollars. Many are fabricated from PVC components and some are modular and freestanding so that any number of units can easily be plumbed into existing aquaculture systems using commonly available fittings. Four types of UV photoreactors are manufactured for aquaculture use: suspended, submerged, jacketed, and a type that encloses the water in fluorocarbon polymer tubes.

Suspended bulb reactors are the oldest type. They are constructed with the UV bulbs suspended over a large, shallow free-flowing water surface. The most common example of this type

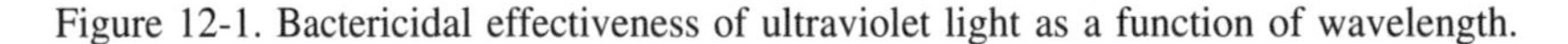

Figure 12-1. Bactericidal effectiveness of ultraviolet light as a function of wavelength.

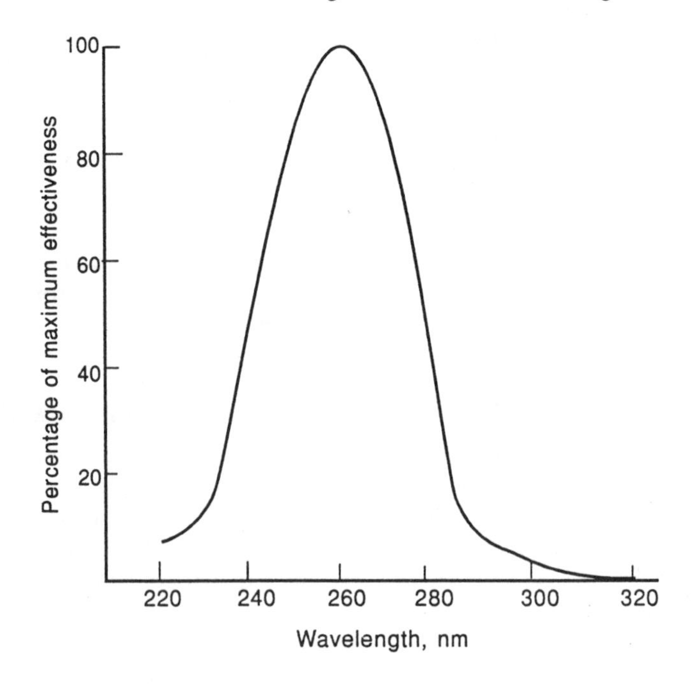

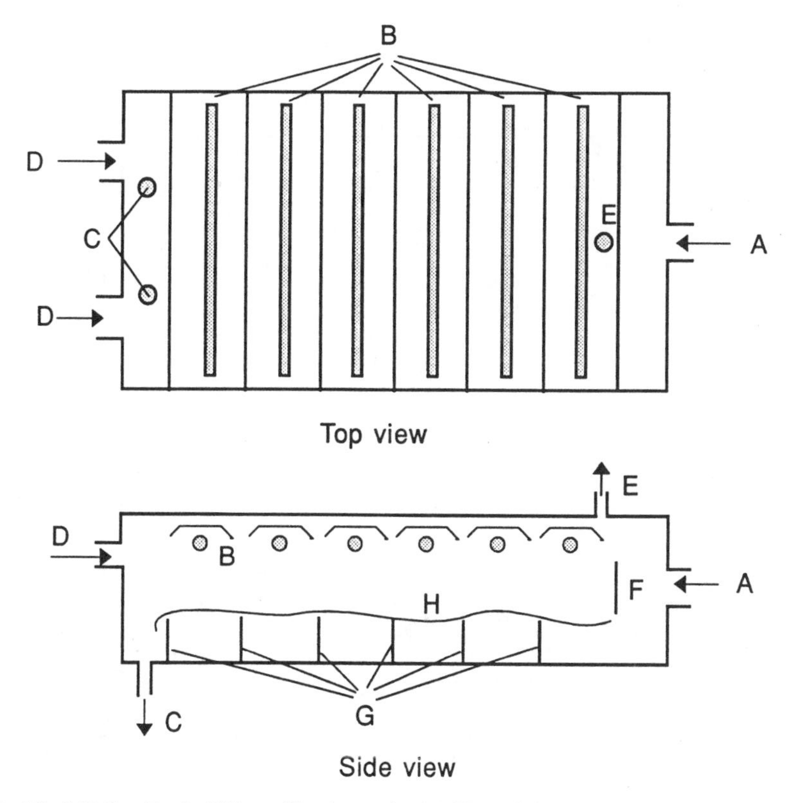

Figure 12-2. Modified Kelley-Purdy UV sterilization unit: (a) Water inlet; (b) UV-bulbs; (c) water outlet; (d) air blower inlets; (e) energy dissipator; (f and g) baffles; and (h) water surface.

reactor is the Kelly-Purdy (K-P) reactor described by Herrington (1991) and illustrated in Figure 12-2. Water enters one end of the unit (A), flows over low baffles (G), and exits the reactor at the opposite end (C). The baffles (usually only about 25 mm in height) provide turbulence and prevent hydraulic short-circuiting through the reactor. The water flows by gravity through the unit creating no back-pressure. The water is exposed to the UV light (B) in a thin sheet. Reflectors above the bulbs (I) are important since, without them, there is considerably less light. The bulbs are suspended about 10–20 cm above the water surface. If the bulbs are too close to the water the light intensity may vary at different points in the unit, creating hot spots directly underneath. Also, water splashing onto the bulbs enhances the growth of biological slime, seriously decreasing light intensity. The reactor should be constructed with vents (D and E) so that ventilating air can be circulated through the unit and excess heat removed.

In K-P units the bulbs are not protected from the corrosive effects of salt in seawater units and should cleaned at regular intervals. This type of unit should also be manufactured with a hinged lid for ease of cleaning and bulb replacement. The water contact region should be cleaned regularly to prevent buildup of algae and organic deposits that can decrease the water volume in the contact area. The resulting increased water flow rate will prevent adequate exposure time of the UV light to the waterborne organisms.

In *submerged bulb reactors* the UV bulb is contained within a PVC tube, and the water flows between the bulb and the tube (Figure 12-3). Hence, the bulb is directly immersed in water. Most small aquarium units are of this type of construction. A major disadvantage is

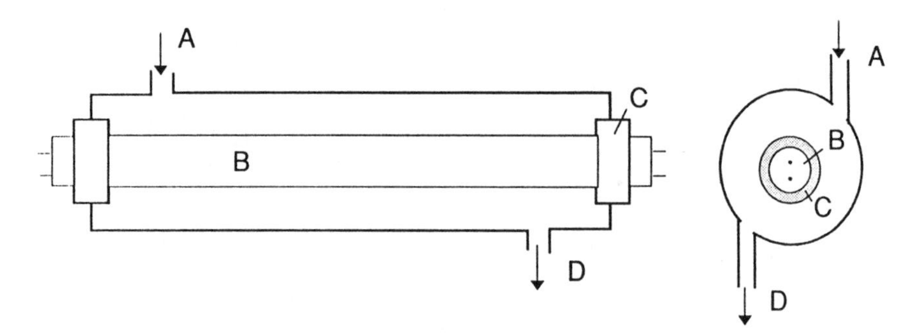

Figure 12-3. A submerged bulb UV sterilizer: (a) Water inlet; (b) UV bulb; (c) watertight seal; (d) outlet; and (e) PVC tube.

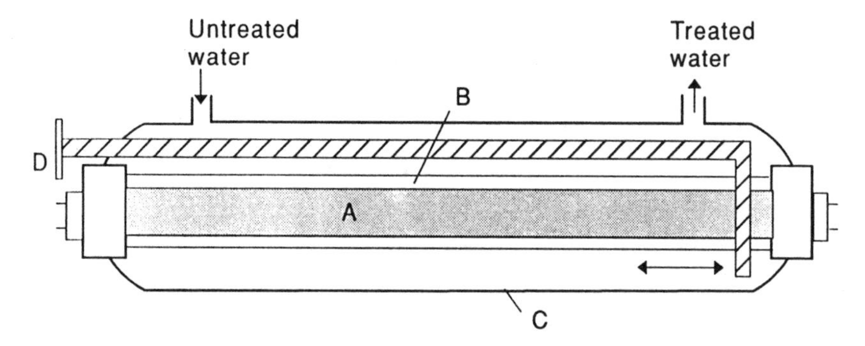

Figure 12-4. Typical quartz-jacketed UV sterilization unit: (a) UV bulb; (b) quartz tube; (c) stainless steel photoreactor; and (d) push rod for cleaning.

the temperature at which the UV bulb operates since the effectiveness of UV irradiation is a function of temperature. The bulb operates at the same temperature as the water. The bulb will operate at its maximum efficiency only if the water temperature is within the range 37.8–49°C. The optimum range is 40–40.6°C (Flatow 1981). At lower temperatures bulb efficiency decreases. For example, at 21°C (70°F), bulbs function at only about 50% efficiency (Spotte 1979).

Jacketed bulb reactors have a quartz sleeve around the UV bulb (Figure 12-4). Both are sealed within a PVC tube, and the water passes between the tube and the quartz jacket. The reason for the quartz sleeve is twofold: (1) Quartz glass has superior light transmission properties to other types of glass, allowing 90–95% of the UV rays through; and (2) the jacket forms an air space around the UV bulb, allowing it to operate at its most efficient temperature without the effects of the outside water temperature (Spotte 1979; Harris et al. 1987). A major advantage of jacketed bulb reactors over submerged reactors is that bulb replacement is much easier. No watertight seal is required around the bulb as in submerged units. A disadvantage is that biological slime can build up on the water side of the sleeve and greatly reduce UV light transmittance. Some jacketed bulb reactors are manufactured with a push rod (D) that slides over the bulb and removes deposits.

A fourth type of UV reactor is illustrated in Figure 12-5. In this type the water is contained in clear fluorescent polymer (Teflon) tubes (A) adjacent to UV bulbs (B). Teflon tubes are used because they allow 70–85% transmission of UV light (Harris et al. 1987). Quartz is not used because construction costs would be too high. The Teflon tube units are available in a variety

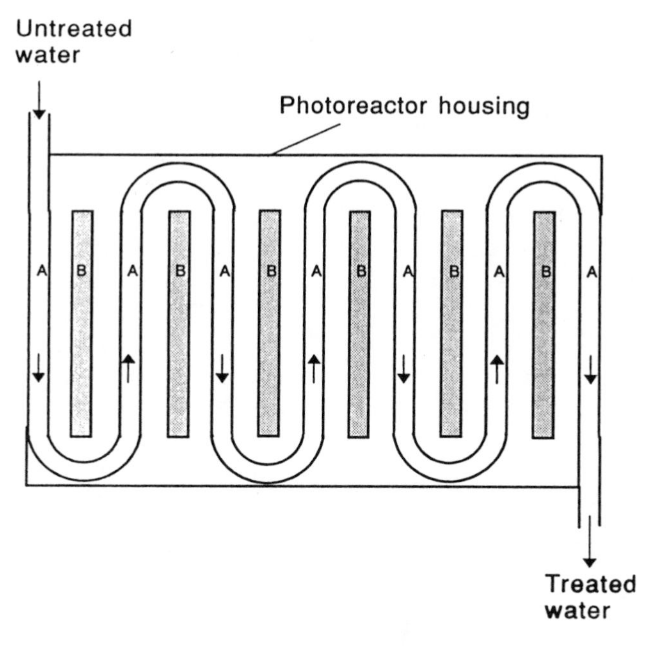

Figure 12-5. A UV reactor using fluorocarbon polymer (Teflon) tubes (A) exposed to UV bulbs (B).

of commercial sizes. Some are wall-mount units while others are freestanding floor units. On the down side, Teflon units are more difficult to clean and maintain than other types. Tube removal and replacement is difficult, and the tubes are easily damaged. It is as important to clean Teflon tubes as quartz-jacketed bulbs. Kreft et al. (1986) demonstrated that tube UV light transmission can increase from 5% before cleaning to 72% after cleaning.

Maintenance. Maintenance of UV photoreactors can be a significant portion of the overall operating costs of large commercial aquaculture facilities. The biological slime, salt, or mineral deposits that accumulate on UV bulbs exposed to culture water should be removed at regular intervals to maintain maximum efficiency. UV bulbs normally lose about 10% of their radiation output during the first 100 hours of operation (Flatow 1981), but efficiency is also reduced because of degradation of the electrodes, caused by switching on and off, and because of the gradual darkening of the inner glass surface of the bulb, called *solarization*. The darkening effect is due to mercury that has been vaporized and settled on the inner surfaces of the glass. The National Shellfish Sanitation Program recommends replacing bulbs after about 7,500 hours of use; however, bulbs do not necessarily last this long. Replacement intervals vary, depending on site conditions and frequency of use. Because of solarization, effective bulb life may be only 6–12 months. In-line metering clocks should be installed to measure bulb use, and records should be kept on each bulb in the system. Bulb intensity should be checked regularly, and when intensity falls below 60% of the bulb's output when new, it should be replaced (Herrington 1991).

Factors Affecting UV Light Effectiveness UV light effectiveness is largely determined by the degree to which the UV rays can penetrate water. Water penetration is a function of water depth, turbidity, and ions dissolved into solution (Spotte 1979).

WATER DEPTH. UV rays can penetrate water only to a depth of about 50 mm (2 in.) (Spotte 1979), and penetration is greater in freshwater than in salt water. Therefore, the culture water should pass under or around bulbs in a thin film to provide the greatest exposure to the UV radiation. Moe (1989) recommended that the flow should be no more than 25 mm (1 in.) deep to ensure complete penetration. Many manufacturers of commercial UV equipment limit water depth in photoreactors to 1–3 mm of prefiltered water in order to achieve a 99.9% kill of microorganisms. High flow rates are discouraged in these units. For this reason, large-volume UV sterilizers are bulky and expensive. The size of an aquaculture system may then become somewhat limited by the amount of water that can be moved through the UV sterilizer.

TURBIDITY. UV light penetration is affected by water clarity. The effect of organic matter, color, turbidity, algae, debris, and suspended particles is to absorb a portion of the light waves before they reach disease organisms, reducing the UV dose and, hence, effectiveness of kill. Also, particulate matter in the water shields disease organisms from the disinfecting effect of the light waves. In shellfish depuration systems receiving UV treatment, turbidity must not exceed 20 nephelometric turbidity units (NTUs) (APHA 1985; NSSP 1990). Presnell and Cummins (1972) reported that the critical upper turbidity limit was 82 Jackson turbidity units (JTUs) in a K-P unit with a 145-L/min flow rate. Color also inhibits light penetration, if the color is of organic origin. For the reasons discussed water to receive UV disinfection may require prefiltration, an expensive proposition in large systems.

CHLORINE. UV light effectiveness is inhibited in systems where chlorine is used for disinfection along with the UV light. If sodium thiosulfate is used for dechlorination, a sodium thiosulfate residual can inhibit the effectiveness of the light.

IRON. If the concentration of dissolved iron is less than 5 mg/L, its inhibition of UV light penetration can be offset by increased dosage, whereas higher concentrations will require iron removal by other methods since this in undesirable for most aquatic species (Flatow 1981).

Determining UV Dosage. The minimum lethal dose (MLD) required to inactivate microorganisms is determined by the size of the target organism, species, and strain of a particular species (Spotte 1979). As a general rule, the larger the organism, the higher the dosage required to kill it. Spotte (1973) suggested an MLD of 35,000 $\mu Ws/cm^2$ for general disinfection of aquarium water. Dosages required to kill most of the disease-causing microorganisms that typically trouble aquaculture systems range from 35,000 to 156,000 $\mu s/cm^2$. Typically, those organisms 40 μ and smaller in size are killed with dosages at the lower end of this range, while larger and flagellated organisms are more resistant (Flatow 1981). Yeasts appear to be more resistant than bacteria (Huguenin and Colt 1989). The most resistant microorganisms may require dosages of over 1,000,000 $\mu Ws/cm^2$. MLDs for some microorganisms and their relative sizes are shown in Table 12-3. Other species are discussed in greater detail in Hoffman (1974), Wheaton (1977), and Spotte (1979).

Before an effective UV dosage can be estimated the size of the target microorganism(s) must be accurately determined with a microscope equipped with measuring device. Once the size is determined, the MLD required to kill the organism can be estimated from Figure 12-6. For example, suppose a fish is infected with an unidentified microorganism that mea-

Table 12-3. Sizes and mean lethal dose (MLD) of ultraviolet radiation for free-living and parasitic organisms in aquarium or hatchery water.

Microorganism	Life stage	Size (μm)	MLD ($\mu Ws/cm^2$)
Trichodina sp.	-	16 × 20	35,000
Trichodina nigra	-	22 × 70	159,000
Saprolegnia sp.	zoospore	4 × 12	35,000
Saprolegnia sp.	hypha	8 × 24	10,000
Sarcina lutea	-	1.5	26,400
Icthyophthirius sp.	tomite	20 × 35	336,000
Icthyophthirius sp.	tomite	20 × 35	100,000
Chilodonella cyprini	-	35 × 70	1,008,400
Paramecium sp.	-	70 × 80	200,000

Source: Hoffman (1974).

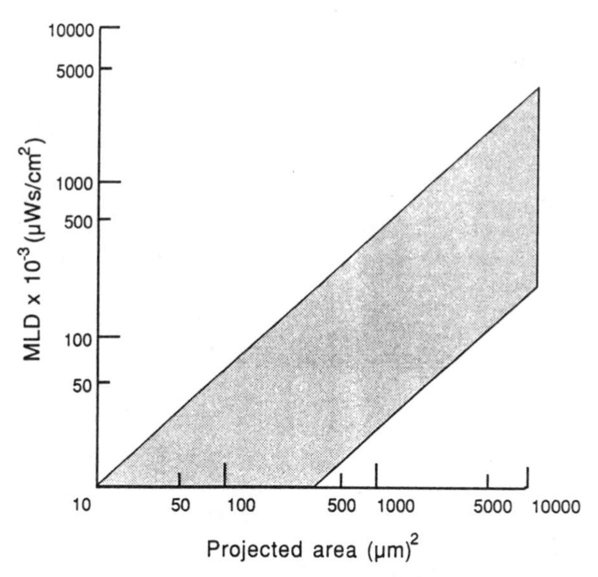

Figure 12-6. Zone of effectiveness of UV light in relation to the projected area of the microorganism (data from Hoffman (1974)).

sures 50 μ in diameter. The cross-sectional area of the organism is $\pi(50/2)^2 \mu m^2 = 1{,}960\ \mu m^2$. From Figure 12-6, a conservative MLD is approximately 100,000 $\mu Ws/cm^2$. Note that this is only an estimate since MLD data is not available for many microorganisms.

Once the MLD is estimated, the flow rate through the UV reactor of known light intensity must be calculated. As an example, the specifications for UV photoreactors manufactured by XYZ Company are shown in Table 12-4. Suppose that a model A reactor having a radiation dosage of 36,000 $\mu Ws/cm^2$ at a flow rate (Q) of 7.6 L/min is selected to disinfect a 1,000-L aquaculture system. The flow rate required to achieve the 100,000 $\mu Ws/cm^2$ MLD is calculated as follows:

$$Q \times 100{,}000 = 7.6 \times 36{,}000$$

$$Q = \frac{(7.6 \times 36{,}000)}{100{,}000}$$

$$Q = 2.7\ L/min$$

The turnover rate through the reactor is calculated from the flow rate:

Table 12-4. Specifications of UV sterilizers used in aquaculture.

Model	Flow rate (L/min/cm bulb)	Dosage level ($\mu Ws/cm^2$)
A	7.6	36,000
B	18.9	39,400
C	37.8	45,200
D	189.3	36,200

$$TR = (2.7\ L/min \times 1{,}440\ min/day)/1{,}000\ L$$

$$TR = 3.9 \text{ or } 4/day$$

This particular example shows that, in order to achieve the desired disinfection, the water flow rate through the reactor must be reduced by approximately 64%, and the water must make nearly four passes daily through the unit. For larger aquaculture systems, a larger UV reactor may be necessary to produce the same effect, or the required water flow rate through the reactor may be so low as to make the process unfeasible. The selection of a larger UV reactor is based purely on economics, that is, capital investment required and operating costs.

Safety Considerations. UV photoreactors should be turned off before cleaning or replacing bulbs since UV rays can damage the eyes and skin (Spotte 1979; Herrington 1991). Care should be taken not to look directly at the UV bulbs while in operation, and appropriate safety warning signs should be posted to alert personnel working in the area. An automatic shutoff switch should be installed on all photoreactors so that the power is turned off when UV units are opened. There is danger of electric shock in addition to radiation exposure.

Ozone

Ozone (O_3) is a three-atom allotrope of oxygen in which the oxygen atoms are held together in an unstable bond. A loosely bound single oxygen atom is quick to break away and reacts with most of the organic and inorganic molecules it comes into contact with. Hence, ozone is a powerful oxidizing agent, second in strength only to fluorine. Fluorine is not used because of the expense. Table 12-5 shows the relative strengths of known oxidizing agents.

Ozone has been used in the sewage treatment industry in Europe since the early 1900s. It was later used for sterilizing potable water supplies. Two decades ago it came into widespread use in the aquarium trade. In the 1970s about 60% of European aquarists were using ozone (Stopka 1975). It has been slow to be adopted in the United States, however. A flurry of research expounded the use of ozone in the 1970s and early 1980s, but it was considered to be too expensive. Therefore, interest waned until around 1990 when the aquaculture industry began to take another look at ozone disinfection.

Table 12-5. Relative oxidation potentials of known oxidants.

Material	Volts	Oxidation potential
Fluorine (F_2)	2.87	Most reactive
Ozone (O_3)	2.07	
Hydrogen peroxide (H_2O_2)	1.78	
Potassium permanganate ($KMnO_4$)	1.70	
Hypobromus acid (HOBr)	1.59	
Hypochlorous acid (HOCl)	1.49	
Chlorine (Cl_2)	1.36	
Chlorine dioxide (ClO_2)	1.27	
Oxygen (O_2)	1.23	
Bromine (Br_2)	1.09	
Iodine (I_2)	0.54	Least reactive

Source: Capitol Controls Co., Inc., (1986).

Ozone is reported to provide many benefits and has distinct advantages over other disinfection methods (Rosenthal 1981; Wedemeyer at al. 1979; Williams at al. 1982; Rosen 1972). Ozone use in aquaculture systems has been cautious due to its potential toxic effects on fish, invertebrates, and bacteria in biofilters (MacLean et al. 1973; King and Spotte 1974). Ozone toxicity has been reviewed by Rosenthal (1981), Wedemeyer et al. (1979) and other researchers. These studies conclude that, if residual concentrations remain low, ozone can be safely used with significant advantages.

The effectiveness of ozone as a disinfecting agent is a function of dosage and contact time. Contact time is acquired by dispersing ozone throughout the culture water. Contact time and required ozone concentration vary with the target microorganism and water quality. It effectively destroys bacteria, viruses, fungi, algae, and protozoa (Lohr and Gratzek 1986). Ozone kills by burning delicate cell membranes. It actually enters the cells and destroys the nuclear chemistry of the cell (Moe 1989). Effects of ozone on various organisms are reviewed by Hoffman (1974), Colberg and Lingg (1978), Wedemeyer at al. (1978), Spotte (1979), and Lohr and Gratzek (1986).

In addition to its disinfecting properties, ozone is reported to reduce nitrite (Evans 1972; Colberg and Lingg 1978), reduce BOD and COD (Colberg and Lingg 1978), and cause the precipitation of certain heavy metals like iron and manganese (Wheaton 1977). Ozone breaks long-chain organic molecules into shorter-chain molecules that are more easily biodegraded. Ozone residuals reduce the gelatinous coating on tank walls and in pipes and sumps, lessening the chances for clogging and short-circuiting in these components. Total system solids loading is reduced with ozone. Its oxidative potential is also a disadvantage, however, in that ozone is very corrosive. All systems components contacting ozone must be fabricated from ozone-resistant material.

Ozone Dissociation in Water. The dissociation of ozone in water results in a chain reaction with hydroxyl ions (OH^-) serving as initiators (Spotte 1979). Other free radicals that are formed include hydroperoxyl (HO_2), hydroxy (OH), oxide (O^-) and ozonide (O_3^-). Peroxide (H_2O_2) and molecular oxygen (O_2) are also formed. The professional literature suggests that free radicals in aquaculture systems pose problems in that they may be as injurious or even more injurious to aquatic life than ozone itself. The most important dissociation product is hydroxyl, the oxidation potential of which is 2.8v at $H^+ = 1M$. This is even higher than ozone, as shown in Table 12-5.

Factors Affecting Ozone Effectiveness. Factors that influence the dissociation of ozone in water also affect the percent kill of free-floating microorganisms. These factors are: (1) Concentration of dissolved organics; (2) concentration of particulate organics; (3) species and concentration of inorganic ions; (4) pH; and (5) temperature. Factors 1–4 are sometimes referred to as *ozone demand.* The higher the ozone demand, the lower will be the kill rate for a given ozone dose. Increases in temperature generally result in higher kill rates (Spotte 1979).

DISSOLVED AND PARTICULATE ORGANICS. When dissolved and particulate organics are present together, hydroxyl radicals react preferentially toward dissolved organic carbon (Hoigne and Bader 1976). For ozonation purposes, microorganisms are considered part of the total particulate matter load. Thus, in water high in dissolved organics, the ozone demand may be so high that free radicals formed by ozone dissociation, or ozone itself, may be used up before microorganisms are killed. Factors enhancing the reactivity of ozone with dissolved organic materials often hinder its ability to oxidize particulate organic material, and vice versa. Particulate organics in general reduce the effectiveness of ozone. Therefore, the higher the density of microorganisms, the lower the disinfection rate, since microorganisms contribute to the particulate organic load, as previously described (Spotte 1979).

Water containing high organic loads must be treated with higher dosages of ozone. Since color, odor, and turbidity can be removed by ozone, this indicates that they also have an ozone demand (Dupree 1981). As organic matter is oxidized, more free ozone becomes available. Therefore, in older installations it is wise to cut back on the ozone output a few days after installation in order to prevent development of high ozone concentrations (Moe 1989).

EFFECT OF PH. The decomposition of ozone is strongly influenced by pH. Decomposition is accelerated in the presence of hydroxyl ions, therefore, rapid reduction of ozone is reported at pH 8.0 and above (Weiss 1935). Colberg and Lingg (1978) reported that, in their experience, even though ozone is more rapidly reduced and its concentration is lowest at higher pH values, its oxidation capacity is greater.

EFFECT OF SALINITY. Evidence supports the use of ozone in marine systems (Honn and Chavin 1976; Sutterlin at al. 1984; Moe 1989). However, since marine systems normally have higher pH values, ozone decomposition is more rapid. Ozone may have detrimental effects on seawater. It is reported to deplete certain trace elements, particularly manganese (Spotte 1970) and calcium (Moe 1989). Ozone reacts with the chloride and bromide ions, forming toxic hypochlorites and hypobromites (Moe 1989; Blogoslawski 1991).

Generation Of Ozone. Ozone is a very unstable gas and must therefore be generated on-site and used immediately. Ozone production is measured as mg ozone per hour. Most ozone generators used in the aquaculture industry use one of two methods to produce ozone: UV irradiation or silent electrical (corona) discharge.

The UV irradiation method is the less efficient of the two and is used primarily in small ozone generators used in the aquarium trade. In this method air is passed through a chamber containing a UV bulb that emits light in the 1,000–2,000 Å range (Klein et al. 1985). The UV light splits some of the oxygen molecules. The single oxygen atoms then attach to other O_2 molecules and form ozone (O_3). The UV approach is only applicable to small systems because of the low-concentration ozone produced (Honn et al. 1976). The feed gas must be dry since residual water vapor decreases the amount of ozone produced and accelerates the formation of corrosive nitrous oxides. Air compressors must be free of oil since any hydrocarbons present in the feed gas will reduce the amount of ozone produced (Honn et al. 1976).

Corona discharge generators are capable of large-scale ozone production. A high voltage is impressed across two plates through which air or pure oxygen passes. The oxygen molecules are excited by the electrical charge and form ozone. The process is energy-consuming and produces heat.

Either air or pure oxygen can be used to produce ozone. The same ozone generator uses twice the power to produce ozone from air as it does to produce ozone from pure oxygen. The cost of producing ozone using pure oxygen is usually less than ozone production using air (Wheaton 1977).

Ozone Dosage. Sufficient time must be allowed for ozone to be dispersed through the water. This brings about the required contact between the ozone and the target organism(s). Contact time and ozone concentration vary with the target organism and water quality.

The scientific literature indicates that ozone concentrations of 0.6–1.0 mg O_3/L and contact times of 1–3 minutes are sufficient to kill most pathogens in aquaculture systems (Dupree 1981), however this varies considerably. Wedemeyer et al. (1978) reported that the minimum dosage required to control *Saprolignia* on fish eggs without damage to eggs or fry was 0.03 mg/L. Ozone was toxic to eggs at 0.3 mg/L. Wedemeyer and Nelson (1977) reported that 0.01 mg/L ozone caused complete inactivation of enteric red mouth (ERM) bacterium in 30 seconds while 10 minutes was required to inactivate *A. salmonicida*. It was felt that pH and other water quality factors could have caused the differences. At the Dworshak National Fish Hatchery in Idaho an ozone residual of 0.20 mg/L for 10 minutes achieved control over the virus infectious hematopoietic necrosis (IHN) (Owsley 1991).

Ozone treatment normally does not leave a residual in the treated water. This is an advantage since it is toxic in low quantities to aquatic animals. However, for disease protection some researchers have advised that a very low residual is not harmful to fish and may prevent disease from reoccurring. Toxicity effects vary considerably in the literature. Arthur and Mount (1975) reported that ozone was toxic to fathead minnows at 0.2–0.3 mg/L. Roselund (1974) and Honn et al. (1976) reported gill damage to rainbow trout at 0.01–0.06 mg/L. In other toxicity testing Wedemeyer et al. (1978) observed no differences between salmon exposed for three months to 0.023 mg O_3/L and the control; however, damage was observed at 0.05 mg/L. Sutterlin et al. (1984) reported that 1.62 g O_3 per kg fish with a one-hour exposure time was sufficient to maintain water clarity with no ill effects on Atlantic salmon smolts. Treatment still had no effects when extended to eight hours, but high mortalities occurred when the ozone generator was accidentally left running for 15 hours. Honn and Chavin (1976) noted no deleterious effects of 0.132 mg O_3/h/L in a seawater system containing nurse sharks.

Wedemeyer et al. (1979) determined that a safe permissible exposure level of ozone to fish was 0.002 mg/L. However, with so much variability being reported, the only positive method to determine ozone dosages and contact times is with on-site bench-scale testing.

Analytical Methods. Because of its recommended limitations, ozone measurement at small dosages is at times difficult. Instrumentation to directly measure ozone in water is commercially available, however, it is expensive, and the sensitivity is usually not great enough to measure ozone in 0.001 mg/L increments.

Several wet chemistry methods are available (APHA 1988; Owsley 1991). The DPD (N,diethyl-P-phenylenediamine) method is relatively simple but has limitations (Paulin 1967). The test must be completed within six minutes after fixing the sample and has many interferences. A modified DPD method showed good results and increased accuracy (Wedemeyer et al. 1978).

The standard accepted procedure for measuring ozone in water is the Indigo Blue method (Bader and Hoigne 1982). It is stable for up to four hours and has the least interferences of any of the analytical methods available (Owsley 1988). Another method that can be used as an indicator is *redox potential.* Redox potential is a measure of the relative amount of the positive and negative charges on the oxidized and reduced molecules in solution and is reported in mv (millivolts). A high redox potential is necessary in culture water for optimum cell respiration, especially important in animals with limited circulatory systems. Redox potential is depressed by accumulated reduced molecules, the result of organic matter addition. Most aquaculture waters have a low redox potential. The addition of ozone will oxidize more of the reduced molecules, thus elevating the redox potential. The higher the redox potential, the more pure the water.

Healthy aquaculture systems should have a redox potential ranging from 200 to 350 mv. Levels below 200–250 mv indicate the presence of toxic, reduced compounds, while levels above 400–450 mv indicates too active an oxidative environment, which can potentially damage plant and animal tissues and cells. Overtreatment with ozone can produce too high a redox potential (Moe 1989). Redox potential can be maintained at a comfortable 300–350 mv with careful ozone regulation. Commercial automatic control units are available that continuously monitor redox potential and adjust ozone output to maintain redox potential at specified levels.

Applications. Ozone units can be incorporated into aquaculture systems in a number of schemes. The choice is left to the aquaculturist and is dependent on system design, site conditions, and economics.

Aquaculturists generally agree that ozonation should precede solids removal and/or biological filtration (Figure 12-7). Since dissolved and solids organic materials are rapidly oxidized by ozone, the load going to settling basins, tube settlers, or sand filters (whichever method is used for solids removal) is reduced, lessening the chance for clogging or short-circuiting in these units. Due to the decrease in solids it may be possible to decrease the size of these units, reducing the initial capital investment and operating costs.

Organic matter reduction prior to biological

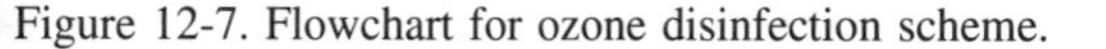
Figure 12-7. Flowchart for ozone disinfection scheme.

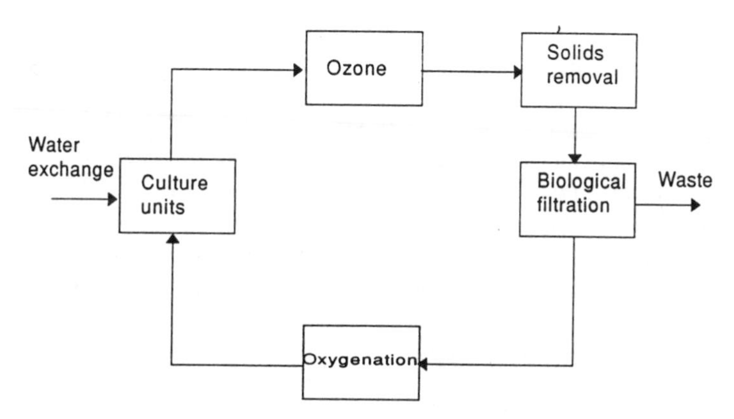

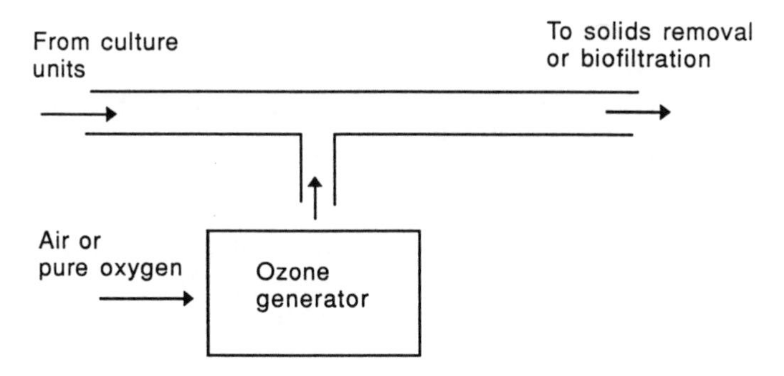

Figure 12-8. Illustration of sidestream method of ozone injection.

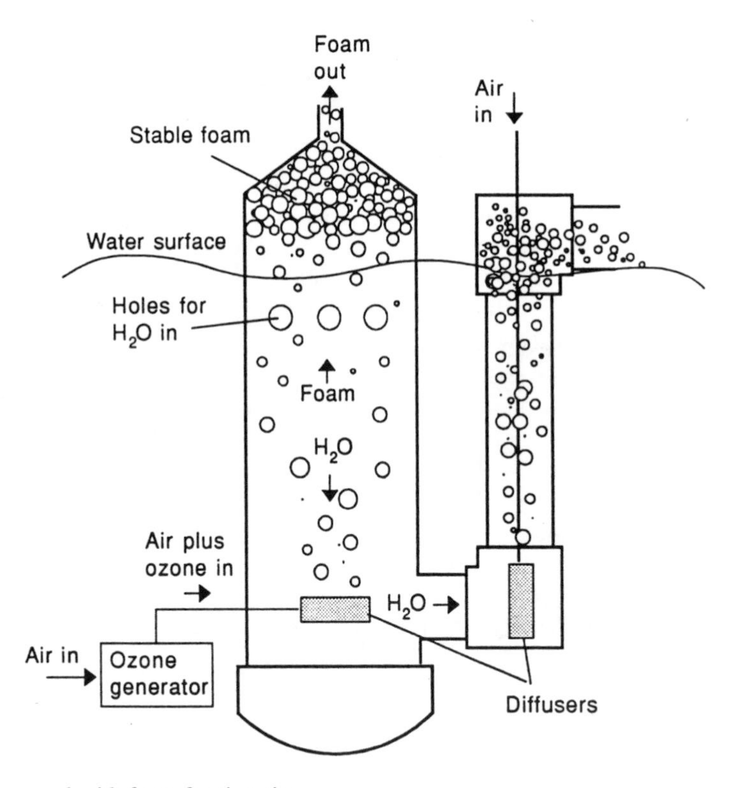

Figure 12-9. Ozone used with foam fractionation.

filtration has the effect of reducing the population of heterotrophic bacteria in the filters. Heterotrophs feed on organics and are the first to become established in biofilters. They compete for space on biofilter substrate with nitrifying bacteria. Too high a population of heterotrophs may inhibit nitrification (Paller and Lewis 1988). A reduction of organics by ozone is followed by a reduction of heterotrophs, leaving more room for the nitrifiers. Ozone also has the effect of cracking large organic molecules into smaller, more biodegradable materials that are more easily removed by heterotrophic bacteria.

The direct ozonation of water in the culture unit is the most simple method of ozone application. This method is most frequently used in home aquaria and small recirculating systems; however, caution must be used to protect the animals from the potentially toxic effects of ozone residuals. Ozone can be applied as a batch dose, lasting just a few hours. Continuous direct ozonation is not recommended. The most simple method of application is to introduce the ozone through air stones or diffusers at the bottom of an air lift. In lieu of direct application in culture units ozone can be applied through a separate mixing chamber following the culture unit and preceding solids removal/biological filtration. In this technique, large doses can be applied to destroy infectious microbes and/or organics while not directly contacting the fish. Contact time is governed only by the size of the chamber. The mixing chamber should be large enough to allow the ozone sufficient time to degrade to oxygen.

Some aquaculture systems are designed for easy retrofit of an ozone generator. For existing systems the sidestream method of injection (Figure 12-8) generally works best since it requires a minimum amount of plumbing changes. However, this method may require an additional air compressor or pure oxygen supply. Smaller ozone generators are fabricated with their own air compressor. New systems may be designed with ozone as an integral part of the waste treatment scheme.

An alternate method, which is reported to be very effective, is to combine ozone with foam fractionation (Rosenthal 1981). The benefits of foam fractionation in aquaculture systems are well known (Wheaton et al. 1979; Lawson and Wheaton 1980). The benefits of ozone and foam fractionation can be realized in one unit. Figure 12-9 illustrates a counterflow foam fractionator with ozone injection. The water enters the foam column near the water surface and flows downward against the upward flow of air (or pure oxygen) and ozone, which is introduced through a fine bubble diffuser near the bottom of the column. The water current is generated by a second diffuser placed in the smaller tube attached to the foam column. Foam is collected in the conical-shaped foam collector located at the top of the foam column. Foamate can be collected in a container and discarded. Foam units can be used individually or in multiple units, depending on system loading. They may be hung on the inside or outside of culture units, or they may be installed into a separate mixing chamber. Several models of foam fractionators using ozone are commercially available.

Ozone Removal. Ozone must be removed from culture water because of its potential toxicity to fish and invertebrates. Since ozone decomposes to oxygen in a short period of time it can be removed by allowing culture water to be held in an aeration chamber for a period of time. This may not be practical in some systems. Carbon filters are also very effective for removing ozone (Owsley 1991). However, for reasons previously discussed, carbon adsorption is not normally economical in large recirculating systems.

A faster, more economical method for ozone removal is to strip it from solution using a packed column (Owsley 1981). Complete removal occurs in properly designed units. Figure 12-10 illustrates a packed column. In this method ozone-laden culture water enters the top of the tower. Column packing material may consist of plastic modules or other rigid, nontoxic material that has a high surface area. Air is pumped in at the bottom of the column and flows upward through the packing material. Ozone is stripped from the water by the countercurrent action of the process. If desired, the unit can be covered and the ozone off-gas vented to the outside. Clean culture water collects in the bottom of the tower and is returned to the culture unit. A word of caution is in order here; the designer should ensure that the level of the water at the bottom of the tower is below the level of the air inlet so that water does not back up into the air compressor when power is off.

Safety Considerations. Gaseous ozone has a very sharp odor that has been described as fresh or grassy. It can easily be detected by humans at low concentrations: 0.02–0.05 mg/L (Spotte 1979). The U.S. Federal Government has estab-

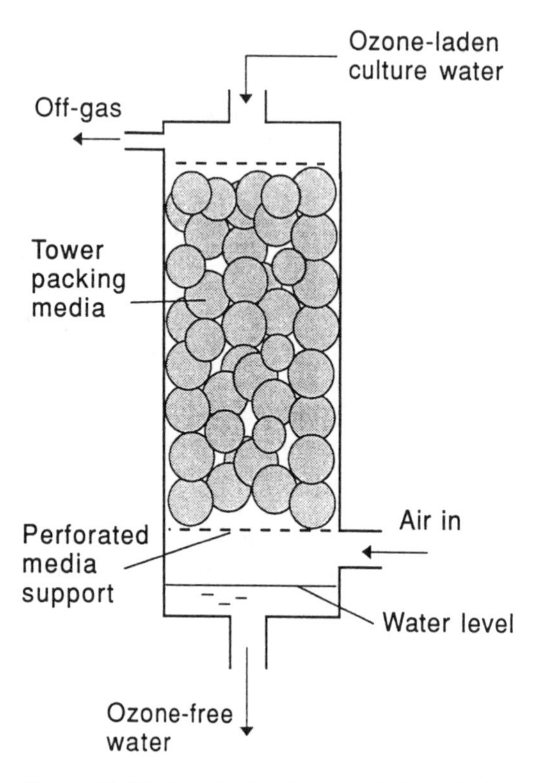

Figure 12-10. A packed column used for stripping ozone from culture water.

lished a permissible exposure of 0.1 mg/L O_3 in an enclosed area over an eight-hour work shift (Capital Controls Co. 1986). At low concentrations ozone causes irritation to the eyes, nose, and throat. Medium-level exposure results in eye irritation, dryness of the nose and throat, and coughing. Severe effects of ozone exposure are depression, cyanosis (a bluish discoloration to the skin caused by lack of oxygen), nausea, severe headache, tightness in the chest, and pulmonary edema. As the concentration of ozone in the atmosphere increases, there is a desensitized reaction on the human body, posing an even greater threat. Figure 12-11 shows a relationship between ozone concentration and toxicity to humans.

When there is a potential for ozone poisoning in certain rooms, buildings or other enclosed areas, the room air should be vented to the outside. Another technique to limit human exposure is to use ozonation equipment during off-hours, such as nighttime and weekends, when personnel are not present in the area.

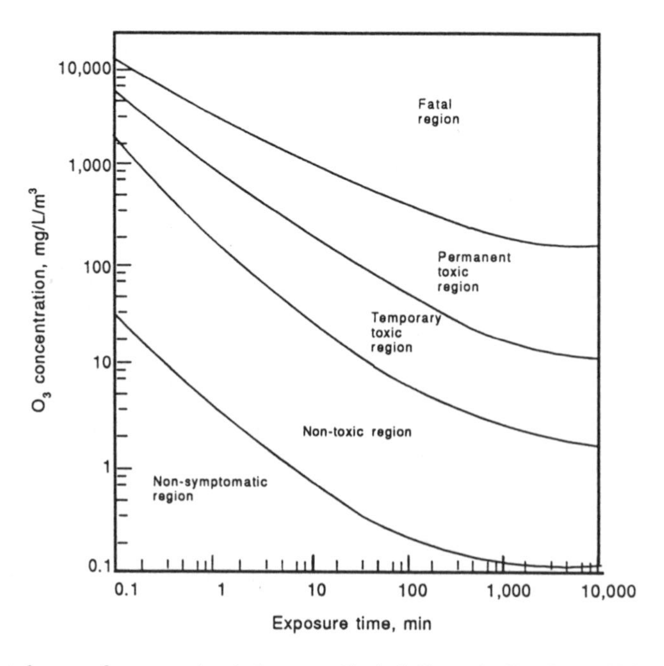

Figure 12-11. Human tolerance for ozone in air (source: Capitol Controls Co., Inc. 1986).

COMPARISON OF DISINFECTION METHODS

Ozone Versus Chlorine Disinfection

Ozone does not leave harmful residuals (Hewes and Darrison 1971), as chlorine does, since it rapidly degrades to molecular oxygen (Layton 1972). The half-life of ozone is about 15 minutes (Honn et al. 1976). Ozone has twice the chemical oxidative capacity of the hypochlorite ion and oxidizes on contact (Rosen 1972; Venosa 1972). Chlorine must first dissociate into the water and then diffuse through cell walls of microorganisms before killing them. Ozone kills by inactivating certain enzymes. Also, ozone does not react with ammonia like chlorine, allowing it to react much quicker. Ozone is not as affected by pH and temperature changes as chlorine.

Ozone Versus UV Irradiation

Ozone rapidly oxidizes turbidity, organics, and color-causing agents, and, therefore, it's efficiency is not affected by these materials. Impurities in the water severely reduce the efficiency of UV light since they absorb the light rays and shield microorganisms from the killing effects of the light. Thus, ozone does not require costly prefiltration like UV light systems to be effective (Lohr and Gratzek 1986).

Another drawback to UV photoreactors is that, where UV light leaves no residual, and the water may be used immediately, a power failure will cause instant disruption of the disinfection process, and the culture unit will begin to receive untreated water in a matter of seconds (Huguenin and Colt 1989). In addition, in recirculating systems or in other systems having a long residence time, the lack of a residual may have little effect on control of the microorganism population in the culture unit. Even when functioning properly, a UV light system in large recirculating systems may have little effect on the growth of microorganisms inside pipes and on the walls of culture units. Populations may reach unacceptable levels in these areas. This suggests that it may be impossible to prevent reinfection in recirculating systems using UV light as the sole means of disinfection. UV light should therefore be used only in smaller facilities employing single-pass application.

REFERENCES

AHMAD, T. and C. E. BOYD. 1988. Design and performance of paddle wheel aerators. *Journal Aquacultural Engineering* 7:39–62.

AKAI, D., O. MIKI and S. OHGAKI. 1983. Nitrification model with inhibitory effect of sea water. *Ecological Modeling* 19:189–198.

ALABASTER, J. S. and R. LLOYD, ed. 1982. *Water Quality Criteria for Freshwater Fish. Second Edition.* London: Butterworth Scientific.

ALBRECHT, A. B. 1964. Some observations on factors associated with survival of striped bass eggs and larvae. *California Fish and Game* 50:100.

ALDERSON, R. 1974. Seawater chlorination and the survival and growth of the early developmental stages of plaice and Dover sole. *Aquaculture* 4:41–53.

ALEEM, M. I. H. 1959. The physiology and chemoautotrophic metabolism of *Nitrobacter agilis.* M. S. Thesis, Cornell University. (Original not available. Cited in Painter (1970)).

AMERICAN PUBLIC HEALTH ASSOCIATION (APHA). 1980. *Standard Methods for the Examination of Water and Wastewater,* 15th Ed., Washington, DC: American Water Works Association, Water Pollution Control Federation.

AMERICAN PUBLIC HEALTH ASSOCIATION (APHA). 1985. *Standard Methods for the Examination of Water and Wastewater,* 16th Ed., Washington, DC: American Water Works Association, Water Pollution Control Federation.

AMERICAN PUBLIC HEALTH ASSOCIATION (APHA). 1988. *Standard Methods for the Examination of Water and Wastewater.* 17th Ed. Washington, DC: American Water Works Association, Water Pollution Control federation.

AMERICAN SOCIETY OF AGRICULTURAL ENGINEERS (ASAE). 1985. *ASAE Standards 1985.* 32nd Ed. St. Joseph, MI: American Society of Agricultural Engineers.

AMERICAN SOCIETY OF AGRICULTURAL ENGINEERS (ASAE). 1991. *Aquaculture Systems Engineering.* St. Joseph, MI: ASAE.

AMERICAN SOCIETY OF CIVIL ENGINEERS (ASCE). 1983. Development of standard procedures for evaluating oxygen transfer devices. EPA-600-2-83-102, Cincinnati: U. S. Environmental Protection Agency.

ANDREWS, J. W., T. MURAI, and G. GIBBONS. 1973. The influence of dissolved oxygen on the growth of channel catfish. *Transactions American Fisheries Society* 102:835–838.

ANTHONISEN, A. C., R. C. LOEHR, T. B. S. PRAKASAM, and E. G. SRINATH. 1976. Inhibition of nitrification by ammonia and nitrous acid. *Journal of the Water Pollution Control Federation* 48:835–852.

ARMSTRONG, D. A., D. CHIPPENDALE, A. W. KNIGHT, and J. E. COLT. 1978. Interaction of ionized and un-ionized ammonia on short-term survival and growth of prawn larvae, *Macro-*

brachium rosenbergii. Biological Bulletin 154:15–31.

ARTHUR, J. W. and D. I. MOUNT. 1975. Toxicity of a dissinfected effluent to aquatic life. *First International Symposium on Ozone in Water and Wastewater Treatment.* International Ozone Institute, Syracuse, NY.

ATLAS, R. M. and R. BARTHA. 1987. *Microbial Ecology: Fundamentals and Applications.* Menlo Park, CA: Benjamin Cummings.

AVAULT, J. W. 1971. Water temperature: An important factor in producing catfish. *Fish Farming Industries* 2:28–32.

AVAULT, J. W. 1989. How will fish be grown in the future? *Aquaculture Magazine* 15:57–59.

AVAULT, J. W. 1991. Some thoughts on marketing aquaculture products. *Aquaculture Magazine* 17:62–67.

AVAULT, J. W., L. W. DE LA BRETONNE, and J. V. HUNER. 1974. Two major problems in culturing crawfish in ponds: Oxygen depletion and overcrowding. *Proceedings of the 2nd International Crayfish Symposium,* Baton Rouge, LA: Louisiana State University/Division of Continuing Education.

BADER, H. and J. HOIGNE. 1982. Determination of ozone in water by the Indigo Method: A submitted standard method. *Ozone: Science and Engineering* 4:169–176.

BAKER, F. E. and J. D. BANKSTON. 1988. Selecting a pumping plant for aquacultural production. Pub. No. 7/88 (4M). Louisiana State University Agricultural Center, Louisiana State University, Baton Rouge, LA.

BALL, I. R. 1967. The relative susceptibility of some species of freshwater fish to poisons—I. Ammonia. *Water Research* 1:767–775.

BARNABÉ, G., ed. 1990a. *Aquaculture: Vol. 1.* New York: Ellis Horwood.

BARNABÉ, G., ed. 1990b. Aquaculture: Vol. 2. New York: Ellis Horwood.

BASS, M. L. and A. G. HEATH. 1976. Toxicity of intermittent chlorination to bluegills: Interaction with temperature. *Bulletin Environmental Contamination and Toxicity* 17:416–423.

BASU, S. P. 1959. Active respiration of fish in relation to ambient concentrations of oxygen and carbon dioxide. *Journal Fisheries Research Board of Canada* 16:175–179.

BEACH, V. 1989. Aquaculture = Water farming = A South Carolina tradition. *Coastal Heritage* 4:2.

BEAN, R. M., R. G. RILEY, and P. W. RYAN. 1978. Investigations of halogenated components formed by chlorination of marine water. (Original not available. Cited in Rosenthal (1981)).

BEDELL, G. W. 1971. Eradicating *Ceratomyxa shasta* from infected waters by chlorination and ultraviolet irradiation. *Progressive Fish Culturist* 33:51–54.

BELLPORT, B. P. and G. E. BURNETT, eds. 1967. *Water Management Manual.* U. S. Department of the Interior, Washington, DC. Benefield, L. D. and C. W. Randall. 1980. *Biological Process Design in Wastewater Treatment.* Englewood Cliffs, NJ: Prentice-Hall.

BERKA, R., B. KUJAL, and K. LAVICKY. 1980. Recirculation systems, *World Symposium on New Developments in the Utilization of Heated Effluents and Recirculation Systems for Intensive Aquaculture.* EIFAC 11th Session, Stavanger, Norway: European Inland Fisheries Advisory Council.

BEVERIDGE, M. C. M. 1984. Cage and pen fish farming: Carrying capacity models and environmental impact. FAO Fisheries Technical Paper No. 255, Rome: Food and Agricultural Organization of the United Nations.

BEVERIDGE, M. C. M. 1987. *Cage Aquaculture.* New York: Scholium International.

BEVERIDGE, M. C. M. and J. F. MUIR. 1987. An evaluation of proposed cage culture in Loch Lomond, an important reservoir in central Scotland. *Canadian Water Resources Journal* 7:107–113.

BISOGNI, J. J. and M. B. TIMMONS. 1991. Control of pH in closed cycle aquaculture systems. *Engineering Aspects of Intensive Aquaculture, Proceedings from the Aquaculture Symposium,* Cornell University, Ithaca, NY, April 4–6, 1991.

BLOCK, R. M., G. R. HELZ, and W. P. DAVIS. 1977. The fate and effects of chlorine in coastal waters: Summary and recommendations. *Chesapeake Science* 18:97–101.

BLOGOSLAWSKI, W. J. 1991. Enhancing shellfish depuration. *Molluscan Shellfish Depuration,* Otwell, W. S., G. E. Rodrick, and R. E. Martin, eds. Boca Raton, FL: CRC Press.

BONN, E. W. and B. J. FOLLIS. 1967. Effects of hydrogen sulfide on channel catfish *Ictalurus punctatus. Transactions American Fisheries Society* 96:31–36.

BONN, E. W., W. M. BAILEY, J. D. BAYLESS, K. E. ERICKSON, and R. E. STEVENS. 1976. *Guidelines for Striped Bass Culture*. Bethesda, MD: American Fisheries Society.

BOON, B. and H. LAUDELOUT. 1962. Kinetics of nitrite oxidation by *Nitrobacter winogradskyi. Biochemical Journal* 85:440–447.

BOUCK, G. R. and R. E. KING. 1983. Tolerance to gas supersaturation in freshwater and seawater by steelhead trout. *Journal Fisheries Biology* 23: 293–300.

BOWER, C. E. and D. T. TURNER. 1981. Accelerated nitrification in new seawater culture systems: Effectiveness of commercial additives and seed media from established systems. *Aquaculture* 24:1–9.

BOWER, C. E. and D. T. TURNER. 1984. Evaluation of two commercial nitrification accelerators in closed seawater culture systems. *Aquaculture* 41:155–159.

BOYD, C. E. 1973. Summer algal communities and primary production in fishponds. *Hydrobiologia* 41:357–390.

BOYD, C. E. 1974. Lime requirements of Alabama fish ponds. Bulletin No. 459. Auburn, AL: Auburn University/Alabama Agricultural Experiment Station.

BOYD, C. E. 1979. *Water Quality in Warmwater Fish Ponds.* Auburn, AL: Auburn University/Alabama Agricultural Experiment Station.

BOYD, C. E. 1982. *Water Quality Management for Pond Fish Culture.* Amsterdam: Elsevier.

BOYD, C. E. 1985a. Hydrology and pond construction. *Channel Catfish Culture,* C. S. Tucker, ed. Amsterdam: Elsevier.

BOYD, C. E. 1985b. Pond evaporation. *Transactions American Fisheries Society* 114:299–303.

BOYD, C. E. 1990. *Water Quality in Ponds for Aquaculture.* Auburn, AL: Auburn University/Alabama Agricultural Experiment Station.

BOYD, C. E. 1992. *Water Quality and Pond Soil Analyses for Aquaculture*. Auburn, AL: Auburn University/Alabama Agricultural Experiment Station.

BOYD, C. E. and T. AHMAD. 1987. Evaluation of aerators for channel catfish farming. Bulletin No. 584. Auburn, AL: Auburn University/Alabama Agricultural Experiment Station.

BOYD, C. E. and D. J. MARTINSON. 1984. Evaluation of propeller aspirator pump aerators. *Aquaculture* 36:283–292.

BOYD, C. E. and W. W. WALLEY. 1975. Total alkalinity and hardness of surface waters in Alabama and Mississippi. Bulletin No. 465. Auburn, AL: Auburn University/Alabama Agricultural Experiment Station.

BOYD, C. E. and B. J. WATTEN. 1989. Aeration systems in aquaculture. *Reviews in Aquatic Sciences* 1(3):425–472.

BOYD, C. E., R. P. ROMAIRE, and E. JOHNSTON. 1978. Predicting early morning dissolved oxygen concentrations in channel catfish ponds. *Transactions American Fisheries Society* 107: 484–492.

BRETT, J. R. and T. D. D. GROVES. 1979. Physiological energetics. *Fish Physiology,* W. S. Hoar, D. J. Randall, and J. R. Brett, eds. New York: Academic Press.

BRETT, J. R. and C. A. ZALA. 1975. Daily pettern of nitrogen excretion and oxygen consumption of sockeye salmon *Oncorhynchus nerka* under controlled conditions. *Journal Fisheries Research Board of Canada* 32:2479–2486.

BROOKS, A. S. and G. L. SEEGERT. 1977. The effects of intermittent chlorination on selected species of great lakes fishes: Effects of 5 and 30 minute exposures on rainbow trout and yellow perch. *Transactions American Fisheries Society* 106:278–286.

BROWN, E. E. 1983. *World Fish Farming: Cultivation and Economics.* 2nd ed., Westport, CT: AVI Publishing.

BROWN, L. C. and C. R. BAILLOD. 1982. Modeling and interpretation of oxygen transfer data. *Journal Environmental Engineering Division (ASCE)* 108:607–628.

BRUNE, D. E. and C. M. DRAPCHO. 1991. Fed pond aquaculture. *Aquaculture Systems Engineering,* St. Joseph, MI: American Society of Agricultural Engineers.

BRUNE, D. E. and J. R. TOMASSO, eds. 1991. *Aquaculture And Water Quality.* Baton Rouge, LA: World Aquaculture Society.

BRUNGS, W. A. 1973. effects of residual chlorine on aquatic life. *Journal Water Pollution Control Federation* 45:2180–2193.

BURKHALTER, D. E. and C. M. KAYA. 1977. Effects of prolonged exposure to ammonia on fertilized eggs and sac fry of rainbow trout (*Salmo gairdneri*). *Transactions American Fisheries Society* 106:470–475.

BURLEY, R. and A. KLAPSIS. 1985. Flow distribution studies in fish rearing tanks. Part 2—Analysis of hydraulic performance in 1m square tanks. *Journal Aquacultural Engineering* 4:113–134.

BURROWS, R. E. and H. H. CHENOWETH. 1955. Evaluation of three types of fish rearing ponds. Records Report No. 39. Washington, DC: U. S. Department of the Interior.

BURROWS, R. E. and H. H. CHENOWETH. 1970. The rectangular circulating rearing pond. *Progressive Fish Culturist* 32:67–80.

BUSCH, R. L. 1985. Channel catfish culture. *Channel Catfish Culture,* Tucker, C. S., ed. Amsterdam, Elsevier.

BUSS, K., D. R. GRAFF, and E. R. MILLER. 1970. Trout culture in vertical units. *Progressive Fish Culturist* 32:187–191.

BUSWELL, A. M., T. SHIOTA, N. LAWRENCE, and I. VAN METER. 1954. Laboratory studies on the kinetics of the growth of *Nitrosomonas* with relation to the nitrification phase of the BOD test. *Applied Microbiology* 2:21–25.

CALSON, D. A., R. W. SEABLOOM, F. B. DEWALLE, T. F. WETZLER, J. ENGESET, R. BUTLER, S. WANGSUPHACHART, and S. WANG. 1985. Ultraviolet disinfection of water for small water supplies. Rep. No. EPA/600/S2–85/092, Cincinnati, OH: U. S. Environmental Protection Agency., Water Engineering and Research Laboratory.

CAMPBELL, J. W. 1973. Nitrogen excretion. *Comparative Animal Physiology,* Prosser, C. L., ed. Philadelphia, PA: W. B. Saunders.

CAPITAL CONTROLS CO. 1986. Application: Ozone and its properties. Bulletin A2.3401.0. Colmar, PA: Capital Controls Co.

CARMIGIANI, G. M. and J. B. BENNETT. 1977. Rapid start-up of a biological filter in a closed aquaculture system. *Aquaculture* 11:85–88.

CARPENTER, P. L. 1967. *Microbiology*. Philadelphia: W. B. Saunders.

CARRICK, T. R. 1979. The effect of acid water on the hatching of salmonid eggs. *Journal Fisheries Biology* 14:165–172.

CASTRO, W. E., P. B. ZIELINSKI, and P. A. STANDIFER. 1975. Performance characteristics of airlift pumps of short length and small diameter. *Journal World Mariculture Society* 6:451–461.

CHAMBERLAIN, G. and K. STRAWN. 1977. Submerged cage culture of fish in supersaturated thermal effluent. *Journal World Mariculture Society* 8:625–645.

CHASTON, I. 1983. *Marketing in Fisheries and Aquaculture*. Surrey, England: Fishing News Books, Ltd.

CHEN, F. Y. 1979. Progress and problems of net cage culture of grouper (*Epinaphelus taurina*) in Singapore. *Journal World Mariculture Society* 10:260–271.

CHEN, S. 1991. Theoretical and experimental investigation of foam separation applied to aquaculture. Ph.D. Dissertation, Cornell University.

CHEN, S. and R. F. MALONE. 1991. Suspended solids control in recirculating aquaculture systems. *Engineering Aspects of Intensive Aquaculture, Proceedings from the Aquaculture Symposium,* Ithaca, NY: Cornell University, April 4–6, 1991.

CHEN, S., M. B. TIMMONS, J. J. BISOGNI, and D. J. ANESHANSLEY. 1993. Modeling surfactant removal in foam fractionation I: Theoretical development. *Journal Aquacultural Engineering,* In press.

CHESNESS, J. L. and J. L. STEPHENS. 1971. A model study of gravity flow aerators for catfish raceway systems. *Transactions American Society of Agricultural Engineers* 14:1167–1169.

CHILDRESS, W. 1990. Leaping lizards! Alligator farming flourishes in Louisiana. *Seafood Leader* 10:84–88.

CHOW, V. T. 1959. *Open Channel Hydraulics*. New York: McGraw-Hill.

CLAUSEN, R. G. 1936. Oxygen consumption in freshwater fish. *Ecology* 17:216–226.

COCHE, A. G. 1982. Cage culture of tilapias. *The Biology And Culture of Tilapias* R. S. V. Pullin and R. H. Lowe-McConnell, eds., Philippines: ICLARM.

COCHE, A. G. 1983. The cultivation of fish in cages. A general bibliography. FAO Circ. No. 714 (Rev. 1).

COLBERG, P. J. and A. J. LINGG. 1978. Effects of ozonation on microbial fish pathogens, ammonia, nitrate, nitrite and BOD in simulated reuse hatchery water. *Journal Fisheries Research Board of Canada* 35:1290–1296.

COLT, J. 1984. Computation of dissolved gas concentrations in water as functions of temperature, salinity and pressure. Pub. No. 14, Bethesda, MD: American Fisheries Society.

COLT, J. 1987. An introduction to water quality management in intensive aquaculture. Oxygen Supplementation: A New Technology In Fish Culture, L. Visscher and W. Godby, eds., Denver, CO: U. S. Department of the Interior, Fish and Wildlife Service.

COLT, J. E. and D. A. ARMSTRONG. 1981. Nitrogen toxicity to crustaceans, fish and molluscs. *Proceedings of the Bio-Engineering Symposium for Fish Culture,* Bethesda, MD: American Fisheries Society.

COLT, J. and G. BOUCK. 1984. Design of packed columns for degassing. *Journal Aquacultural Engineering* 3:251–273.

COLT, J. and C. ORWICZ. 1991. Aeration in intensive culture. *Aquaculture And Water Quality,* Brune, D. E. and J. R. Tomasso, eds., Baton Rouge, LA: The World Aquaculture Society.

COLT, J. E. and G. TCHOBANOGLOUS. 1976. Evaluation of the short-term toxicity of nitrogenous compounds to channel catfish, *Ictalurus punctatus. Aquaculture* 8(3):209–224.

COLT, J. and G. TCHOBANOGLOUS. 1981. Design of aeration systems for aquaculture. *Proceedings of the Bio-Engineering Symposium for Fish Culture*. Allen, L. J. and E. C. Kinney, eds., Bethesda, MD: American Fisheries Society.

COLT, J. and B. J. WATTEN. 1988. Applications of pure oxygen in fish culture. *Journal Aquacultural Engineering* 7:397–441.

COLT, J. and H. WESTERS. 1982. Production of gas supersaturation by aeration. *Transactions American Fisheries Society* 111:342–360.

COLT, J., MITCHELL, S., TCHOBANOGLOUS, G., and A. KNIGHT. 1979. The environmental requirements of fish. *The Use and Potential of Aquatic Species for Wastewater Treatment.* Sacramento, CA: California State Water Resources Control Board.

CONTE, F. S., S. I. DOROSHOV, P. B. LUTES, and E. M. STRANGE. 1988. Hatchery Manual For The White Sturgeon, *Acipenser transmontanus* Richardson: With application to other North American Acipenseridae. Publication No. 3322, California Oakland, CA: University of California, Cooperative Extension Service, Division of Agriculture and Natural Resources.

CORBITT, R. A. 1990. *Standard Handbook Of Environmental Engineering*. New York: McGraw-Hill.

CORMORANT FEEDING RATES ON COMMERCIALLY GROWN CATFISH. 1991. *Aquaculture Magazine* 17:89–90.

CRESWELL, R. L. 1993. *Aquaculture Desk Reference*. New York: Van Nostrand Reinhold.

CUENCA, R. H. 1989. *Irrigation System Design*. Englewood Cliffs, NJ: Prentice Hall.

CULLEY, D. D. and L. F. DUOBINIS-GRAY. 1990. Culture of the Louisiana soft crawfish. Baton Rouge, LA: Louisiana Sea Grant College Program, Louisiana State University.

CUTLER, D. W. and L. M. CRUMP. 1933. Somer aspects of physiology of certain nitrite-forming bacteria. *Annals of Applied Biology* 20:291–296.

DARDIGNAC-CORBEIL, M. 1990. Traditional mussel culture. *Aquaculture: Vol. 1,* G Barnabé, ed. New York: Ellis Horwood.

DAYE, P. G. and E. T. GARSIDE. 1976. Histopathological changes in superficial tissues in brook trout *Salvelinus fontinalis* (Mitchell) exposed to acute and chronic levels of pH. *Canadian Journal of Zoology* 54:2140–2155.

DEERING, R. A. 1962. Ultraviolet radiation and nucleic acid. *Scientific American* 207(6):135–144.

DE LA BRETONNE, L. W. and R. P. ROMAIRE. 1989. Commercial crawfish cultivation practices: A review. *Journal Shellfish Research* 8:267–276.

DE LA BRETONNE, L., J. W. AVAULT, and R. O. SMITHERMAN. 1969. Effects of soil and water hardness on survival and growth of the red swamp crawfish, *Procambarus clarkii,* in plastic pools. *Proceedings Southeastern Association of Game and Fish Commissioners* 23: 626–633.

DEPPE, K. and J. ENGLE. 1960. (Original not available. Cited in Painter (1970)).

DOUMENGE, F. 1990. Aquaculture in Japan. 1990b. *Aquaculture, Vol. 2,* G. Barnabé, ed. New York: Ellis Horwood.

DOWNING, K. M. and J. C. MERKENS. 1955. The influence of dissolved oxygen concentrations on the toxicity of un-ionized ammonia to rainbow trout (*Salmo gairdneri*) *Annals of Applied Biology* 43:243–246.

DRAPCHO, C. M. and D. E. BRUNE. 1984. Application of a retained biomass surface in a rotating biological contactor. *Proceedings Second International Conference on Fixed Film Processes,* Pittsburgh, PA.

DUPREE, H. K. 1981. An overview of the various techniques to control infectious diseases in water supplies and in water reuse aquacultural systems. *Bioengineering Symposium for Fish Culture,* Allen, L. J. and E. C. Kinney, eds. Bethesda, MD: American Fisheries Society.

DWIVEDY, D. 1973. Removal of dissolved organics through foam fractionation in closed cycle systems for oyster culture. ASAE Tech. Pap. No. 73–561, St. Joseph, MI: American Society of Agricultural Engineers.

DYER, K. R. 1973. *Estuaries—A Physical Introduction.* New York: Wiley Interscience.

ECKENFELDER, W. W. 1991. *Principles of Water Quality Management. Malabar, FL: Krieger Publishing.*

ELEKES, K. 1984. Principles of designing inland fish farms. *Inland Aquaculture Engineering.* Rome: Food and Agricultural Organization of the United Nations.

EMERSON, K., R. C. RUSSO, R. E. LUND, and R. V. THURSTON. 1975. Aqueous ammonia equilibrium calculations: Effect of pH and temperature. *Journal Fisheries Research Board of Canada* 32:2379–2383.

ENGLE, M. S. and M. ALEXANDER. 1958. Growth and autotrophic metabolism of *Nitrosomonas eoropea. Journal of Bacteriology* 76:217–222.

EPIFANIO, E. C. and R. F. SRNA. 1975. Toxicity of ammonia, nitrite ion and orthophosphate to *Mercenaria mercanaria* and *Crassostrea virginica. Marine Biology* 33:241–246.

ESCHBACH, O. W., ed. 1952. *Handbook Of Engineering Fundamentals.* New York: John Wiley and Sons.

EIFAC. 1973. Water quality criteria for European freshwater fish. EIFAC Tech. Rep. No. 20, Rome: European Inland Fisheries Advisory Council (EIFAC).

EIFAC. 1986. Report of the working group on terminology, format, and units of measurement as related to flow-through and recirculation systems. EIFAC Tech. Pap. No. 49. Rome: European Inland Fisheries Advisory Commission (EIFAC).

EVANS, F. L. 1972. Ozone technology: Current status. *Ozone In Water And Wastewater Treatment,* F. L. Evans, ed. Ann Arbor, MI: Ann Arbor Science Publishers, Inc.

FINKEL, H. J. 1982. *Handbook Of Irrigation Technology,* Boca Raton, FL: CRC Press.

FINSTEIN, M. S. and C. C. DELWICHE. 1965. Molybdenum as a micro-nutrient for *Nitrobacter. Journal of Bacteriology* 89:123–128.

FITZGERLAD, R. 1990. Louisiana—on the brink of a boom. *Seafood Leader* 10:111 and 112.

FIVELSTAD, S. 1988. Waterflow requirements for salmonids in single pass and semi-closed land-based seawater and freshwater systems. *Journal Aquacultural Engineering* 7:183–200.

FIVELSTAD, S. and M. J. SMITH. 1991. The oxygen consumption rate of Atlantic salmon (*Salmo salar* L.) reared in a single pass landbased seawater system. *Journal Aquacultural Engineering* 10:227–235.

FLATOW, R. E. 1981. High dosage ultraviolet water purification: An indispensable tool for recycling, fish hatcheries, and heated effluent aquaculture. *World Symposium on New Developments in the Utilization of Heated Effluents and Recirculation Systems for Intensive Aquaculture,* Stavanger, Norway: European Inland Fisheries Advisory Council.

FOLSOM, W. B. and E. A. SANBORN. 1992. World salmon culture 1991–92. *Aquaculture Magazine* 18:55–65.

FAO. 1989. *Yearbook of Fishery Statistics: 1988,* Volume 67, Rome: Food and Agricultural Organization of the United Nations (FAO).

GALTSOFF, P. S. 1964. The American Oyster. Bulletin No. 64. Milford, CT: U. S. Fish and Wildlife Service.

GARLING, D. L. and R. P. WILSON. 1976. Optimum dietary protein to energy ratio for channel catfish fingerlings, *Ictalurus punctatus. Journal of Nutrition* 106:1368–1375.

GILES, M. A. and R. DANELL. 1983. Water dechlorination by activated carbon, ultraviolet radiation and sodium sulfite. *Water Research* 17(6): 667–676.

GOULD, G. W. and H. LEES. 1960. The isolation and culture of the nitrifying organisms. Part I. *Nitrobacter. Canadian Journal of Microbiology* 6:299–307.

GRIEVE, F. W. 1928. Measurement of pipe flow by the coordinate method. *Water Measurement Manual,* Bellport, B. P. and G. E. Burnett, eds. Washington, DC: U. S. Department of the Interior.

GUJER, W. and M. BOLLER. 1986. Design of a nitrifying tertiary trickling filter based on theoretical concepts. *Water Research* 20:1353–1362.

HACKNEY, G. E. and J. COLT. 1982. The performance and design of packed column aerators for aquaculture. *Journal Aquacultural Engineering* 1:275–295.

HANEY, P. D. 1954. Theoretical principles of aeration. *Journal of the American Water Works Association* 46:353–376.

HARRIS, G. D., V. D. ADAMS, D. L. SORENSEN, and R. R. DUPONT. 1987. The influence of photoreactivation and water quality on ultraviolet disinfection of secondary municipal wastewater. *Journal Water Pollution Control Federation* 59:781–787.

HAUG, R. T. and P. L. MCCARTY. 1972. Nitrification with submerged filters. *Journal Water Pollution Control Federation* 44:1086–2102.

HAZEL, C. R., W. THOMSEN, and S. J. MEITH. 1971. Sensitivity of striped bass and stickleback to ammonia in relation to temperature and salinity. *California Fish and Game* 57:154.

HEATH, A. G. 1977. Toxicity of intermittent chlorination to freshwater fish: influence of temperature and chlorine form. *Hydrobiologia* 56(1): 39–47.

HEDGECOCK, D., K. NELSON, and R. A. SHLESER. 1976. Growth differences among families of the lobster *Homerus americanus. Proceedings World Mariculture Society* 7:347–361.

HEDSTROM, C. E. and E. LYCKE. 1964. An experimental study on oysters as virus carriers. *American Journal of Hygiene* 79:134–142.

HERRINGTON, T. L. 1991. use of ultraviolet light in depuration. *Molluscan Shellfish Depuration,* Otwell, W. S., G. E. Rodrick, and R. E. Martin, eds. Boca Raton, FL: CRC Press.

HEWES, C. G. and R. P. DARRISON. 1971. Kinetics of ozone decomposition and reaction with organics in water. (Original not available. Cited in Lohr and Gratzek (1986)).

HIRAYAMA, K. 1974. Water control by filtration in closed systems. *Aquaculture* 4:369–385.

HOCHHEIMER, J. N. 1990. Trickling filter model for closed system aquaculture. Ph.D. Dissertation, Univ. of Maryland.

HOCHHEIMER, J. N. and F. W. WHEATON. 1991. Understanding Biofilters: Practical microbiology for ammonia removal in aquaculture. *Engineering Aspects of Intensive Aquaculture, Proceedings from the Aquaculture Symposium,* Cornell University, Ithaca, NY, April 4–6, 1991.

HOFMAN, T. and H. LEES. 1952. The biochemistry of the nitifying organisms. *Biochemical Journal* 52:140–142.

HOFFMAN, G. L. 1974. Disinfection of contaminated water by ultraviolet radiation, with emphasis on whirling disease (*Myxosoma cenebralis*) and its effects on fish. *Transactions American Fisheries Society* 103:541–550.

HOIGNE, J. and H. BADER. 1976. The role of hydroxyl radical reactions in ozonation processes in aqueous solutions. *Water Research* 10:377–386.

HOLLAND, G. A., J. E. LASATER, E. D. NEWMAN, and W. E. ELRIDGE. 1960. Toxic effects of organic and inorganic pollutants on young salmon and trout. Res. Bull. No. 5. Dept. of Fisheries, State of Washington. (Original not available. Cited in Heath et al. (1977)).

HOLLERMAN, W. D. and C. E. BOYD. 1980. Nightly aeration to increase production of channel catfish. *Transaction American Fisheries Society* 109:446–452.

HONN, K. V. and W. CHAVIN. 1976. Utility of ozone treatment in the maintenance of water quality in a closed marine system. *Marine Biology* 34: 201–209.

HONN, K. V., G. M. GLEZMAN and W. CHAVIN. 1976. A high capacity ozone generator for use in aquaculture and water processing. *Marine Biology* 34:211–216.

HORA, S. L. and T. V. R. PILLAY. 1962. Handbook of fish culture in the Indo-Pacific region. Fisheries Technical Report 14. Rome: Food and Agricultural Organization of the United Nations.

HORRIGAN, S. G., A. F. CARLUCCI, and P. M. WILLIAMS. 1981. Light inhibition of nitrification in seawater films. *Journal Marine Research* 39:557–565.

HUET, M. 1970. *Textbook of Fish Culture, Breeding and Cultivation of Fish*. London: Fishing News Books.

HUGUENIN, J. E. and F. J. ANSUINI. 1978. A review of the technology and economics of marine fish cage systems. *Aquaculture* 15:151–170.

HUGUENIN, J. E. and J. COLT. 1989. *Design And Operating Guide For Aquaculture Seawater Systems*. Amsterdam: Elsevier.

HUGUENIN, J. E., S. C. FULLER, F. J. ANSUINI, and W. T. DODGE. 1981. Experiences with a fouling-resistant modular marine fish cage system. *Proceedings of the Bio-Engineering Symposium for Fish Culture,* Allen, L. J. and E. C. Kinney, eds. Bethesda, MD: American Fisheries Society.

HUNER, J. V. and J. E. BARR. 1984. Red swamp crawfish: Biology and exploitation. Baton Rouge, LA: Louisiana Sea Grant College Program, Center for Wetland Resources, Louisiana State University.

HYMEL, T. M. 1985. Water quality dynamics in commercial crawfish ponds and toxicity of selected water quality variables to *Procambarus clarkii*. M. S. Thesis, Louisiana State University.

JAMES, L. G. 1988. *Principles Of Farm Irrigation System Design*. New York: John Wiley and Sons.

JASPERS, E. J. 1969. Environmental conditions in burrows and ponds of the red swamp crawfish, *Procambarus clarkii* (Girard), near Baton Rouge, La. M. S. Thesis, Louisiana State University.

JENSEN, G. L. and J. D. BANKSTON. 1988. Guide To Oxygen and Aeration in Commercial Fish Ponds. Baton Rouge, LA: Louisiana Cooperative Extension Service, Louisiana State University Agricultural Center.

JOHANSEN, P. 1991. Heat pumps in salmon smolt hatcheries. *IEA Heat Pump Centre Newsletter* 9(3):16–19.

JONES, F. R. and W. H. ALFRED. 1980. *Farm Power and Tractors*. New York: McGraw-Hill.

JONES, R. D. and R. Y. MORITA. 1985. Low temperature growth and whole cell kinetics of a marine ammonium oxidizer. *Marine Ecology Progress Series* 21:239–243.

JORGENSEN, S. F., O. LIBOR, K. BARKACS, and K. LUNA. 1979. Equilibrium and capacity data of clinoptilolite. *Water Research* 13:159–165.

KAWAI, A., Y. YOSHIDA, and M. KIMATA. 1965. Biochemical studies on the bacteria in the aquarium with a circulation system—II: Nitrifying activity of the filter sand. *Bulletin of the Japanese Society of Scientific Fisheries* 31(1):65–71.

KAWAKAMI, T. 1964. The theory of designing and testing fishing nets in model. *Modern Fishing Gear of the World, Vol. 2*. H. Kristjansson, ed. Surrey, England: Fishing News Books, Ltd.

KELLEY, K. 1992. Zebra mussels gain a foothold. *National Fisherman* 73(6):32–33.

KERBY, J. H. 1986. Striped bass and striped bass hybrids. *Culture of Nonsalmonid Freshwater Fishes*. R. R. Stickney, ed. Boca Raton, FL: CRC Press.

KERBY, J. H. and E. B. JOSEPH. 1979. Growth and survival of striped bass and striped bass X white bass hybrids. *Proceedings Southeastern Association of Game and Fish Commissioners* 32:715.

KERR, N. M., M. J. GILLESPIE, S. T. HULL, and S. J. KINGWELL. 1980. The design, construction and location of marine floating cages. *Proceedings The Institute of Fisheries Management Cage Fish Rearing Symposium,* March 26–27, 1980. University of Reading. London: Janssen Services.

KHOLDEBARIN, B. and J. J. OERTLI. 1977a. Effects of pH and ammonia on the rate of nitrification in surface water. *Journal Water Pollution Control Federation* 49:1688–1690.

KHOLDEBARIN, B. and J. J. OERTLI. 1977b. Effect of suspended particles and their sizes on nitrification in surface water. *Journal Water Pollution Control Federation* 49:1691–1697.

KILAMBI, R. W., J. NOBLE, and C. E. HOFFMAN. 1970. Influence of temperature and photoperiod on growth, food consumption and food conversion efficiency of channel catfish. *Pro-*

ceedings Southeastern Association of Game and Fish Commissioners 24:519–531.

KING, J. M. and S. SPOTTE. 1974. Marine aquariums in the research laboratory. (Original not available. Cited in Honn et al. (1976)).

KINNE, O. and H. ROSENTHAL. 1977. Cultivation of animals: Commercial cultivation (aquaculture). *Marine Ecology: A Comprehensive Integrated Treatise On Life In Oceans and Coastal Waters,* Kinne, O. and U. K. Chichester, eds. New York: John Wiley and Sons.

KISSEL, J. C., P. L. MCCARTY and R. L. STREET. 1984. Numerical simulation of mixed culture biofilters. *Journal of Environmental Engineering* 110:393–411.

KLAPSIS, A. and R. BURLEY. 1984. Flow distribution studies in fish rearing tanks. Part 1—Design constraints. *Journal Aquacultural Engineering* 3:103–118.

KLEIN, M. S., R. I. BRABETS and L. C. KINNEY. 1985. Generation of ozone. *Proceedings First International Symposium on Ozone for Water and Wastewater Treatment,* Waterbury, CT: International Ozone Institute.

KNEPP, G. L. and G. F. ARKIN. 1972. Ammonia toxicity levels and nitrate tolerance for channel catfish (*Ictalurus punctatus*). Presented at Annual Meeting of the American Society of Agricultural Engineers, Hot Springs, AR.

KNOWLES, G., A. L. DOWNING, and M. J. BARRETT. 1965. Determination of kinetic constants for nitrifying bacteria in mixed culture with the aid of an electronic computer. *Journal General Microbiology* 38:263–278.

KOLLER, L. R. 1965. *Ultraviolet Radiation.* New York: John Wiley and Sons.

KÖVÁRI, J. 1984. Considerations in the selection of sites for aquaculture. *Inland Aquaculture Engineering.* Rome: Food and Agricultural Organization of the United Nations.

KREFT, P., O. K. SCHEIBLE, and A. VENOSA. 1986. Hydraulic studies and cleansing evaluations of disinfection units. *Journal Water Pollution Control Federation* 58:1129–1137.

KUGELMAN, I. J. and S. VAN GORDER. 1991. Water and energy recycling in closed aquaculture systems. *Engineering Aspects of Intensive Aquaculture, Proceedings from the Aquaculture Symposium,* April 4–6, Ithaca, NY: Cornell University.

LAI-FA, Z. and C. E. BOYD. 1988. Nightly aeration to increase the efficiency of channel catfish production. *Progressive Fish Culturist* 50:237–242.

LAL, K., R. LASKER, and A. KLUJIS. 1977. Acclimation and rearing of striped bass larvae in sea water. *California Fish and Game* 63:210.

LARSON, G. L., F. E. HUTCHINS, and L. P. LAMPERTI. 1977. Laboratory determination of acute and sublethal toxicities of inorganic chloramines to early life stages of soho salmon (*Oncorhynchus kisutch*). *Transactions American Fisheries Society* 106:268–277.

LASKIN, A. I. and H. A. LECHEVALIER. 1974. *Handbook Of Microbiology.* Cleveland, OH: CRC Press.

LAUDELOUT, H. and L. VAN TICHELEN. 1960. Kinetics of the nitrite oxidation by *Nitrobacter winogradsky. Journal of Bacteriology* 79:39–42.

LAUDELOUT, H., P. C. SIMONART, and R. VAN DROOGANBROEK. 1968. Calorimetric measurement of free energy utilization by *Nitrosomonas* and *Nitrobacter. Archive fur Mikrobiologic* 63: 256–277.

LAWRENCE, F. E. and P. L. BRAUNWORTH. 1906. Fountain flow of water in vertical pipes. *Transactions American Society of Civil Engineers* 57:265–306.

LAWSON, T. B. 1989. Unpublished data. Department of Biological and Agricultural Engineering, Louisiana State University, Baton Rouge, LA.

LAWSON, T. B. and F. W. WHEATON. 1980. Removal of organics from fish culture water by foam fractionation. *Journal World Mariculture Society* 11:128–134.

LAWSON, T. B., F. E. BAKER, and J. D. BANKSTON. 1989. Recirculating ponds for crawfish culture. American Society of Agricultural Engineers Paper No. 89–7012, St. Joseph, MI: ASAE.

LAWSON, T. B., C. M. DRAPCHO, S. MCNAMARA, H. J. BRAUD, and W. R. WOLTERS. 1989. A heat exchange system for spawning red drum. *Journal Aquacultural Engineering* 8:177–191.

LAWSON, T. B., J. D. BANKSTON, and F. E. BAKER. 1994. Engineering evaluation of paddlewheel aerators to enhance crawfish production. *Proceedings 8th International Symposium of the*

International Association of Astacology, Baton Rouge, LA, 1990, In press.

LAYTON, R. F. 1972. Analytical methods for ozone in water and wastewater applications. *Ozone in Water and Wastewater Treatment,* Evans, F. L., ed. Ann Arbor, MI: Ann Arbor Science Publishers.

LEES, H. 1952. The biochemistry of the nitrifying organism. *Biochemical Journal* 52:134–139.

LEES, H. and J. R. SIMPSON. 1957. The biochemistry of the nitrifying organisms. *Biochemical Journal* 65:297–305.

LEMLICH, R. 1966. A theoretical approach to non-foaming absorptive bubble fractionation. *American Institute of Chemical Engineering Journal* 12(4):802–804.

LEWIS, W. K. and W. C. WHITMAN. 1924. Principles of gas adsorption. *Journal Industrial Engineering* 16:1215–1220.

LEWIS, W. M. and G. L. BUYNAK. 1976. Evaluation of revolving plate type biofilter for use in recirculated fish production and holding systems. *Transactions American Fisheries Society* 105:704–708.

LIAO, P. B. 1971. Water requirements of salmonids. *Progressive Fish Culturist* 33(4):210–215.

LIAO, P. B. 1974. Ammonia production rate and its application to fish culture system planning and design. (Original not available. Cited in Meade (1991)).

LIAO, P. B. 1980. Treatment units used in recirculation systems for intensive aquaculture. *World Symposium on New Developments in the Utilization of Heated Effluents and Recirculation Systems.* Stavanger, Norway: European Inland Fisheries Advisory Council.

LIAO, P. B. and R. B. MAYO. 1974. Intensified fish culture combining water recirculation with pollution abatement. *Aquaculture* 3:61–85.

LIBEY, G. S. 1992a. Maximizing nitrification with rotating biological contactors (RBC). *Proceedings Workshop On Design Of High density Recirculating Aquaculture Systems,* Baton Rouge, LA: Louisiana State University, September 25–27, 1991.

LIBEY, G. S. 1992b. Recirculating system for the production of hybrid striped bass: Description and design rationales. Presented at Aquaculture Expo V, New Orleans, LA, January 12–16, 1992.

LLOYD, R. and L. D. ORR. 1969. The diuretic response of rainbow trout to sublethal concentrations of ammonia. *Water Research* 3:335–344.

LOHR, A. L. and J. B. GRATZEK. 1986. Bactericidal and parasitic effects of an activated air oxidant in a closed aquatic system. *Journal Aquaculture and Aquatic Sciences* 4:1–8.

LOMAX, K. M. 1976. Nitrification with waste pretreatment on a closed cycle catfish culture system. Ph.D. Dissertation, University of Maryland.

LORIO, W. 1992. Personal communication. Louisiana Cooperative Extension Service, Louisiana State University, Baton Rouge, LA.

LOSORDO, T. M. 1991. An introduction to recirculating production systems design. *Engineering Aspects of Intensive Aquaculture, Proceedings from the Aquaculture Symposium,* Ithaca, NY: Cornell University, April 4–6, 1991.

LOVELESS, J. E. and H. A. PAINTER. 1968. The influence of metal ion concentration and pH value on the growth of a *Nitrosomonas* strain isolated from activated sludge. *Journal General Microbiology* 52:1–14.

LOYACANO, H. 1967. Some effects of salinity on two populations of red swamp crawfish, *Procambarus clarkii. Proceedings Southeastern Association of Game and Fish Commissioners* 21: 423–434.

LOYACANO, H. A. and D. C. SMITH. 1976. Attraction of native fishes to catfish culture cages in reservoirs. *Proceedings 29th Annual Conference of the Southeastern Association of Game Fisheries Commissioners,* Columbia, SC.

MACALADY, D. L., J. H. CARPENTER and C. A. MOORE. 1977. Sunlight-influenced bromate formation in chlorinated seawater. *Science* 195:1335–1337.

MACEINA, M. J. and J. V. SHINEMAN. 1979. Grass carp: Effects of salinity on survival, weight loss and muscle tissue water content. *Progressive Fish Culturist* 41:69–73.

MACLEAN, S. A., A. C. LONGWELL, and W. J. BLOGOSLAWSKI. 1973. Effects of ozone-treated seawater on the spawned, fertilized, meiotic and cleaving eggs of the commercial American oyster. *Annals of Applied Biology* 45:521–527.

MALONE, R. F. and D. G. BURDEN. 1988a. Design of recirculating soft crawfish shedding systems. Baton Rouge, LA: Louisiana Sea Grant College Program, Louisiana State University.

MALONE, R. F. and D. G. BURDEN. 1988b. Design of recirculating blue crab shedding systems. Baton Rouge, LA: Louisiana Sea Grant College Program, Louisiana State University.

MALONE, R. F., B. S. CHITTA, and D. G. DRENNAN. 1993. Optimizing nitrification in bead filters for warmwater recirculating aquaculture systems. *Techniques For Modern Aquaculture,* Wang, J., ed. St. Joseph, MI: American Society of Agricultural Engineers.

MANTHE, D. P. and R. F. MALONE. 1987. Chemical addition for accelerated biofilter acclimation in closed blue crab shedding systems. *Journal Aquaculture Engineering* 6:227–236.

MARCELLO, R. A., JR. and K. STRAWN. 1973. Cage culture of some marine fishes in the intake and discharge canals of a steam-electric generating station, Galveston Bay, Texas. *Journal World Mariculture Society* 4:97–112.

MARGERIT, C. and J. P. MARION. 1990. Culturing Tilapia in sea water in Martinique. *Aquaculture, Vol. 2,* Barnabé, G., ed. New York: Ellis Horwood.

MARUYAMA, T. and R. ISHIDA. 1976. Effect of water depth in net cages on the growth and body shape of *Tilapia mossambica*. (Original not available. Cited in Beveridge (1987)).

MARUYAMA, T. and R. ISHIDA. 1977. Effects of water depth on the growth and body shape of common carp raised in net cages. (Original not available. Cited in Beveridge (1987)).

MATHIAS, J. A. and J. BARCIS. 1985. Gas supersaturation as a course of early spring mortality of stocked trout. *Canadian Journal of Fisheries and Aquatic Sciences* 42:268–279.

MAYO, R. D. 1976. A technical and economic review of the use of reconditioned water in aquaculture. Rep. 30. Rome: Food and Agricultural Organization of the United Nations.

MCKAY, L. R. and B. GJERDE. 1985. The effect of salinity on growth of rainbow trout. *Aquaculture* 49:325–331.

MCKEE, J. E. and H. W. WOLF, eds. 1963. *Water Quality Criteria,* 2nd Ed. Sacramento, CA: California State Water Quality Control Board.

MCLARNEY, W. 1984. *The Freshwater Aquaculture Book*. Vancouver: Hartley and Marks.

MCLAUGHLIN, T. W. 1981. Hatchery effluent treatment U. S. Fish and Wildlife Service. *Proceedings of the Bioengineering Symposium For Fish Culture,* Allen, L. J. and E. C. Kinney, eds. Bethesda, MD: American Fisheries Society.

MEADE, J. W. 1989. *Aquaculture Management.* New York: Van Nostrand Reinhold.

MEADE, J. W. 1990. Personal communication. Tunnison Laboratory of Fish Nutrition, Cortland, New York.

MEADE, J. W. 1991. Intensity of aquaculture production, definitions, meanings and measures. *Engineering Aspects of Intensive Aquaculture, Proceedings from the Aquaculture Conference,* Ithaca, NY: Cornell University, April 4–6, 1991.

METCALF and EDDY. 1972. *Wastewater Engineering: Collection, Treatment, Disposal.* New York: McGraw-Hill.

METCALF and EDDY. 1991. *Wastewater Engineering.* 3rd Ed. Revised by G. Tchobanoglous and F. L. Burton, New York: McGraw-Hill.

MEVEL, G. and S. CHAMROUX. 1981. A study of nitrification in the presence of prawns (*Peneaux japonicus*) in marine closed systems. *Aquaculture* 123:29–43.

MEYERHOF, O. 1917. (Original not available. Cited in Painter (1970)).

MEZAINIS, V. E. 1977. Metabolic rates of pond ecosystems under intensive catfish cultivation. M. S. Thesis, Auburn University. (Original not available. Cited in Boyd (1990)).

MILLER, G. E. and G. S. LIBEY. 1985. Evaluation of three biological filters suitable for aquaculture applications. *Journal World Mariculture Society* 16:158–168.

MILNE, P. H. 1972. *Fish and Shellfish Farming in Coastal Waters*. Surrey, England: Fishing News Books, Ltd.

MOE, M. A. 1989. *The Marine Aquarium Reference: Systems and Invertebrates*. Plantation, FL: Green Turtle Publications.

MOLLER, D. 1979. Recent developments in cage and enclosed aquaculture in Norway. *Advances In Aquaculture,* Pillay, T. V. R. and W. A. Dill, eds. Surrey, England: Fishing News Books, Ltd.

MONOD, J. 1949. The growth of bacterial cultures. *Annual Review of Microbiology* 3:371.

MOODY, L. F. 1944. Friction factors for pipe flow. *Transactions of the American Society of Mechanical Engineers* 66:671–678.

MOORE, J. M. and C. E. BOYD. 1984. Comparisons of devices for aerating inflow of pipes. *Aquaculture* 38:89–96.

MOREL, F. M. M. 1983. *Principles of Aquatic Chemistry*. New York: Wiley-Interscience.

MUIR, J. F. 1982. Recirculated water systems in aquaculture. *Recent Advances In Aquaculture,* Muir, J. F. and R. J. Roberts, eds. London: Croom Helm.

MUIR, J. F. and R. J. ROBERTS, eds. 1982. *Recent Advances In Aquaculture,* London: Croom Helm.

MÜLLER, F. and VÁRADI, L. 1984. Freshwater cage fish farming. *Inland Aquacultural Engineering*. Rome: Food and Agricultural Organization of the United Nations.

MULLER-FEUGA, A., A. PETIT, and J. J. SABBAUT. 1978. The influence of temperature and wet weight on the oxygen demand of rainbow trout (*Salmo gairdneri*) in freshwater. *Aquaculture* 14:355–359.

NASH, C. E. 1988. A global overview of aquaculture production. *Journal World Aquaculture Society* 19:51–58.

NATIONAL SHELLFISH SANITATION PROGRAM (NSSP). 1990. *Manual of Operations, Part II: Sanitation of the Harvesting, Processing and Distribution of Shellfish.* Washington, DC: Public Health Service, U.S. Food and Drug Administration.

NATIONAL MARINE FISHERIES SERVICE (NMFS). 1990. *Fisheries of the United States, 1989.* Washington, DC: U.S. Department of Commerce.

NEILL, W. H. and J. D. BRYAN. 1991. Responses of fish to temperature and oxygen, and response integration through metabolic scope. *Aquaculture and Water Quality,* Brune, D. E. and J. R. Tomasso, eds. Baton Rouge: World Aquaculture Society.

NELSON, D. H. 1931. (Original not available. Cited in Painter (1970)).

NUNLEY, C. E. and G. S. LIBEY. 1991. The performance of hybrid striped bass in a recirculating system. Original not available. Cited in Livey. (1992a)).

OLSON, R. J. 1981. Differential photo inhibition of marine nitrifying bacteria. *Journal Marine Research* 39:227–238.

OWSLEY, D. E. 1981. Nitrogen gas removal using packed columns. *Bionegineering Symposium for Fish Culture,* Allen, L. J. and E. C. Kinney, eds. Bethesda, MD: American Fisheries Society.

OWSLEY, D. E. 1991. Ozone for disinfecting hatchery rearing water. *Fisheries Bioengineering Symposium,* Colt, J. and R. J. White, eds. Bethesda, MD: American Fisheries Society.

PAINTER, H. A. 1970. A review of literature on inorganic nitrogen metabolism in microorganisms. *Water Research* 4:393–450.

PALLER, M. H. and W. M. LEWIS. 1988. Use of ozone and fluidized bed biofilters for increased ammonia removal and fish loading rates. *Progressive Fish Culturist* 50:141–147.

PAPKO, S. I. 1957. Action of certain heterogeneous catalysts on the oxidation of ammonia in aqueous solution by ozonated oxygen. *Journal Applied Chemistry* 30:1361–1367.

PARKER, N. C. 1980. External standpipe drain system for fish tanks. *Progressive Fish Culturist* 42: 52–54.

PARKER, N. C. 1984. Culture requirements for striped bass. *The Aquaculture of Striped Bass: A Proceedings.* J. P. McCraren, ed. College Park, MD: Sea Grant College Program, University of Maryland.

PARKER, N. C. 1989. History, status, and future of aquaculture in the United States. *Aquatic Sciences* 1:97–109.

PARKER, N. C. and M. A. SUTTLE. 1987. Design of air-lift pumps for water circulation and aeration in aquaculture. *Journal Aquacultural Engineering* 6:97–110.

PAULIN, A. T. 1967. Procedures for measuring ozone in water. *Journal Institute Water Engineering* 21:537.

PAZ, J. D. 1984. The effects of borderline alkalinity on the nitrification rate in natural water systems. Ph.D. Dissertation, Polytechnical Institute of New York.

PECOR, C. H. 1979. Experimental intensive culture of tiger muskellunge in a water reuse system. *Progressive Fish Culturist* 41:103–108.

PERRY, W. G. 1971. Salt tolerance and factors affecting crawfish production in coastal marshes. Presented at the 2nd Annual Meeting of the Louisiana Crawfish Farmers Association, Lafayette, LA.

PERRY, W. G. and J. W. AVAULT. 1970. Culture of blue, channel and white catfish in brackish water ponds. *Proceedings of the Southeastern Association of Game and Fish Commissioners* 23:592–597.

PETIT, J. 1990. Water supply, treatment and recycling in aquaculture. *Aquaculture, Vol. 1,* G. Barnabé, ed. New York: Ellis Horwood.

PHILLIPS, M. J. 1985. Behavior of rainbow trout, *Salmo gairdneri* Richardson, in marine cages. *Aquaculture Fisheries Management* 16:223–232.

PIEDRAHITA, R. H. 1991. Engineering aspects of warmwater hatchery design. *Aquaculture Systems Engineering,* St. Joseph, MI: American Society of Agricultural Engineers.

PILLAY, T. V. R. 1990. *Aquaculture: Principles and Practices*. Surrey, England: Fishing News Books, Ltd.

PIPER, R. G., I. B. MCELWAIN, L. E. ORME, J. O. MCCRARAN, L. G. FOWLER, and J. R. LEONARD. 1982. *Fish Hatchery Management.* Washington, DC: U. S. Fish and Wildlife Service.

PLEMMONS, B. and J. W. AVAULT. 1980. Six tons of catfish per acre with constant aeration. *Louisiana Agriculture* 23:6–8.

POND, S. and G. L. PICKARD. 1978. *Introductory Dynamic Oceanography*. Oxford, England: Pergamon Press.

POPE, P., J. E. COLT, AND and R. LUDWIG. 1981. The environmental requirements of crustaceans. *The Use and Potential of Aquatic Species for Wastewater Treatments.* Sacramento, CA: California State Water Resources Control Board.

POSTON, H. A. 1983. Effect of population density of lake trout in cylindrical jars on growth and oxygen consumption. *Progressive Fish Culturist* 45:8–13.

POXTON, M. G. and S. B. ALLOUSE. 1982. Water quality criteria for marine fisheries. *Journal Aquacultural Engineering* 1:153–192.

PRESNELL, M. W. and J. M. CUMMINS. 1972. Effectiveness of ultraviolet radiation units in the bactericidal treatment of seawater. *Water Research* 6:1203–1212.

PURSLEY, M. G. and W. R. WOLTERS. 1989. Water quality affects growth of young redfish. *Louisiana Agriculture* 32:14–15.

RANDALL, D. 1991. The impact of variations in water pH on fish. *Aquaculture And Water Quality,* Brune, D. E. and J. R. Tomasso, eds. Baton Rouge, LA: World Aquaculture Society.

RAPPAPORT, U., S. SARIG, and M. MAREK. 1976. Results of tests of various aeration systems on the oxygen regime in the Genosar experimental ponds and growth of fish there in 1975. *Bamidgeh* 28:35–49.

RAY, L. 1981. Channel catfish production in geothermal water. *Bioengineering Symposium for Fish Culture,* Allen, L. J. and E. C. Kinney, eds. Bethesda, MD American Fisheries Society.

RILEY, J. 1991. Open net-pen culture of salmonids: Engineering considerations. *Engineering Aspects Of Intensive Aquaculture, Proceedings from the Aquaculture Symposium,* Ithaca, NY: Cornell University, April 4–6, 1991.

ROBERSON, J. A. and C. T. CROWE. 1990. *Engineering Fluid Mechanics*. Boston: Houghton Mifflin.

ROGERS, G. L. and S. L. KLEMETSON. 1981. Evaluation of a biofilter water reuse system for *Machrobrachium rosenbergii* prawn aquaculture ponds. *Second Water Reuse Symposium,* Washington, DC: American Water Works Association Research Foundation.

ROGERS, G. L. and S. L. KLEMETSON. 1985. Ammonia removal in selected aquaculture water reuse biofilters. *Journal Aquacultural Engineering* 4:135–154.

ROMAIRE, R. P. 1985. Water quality. *Crustacean And Mollusk Aquaculture In The United States,* Huner, J. V. and E. E. Brown, eds. Westport, CT: AVI Publishing.

ROMAIRE, R. P. and C. E. BOYD. 1978. Predicting nighttime oxygen depletions in catfish ponds. Bulletin No. 505, Auburn, AL: Auburn University/Alabama Agricultural Experiment Station.

ROMAIRE, R. P. and C. E. BOYD. 1979. Effects of solar radiation on the dynamics of dissolved oxygen in channel catfish ponds. *Transactions American Fisheries Society* 108:473–478.

ROMAIRE, R.P, C. E. BOYD and W. J. COLLINS. 1978. Predicting nighttime dissolved oxygen decline in ponds used for tilapia culture. *Transactions American Fisheries Society* 107:804–808.

ROSELUND, B. 1974. Disinfection of hatchery effluent by ozonation and the effect of ozonated water on rainbow trout. *Aquatic Applications of Oxone,* Blogoslawski, W. J. and R. G. Rice, eds. Syracuse, NY: International Ozone Institute.

ROSEN, H. M. 1972. Ozone generation and its relationship to the economical application of ozone in wastewater treatment. *Ozone in Water and Wastewater Treatment,* Evans, F. L., ed. Ann Arbor, MI: Ann Arbor Science Publishers.

ROSENTHAL, H. 1981. Ozonation and Sterilization. *World Symposium on New Developments in the Utilization of Heated Effluents and Recirculation Systems.* Stavanger, Norway: European Inland Fisheries Advisory Council, 28–30 May, 1980.

ROSENTHAL, H. 1985. Constraints and perspectives in aquaculture development. (Original not available. Cited in Beveridge (1987)).

ROSS, D. A. 1970. *Introduction to Oceanography.* New York: Appleton-Century-Crofts.

RUSSO, R. C. and R. V. THURSTON. 1977. The acute toxicity of nitrites to fishes. *Recent Advances In Fish Toxicology,* Tubb, R. A. ed. Corvalis, OR: U. S. Environmental Protection Agency, Ecological Research Service.

RUSSO, R. C. and R. V. THURSTON. 1991. Toxicity of ammonia, nitrite and nitrate to fishes. *Aquaculture And Water Quality,* Brune, D. E. and J. R. Tomasso, eds. Baton Rouge, LA: World Aquaculture Society.

SANDER, E. and H. ROSENTHAL. 1975. Application of ozone in water treatment for home aquaria, public aquaria and for aquaculture purposes. *Aquatic Applications of Ozone,* Blogoslawski, W. J. and R. G. Rice, eds. Syracuse, NY: International Ozone Institute.

SAWYER, C. N. and P. L. MCCARTY. 1978. *Chemistry For Environmental Engineering.* New York: McGraw-Hill.

SCHMIDT-NIELSEN, K. 1979. *Animal Physiology: Adaptation and Environment.* New York: Cambridge University Press.

SCHROEDER, G. L. 1975. Nighttime material balance for oxygen in fish ponds receiving organic wastes. *Bamidgeh* 27:65–74.

SCHWAB, G. O., R. K. FREVERT, T. W. EDMINSTER, and K. K. BARNES. 1966. *Soil and Water Conservation Engineering.* New York: John Wiley and Sons.

SCHWEDLER, T. E. and C. S. TUCKER. 1983. Empirical relationship between percent methemoglobin in channel catfish and dissolved nitrite and chloride in ponds. *Transactions American Fisheries Society* 112:117–119.

SCHWEDLER, T. E., C. S. TUCKER, and M. H. BELEAU. 1985. Non-infectious diseases. *Channel Catfish Culture,* Tucker, C. S., ed. Amsterdam: Elsevier.

SEEGERT, G. L. and A. S. BROOKS. 1978. Dechlorination of water for fish culture: Comparison of activated carbon, sulfite reduction and photochemical methods. *Journal Fisheries Research Board of Canada* 35:88–92.

SHARP, J. 1976. The effects of dissolved oxygen, temperature and weight on respiration of *Machrobrachium rosenbergii.* Pap. No. 4501. Davis, CA: Department of Water Sciences and Engineering, University of California.

SHARP, R. W. 1951. Chlorine removal unit at full-scale operating hatchery. *Progressive Fish Culturist* 13:146–148.

SHAW, S. and J. F. MUIR. 1986. *Salmon Economics and Marketing.* London: Croom Helm.

SCHECHMEISTER, I. L. 1977. Sterilization by ultraviolet radiation. *Disinfection, Sterilization and Preservation,* Buck, S. S., ed. London: Lee and Febiger.

SHELTON, J. L. and C. E. BOYD. 1983. Correction factors for calculating oxygen transfer rates of pond aerators. *Transactions American Fisheries Society* 120:120–122.

SIMON, A. L. 1981. *Practical Hyraulics* New York: John Wiley and Sons.

SKINNER, F. A. and N. WALKER. 1961. (Original not available. Cited in Hochheimer and Wheaton (1991)).

SKOGERBOE, G. V., M. L. HYATT, J. D. ENGLAND, and J. R. JOHNSON. 1967. Design and calibration of submerged open channel flow measurement structures: Part 2. Parshall flumes. Report WG31-2, Logan, UT: Utah Water Research Laboratory, Utah State University.

SMART, G. R. 1978. Investigation of the toxic mechanisms of ammonia to fish—Gas exchange in rainbow trout (*Salmo gairdneri*) exposed to acutely lethal concentrations. *Journal Fisheries Biology* 12:93–104.

SMART, G. R. 1981. Aspects of water quality producing stress in intensive culture. *Stress and Fish,* Pickering, A. D., ed. London: Academic Press.

SMITH, C. E. and R. G. PIPER. 1975. Lesions associated with chronic exposure to ammonia. *The Pathology of Fishes,* Ribelin, W. E. and H. Migaki, eds. Madison, WI: University of Wisconsin Press.

SMITH, L. L., P. M. OSEID, L. L. KIMBALL, and S. M. EL-KAUDELGY. 1976 Toxicity of hydrogen sulfide to various life history stages of bluegill (*Lepomis macrochirus*). *Transactions American Fisheries Society* 105:442–449.

SODERBERG, R. W. 1982. Aeration of water supplies for fish culture in flowing water. *Progressive Fish Culturist* 44(2):89–93.

SOIL CONSERVATION SERVICE (SCS). 1969. *Engineering Field Manual.* Handbook No. 590. Washington, DC: U. S. Department of Agriculture, Soil Conservation Service.

SOIL CONSERVATION SERVICE (SCS). 1982. *Ponds—Planning, Design and Construction.* Handbook No. 590. Washington, DC: U. S. Department of Agriculture, Soil Conservation Service.

SPEECE, R. E. 1973. Trout metabolism characteristics and the rational design of nitrification filters for water reuse in hatcheries. *Transactions American Fisheries Society* 2:323–334.

SPOTTE, S. H. 1970. *Fish and Invertebrate Culture: Water Management In Closed Systems.* New York: John Wiley and Sons.

SPOTTE, S. H. 1973. *Marine Aquarium Keeping.* New York: John Wiley and Sons.

SPOTTE, S. H. 1979. *Seawater Aquariums: The Captive Environment.* New York: John Wiley and Sons.

SPOTTE, S. H. 1992. *Captive Seawater Fishes: Science and Technology.* New York: John Wiley and Sons.

STEEBY, J. A. and C. S. TUCKER. 1988. Comparison of nightly and emergency aeration of channel catfish ponds. Res. Rep. Vol. 13 No. 8. Starkville, MS: Mississippi State University/ Mississippi Agricultural Experiment Station.

STEELS, I. H. 1974. Design basis for the rotating disc process. *Effluent Water Treatment Journal* 14: 431–445.

STENSTROM, M. K. and R. G. GILBERT. 1981. The effects of alpha, beta and theta factors upon the design, specification and operation of aeration systems. *Water Research* 15:643–654.

STICKNEY, R. R. 1979. *Principles of Warmwater Aquaculture.* New York: Wiley-Interscience.

STICKNEY, R. R. 1986. Tilapia tolerance of saline waters: A review. *Progressive Fish Culturist* 48:161–167.

STICKNEY, R. R. 1991. Effects of salinity on aquaculture production. *Aquaculture And Water Quality,* Brune, D. E. and J. R. Tomasso, eds. Baton Rouge, LA: World Aquaculture Society.

STOPKA, K. 1975. European and Canadian experiences with ozone in controlled closed circuit fresh and salt water systems. *Aquatic Applications of Ozone,* Blogoslawski, W. J. and R. G. Rice, eds. Syracuse, NY: International Ozone Institute.

SUTTERLIN, A. M., C. Y. COUTERIER, and T. DEVEREAUX. 1984. A recirculation system using ozone for the culture of Atlantic salmon. *Progressive Fish Culturist* 46:239–244.

SUTTERLIN, A. M., K. J. JOKOLA, and B. HOLTE. 1979. Swimming behavior of salmonid fish in ocean pens. *Journal Fisheries Research Board of Canada* 36:948–954.

SWIFT, D. R. 1963. Influence of oxygen concentration on growth of brown trout *Salmo trutta* L. *Transactions American Fisheries Society* 92: 300–304.

SWINGLE, H. S. 1969. Methods of analysis for waters, organic matter and pond bottom soils used in fisheries research. Auburn, AL: Auburn University.

SZYPER, J. P. and C. K. LIN. 1990. Techniques for assessment of stratification and effects of mechanical mixing in tropical fish ponds. *Journal Aquacultural Engineering* 9:151–165.

TANSAKUL, R. 1983. Progress in Thailand rearing larvae of the giant prawn, *Macrobrachium rosenbergii* DeMan, in salted water. *Aquaculture* 31:95–98.

TCHOBANOGLOUS, G. and E. D. SCHROEDER. 1985. (Revised 1987). *Water Quality: Characteris-*

tics, Modeling, Modification. Reading, MA: Addison-Wesley.

THOMAS, H. A. and R. S. ARCHIBALD. 1952. Longitudinal mixing measured by radioactive tracers. *Transactions American Society of Civil Engineers* 117:839.

THOMAS, W. A., H. A. SPALDING and Z. PAVLOVICH. 1967. *The Engineer's Vest Pocket Book*. Philippines: National Book Store.

TIMMONS, M. B. and W. D. YOUNGS. 1991. Considerations on the design of raceways. *Aquaculture Systems Engineering,* St. Joseph, MI: American Society of Agricultural Engineers.

TIMMONS, M. B., W. D. YOUNG, J. M. REGENSTEIN, G. A. GERMAN, P. R. BOSWER, and C. A. BISOGNI. 1987. A systems approach to the development of an integrated trout industry for New York State; Final Report Presented to New York State Department of Agric. and Markets. Ithaca, NY: Cornell University.

TOMASSO, J. R., B. A. SIMCO, and K. DAVIS. 1979. Chloride inhibition of nitrite induced methemoglobinemia in channel catfish (*Ictalurus punctatus*). *Journal of the Fisheries Research Board of Canada* 36:1141–1144.

TSIVOGLOU, E. C., R. L. O'CONNELL, C. M. WALTER, P. J. GODSIL, and G. S. LOGSDON. 1965. Tracer measurements of atmospheric respiration. 1. Laboratory studies. *Journal Water Pollution Control Federation* 37:1343–1362.

TUCKER, C. S. ed. 1985. *Channel Catfish Culture*. Amsterdam: Elsevier.

TUCKER, C. S. and C. E. BOYD. 1985. Water quality. *Channel Catfish Culture,* Tucker, C. S., ed. Amsterdam: Elsevier.

TUCKER, C. S. and E. H. ROBINSON. 1990. *Channel Catfish Farming Handbook*. New York: Van Nostrand Reinhold.

TURNER, J. L. and T. C. FARLEY. 1971. Effects of temperature, salinity, and dissolved oxygen on the survival of striped bass eggs and larvae. *California Fish and Game* 57:268.

ULKEN. A. 1963. (Original not available. Cited in Painter (1970)).

U. S. DEPARTMENT OF AGRICULTURE (USDA). 1989a. Aquaculture: Situation and Outlook Report. Pub. No. AQUA-2, March, 1989. Washington, DC: Economic Research Service.

U. S. DEPARTMENT OF AGRICULTURE (USDA). 1989b. Aquaculture: Situation and Outlook Report. Pub. No. AQUA-3, September, 1989. Washington, DC: Economic Research Service.

U. S. DEPARTMENT OF AGRICULTURE (USDA). 1990. Aquaculture: Situation and Outlook Report. Pub. No. AQUA-4, March, 1990. Washington, DC: Economic Research Service.

U. S. DEPARTMENT OF AGRICULTURE (USDA). 1992a. Aquaculture: Situation and Outlook Report. Pub. No. AQUA-8, March, 1992. Washington, DC: Economic Research Service.

U. S. DEPARTMENT OF AGRICULTURE (USDA). 1992b. Aquaculture: Situation and Outlook Report. Pub. No. AQUA-9, September, 1992. Washington, DC: Economic Research Service.

U. S. DEPARTMENT OF THE INTERIOR. 1967. *Water Measurement Manual,* Bellport, B. P. and G. E. Burnett, eds. Washington, DC: U. S. Department of the Interior.

U. S. ENVIRONMENTAL PROTECTION AGENCY (USEPA). 1975. Process design manual for nitrogen control. Washington, DC: USEPA, Office of Technology Transfer.

VAN OLST, J. C., R. F. FORD, J. M. CARLBERG, and W. R. DORBAND. 1976. Use of therma effluent in culturing the American lobster. Power Plant Waste Heat Utilization in Aquaculture Workshop, Trenton, NJ, November 6–7, 1975.

VENOSA, A. D. 1972. Ozone as a water and wastewater disinfectant: a literature review. *Ozone in Water and Wastewater Treatment,* Evans, F. L., ed. Ann Arbor, MI: Ann Arbor Science Publishers.

VIESSMAN, W. and M. J. HAMMER. 1985. *Water Supply and Pollution Control*. New York: Harper and Row.

WATER POLLUTION CONTROL FEDERATION (WPCF). 1983. Nutrient control. Manual of Practice. Pub. No. FD-7, Washington, DC: Water pollution Control Federation.

WATTEN, B. J. 1991. Application of pure oxygen in raceway culture systems. *Engineering Aspects of Intensive Aquaculture, Proceedings from the Aquaculture Symposium,* Ithaca, NY: Cornell University, April 4–6, 1991.

WEDEMEYER, G. A. and N. C. NELSON. 1977. Survival of two bacterial fish pathogens (*Aeromonas salmonicida* and the enteric Redmouth Bacterium) in ozonated, chlorinated and untreated waters. *Journal Fisheries Research Board of Canada* 34(3):429–432.

WEDEMEYER, G. A. and W. T. YASUTAKE. 1978. Prevention and treatment of nitrite toxicity in juvenile steelhead trout (*Salmo gairdneri*). *Journal Fisheries Research Board of Canada* 35: 822–827.

WEDEMEYER, G. A., N. C. NELSON, and C. A. SMITH. 1978. Survival of the salmonid viruses infectious hematopoietic necrosis (IHNV) and infectious pancreatic necrosis (IPNV) in ozonated, chlorinated and untreated waters. *Journal Fisheries Research Board of Canada* 34:433–437.

WEDEMEYER, G. A., N. C. NELSON, and W. T. YASUTAKA. 1979. Potentials and limits for the use of ozone as a fish disease control agent. *Ozone: Science and Engineering* 1:295–318.

WEISS, J. 1935. The radical HO_2 in solution. *Transactions Farraday Society* 31:668–681.

WEISS, R. F. 1970. The solubility of nitrogen, oxygen and argon in water and seawater. *Deep Sea Research* 17:721–735.

WESTERS, H. 1981. *Fish Culture manual for the State of Michigan.* Lansing, MI: Michigan Department of Natural Resources.

WESTERS, H. 1987. The use of pure oxygen in fish hatcheries. *Oxygen Supplementation: A New Technology In Fish Culture.* Visscher, L. and W. Godby, eds. Information Bulletin No. 1, Denver, CO: U. S. Department of the Interior, Region 6.

WESTERS, H. 1991. Modes of operation and design relative to carrying capacities of flow-through systems. *Engineering Aspects of Intensive Aquaculture, Proceedings from the Aquaculture Symposium,* Ithaca, NY: Cornell University, April 4–6, 1991.

WESTERS, H. and K. M. PRATT. 1977. Rational design of hatcheries for intensive culture based on metabolic characteristics. *Progressive Fish Culturist* 39:157–165.

WHEATON, F. W. 1977. *Aquacultural Engineering.* New York: Wiley-Interscience.

WHEATON, F. W., J. N. HOCHHEIMER, G. E. KAISER and M. J. KRONES. 1991. Principles of biological filtration. *Engineering Aspects of Intensive Aquaculture, Proceedings from the Aquaculture Symposium,* Ithaca, NY: Cornell University, April 4–6, 1991.

WHEATON, F. W., T. B. LAWSON, and K. M. LOMAX. 1979. Foam fractionation applied to aquacultural systems. *Journal World Mariculture Society* 10:795–808.

WHITE, W. R. 1966. The effect of low-level chlorination on mussels at Poole Power Station. (Original not available. Cited in Wickens (1981)).

WICKENS, J. F. 1981. Water quality requirements for intensive aquaculture: A review. *World Symposium on New Developments in the Utilization of Heated Effluents and Recirculation Systems,* Stavanger, Norway: European Inland Fisheries Advisory Council.

WICKINS, J. F. 1976. The tolerance of warmwater prawns to recirculated water. *Aquaculture* 9:19–37.

WILLIAMS, R. C., S. G. HUGHES, and G. L. RUMSEY. 1982. Use of ozone in a water reuse system for salmonids. *Progressive Fish Culturist* 44:102–105.

WILLIAMSON, K. and P. L. MCCARTY. 1976. Verification studies of the biofilm model for bacterial substrate utilization. *Journal Water Pollution Control federation* 48:281–296.

WILLOUGHBY, H. 1968. A method for calculating carrying capacities of hatchery troughs and ponds. *Progressive Fish Culturist* 30:173–175.

WIMBERLY, D. M. 1990. Development and evaluation of a low-density media biofilter unit for use in recirculating finfish culture systems. M. S. Thesis, Louisiana State University.

WINOGRADSKY, S. and H. WINOGRADSKY. 1933. (Original not available. Cited in Painter (1970)).

WITTENBERG, J. B. and B. A. WITTENBERG. 1974. The choroid rete mirabile of the fish eye. I. Oxygen secretion and structure: comparison with the swimbladder rete mirabile. *Biological Bulletin* 146:116–136.

WONG, G. T. F. 1980. The effects of light on the dissipation of chlorine in seawater. *Water Research* 14:1263–1268.

WONG, G. T. F. and J. A. DAVIDSON. 1977. The fate of chlorine in seawater. *Water Research* 11: 971–978.

WORTMAN, B. and F. W. WHEATON. 1991. Temperature effects on biodrum nitrification. *Journal Aquacultural Engineering* 10:183–205.

APPENDIX

Table A-1. Celsius–Fahrenheit temperature conversions.

°C		°F	°C		°F	°C		°F
−17.8	0	32	1.67	35	95.0	20.6	69	156.2
−17.2	1	33.8	2.22	36	96.8	21.1	70	158.0
−16.7	2	35.6	2.78	37	98.6	21.7	71	159.8
−16.1	3	37.4	3.33	38	100.4	22.2	72	161.6
−15.6	4	39.2	3.89	39	102.2	22.8	73	163.4
−15.0	5	41.0	4.44	40	104.0	23.3	74	165.2
−14.4	6	42.8	5.00	41	105.8	23.0	75	167.0
−13.9	7	44.6	5.56	42	107.6	24.4	76	168.8
−13.3	8	46.4	6.11	43	109.4	25.0	77	170.6
−12.8	9	48.2	6.67	44	111.2	25.6	78	172.4
−12.2	10	50.0	7.22	45	113.0	26.1	79	174.2
−11.7	11	51.8	7.78	46	114.8	26.7	80	176.0
−11.1	12	53.6	8.33	47	116.6	27.2	81	177.8
−10.6	13	55.4	8.89	48	118.4	27.8	82	179.6
−10.0	14	57.2	9.44	49	120.2	28.3	83	181.4
−9.44	15	59.0	10.0	50	122.0	28.9	84	183.2
−8.89	16	60.8	10.6	51	123.8	29.4	85	185.0
−8.33	17	62.6	11.1	52	125.6	30.0	86	186.8
−7.78	18	64.4	11.7	53	127.4	30.6	87	188.6
−7.22	19	66.2	12.2	54	129.2	31.1	88	190.4
−6.67	20	68.0	12.8	55	131.0	31.7	89	192.2
−6.11	21	69.8	13.3	56	132.8	32.2	90	194.0
−5.56	22	71.6	13.9	57	134.6	32.8	91	195.8
−5.00	23	73.4	14.4	58	136.4	33.3	92	197.6
−4.44	24	75.2	15.0	59	138.2	33.9	93	199.4
−3.89	25	77.0	15.6	60	140.0	34.4	94	201.2
−3.33	26	78.8	16.1	61	141.8	35.0	95	203.0
−2.78	27	80.6	16.7	62	143.6	35.6	96	204.8
−2.22	28	82.4	17.2	63	145.4	36.1	97	206.6
−1.67	29	84.2	17.8	64	147.2	36.7	98	208.4
−1.11	30	86.0	18.3	65	149.0	37.2	99	210.25
−0.56	31	87.8	18.9	66	150.8	37.8	100	212.0
0	32	89.6	19.4	67	152.6	37.8	100	212
0.56	33	91.4	20.0	68	154.4	43	110	230
1.11	34	93.2	20.6	69	156.2			

Note: The whole numbers refer to the temperature in either degrees Celsius or Fahrenheit which is to be converted to the other scale. If converting from degrees Celsius, the center column is the temperature converting from, and the right column is its equivqlent in degrees Fahrenheit. When converting fron degrees Fahrenheit, the center column is the temperature converting from, and the left column is its equivalent in degrees Celsius.

Table A-2. Frictional head losses in schedule 80 plastic (PVC) pipe.

	Nominal pipe size (in.)											
Flow	1		2		4		6		8		12	
gpm	Vel	Loss	Vel	Loss	Vel	Loss	Vel	Loss	Vel	Loss	Vel	Loss
10	4.4	8.8	1.1	0.3								
20	8.9	32.0	2.2	1.1	0.6	0.0						
30	13.3	72.1	3.2	2.3	0.8	0.1						
50			5.4	5.6	1.4	0.2	0.6	0.0				
70			7.6	10.7	2.0	0.4	0.9	0.1				
100			10.8	20.2	2.8	0.7	1.2	0.1	0.7	0.0		
150			16.2	42.8	4.2	1.5	1.9	0.2	1.1	0.1	0.5	0.01
200					5.6	2.6	2.5	0.4	1.4	0.1	0.6	0.01
250					7.0	4.0	3.1	0.5	1.8	0.1	0.8	0.02
300					8.4	5.5	3.7	0.7	2.1	0.2	1.0	0.03
350					9.8	7.4	4.3	1.0	2.5	0.3	1.1	0.04
400					11.2	9.7	4.9	1.2	2.8	0.3	1.3	0.05
500					13.9	14.5	6.2	1.9	3.5	0.5	1.6	0.07
600					16.7	21.5	7.4	2.7	4.2	0.7	1.9	0.10
700					19.5	26.9	8.6	3.6	4.9	0.9	2.2	0.13
800							9.9	4.6	5.6	1.2	2.5	0.17
900							11.1	5.7	6.3	1.4	2.8	0.20
1,000							12.3	6.9	7.0	1.8	3.2	0.24
1,200							14.8	9.7	8.4	2.4	3.8	0.34

Note: The above values were calculated using the Darcy-Weisbach equation with the following assumptions: Clean, used PVC pipe; seawater at 70°F (21°C); equivalent sand roughness = 0.000042 ft^2/s.

gpm = Gallons per minute.

Lpm = Liters per minute.

Vel = Velocity in ft/s.

Loss = Frictional head loss in ft.

Source: Huguenin and Colt (1989).

Table A-3. Conversion factors.

Multiply	By	To obtain
in.	25.4	mm
in.	0.0254	mm
in.	2.54	cm
in.	0.0254	m
ft	0.3048	m
ft	304.8	mm
ft	30.48	cm
ft^2	0.0929	m^2
$1n^2$	0.0006452	m^2
acre	4,047.0	m^2
acre	0.4047	ha
ft^3	0.0283	m^3
yd^3	0.7646	m^3
gal	3.785	L
gal	0.003785	m^3
ft/s	0.3048	m/s
ft^3/s	0.0283	m^3/s
ft^3/s	28.3	L/s
gpm	0.00006308	m^3/s
gpm	0.06308	L/s
lbf	4.448	N
ton (2,000 lbf)	8,896.0	N
ft-lbf	1.356	N-m
slug	14.49	kg
lbm	0.4536	kg
lbm/ft^3	16.02	kg/m^3
lbf/ft^2	47.88	N/m^2
lbf/in^2	6,895.0	N/m^2
lbf/ft^3	157.1	N/m^3
ft-lbf	1.356	J
ft^2/s	0.0929	m^2/s
$lbf\text{-}s/ft^2$	47.88	$N\text{-}s/m^2$
hp	0.747	kW
BTU	1,055.0	J

INDEX